Patterns and Processes in Forest Landscapes

Raffaele Lafortezza · Jiquan Chen ·
Giovanni Sanesi · Thomas R. Crow
Editors

Patterns and Processes in Forest Landscapes

Multiple Use and Sustainable Management

Foreword by Thomas A. Spies

Springer

Editors

Dr. Raffaele Lafortezza
University of Bari
Dept. of Scienze
delle Produzioni
Vegetali
Via Amendola 165/A
70126 Bari BA
Italy
r.lafortezza@agr.uniba.it
www.greenlab.uniba.it

Dr. Giovanni Sanesi
University of Bari
Dept. of Scienze
delle Produzioni
Vegetali
Via Amendola 165/A
70126 Bari BA
Italy
sanesi@agr.uniba.it

Dr. Jiquan Chen
University of Toledo
Dept. Earth, Ecological &
Environmental Sciences
2801 W. Bancroft St.
Toledo OH 43606-3390
USA

Dr. Thomas R. Crow
USDA Forest Service
Environmental Sciences
Research
WFWAR - Stop Code 1113
1400 Independence Avenue
SW Washington DC 20250-1113
USA

ISBN: 978-1-4020-8503-1 e-ISBN: 978-1-4020-8504-8

Library of Congress Control Number: 2008930786

Cover photograph by Claudia Cotugno

Printed on acid-free paper

9 8 7 6 5 4 3 2 1

springer.com

Contents

Foreword

Landscape ecology has been at the center of a sea change in how we think about forest ecosystems and approach the problem of managing forests for multiple goals. Instead of using homogeneity and stability as a basis for our scientific and management models as was done in the recent past, we now seek to understand forest systems in terms of heterogeneity and dynamics. Rather than thinking in terms of single resources or scales, scientists and managers are now focused on integration and holistic approaches to forestry across multiple scales. Instead of avoiding places were humans have changed nature, we now study those places to learn how we have changed ecosystems and how we could restore them. Landscape ecology is now poised to help us meet new management challenges such as globalization and global climate change.

We have learned much in the last 25 years especially about landscape pattern, its origins, its dynamics, and its role in structuring the flow of materials, processes and species. We have learned that pattern matters. We are increasingly learning, however, that the way pattern affects systems varies with species, process, and scale. For example, forest fragmentation, which is typically seen in a negative light, can have positive or negative effects, as illustrated by the case of the Northern Spotted Owl (*Strix occidentalis caurina*) in the Pacific Northwest, USA. This older-forest species was listed because its populations were threatened by the loss and fragmentation of its dense older forest habitat from logging. But, new threats to the populations have become apparent and these appear to indirectly result from increasing forest connectivity. Barred owls (*Strix varia*), a more aggressive competitor species from eastern North America have arrived in the Pacific Northwest. Their arrival may be a result of increased connectivity of forests across the plains, whose grasslands were a barrier that separated the two species in the past. Threats of loss of owl habitat from wildfire have also increased in parts of the Northern Spotted Owl's range. Increasing homogeneity of forest fuels resulting from fire suppression by humans may be leading to larger, high-severity fires now than in the past when fuels were patchier. This case illustrates that as our understanding of landscape ecology and the application to management have matured we have moved from simple conceptual models of pattern effects to complex and more realistic models that include scale, dynamics, and interactions.

The science of landscape ecology is diverse and we have learned much, as this book illustrates, but we have a long way to go yet. In some ways landscape ecologists are like the naturalists of the 19th century, collecting and describing different landscape specimens. Given the enormous diversity of biophysical settings and human cultures, our collections are meager and our theories are young. In other ways, landscape ecologists are like the most advanced theoretical physicists, using large, complex computer simulations to search for rules that govern the flows of organisms, fire, disease or water across the fabric of landscapes. Some of the "rules" appearing in the form of hierarchy theory, metapopulation dynamics, or disturbance regimes form the basis of new approaches to forest management and planning. An increasing number of landscape ecologists are able to act as experimentalists with treatment and controls that span hundreds of hectares. The results from these experiments will provide the empirical sparks to advance our theories, test our models, and improve our management practices.

Increasingly the boundaries between management and research in landscape ecology are hard to see. Scientists have become more involved in informing policy and evaluating alternative scenarios for policy makers and stakeholder groups. Managers and planners with Ph.D.s now often use sophisticated landscape models to inform management decisions on the ground. For a science with a strong applied vein this blurring of the boundaries is a good thing and can lead to greater mutual learning and more sustainable and adaptive natural and human systems.

The Landscape Ecology Working Party of the IUFRO (08.01.02), primarily through its bi-annual conferences, has been expanding its diversity of scientists and perspectives to facilitate the growth of landscape ecology. This book based on its conference in Locorotondo, Bari (Italy) in 2006 provides many valuable perspectives on the state of theory and application of landscape ecology in forests. I believe that one of its main contributions is to demonstrate the global nature of the discipline, as the authors of this text come from many different countries and studied a wide range of ecosystems and landscapes. This is particularly important because processes such as climate change, land use change, and the flow of wood products and organisms are global in scale. It is also important because comparative analysis of different ecosystems and cultures and their interactions may be one of the best ways that we can take the specimens of landscapes we analyze and discover broader patterns that can help both advance theory and enable application of the science to some of the earth's most challenging environmental and human problems.

Corvallis, Oregon, USA Thomas A. Spies

Preface

In the last two decades, landscape ecology has advanced with rapid developments in technology, increased knowledge exchange among different scientific arenas, and the growing number of modern scientists across disciplines. Landscape ecology has now reached a stage of maturity that principles, methods, and models can be applied with confidence in real situations and field studies. Agencies and organization worldwide (e.g., USDA-FS, EEA, FAO, etc.) are now embracing integrated approaches to landscape planning and management in the hope to recover from a long period of environmental "consumption". In particular, natural resource managers are progressively shifting their emphasis from management of separate resources (e.g., riparian habitat, wetlands, old-growth forests, etc.) to the focus on the integrity of entire landscapes.

The demand for input and guiding principles from landscape ecologists for resource management decision at all scales is therefore very high. Forest resources are within this context as they constitute fundamental parts of our living environment. From a landscape ecological perspective, such resources can be approached as part of an overall forest landscape whose patterns interact with ecological processes (e.g., energy flows, nutrient cycling, and flora/fauna dispersal) across dimensions of common time and space. The application of landscape ecology in forest landscapes requires focusing on mosaics of patches and long-term changes in these mosaics to integrate ecological values, such as the maintenance of forest ecosystem health, biodiversity conservation, and carbon sequestration, with economic and social purposes, such as timber and recreation. A daunting challenge in this application is to balance natural disturbances with human-induced changes, thus determining how forest harvesting or other management plans may affect the mechanisms underlying these changes and, ultimately, maintain the high function stability and productivity of forest landscapes. Indeed, depending on the geographical region and the silviculture tradition, human intervention is critical to the maintenance of forest landscapes and to achieve a sustainable and multiple use of the land.

Managing forest landscapes is therefore a complex practice of understanding the critical patterns of the landscape and their reciprocal interrelationship through processes. Managing forests at a landscape level implies focusing on mosaics of patches and long-term changes in these mosaics to integrate ecological values (e.g., the maintenance of forest ecosystem health and biodiversity conservation), with economic and social purposes (e.g., timber and recreation).

Prior texts on landscape ecology have focused largely on analyzing and quantifying landscapes with a dominance of natural disturbances rather than human-induced changes. Consequently, the large part of the available literature in landscape ecology has a focus on large and "uncontaminated" forest landscapes. Less emphasis has been placed on the interplay between forest landscapes and human societies and, specifically, on the ecology of human-dominated landscapes where natural and forest habitats have been extensively cleared, fragmented or modified. Urban forest landscapes, for example, were received as one of the hot topics at the of the 2006 IUFRO Landscape Ecology Workshop held in Locorotondo, Bari, Italy in Sept. 2006, but significant less attention had been made in landscape ecology studies.

The proposed book attempts to lay the theoretical and applicative foundations for such a comprehensive approach by a group of international landscape ecologists to broaden the potential impacts of this product. Withdrawing from a very diverse international community and many unique landscapes in Europe, Asia, North America, Africa and Australia, this book provides advanced knowledge into some of the applicable landscape ecological theory that underlies forest management with a specific focus on how forest management can benefit from landscape ecology, and how landscape ecology can be advanced by tackling challenging problems in forest landscape management.

The book offers examples of successful ecological research and case-studies from an international perspective that are conducted at landscape level and subsequent implications on which to base forest ecosystems and landscapes management. The objectives of the book are to:

(1) Understand the dynamics of forest ecosystems in a landscape context across multiple spatial and temporal scales.
(2) Introduce effective methods for linking landscape ecology with remotely-sensed data in geographic information systems (GIS) to extract specific, user-oriented information at multiple scales on forest ecosystems and landscapes.
(3) Explore management strategies and policies that enable a sustainable and multiple use of forest resources in different geographical regions.

This book is introduced by a synopsis of some relevant research findings in landscape ecology and may be relevant to international forest management as our contributors are positioned in leading science and management from many countries where land use history and current management plans may be greatly associated with the local political systems and culture.

The book consists of four sections, with chapters for each section, focusing on:

(1) Underlying concepts and applicative approaches
(2) Consequences of management across regions and scales
(3) Landscape-scale indicators and projection models
(4) Long-term sustainable plans and management actions.

The book provides valuable information to the existing international literature that can be used for expanding the scope of environmental education in upper-level undergraduate and graduate classes of Landscape Ecology, Conservation Biology, Forest ecology and Natural Resource Management.

Most of the chapters have been authored by participants of the 2006 IUFRO Landscape Ecology Workshop, but not excluding other interested parties from participating. The workshop would not be successful without support of many organizations (e.g., University of Bari, CRSA-Basile Caramia, Urban Forestry Working Party of IUFRO, USDA-Forest Service, University of Toledo) and many individuals (Claudia, Marco, Giuseppe, Leonardo, Zaira, etc.). We thank people of Springer for their consistent and positive support for considering this book.

The following reviewers provided their timely and valuable for some chapters of the book João Azevedo, Jan Bogaert, Jim Bouldin, Robert Brown, Gherardo Chirici, Robert Corry, Igor Danilin, Chao Li, Richard Lucas, Sari Saunders, and Santiago Saura.

Raffaele Lafortezza
Jiquan Chen
Giovanni Sanesi and
Thomas R. Crow
The Editors

Contributors

Arnon Accad

Queensland Herbarium, Environmental Sciences, EPA Brisbane Botanic Gardens, Mt Cool-tha, 4066 Australia

John Armston

Natural Resource Sciences, Department of Natural Resources and Water Climate, QLD 4068, Australia

João C. Azevedo

Departamento de Ambiente e Recursos Naturais/CIMO-Centro de Investigação de Montanha, Escola Superior Agrária, Instituto Politécnico de Bragança, 5301-854 Bragança, Portugal, jazevedo@ipb.pt

Issouf Bamba

Université libre de Bruxelles, Laboratoire d'Ecologie du Paysage, B-1050 Bruxelles, Belgium

Daniel S. Banfai

School for Environmental Research, Institute of Advanced Studies, Charles Darwin University, Darwin, Australia, daniel.banfai@earthsystems.com.au

Anna Barbati

Department of Forest Environment and Resources, University of Tuscia, Italy

Jean-Marc Boffa

ICRAF, World Agroforestry Centre, Nairobi, Kenya

Jan Bogaert

Université libre de Bruxelles, Laboratoire d'Ecologie du Paysage, B-1050 Bruxelles, Belgique, jan.bogaert@ulb.ac.be

Janine Bolliger

Swiss Federal Research Institute WSL, Birmensdorf, Switzerland.

David M.J.S. Bowman
School of Plant Science, University of Tasmania, Hobart, Australia

Don C. Bragg
U.S. Department of Agriculture, Forest Service, Southern Research Station, Monticello, Arkansas, USA

Kimberley D. Brosofske
School of Forest Resources and Environmental Science, Michigan Technological University, USA

Robert D. Brown
School of Environmental Design and Rural Development, University of Guelph, Guelph, Ontario, Canada N1G 2W1

Peter Bunting
Institute of Geography and Earth Sciences, Aberystwyth University, Aberystwyth, Ceredigion, SY23 3DB, United Kingdom.

Francisco Carabelli
Patagonia National University, Esquel, Chubut Province, Argentina, francisco.carabelli@gmail.com

Dominique Champluvier
National Botanic Garden of Belgium, Domein van Bouchout, B-1860 Meise, Belgium

Jiquan Chen
Landscape Ecology and Ecosystem Science (LEES) Lab, University of Toledo, Department of Environmental Sciences, Toledo OH 43606-3390 USA, Jiquan.Chen@utoledo.edu

Gherardo Chirci
Department of Scienze e Tecnologie per l'Ambiente e il Territorio, University of Molise, Italy

Piermaria Corona
Department of Forest Environment and Resources, University of Tuscia, Italy, piermaria.corona@unitus.it

Robert C. Corry
School of Environmental Design and Rural Development, University of Guelph, Guelph, Ontario, Canada N1G 2W1

Thomas R. Crow
Research and Development, Environmental Sciences Staff, USDA Forest Service, Washington DC 20250-1113 USA, tcrow@fs.fed.us

Igor M. Danilin
V.N. Sukachev Institute of Forest, Russian Academy of Sciences, Siberian Branch Akademgorodok, 50/28, Krasnoyarsk, 660036, Russia, danilin@ksc.krasn.ru

Charles De Cannière
Université libre de Bruxelles, Laboratoire d'Ecologie du Paysage, B-1050 Bruxelles, Belgium

Jean-Pierre Kabulu Djibu
Université libre de Bruxelles, Laboratoire d'Ecologie du Paysage, B-1050 Bruxelles, Belgium; Université de Lubumbashi, Faculté des Sciences, BP 1825, Lubumbashi, Democratic Republic of the Congo

Amy E. Eycott
Forest Research, Alice Holt, Wrecclesham, Farnham, Surrey, England, UK

Richard F. Fisher
Texas A&M University, College Station, TX 77843-2135, USA

Assu Gil-Tena
Department of Agroforestry Engineering, Higher Technical School of Agrarian Engineering, University of Lleida, Lleida, Spain

Mark H. Hansen
U.S. Department of Agriculture, Forest Service, Northern Research Station, Forest Inventory and Analysis Program, St. Paul, Minnesota, USA

Mark A. Hatfield
U.S. Department of Agriculture, Forest Service, Northern Research Station, Forest Inventory and Analysis Program, St. Paul, Minnesota, USA

Jonathan W. Humphrey
Forest Research, Northern Research Station, Roslin, Midlothian, Scotland, UK

William S. Keeton
Rubenstein School of Environment and Natural Resources, University of Vermont, Burlington, VT, USA

Kouao J. Koffi
Université libre de Bruxelles, Laboratoire d'Ecologie du Paysage, B-1050 Bruxelles, Belgium

Piia Koponen
CIFOR, Center for International Forestry Research, Bogor, Indonesia

Ashish Kumar
Gujarat Institute of Desert Ecology, 370 001, Kachchh, Gujarat, India, ashishkumarphd@gmail.com

Raffaele Lafortezza

Department of Scienze delle Produzioni Vegetali, Università degli Studi di Bari, Italy, r.lafortezza@agr.uniba.it

Chao Li

Canadian Wood Fibre Centre, Canadian Forest Service, Natural Resources Canada, Edmonton, Alberta, Canada T6H 3S5, cli@nrcan.gc.ca

Shirong Liu

Chinese Academy of Forestry, Wan Shou Shan, Beijing, P.R. China 100091

Jianwei Liu

Forestry Branch, Manitoba Conservation, Winnipeg, Manitoba, Canada R3J 3W3

Richard Lucas

Institute of Geography and Earth Sciences, Aberystwyth University, Aberystwyth, Ceredigion, SY23 3DB, United Kingdom, rml@aber.ac.uk

Chuanwen Luo

School of Forestry, Northeast Forestry University, Heilongjiang Province, P.R. China 150040

Sandra Luque

Cemagref, – Institute for Agricultural and Environmental Engineering Research. UR Ecosystèmes Montagnards, Grenoble, F-38402 Saint-Martin-d'Hères cedex, France, sandra.luque@cemagref.fr

Bruce G. Marcot

Ph.D. Research Wildlife Ecologist USDA Forest Service, Pacific Northwest Research Station, Portland, Oregon 97205, USA

Michael G. Messina

Department of Ecosystem Science and Management, Texas A&M University, College Station, TX 77843-2138, USA

Darren Moseley

Forest Research, Northern Research Station, Roslin, Midlothian, Scotland, UK

W. Keith Moser

U.S. Department of Agriculture, Forest Service, Northern Research Station, Forest Inventory and Analysis Program,, St. Paul, Minnesota, USA, wkmoser@fs.fed.us

Timothy A. Nigh

Missouri Department of Conservation, Columbia, Missouri, USA

Malcolm P. North
Sierra Nevada Research Center, Davis, CA 95618, mpnorth@ucdavis.edu

Trudy O'Connor
ICRAF, World Agroforestry Centre, Bogor, Indonesia

Lucía Pascual-Hortal
Department of Agroforestry Engineering, Higher Technical School of Agrarian Engineering, University of Lleida, Lleida, Spain

Jean-Laurent Pfund
CIFOR, Center for International Forestry Research, Bogor, Indonesia, j.pfund@ cgiar.org

Christopher P. Quine
Forest Research, Northern Research Station, Roslin, Midlothian, Scotland, UK

Lucy Randall
Land Management Sciences, Bureau of Rural Sciences, Canberra ACT 2601 Australia

Duncan Ray
Forest Research, Northern Research Station, Roslin, Midlothian, Scotland, UK

Elmar Robbrecht
National Botanic Garden of Belgium, B-1860 Meise, Belgium

P.S. Roy
National Remote Sensing Agency (Govt. of India) Balanagar, Hyderabad - 500 037, India

Riccardo Salvati
Department of Forest Environment and Resources, University of Tuscia, Italy

Giovanni Sanesi
Università degli Studi di Bari, University of Bari, Department of Scienze delle Produzioni Vegetali Bari Italy, sanesi@agr.uniba.it

Santiago Saura
Department of Agroforestry Engineering, Higher Technical School of Agrarian Engineering, University of Lleida, Lleida, Spain; Forest Technology Centre of Catalonia, Solsona, Lleida, Spain, ssaura@eagrof.udl.es

Lisa A. Schulte
Iowa State University, Department of Natural Resource Ecology and Management, Ames, Iowa, U.S.A

Roberto Scoz
Department of Forest Biometry, Albert-Ludwigs University, Freiburg im Breisgau, Germany

Serge Sibomana
Université libre de Bruxelles, Laboratoire d'Ecologie du Paysage, B-1050 Bruxelles, Belgium; Université du Burundi, Faculté des Sciences, B.P. 2700, Bujumbura, Burundi

Jean-Pierre Sorg
ETH Zurich, Swiss Federal Institute of Technology, Switzerland

Olga Torras
Department of Agroforestry Engineering, Higher Technical School of Agrarian Engineering, University of Lleida, Lleida, Spain

Nina Vainikainen
Finnish Forest Research Institute, 00170 Helsinki, Finland

Meine van Noordwijk
ICRAF, World Agroforestry Centre, Bogor, Indonesia

Marjolein N. Visser
Université libre de Bruxelles, Laboratoire d'Ecologie du Paysage, B-1050 Bruxelles, Belgium

Kevin Watts
Forest Research, Alice Holt, Wrecclesham, Farnham, Surrey, England, UK, kevin.watts@forestry.gsi.gov.uk

Jimmy R. Williams
Texas A&M University, Blackland Agricultural Research Center, Temple, TX 76502, USA

X. Ben Wu
Department of Ecosystem Science and Management, Texas A&M University, College Station, TX 77843-2138, USA

Yuandong Zhang
Research Institute of Forest Ecology and Environment, Chinese Academy of Forestry, Wan Shou Shan, Beijing, P.R. China 100091

Authors' Profiles

João C. Azevedo Professor at the Department of Environment and Natural Resources of the School of Agriculture, Polytechnic Institute of Bragança, Portugal, where he teaches landscape ecology, conservation, and GIS. João is the principal investigator and a researcher in the Marginal Land Ecosystem Services research group at the Mountain Research Center (CIMO), Portugal. He is a local co-coordinator of the MS program in Nature Conservation and Management from the Azores University. He received a Ph.D. degree in Forestry from Texas A&M University, EUA. His current research interests include landscape change and fire behavior, forested landscape management and conservation, and sustainable systems for biomass production.

Daniel Banfai Ecologist, has over six years research and consulting experience in vegetation and wildlife ecology, including involvement with mining and hydropower projects. Daniel has a broad range of applied skills with biological surveys, research design, GIS, and statistical analyses. He holds a PhD on the ecology and management of rainforest in Kakadu National Park which he completed at Charles Darwin University, Australia. He has presented his research at both national and international conferences, and is the author of several scientific journal papers.

Jan Bogaert Landscape ecologist, is a professor of landscape ecology and agronomy at the Université Libre de Bruxelles, Belgium. He holds Master degrees in Botany (University of Antwerp, Belgium) as well as in Land and Forest Management (University of Ghent, Belgium), and a PhD in Biology (University of Antwerp, Belgium). In 2001, he was a postdoctoral visiting scholar at Boston University, MA, USA. Research fields: development of landscape metrics to characterize landscape composition and configuration; analysis of drivers and ecological consequences of landscape dynamics in West and Central Africa.

Francisco Carabelli Specialist in territorial planning, is Professor at the Patagonia National University in Argentina. He holds a PhD in land use planning from the University of Munich, Germany. He had been research fellow of the National Council of Scientific and Technical Research and Professor at the Doctoral Program of the Iberoamerican Center for Biodiversity in the University of Alicante, Spain. His research interests are the study of temporal and spatial changes of forests, being founder and coordinator of the Group for Studies on Fragmentation of the

Patagonian Andes Native Forests. Complementary research is focused on territorial planning and forest uses. He has authored and co-authored a significant number of articles on these subjects.

Jiquan Chen Landscape and ecosystem ecologist, is a Professor of ecology at the University of Toledo (USA). He holds a PhD in ecosystem analysis from the University of Washington in 1991 and MS degree in forest ecology from the Chinese Academy of Sciences. His main research interests are in the fields of ecosystem process and community structure-function at landscape level. He is the Co-Chair of the Landscape Ecology Working Party of the IUFRO (08.01.02) and the chief scientist for the US-China Carbon Consortium (USCCC). He has authored and co-authored more than 130 scientific papers and four books on edge effects, landscape fragmentation, biological diversity, microclimate, carbon-water cycle, landscape analysis, remote sensing, and ecosystem modelling.

Piermaria Corona Full Professor (chair of Forest Management and Planning) at the University of Tuscia (Italy), where he is also the Dean of the School of Forest and Environmental Sciences. His main research fields are forest resources inventory and management, and geomatics for natural resources assessment. He is the author of over 270 scientific and technical publications, and five books dealing with forest monitoring issues. Coordinator of the 4.02.06 IUFRO working unit (Resource Data in Boreal and Temperate Regions).

Thomas R. Crow National Program Leader and Assistant Director for Ecology on the Environmental Sciences Research and Development Staff, USDA Forest Service, Washington, DC. He is the past president of the U.S. Section of the International Association of Landscape Ecology, past chair of the SAF Forest Ecology technical group, and currently is the co-chair of the International Union for Forestry Research Organizations (IUFRO) Working Party for Landscape Ecology. His research has focused largely on applying principles and concepts from landscape ecology to managing natural resources.

Igor Danilin Forest ecologist, is a Leading Research Scientist at the V.N. Sukachev Institute of Forest, Russian Academy of Sciences, Siberian Branch, Professor for Land Management Department at Krasnoyarsk State Agricultural University (Russia). He holds a PhD in Forest Inventory and Forest Planning from the Russian Academy of Sciences. His main research interests are in the fields of forest inventory, forest/landscape planning and ecology, land management, remote sensing methods. He has authored and co-authored a large number of articles on forest inventory and planning, landscape ecological and GIS-based modelling, remote sensing methods of forest landscapes' and land monitoring.

Ashish Kumar Forest ecologist, is a Scientist at the Gujarat Institute of Desert Ecology, Bhuj, India. He holds a PhD in Forestry (Forest Ecology and Environment) from the Forest Research Institute and Deemed University, Dehradun, India. His expertise is in forest biodiversity conservation, wildlife management, landscape ecology and regional environmental planning. He has provided consultancy to the National Tiger Conservation Authority, a statutory body of the Ministry of

Environment and Forests, Government of India for landscape level conservation of tigers and their habitats in India, and is a registered member of IUCN World Commission on Protected Areas and IUCN Commission on Ecosystem Management. He has authored articles and book chapters on forest biodiversity conservation and wildlife management.

Raffaele Lafortezza Landscape ecologist, is a Research Associate at the University of Cambridge, UK and University of Bari, Italy. Contract Professor of Landscape Ecology at the Polytechnic of Bari. He holds a PhD in landscape analysis from the University of Bari and has accumulated a wide-range of experience in landscape issues through numerous research projects and scientific collaborations conducted worldwide. His main research interests are in the fields of spatial analysis and (forest) landscape ecology. He is the European coordinator of the Landscape Ecology Working Party of the IUFRO (08.01.02) and the IUFRO representative for the newly established IALE-IUFRO Working Group on Forest Landscape Ecology. He has published in numerous international journals on land and planning related topics.

Chao Li Research Scientist (Landscape Dynamics) of the Canadian Forest Service in Edmonton, Alberta, Canada. He holds a PhD in Insect Ecology from the Simon Fraser University in Burnaby, BC. His extensive ecological modelling research has been published in a large number of articles in various research fields. His current research interests are in the fields of modelling spatial forest fire regimes and associated forest landscape dynamics, integrated land-based management, integration and optimization of forest wood value chain, and artificial intelligent and decision support systems. He is the North America coordinator of the Landscape Ecology Working Party of the IUFRO.

Richard Lucas Biogeographer, is a Reader at Aberystwyth University He holds a PhD in remote sensing of snow and vegetation from the University of Bristol. His main research interests are in the integration of remote sensing data for understanding and quantifying the dynamics of terrestrial vegetation in response to natural and anthropogenic change. He has authored and co-authored over 100 articles on remote sensing of tropical forests, wooded savannas, mangroves, and temperate/high mountain environments.

Sandra S. Luque PhD (Rutgers, USA), Landscape ecologist working in the development of rapid assessment methods for biodiversity evaluation and comprehensive landscape monitoring and modelling. For 17 years she has been working on remote sensing/GIS and landscape ecology issues in relation to change detection, forest ecology, biodiversity indicators, and habitat models. At present she is a trained and active project manager working as a Research Director at Cemagref, France. She serves as one of the elected Vice-presidents of IALE (International Association of Landscape Ecology). She is also the IALE representative for the newly established IALE-IUFRO Working Group on Forest Landscape Ecology.

W. Keith Moser Research forester with the U.S.D.A., Forest Service, Northern Research Station Forest Inventory and Analysis (FIA) Program in St. Paul, Minnesota.

He holds a Doctor of Forestry in silviculture and forest stand dynamics from Yale University. His research focuses on disturbance ecology, vegetation analysis, and management activities and implications using FIA data. Before assuming his current position, Dr. Moser was the silviculturist with the Missouri Department of Conservation (2000–2003), the ecological forestry research scientist at Tall Timbers Research Station in Tallahassee, Florida (1995–2000), and a visiting fellow at Ludwig Maximilian University in Munich, Germany (1994). He also managed the Yale University Forests from 1989 through 1993.

Malcolm North Forest ecologist, is a Research Scientist with the U.S. Forest Service Pacific Southwest Research Station and an Associate Professor at the University of California, Davis. His PhD in Forest Ecosystem Analysis is from the College of Forest Resources, University of Washington. His main research interests are the effects of disturbance on forest ecosystem structure and function. He has authored many papers and a DVD movie on ecosystem dynamics in the forests of California's Sierra Nevada Mountains.

Jean-Laurent Pfund Forest Engineer from the Federal Institute of Technology of Zürich (ETHZ) specialized in research for development, is currently a scientist seconded by SDC and Intercooperation (Switzerland) at the headquarters of CIFOR, Indonesia. He holds a PhD in Natural Sciences. His main research interests are in the fields of biodiversity conservation and sustainable use of natural resources in forest landscapes. Coordinator of the joint CIFOR-ICRAF Biodiversity Platform, he has been involved in several development and conservation projects as well as in transdisciplinary research activities.

Giovanni Sanesi Associate Professor of Urban Forestry and Forest Ecology and Planning at the University of Bari, Italy. His research has been related to many aspects of trees, forests and open spaces in urban and periurban landscapes. He is the national delegate and member of the Management Committee for the implementation of the European research action – COST Action E39: "Forests, trees and human health and wellbeing" (2004–2008). He has authored and co-authored a large number of peer-reviewed articles on sustainable planning and management of urban forests and integration between urban forestry and landscape ecology.

Santiago Saura Associate Professor at the University of Lleida and head of the Landscape Planning, Forest Management and Biodiversity Conservation research group at this University. He is also a researcher at the Forest Technology Centre of Catalonia. He received his awarded Ph.D. in Forestry at the Universidad Politécnica de Madrid in 2002. His main research interests are quantitative methods in landscape ecology, forest biodiversity indicators, landscape connectivity, scale analysis, forest inventory and forest planning. He has leaded numerous research projects and international publications on these topics.

Kevin Watts Programme Leader for Landscape Ecology research at Forest Research, the research agency of the Forestry Commission. His main research interest is focussed on the development of landscape ecology tools to enable forest and land managers to target and evaluate their actions on biodiversity. These tools are being developed and applied at a range of scales, including the forest, catchment, region and country. He is also working on a number of species-based landscape ecology studies, including landscape genetics, to improve the underpinning scientific evidence base. He is an active member of the IUFRO landscape ecology (08.01.02) and a committee member of IAKE UK.

Part I
Underlying Concepts and Applicative Approaches

Chapter 1

Ecology and Management of Forest Landscapes

Jiquan Chen, Kimberley D. Brosofske and Raffaele Lafortezza

Abstract With an emphasis on ecology as the key word in landscape ecology and landscape as the context, we discuss the importance of ecological theory as the basis for the study of forest landscapes. We identify and discuss the core components and challenges in current forest landscape ecological research with an emphasis on how they relate to forest management. Among these challenges are issues such as scaling, ecosystem interactions, modelling and technology/information transfer, hypothesis development, and experimental design at broader scales. Although an extensive literature exists focusing on many of these issues, few contributions have addressed them in the context of forest landscape ecology and management. We propose a conceptual framework working toward a more complex and realistic analysis of forest landscapes. Finally, we provide our perspective on the key challenges for incorporating landscape ecological principles into forest landscape management by highlighting the missing links between academic researchers who often deal with abstract ideas and decision makers who need realistic information presented in a user-friendly format.

1.1 Ecology as the Keyword

Landscapes, consisting of multiple components coexisting and interacting on a continuous land surface, have become a focus of much modern ecological research and application. Moving from the individual pieces of the landscape (i.e., ecosystems) to the interacting whole is a useful perspective when addressing large-scale ecological questions and management tasks. With varying scales, different ecological features emerge such as landscape patches that are configured in time and space by natural and cultural determinants (see: Lafortezza et al., Chapter 2). Under this perspective, landscapes can be seen as complex assemblages of structural units whose patterns interact with ecological processes in addition to reflecting the

J. Chen
Landscape Ecology and Ecosystem Science (LEES) Lab, University of Toledo, Department of Environmental Sciences 2801 W. Bancroft St. Toledo OH 43606-3390 USA
e-mail: Jiquan.Chen@utoledo.edu

R. Lafortezza et al. (eds.), *Patterns and Processes in Forest Landscapes*,

magnitude and intensity of human activities. Understanding the underlying mechanisms responsible for creating and maintaining landscape mosaics (Forman 1995), as well as the ecological consequences of these mosaics, remains a central issue in landscape ecological research (Chen et al. 2006; Farina 2006; Green et al. 2006; Kienast et al. 2007; Wu and Hobbs 2007).

Traditionally, ecologists have taken a bottom-up approach and defined the various sub-disciplines according to the organismal ladder (e.g., population and community ecology) (Fig. 1.1), which places organisms at the center of ecological research (i.e., the "bio-view" of ecology). Advances in ecosystem studies have since emphasized the importance of the abiotic components of ecosystems such as biogeochemical cycles (Bormann and Likens 1979) and disturbances (Pickett and White 1985), bringing physical processes to the fore in ecological research. Meanwhile, a "geo-view" of ecology has typically focused on large, three-dimensional geographic spaces, with an interest on broad-scale terrain and physical processes (Rowe and Barnes 1994). Interestingly, geoscientists have gradually moved from this broader-scale view to examining the details of the component ecosystems, including the inherent contributions of the resident organisms, i.e., top-down approach. The concept of landscape is intermediate between the bio- and geo-ecological points of view (Rowe and Barnes 1994).

As the broad-scale viewpoint has become more prevalent, it has also become clear that humans and human influences on the landscape (i.e., socio-economic, cultural, and political) are not trivial, especially in our modern world. Humans,

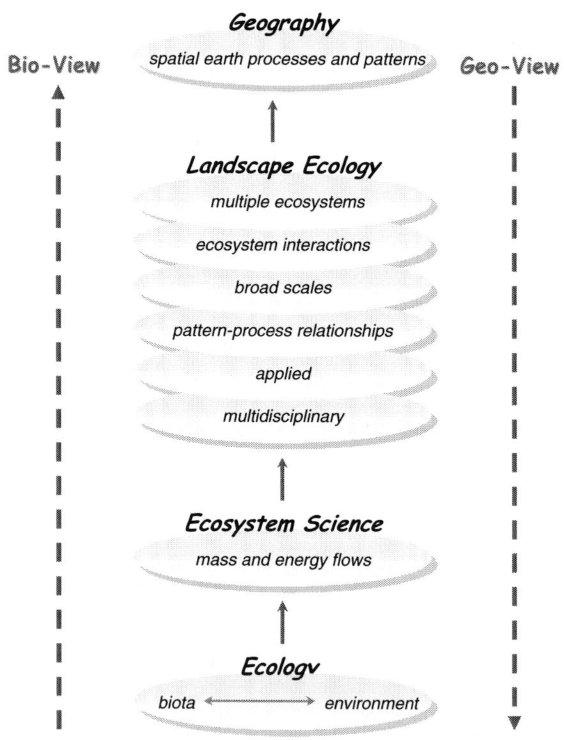

Fig. 1.1 Bio- and geo-ecological views of nature. Traditional ecological research evolved from aggregation of organisms (*bottom up*) into ecosystem and landscape research, while geo-ecologists zoomed in from a large landform perspective to explore details of each land surface unit (i.e., *top down*). The idea that a landscape is composed of multiple ecosystems distinguishes landscape research from other ecological disciplines

while dependent upon and changing with surrounding landscapes (Diamond 1999), are dominant organisms shaping most landscapes, and it is inevitable that full understanding of ecological systems can only come if we specifically address the interactions between humans and nature (Kienast et al. 2007; Liu et al. 2007). Thus, much landscape ecological research is of value for applied uses, such as resource management in many parts of the world. The demands for resources have produced unexpected ecological and social impacts (e.g., loss of biological diversity, loss of valuable agricultural fields, urban sprawl, etc.) that are often associated with landscape-level processes such as fragmentation, spread of infectious diseases and wildfire, urbanization, and species introductions and invasions (Lindenmayer and Franklin 2002; Kienast et al. 2007). These human-related processes further emphasize the importance of ecology as a keyword for researchers and practitioners because of the need for information relevant to conservation, landscape and urban design, and strategic planning for a sustainable environment (Nassauer 1997; Liu and Taylor 2002; Silbernagel 2003; Perera et al. 2007). In this chapter, we discuss some of the main issues associated with the study of landscape ecology and provide a general background for analyzing and managing forest ecosystems within a landscape context.

1.2 Landscape as the Context

Landscapes are fundamental components of the living environment. Because landscapes change across regions and scales, a variety of (conceptual) models and frameworks have been proposed to support research approaches and applicative goals. The most common way to define landscapes is through the lens of landscape ecology that is the study of ecological patterns and processes in heterogeneous environments (Urban et al. 1987; Sanderson and Harris 2000; Turner et al. 2001). Landscapes are explained as heterogeneous entities composed of multiple kinds and spatial arrangements of ecosystems. The arrangement of the different kinds of ecosystems constitutes the spatial pattern of the landscape (i.e., the landscape mosaic), while the interactions among ecosystems represent the underlying mechanisms or processes that create and maintain the landscape pattern, e.g. movement of energy, materials, and species through heterogeneous land mosaics (Forman and Godron 1986; Turner 1989).

As a unique scientific discipline handling biophysical and social processes among multiple ecosystems, landscape ecology requires solid foundations and principles to be of benefit to practitioners and resource managers. In their recent contribution, Chen and Saunders (2006) discussed some of the main facets and core elements of landscape ecology, identifying several critical and emerging issues in the discipline (Fig. 1.1), including:

- Delineation of landscape patterns in relation to processes.
- Interest in up-scaling ecological patterns and processes.
- Understanding ecosystem interactions (areas-of-edge influence, AEI, as landscape elements).
- Development of hypothesis-oriented ecological investigations.

Despite the extensive literature on the subject, few contributions have attempted to place these elements in the context of forest landscape ecology and management. In the following sections, we describe some of the main implications for incorporating these elements into current landscape ecological research, with particular emphasis on the analysis and management of forest ecosystems and landscapes.

1.2.1 Ecological Patterns and Processes

In landscape ecology, a large amount of effort has thus far gone into quantifying spatial patterns and their dynamics through time (e.g., Turner and Gardner 1991; Bresee et al. 2004). The analysis of the landscape pattern generally involves the adoption of quantitative approaches and methods along with dedicated tools based on geographical information systems (GIS) and remote sensing technologies. Once spatial information on landscapes has been made available and/or derived from remotely sensed data, landscape pattern analysis can take place considering each landscape element as part of a discrete patch mosaic: each patch is treated as a structural element of the landscape bounded by other patches or matrix that may be more or less similar (Lafortezza et al. 2005). Landscape patches are then subject to further analysis and computation aimed at determining quantitative measures of landscape pattern. Conversely, less effort has focused on the underlying ecological processes and functions of landscapes, although such efforts are beginning to gain momentum. In order for the science of landscape ecology to progress, it must move toward a more mechanistic understanding of landscape patterns and processes. Farina (2006) identified several categories of emerging landscape processes, including disturbance (e.g., wind-throw, fire, human activities, snow, animals, pathogens), fragmentation, corridors and connectivity (e.g., species movements across landscapes), and nutrient/water movement across landscapes. In addition, ecosystem processes such as biogeochemical cycling and species interactions (e.g., food webs) have great relevance to landscape studies, especially as we try to understand related processes at even broader scales (e.g., global carbon budget). Understanding how landscape patterns influence these processes and how these processes in turn influence the patterns observed is at the core of landscape ecology. Describing and quantifying spatial patterns and their dynamics can give useful information, but landscape ecology is now ripe for a more complex, ecologically-focused phase, emphasizing: (1) reasons and mechanisms for the patterns; and (2) ecological or socio-economic consequences of the patterns (see: Bogaert et al., Chapter 5).

1.2.2 Scaling Issues

Scaling issues are paramount in landscape ecology, particularly because of the difficulties of measuring ecological variables at landscape scales. It is generally accepted in ecology that: (1) scientific investigations of a system (structure and function) need to be conducted at proper scales, and (2) ecological processes need to be explored across a range of scales. Often in ecology, measurements are taken at finer scales and "scaled-up" to give a landscape estimate (e.g., Saunders et al., 2002; Corry and Lafortezza 2007). How to do this effectively is a topic of intense interest in landscape ecology, but it is generally agreed that a multi-scale perspective is necessary; that is, scaling necessitates examining both finer-scale mechanistic processes as well as broader-scale constraints (see: Danilin and Crow, Chapter 4). Because landscapes consist of a collection of ecosystems, ecological processes occurring at the community or ecosystem level are typically measured and aggregated to provide an estimate for the landscape (Peters et al. 2004). Extrapolating ecosystem processes to the landscape level can provide helpful information concerning critical environmental issues, such as global climate change (see: Crow, Chapter 3). However, landscapes are composed of interacting ecosystems, and a simple aggregation approach might well lead to inaccurate landscape estimates if transitional or edge areas are ignored (Noormets et al. 2006). Another problem with scaling-up involves the high likelihood that many small ecosystems or other important landscape elements (e.g., roadsides, small wetlands, riparian zones, etc.) may not be sampled using an ecosystem-based sampling approach. For example, overlooking small, embedded wetlands in an upland matrix might result in a gross underestimate of landscape-level species richness because those species only occurring in the wetlands (wetland-obligate species) will not be sampled. Development of appropriate sampling (or experimental) designs that allows for extrapolation is a common difficulty in landscape ecology. When one is interested in exploring pattern-process relationships at multiple scales in time and space, designing an effective experiment could be even more challenging.

1.2.3 Area of Edge Influence

An important element of this pattern-to-process interdependency is the study of the influence of edges. A conceptual framework for theorizing the importance of edges in fragmented landscapes has been proposed by Chen et al. (1996). Edges and their configurations dominate the overall landscape structure. Virtually all movements of energy, materials, and species across a landscape involve passing through and, thus, being influenced by, a structural edge. Previous studies have shown that ecosystem properties within the area-of-edge-influence (AEI) are much more variable and dynamic than those in the interior, adjacent forests (Harper et al. 2005). The depth-of-edge-influence (DEI, Chen et al. 1992) depends on the variable of interest, time of day (season and year), edge orientation, and edge age (Chen et al. 1995). In the northern hemisphere, edge effects on microclimate, vegetation, and soil are greatest in south- or southwest-facing edges (Chen et al. 1992, 1995). Pronounced increases in temperature, light, vapor pressure deficit, and tree mortality (i.e. lower leaf-area index) have been documented within the AEI.

These changes in vegetation and microclimate collectively determine the underlying processes influencing (positively or negatively) ecosystem production (Fig. 1.2). Because the magnitudes of the independent variables change at different

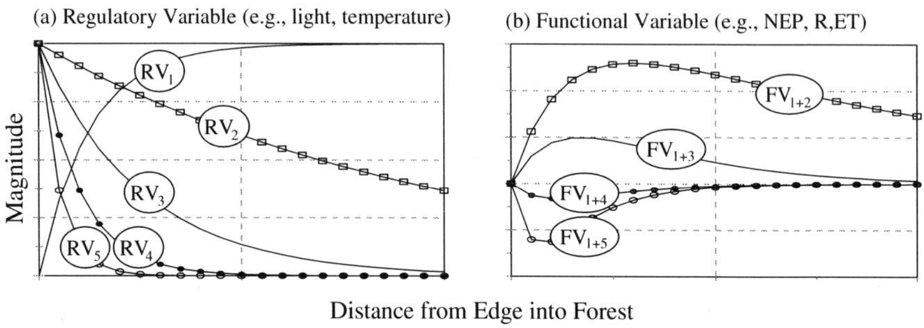

Fig. 1.2 Changes in the biological and physical environment from a structural edge into both forests and harvested areas have been reported (**a**). The combined biophysical effects of regulatory drivers for ecosystem function (e.g., ecosystem production) will produce different gradients (**b**) depending on the importance of each independent variable in controlling the function and their interactions (i.e., linear or not). The illustrated changes in ecosystem production as a functional variable (FV) in (**b**) was simplified with linear combinations of RF1 with others (i.e., RF2–RF5)

rates and sometimes in different directions from the edge into the forest, the combined effects on the dependent variable may be positive or negative (Fig. 1.2). For example, we expect an increase in the ratio of ecosystem respiration (R) to primary production (GPP) within the AEI and, hence, reductions in net ecosystem production (NEP) relative to the forest interior. However, increased light penetration and higher temperatures at north-facing edges may favor plant growth in the AEI in comparison to the forest interior (Chen et al. 1992). The edge influence on the two adjacent patches is also asymmetrical, with greater DEI on the taller side of the edge than on the shorter. Studying ecosystem processes such as carbon and water cycles in relation to landscape patterns is especially timely, given current concerns about global climate change and human influences on the global carbon budget. It has been suggested that human activities, especially involving land-use change, can have large effects on ecosystem productivity and carbon dynamics (e.g., Houghton et al. 1987; Schimel et al. 2000).

We suggest that the interactions between neighboring ecosystems through edge effects modify landscape-level NEP because of altered microclimate and NEP within AEIs. Edge effects on microclimatic parameters affecting NEP are asymmetric across the edge (Euskirchen et al. 2002) and differ with edge orientation. Greater reductions in NEP are expected for old stands than for young ones, and the magnitude of reduction at south-facing edges should be significantly higher than the minor increase at north-facing edges (Fig. 1.3). However, AEIs are relatively narrow strips in fragmented landscapes too small for tower-based flux measurements (Noormets et al. 2007), yet are potentially significant in terms of their contributions to landscape NEP. Therefore, repeated NEP estimates in AEIs of different types, ages, and orientations are needed to test the hypothesis. A potential method might involve sampling the key biometric and physiological input variables for an ecosystem model and combining these with chamber-based measurements of soil respiration and photosynthesis in AEIs of different forest types.

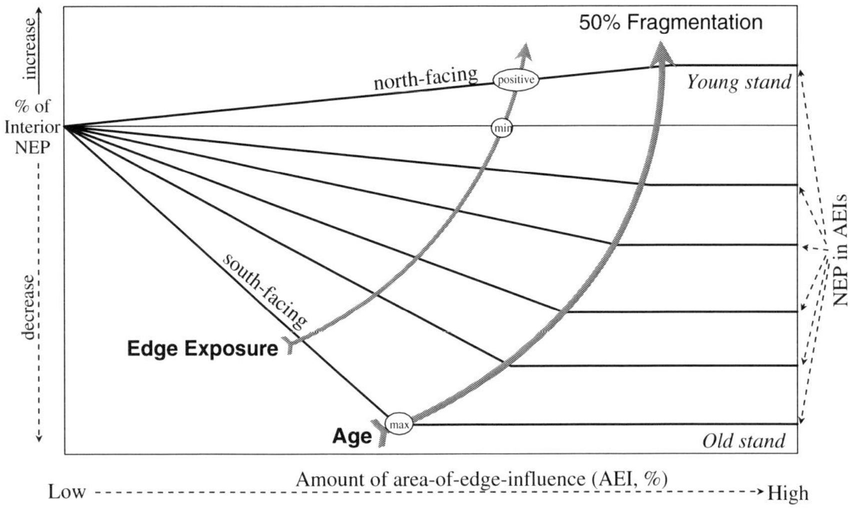

Fig. 1.3 Hypothesized effects of edges on NEP within the area of edge influence (AEI). Edge orientation and edge age are the two most important factors determining the changes in NEP and ecosystem water use within the AEIs

Recent perspectives have emphasized the role of the AEI as a basic landscape element in addition to patches, corridors, and the matrix (e.g., Chen and Saunders 2006), and a generalized theory of edge influence has begun to emerge (Hansen and Rotella 2000; Sanderson and Harris 2000; Harper et al. 2005). However, even more recent works have suggested that edge structures could be even more complex. Areas where multiple edges converge have been recognized as being prevalent on many landscapes, especially where human activities are prominent, and questions have arisen as to whether ecological properties in these "areas-of-multiple-edge-influence" (AMEI) might be affected differently from areas influenced by only a single edge (Li et al. 2007). These questions open up a whole new area for research and could end up being another important consideration in landscape management.

1.2.4 Developing Testable Hypotheses and Sound Experiments

Developing testable hypotheses and manipulative experiments in landscape research is perhaps the most challenging task in the discipline (Chen and Saunders 2006; Green et al. 2006). Rarely do researchers have the opportunity to manipulate multiple landscapes over a long time period to meet the robust requirements of sound statistical analysis (e.g., classical ANOVA). Thus, "true" replications are notoriously difficult to obtain and impossible to guarantee. However, many landscape-scale studies do not attempt any sort of replication (although, certainly, some do). Despite the difficulties in finding independent replicates for a study, theoretical advances in the field rely upon the ability to generalize, which is difficult or impossible from single, unreplicated studies; therefore, significantly greater effort should be expended in developing sound experimental designs with replication where possible. Where true replication is not possible, it is important to recognize the limitations on the statistical interpretation of the results, and it is especially crucial to place the results in the context of other similar work. Ignoring the limitations of the study design does a disservice to the science by masking the appropriate interpretation of the results. Addressing these limitations in a straightforward way does not invalidate or deemphasize the importance of the results, but instead allows them to be used more appropriately to advance theory as well as to better discover improvements in study design.

If handled appropriately, un- or pseudo-replicated studies can contribute new insights to landscape-scale research. Synthesizing the results of many unreplicated studies could identify commonalities that allow generalization and contribute to theoretical development. However, comparing the results of many studies is made more difficult, and sometimes impossible, because of the wide variety of sampling methods employed. A popular approach to overcoming this sampling design issue and achieving study objectives at broader spatial and temporal scales (i.e., those relevant to landscape studies) is to develop a consortium where multiple research teams agree on a unified design and data sharing policy (i.e., to achieve replication). The US-China Carbon Consortium (USCCC, http://research.eeescience.utoledo.edu/lees/research/USCCC/) and The Global Lake Ecological Observatory Network (GLEON, http://gleon.org) were constructed along this line of thought. While developing networked studies with standardized sampling protocols could promote huge advances in the science by making synthesis and generalization a far simpler task, labor and other costs can be prohibitive in such endeavors. For example, the recent development of the National Ecological Observatory Network (NEON, http://www.neoninc.org/) was originally conceived as a way to provide a sound, thorough, consistent sampling design for USA landscapes, but was immediately scaled down because of cost.

An alternative to the difficulties of landscape manipulation in research is to search for existing landscape configurations that can provide an almost experiment-like setting. Although these will not represent "true replications" in the sense of a controlled experiment, they can still provide valuable data. For example, several studies have taken advantage of the contrasting patch patterns and landscape structural elements found in the Chequamegon-Nicolet National Forest in northern Wisconsin, USA, to study plant distribution, microclimatic changes, and carbon and water dynamics (e.g., Brosofske et al. 1999, 2001; Brosofske 2006; Saunders et al. 1998, 1999, 2002; Euskirchen et al. 2001, 2003; Watkins et al. 2003; Noormets et al. 2006). The varying patch patterns of this study area have been largely imposed by forest management activities, while other properties (e.g., soils, topography, historic vegetation) are relatively homogeneous, resulting in a fantastic opportunity for landscape study, even without deliberate manipulation on the part of the researchers. In the absence of experimental manipulation, such geographical areas can be extremely valuable assets to the study of landscapes.

1.3 Knowledge and Technology Transfer

All too often, knowledge and technology transfer in landscape ecology is limited to a passive delivery of an overview of research results and techniques. The best way to assure understanding and competence is through facilitated hands-on practice. Hands-on technology transfer (HOTT) involves formulating strategies for resource managers to improve their current plans and protocols by evaluating options and by using instantaneous models to evaluate the consequences of different scenarios. Both passive and active technology transfer methods need to be exercised to deliver research products effectively. An important issue here is that scientists need to conduct research that is of current interest to managers and others who want answers to timely questions (see: North et al., Chapter 17). This requires strong communication between researchers and managers/stakeholders to identify topics of interest and the format in which the results should be delivered. Developing research proposals in concert with managers and agencies that want to use the research to guide their activities might be one of the best approaches to ensuring the research will be useful. Although basic research is integral to advancing any scientific field, applied research, by definition, must address issues considered to be important within the relevant political, cultural, and social context. Engaging a diverse scientific community, which might include ecological, social, mathematical, physical, geological, economic, and other disciplines, as well as decision-makers, can help to ensure that appropriate questions are addressed and that results are useful for developing a socially- and politically-acceptable management plan.

In addition to conducting relevant research, it is important that the results of that research be made available to resource managers in a user-friendly form. Easy-to-use spatial models that depict spread or spatial pattern of some ecological, social, or economic property or process of interest (e.g., disturbance, invasive species, historic "reference conditions", timber output, etc.) across the landscape have been developed as contributions to resource management and policy making (e.g., Thomas et al. 1990; He and Mladenoff 1999; Gustafson and Rasmussen 2005; Perera et al. 2007). However, many models produced by researchers are complicated and require large amounts of data for parameterization and validation. Such models are generally research tools and not particularly useful to those seeking to develop feasible management strategies. Scenario testing, in which multiple possible events or courses of action are simulated and the spatial consequences for the property or process of interest are assessed, has the potential to be especially helpful. For example, Euskirchen et al. (2002) tested the effects of three different hypothetical landscape and disturbance scenarios on net ecosystem productivity and biomass, finding that the timing of timber harvest significantly affected the degree to which a landscape sequesters carbon. Care must be taken, however, to choose scenarios that are realistic – another reason to get resource managers involved throughout the study. In managed forest landscapes, major considerations for developing scenarios might include spatial allocations of disturbance (e.g., clustered vs. aggregated harvesting), treatment size, amount of landscape to be treated (e.g., 0.5–10% per year), magnitude of treatment (e.g., clearcutting vs. partial cutting), policy or other constraints on harvesting, regeneration and successional pathways, and other disturbances besides timber harvesting. Consultations with those who manage the forest can provide more realistic scenarios for testing that take into account the goals specific to that landscape and the activities that could actually be implemented or alternatives actually under consideration. Testing practical, sound scenarios and developing predictions based on those scenarios are becoming a popular choice of methods in ecological science when facing long time periods or large spatial extents that do not lend themselves well to field inquiry. These modelling and predictive approaches have demonstrated their effectiveness in influencing policy decisions, such as the management of old-growth forests in the Pacific Northwest (Thomas et al. 1990) and strategic planning for global warming (IPCC 2007).

1.4 Sustainable Management as the Priority

Sustainability of forest ecosystems and landscapes has become a pressing problem and important management goal in recent years as recognition of the extent of land use change (e.g., de- or re-forestation) due to accelerated human activities increases, along with concerns about interactions with the changing climate and an increasing rate of exotic species invasion. Forest management has been undergoing a paradigm shift that calls for a more integrated ecosystem management that emphasizes the maintenance of ecosystem functions as equal in importance to economic output. Effects on the landscape mosaic of different forest management activities may be more complex than simple models would suggest, with potential large variability in the outcomes of those management activities when measured by different ecological functions.

Landscape ecology principles suggest that all parts of a landscape work together to maximize functional output. However, one cannot assume that restoration of a habitat, for example, will always be accompanied by sound ecosystem functioning. Likewise, habitat loss due to non-indigenous species can be accompanied by high production (Binkley et al. 2004; Chen et al. 2008). In northern Wisconsin, fragmentation and plantations have greatly altered the successional progression of the forests and landscape, resulting in very complex patch-dynamics of production and biological diversity (Chen et al. 2004; Brosofske 2006; Noormets et al. 2007). Optimal solutions to managing a forest landscape cannot be found without clear definition of functions; the search for optimal outcomes is further complicated by the need to incorporate social and economic perspectives and objectives. It is particularly important for landscape ecologists and managers to bear in mind that a management plan developed solely based on sound science may not work unless it is also socially and politically acceptable. This is vital because it is often the societal culture that decides what plan to adopt, while the political system of a country (or region) constrains the options. In many parts of the world, for example, people have begun considering some historically invasive species (e.g., olives in southern Italy) as "native" (i.e., culturally acceptable for the local landscape). As another example, any landscape design proposal near the Three Gorges Dam site on the Yangtze River that assumes a "no dam" option will receive no consideration regardless of merit because the government has already decided to construct the dam. In order to effectively address the functioning of ecosystems and modern problems such as fragmentation, loss of biological diversity, and species invasions, as well as economic outputs of forests, a broad-scale perspective is necessary in planning activities. Actual management applications, however, typically take place at the stand level, creating a need for a multi-scale perspective in ecosystem management. When implementing ecosystem management, the issue of feasibility in forest planning has become more important because of the often disparate scales at which planners and on-the-ground managers function. Not all management strategies are relevant to all scales, and different outcomes might be evident at different scales. It is understood that cultural, political, economic, and engineering constraints can act to limit or define potential management activities, but broader scales and complex spatial interactions must now be addressed as well.

With the shift to ecosystem management, managers are being asked to manipulate forests and other resources at scales rarely considered previously, but empirical research at landscape scales is sparse, and theory is still developing. Thus, management has often responded by becoming more adaptive, or adopting a "learn-as-you-go" mentality. Several large-scale, long-term ecosystem projects, such as the DEMO project in the Pacific Northwest (Halpern et al. 2005), the Structural Complexity Enhancement (SCE) project in Vermont (Keeton 2006), the Sierran Mixed-Conifer Research in California (North and Chen 2005), and the Missouri Ozark Forest Ecosystem Project (MOFEP) in the southeastern Ozarks of Missouri (Shifley and Kabrick 2002), have been developed based on the adaptive management concept and provide encouraging examples of successful collaborations between scientists and managers (see: North et al., Chapter 17). Such projects provide excellent opportunities for scientists to cooperate with land managers to identify relevant research questions, conduct experimental, manipulative studies at broad scales, and thereby develop the scientific base needed for sound forest landscape management.

References

Bormann FH, Likens GE (1979) Pattern and process in a forested ecosystem. Springer, New York

Bresee M, Le Moine J, Mather S, Crow T, Brosofske KD, Chen J, Crow TR, Radmacher J (2004) Disturbance and landscape dynamics in the Chequamegon National Forest Wisconsin, USA, from 1972 to 2001. Landsc Ecol 19:291–309

Brosofske KD (2006) Spatial structure and plant diversity in hierarchical landscapes. In Chen J, Saunders SC, Brosofske KD, Crow TR (eds) Ecology of hierarchical landscapes: from theory to application. Nova Publishing, Carbondale, IL, pp 35–69

Brosofske KD, Chen J, Crow TR (2001) Plant diversity and local site factors: implications for a managed Wisconsin landscape. Forest Ecol Manag 146:75–87

Brosofske KD, Chen J, Crow TR, Saunders SC (1999) Vegetation responses to landscape structure at multiple scales across a northern Wisconsin, USA, pine barrens landscape. Plant Ecol 143:203–218

Chen J, Brosofske KD, Noormets A, Crow TR, Bresee MK, Le Moine JM, Euskirchen ES, Mather SV, Zheng D (2004) A working framework for quantifying carbon sequestration in disturbed land mosaics. Environ Manage 34(3):210–221
Chen J, Franklin JF, Lowe JS (1996) Comparison of abiotic and structurally defined patch patterns in a hypothetical forest landscape. Conserv Biol 10(3):854–862
Chen J, Franklin JF, Spies TA (1992) Vegetation responses to edge environments in old-growth Douglas-fir forests. Ecol Appl 2:387–396
Chen J, Franklin JF, Spies TA (1995) Growing season microclimatic gradients extending into old-growth Douglas-fir forests from clearcut edges. Ecol Appl 5:74–86
Chen J, Saunders SC (2006) Ecology of multiple ecosystems in time and space. In Chen J, Saunders SC, Brosofske KD, Crow TR (eds) Ecology of hierarchical landscapes: from theory to application, Nova Publishing, Carbondale, IL, pp 1–34
Chen J, Zhao B, Ren W, Saunders SC, Ma Z, Li B, Luo Y, Chen J (2008) Invasive Spartina and reduced sediments: Shanghai's dangerous silver bullet. J Plant Ecol 1:79–84
Corry RC, Lafortezza R (2007) Sensitivity of landscape measurements to changing grain size for fine-scale design and management. Landsc Ecol Eng 3:47–53
Diamond J (1999) Guns, germs, and steel: the fates of human societies. WW Norton & Company Inc., New York
Euskirchen ES, Chen J, Bi R (2001) Effects of edges on plant communities in a managed landscape in northern Wisconsin. Forest Ecol Manag 148:93–108
Euskirchen E, Chen J, Gustafson EJ, Ma S (2003) Soil respiration at dominant patch types within a managed northern Wisconsin landscape. Ecosystems 6:595–607
Euskirchen E, Chen J, Li H, Gustafson EJ, Crow TR (2002) Modeling net carbon flows under alternative management regimes at a landscape level. Ecol Model 154:75–91
Farina A (2006) Principles and methods of landscape ecology: towards a science of the landscape. Springer, The Netherlands
Forman RTT (1995) Land mosaics: The ecology of landscapes and regions. New York, NY: Cambridge University Press
Forman RTT, Godron M (1986) Landscape ecology. John Wiley & Sons, New York
Green DG, Klomp N, Rimmington G, Sadedin S (2006) Complexity in landscape ecology. Springer, New York
Gustafson EJ, Rasmussen LV (2005) HARVEST for Windows v6.1: User's guide. Published on the Internet by the USDA Forest Service, North Central Research Station, St. Paul, MN. Available via: http://www.ncrs.fs.fed.us/4153/harvest/v61/documentation/default.asp Cited 20 Dec 2007
Halpern CB, McKenzie DE, Shelley A, Maguire DA (2005) Initial responses of forest understories to varying levels and patterns of green-tree retention. Ecol Appl 15:175–195
Hansen AJ, Rotella JJ (2000) Bird response to forest fragmentation. In Knigh RL, Smith FW, Buskirk SW, Romme WH, Baker WL (eds) Forest Fragmentation in the Southern Rocky Mountains. University Press of Colorado, Boulder, CO, pp 201–219
Harper K, MacDonald E, Burton P, Chen J, Brosofske K, Saunders S, Euskirchen E, Roberts D, Jaiteh M, Esseen PA (2005) Edge Influence on forest structure and composition in fragmented landscapes. Conserv Biol 19(3):768–782
He HS, Mladenoff DJ (1999) Spatially explicit and stochastic simulation of forest-landscape fire disturbance and succession. Ecology 80:81–99
Houghton RA, Boone RD, Fruci JR, Hobbie JE, Melillo JM, Palm CA, Peterson BJ, Shaver GR, Woodwell GM, Moore B, Skole DL, Meyers N (1987) The flux of carbon from terrestrial ecosystems to the atmosphere in 1980 due to changes in land use: geographic distribution of the global flux. Tellus 39:122–139
IPCC (2007) Climate Change 2007: Impacts, Adaptation and Vulnerability. Contribution of Working Group II to the Fourth Assessment Report of the Intergovernmental Panel on Climate. Parry ML, Canziani OF, Palutikof JP, van der Linden PJ, Hanson CE (eds) Cambridge niversity Press, Cambridge, UK, p 976
Keeton WS (2006) Managing for late-successional/old-growth characteristics in northern hardwood-conifer forests. Forest Ecol Manag 235:129–142

Kienast F, Wildi O, Ghosh S (eds) (2007) A changing world: challenges for landscape research. Springer, The Netherland

Lafortezza R, Corry RC, Sanesi G, Brown RD (2005) Quantitative approaches to landscape spatial planning: clues from landscape ecology. In Brebbia CA, Kungolos A (eds) Sustainable development and planning, Vol 1, WIT Press, Southampton, UK, pp 239–250

Li Q, Chen J, Bresee MK, Rademacher JA, LaCroix JJ (2007) Areas influenced by multiple edges and their implications in fragmented landscapes. Forest Ecol Manag 242:99–107

Lindenmayer DB, Franklin JF (2002) Conserving forest biodiversity: a comprehensive multiscaled approach. Island Press, Washington, DC

Liu J, Dietz T, Carpenter SR, Alberti M, Folke C, Moran E, Pell AN, Deadman P, Kratz T, Lubchenco J, Ostrom E, Ouyang Z, Provencher W, Redman CL, Schneider SH, Taylor WW (2007) Complexity of coupled human and natural systems. Science 317:1513–1516

Liu J, Taylor WW (eds) (2002) Integrating landscape ecology into natural resource management. Cambridge University Press, Cambridge, UK

Nassauer JJ (ed) (1997) Placing nature: Culture and landscape ecology. Island Press, Washington, D.C.

Noormets A, Chen J, Crow TR (2007) Age-dependent changes in ecosystem carbon fluxes in managed forests in northern Wisconsin, USA. Ecosystems 10:187–203

Noormets A, Ewers B, Sun G, Mackay S, Zheng D, McNulty S, Chen J (2006) Linking ecology to land mosaics: an ecosystem perspective. In Chen J, Saunders SC, Brosofske KD, Crow TR (eds) Ecology of hierarchical landscapes: from theory to application, Nova Publishing, Carbondale, IL, pp 89–124

North M, Chen J (2005) Introduction to the special issue on Sierran Mixed-Conifer Research. Forest Sci 51(3):185–186

Perera HP, Buse LJ, Crow TR (eds) (2007). Forest landscape ecology. Springer, New York

Peters DPC, Herrick JE, Urban DL, Gardner RH, Breshears DD (2004) Strategies for ecological extrapolation. Oikos 106:627–636

Pickett STA, White PS (1985) The ecology of natural disturbance and patch dynamics. Academic Press, Inc., Orlando

Rowe JS, Barnes BV (1994) Geo-ecosystems and bio-ecosystems. B Ecol Soc Am 75(1):40–41

Sanderson J, Harris LD (eds) (2000) Landscape ecology: A top-down approach. Lewis Publishers, Washington D.C.

Saunders CS, Chen J, Crow TR, Brosofske KD (1998) Hierarchical relationships between landscape structure and temperature in a managed forest landscape. Landsc Ecol 13:381–395

Saunders SC, Chen J, Drummer TD, Crow TR (1999) Modeling temperature gradients across edges over time in a managed landscape. Forest Ecology and Management 117:1731

Saunders SC, Chen J, Drummer TD, Crow TR, Brosofske KD, Gustafson EJ (2002) The patch mosaic and ecological decomposition across spatial scales in a managed landscape of northern Wisconsin, USA. Basic Appl Ecol 3:49–64

Schimel D, Melillo J, Tian H, McGuire AD, Kicklighter D, Kittel T, Rosenbloom N, Running S, Thornton P, Ojima D, Parton W, Kelly R, Sykes M, Neilson R, Rizzo B (2000) Contribution of increasing CO2 and climate to carbon storage by ecosystems in the United States. Science 287:2004–2006

Shifley SR, Kabrick JM (2002) Proceedings of the second Missouri Ozark Forest Ecosystem Project Symposium: post treatment results of the landscape experiment. FS GTR NC-227. St Paul, MN, USA

Silbernagel J (2003) Spatial theory in early conservation design: examples from Aldo Leopold's work. Landsc Ecol 18:635–646

Thomas JW, Forsmann ED, Lint JB, Meslow EC, Noon BR, Verner J (1990) A conservation strategy for the northern spotted owl. Portland, OR

Turner MG (1989) Landscape ecology: the effect of pattern on process. Annu Rev Ecol Evol S 20:171–197

Turner MG, Gardner RH (eds) (1991) Quantitative methods in landscape ecology. Springer, New York

Turner MG, Gardner RH, O'Neill RV (2001) Landscape ecology in theory and practice: Pattern and process. Springer, New York

Urban DL, O'Neill RV, Shugart HH (1987) Landscape ecology. Bioscience 37:119–127

Watkins Z, Chen J, Pickens J, Brosofske KD (2003) Effects of forest roads on understory plants in a managed hardwood landscape. Conserv Biol 17:411–419

Wu J, Hobbs RH (eds) (2007) Key topics in landscape ecology. Cambridge University Press, Cambridge, UK

Chapter 2

Cultural Determinants of Spatial Heterogeneity in Forest Landscapes

Raffaele Lafortezza, Robert C. Corry, Giovanni Sanesi and Robert D. Brown

Abstract Forests constitute fundamental parts of our living environment and provide a wide range of important benefits and services to society that go far beyond forest products. From a landscape ecological perspective forests can be approached as part of an overall landscape whose pattern affects ecological processes across dimensions of common time and space. Forest landscapes often consist of complex assemblages of forest and non-forest elements (patches, corridors, and matrix) whose arrangement reflects, in part, the magnitude, intensity, and type of human intervention and disturbance. This chapter describes some of the cultural patterns inherent in selected forest landscapes with examples from southern Italy and southern Ontario, Canada. We outline how cultural determinants, such as land tenure systems, forest tenure regimes, silviculture traditions, management plans and practices can affect the way forest landscapes are spatially-arranged and the intrinsic heterogeneity associated with them. We provide illustrative examples of cultural determinants of spatial heterogeneity and conclude by discussing ways for enhancing functional and cultural attributes of forest and non-forest landscape elements within a landscape ecological perspective.

2.1 Introduction

Since its foundation, landscape ecology has been devoted to solve a number of research questions across the gradient between natural and cultural ecosystems (Risser et al. 1984; Forman and Godron 1986). The concept of landscape itself conveys the idea of something that is not totally natural, but somehow modified by humans for cultural needs. Each landscape includes traces of cultural effects that emerge across regions and scales (Nassauer 1997). Some effects are powerful enough to control landscape patterns. Cultural controls markedly emerge in landscapes that have been modified and shaped for productive reasons, like rural and forest landscapes (Brown

R. Lafortezza
Department of Scienze delle Produzioni Vegetali, Università degli Studi di Bari, Italy
e-mail: r.lafortezza@agr.uniba.it

R. Lafortezza et al. (eds.), *Patterns and Processes in Forest Landscapes*,

et al. 2007). This is because the cultural influence on single or multiple ecosystems expands when benefits and commodities are largely tangible and relevant for a given community.

One of the most-evident consequences of this cultural control is the alteration of landscape patterns that, depending on the economic and social relevance of the control, may lead landscapes towards simplified conditions (e.g., landscapes in Iowa, see Corry and Nassauer 2002) or, in some circumstances, diversified and heterogeneous structures (e.g., forested matrix landscapes like in the Mediterranean basin, Fig. 2.1).

In landscape ecology, these myriad possible patterns are described by the term *spatial heterogeneity* which refers to the composition and distribution of patches (i.e., ecosystems) at a given temporal and spatial scale (Forman 1995; Li and Reynolds 1995). Spatial heterogeneity is a reflection of the physical environment, the imprint of past and current land use, and the interaction among these variables (Crow et al. 1999). Landscapes are heterogeneous systems and therefore their study requires a thoughtful understanding of the main causes of spatial variability and the relative consequences for ecological processes (Turner 1989). At the same time cultural effects control and define many landscape patterns and alter ecological processes, while landscape patterns affect local culture (Nassauer 1995). For example, farmers in rolling landscapes change perceptions and attitudes for land stewardship based on suitability of cultivation methods (Nassauer and Westmacott 1987).

Recent studies in landscape ecology reviewed factors that control for landscape patterns and heterogeneity in rural landscapes. For example, Corry and Nassauer (2002), identify three sets of cultural values and traditions that affect the structure and function of rural landscapes in midwestern USA: (a) land division, settlement patterns, and ownership traditions; (b) applied science and technology; and (c) stewardship values and landscape aesthetic values. These values and traditions include things like traditional farm and field size, farm management tools and choices, and norms for caring for the rural landscape, including its appearance.

Fig. 2.1 Complex mosaic of forest and non-forest elements in the Mediterranean region, Apulia Region, southern Italy (photograph by Claudia Cotugno)

Similarly, land tenure systems and forest tenure regimes can be identified as factors controlling for patterns and processes in forest landscapes. For example, Chidumayo (2002) demonstrated how land tenure is responsible for significant structural and functional differences in re-growth following clearing of mature woodland.

Other factors include silviculture traditions and current management plans and practices (e.g., forest harvest and thinning techniques) which are important determinants of cultural patterns and ecological processes variations. These factors contribute to the cultural control of forest landscapes thus acting as drivers of spatial heterogeneity across scales.

In this chapter, we review some of the main determinants of spatial heterogeneity in forest landscapes placing emphasis on the role of humans in shaping patterns and maintaining or altering processes. We stress the concept of forest landscape thus putting it in the broader context of landscape ecology and natural resource management. We provide illustrative examples of cultural determinants of spatial heterogeneity in selected forest landscapes of southern Italy and southern Ontario, Canada. We conclude by discussing ways for enhancing functional and cultural attributes of forest landscapes' components.

2.2 Defining and Understanding Forest Landscapes

Forest landscapes are undoubtedly difficult to define in a general way because they often consist of a mixture between forest and non-forest elements, such as farmlands, roads, water bodies, villages, and different types of vegetation. According to Runesson (2004), forest landscapes can be seen as: *portions of the land that the eye can see in one glance and in which the forest is the most dominant element.* Most forest landscapes are, indeed, land mosaics in which forest management attempts to cope with nature conservation, recreation, water management, and other major societal objectives or multiple uses (Forman 1995). Forest landscapes reflect past and present landscape management activities, and to some degree, the consequences of various types of natural and human disturbances, such as climatic disturbance, fire suppression, and agriculture abandonment (Baker 1993; Boose et al. 2004). In southern Ontario, for example, several different land division systems created a landscape mosaic made by forest patches, crop fields, farmsteads, roads and clustering houses with odd angles, triangular "gores" and an unusual spatial pattern and heterogeneity (Corry et al. 2006; Hart 1998) (Fig. 2.2).

One of the main characteristics of forest landscapes is therefore the presence of forest-type vegetation with patches of various size, shape, and degree of "connectness" (Perera and Baldwin 2000). The structure and spatial arrangement of these patches largely depend on their origin, type and magnitude of human control or natural disturbance of both the forest and surrounding landscape matrix.

Forest patches represent the fundamental elements of these landscapes as they affect many ecological processes, including the movement and persistence of particular species, the susceptibility and spread of disturbances, such as fires or pest

Fig. 2.2 Forest pattern heterogeneity in southern Ontario (photograph by Robert Corry)

outbreaks, and the redistribution of matter and nutrients. In managed landscapes like rural areas, forests are often the most biologically-diverse habitats. Studies of insect outbreaks have shown how the spatial arrangement of forest patches influences the distribution and abundance of organisms (e.g., forest tree leaf-feeding insects) across landscapes, see: Coulson et al. (1999). Complex forest landscape mosaics are considered to provide more locations for different foraging behaviours and for encouraging more boundary-crossing animals through elements of connectivity (Schooley and Wiens 2004).

Other studies have focused on the ecological impacts of management on forest patches and the effects of silvicultural practices on species dynamics (Gustafson and Crow 1994; Michael and Hermy 2002). The most significant impact of forest management is that caused by large-scale clear-cut logging, where forest composition, structure, and function are drastically changed, often for the very long term. This process may lead to forest landscapes made of even-aged patches with an overall reduction in the spatial heterogeneity and landscape functionality. At a finer contextual scale, management may result in multi-aged patches with fine-scale heterogeneous conditions.

2.2.1 Cultural Determinants of Forest Landscapes

The way forest landscapes appear to an observer has to do with many driving or controlling factors that operate simultaneously (Bürgi et al. 2004). As forest landscapes are the results of the interplay between natural disturbances and human interventions, a general perspective is needed. Such perspective provides valuable information on how landscapes are composed and configured and the reasons behind past and current patterns and processes (Brown et al. 2007).

A conceptual model is therefore proposed in order to organize the flow of information and the main cultural factors influencing the spatial heterogeneity of forest landscapes (Fig. 2.3). The model is an adaptation of the so called **DPSIR** framework: **D**riving forces, **P**ressure, **S**tate, **I**mpact, and **R**esponse (Smeets and Weterings 1999).

The model considers *land tenure systems*, *forest tenure regimes*, *silvicultural traditions*, *management plans and practices* as main cultural drivers or determinants (**D**) of **patterns** and **processes** in forest landscapes. These drivers operate through specific pressures (**P**), such as *land-use change* and *landscape disturbance* which in turn determine a change in the state (**S**) of forest landscapes in terms of cultural patterns and ecological processes. Such changes emerge in a series of subsequent and interrelated impacts (**I**) which include alterations of landscape spatial heterogeneity, forest fragmentation, shape complexity as well as impacts on species diversity and natural successions. A wide range of indicators and spatial measures exist to quantify the magnitude and trend of impact and to inform the response (**R**) of forest planners and mangers in setting long-term sustainable plans and management actions (see: North and Keeton, Chapter 17).

For the practical purpose of this chapter, we define *land tenure system* as the institutional framework which society creates to make land ownership, use and management possible and which reflects the level of development in society, economy and technology (Bruce 1998). A land tenure system is the bundle of rights, rules, regulations, and laws that establish the ownership of, and access to a land property, and protect the pattern of ownership (Saastamoinen and Matero 2004). Land tenure, either private or public, is a socio-cultural sensitive issue and generalization to describe global trends is therefore limited by regional factors and local constraints. In forest landscapes, land tenure is one of the main determinants of spatial heterogeneity of the landscape matrix that is the "medium" in which forest patches are embedded (Lindenmayer and Fischer 2006). Where most of the land is owned by smaller, private landowners as opposed to larger, publicly-owned land, the forest landscape appears parcelized into units or patches whose size and shape are related to ownership and land management units. This condition facilitates forest fragmentation and land-use conversion. Where most land is held by very large landowners, the resulting landscape is composed by larger divisions of land into units, like fields or forest blocks (Corry et al. 2006). Ownership parcelization determines spatial structure, with consequences for biodiversity and forest productivity (Crow et al. 1999). In addition, a large number of landowners with a diversity of interests can result in uncoordinated forest management regimes varying in extent, intensities, and spatial implementations. Using timber harvest models, Gustafson and Loehle (2006) predicted the cumulative effects of ownership parcelization and land divestiture on forest landscapes and found significant effects on most measures of forest fragmentation.

Parcelization and subdivision of property into small units are typical driving forces of spatial heterogeneity in Mediterranean forest landscapes, such as those in southern Italy. The patterns of land division and ownership have commonly fragmented primeval forest ecosystems along lines that coincide with roads network, farm boundaries, and settlements. Cyclical disturbances, such as rotational grazing, cutting and coppicing or fire management (i.e., landscape pressures, *P*) have gradually led to complex and heterogeneous cultural patterns characterized by relatively

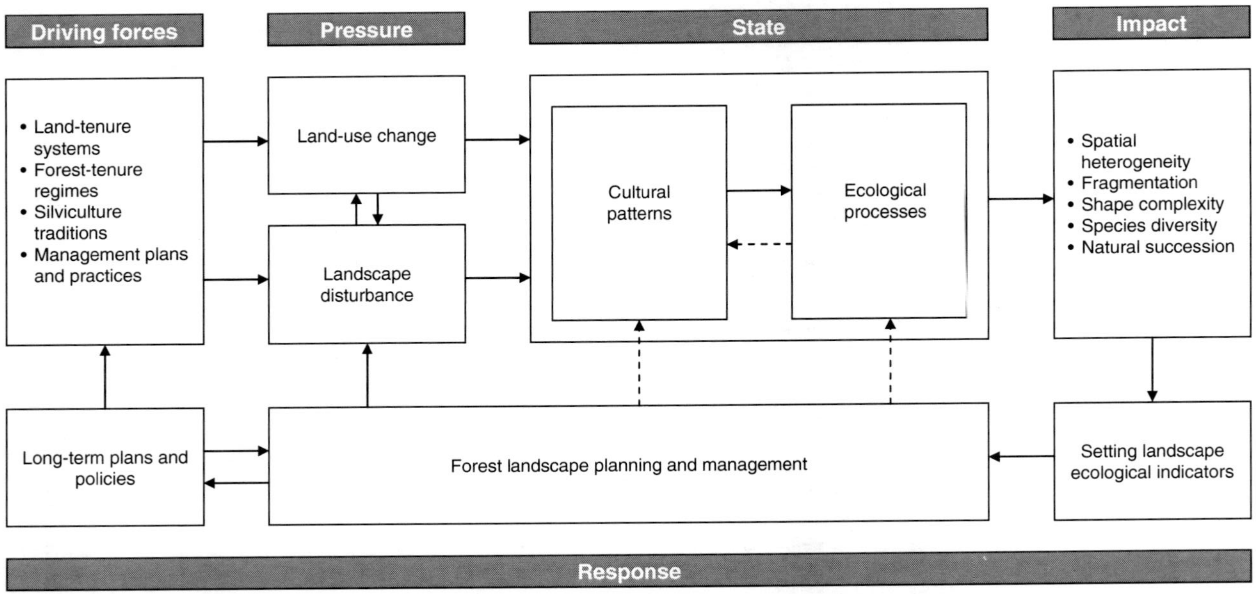

Fig. 2.3 Conceptual model for analyzing the main cultural factors influencing the spatial heterogeneity of forest landscapes

small and regular forest patches. These patterns not only increase spatial heterogeneity in the landscape, they also provide valuable habitat for species that would otherwise disappear or be displaced in the agricultural matrix (Brown et al. 2007).

In southern Ontario forest patches relate to farm and field sizes, and show boundary effects. Since woodlots are sources of fuelwood, construction timber, hunting, trapping, and maple syrup, most farms retain woodlots along back property lines. Farm shapes and layouts affect the shape and alignment of forest patches (see: Fig. 2.2). Farm layout and field sizes in Ontario are generally divisions of five acres (the cultural land division norm, equal to about two hectares), thus common forest patch sizes on Ontario farmland are two, four, six hectares (and so on). Similarly, farm dimensions control forest patch shape, such that a two hectare forest is approximately square, and a four hectare forest is a rectangle with length approximately twice that of the width. Forest patches throughout are small and regular in shape (Pearce 1992; Moss and Strath Davis 1994).

Forest tenure regime is the combination of legally or commonly defined forest ownership rights, responsibilities, and other arrangements for the management and use of forest resources (Romano 2007). Forest tenure directly affects the conduct of forest landowners which in turn affects the way forests are managed and the resulting forest landscape pattern (Gustafson and Crow 1996). In their review, White and Martin (2002) adopt a set of categories to describe global trends in forest tenure and forest ownership rights, starting with the predominant legal categories of public and private ownership (Table 2.1).

Although most forests in the world are in public ownership, trends towards community access and ownership have increased, based on a common view that private ownership may improve economic efficiency and sustainable use of landscapes (FAO 2007). In countries with economies that are in transition, legal transfer of rights to communities or recognition of pre-existing community-based rights have

Table 2.1 Official forest ownership in 24 of the top 30 forest countries

Category	Sub-category	Estimated distribution at global scale	
		Area in $10^6 \times$ ha	Percent of total
Public ownership*	Administered by governments	2,803.20	77
	Reserved for community and indigenous groups	131.40	4
Private ownership**	Owned by indigenous and other local community groups	246.30	7
	Owned by private individuals and firms	443.00	12
Total		3,623.90	100

*All lands owned by central, regional or local governments.
**Rights over a specific area that cannot unilaterally be terminated by a government without some form of due process and compensation (White and Martin 2002).
(see: White and Martin 2002 for more details).

increased in the share of private forests. Securing forest tenure to local communities is indeed a prerequisite to an efficient forest management as it creates common property rights on forest resources (LeMaster and Owubah 2000; FAO 2006). From a landscape ecological perspective, the type of forest tenure regime or ownership may affect the structural pattern of forest patches, thus influencing the degree of spatial heterogeneity of forest landscapes at large. Studies on forest tenure regimes have shown the relationship between different types of ownership and spatial patterns of forest cover (Spies et al. 1994; Turner et al. 1996; Wear et al. 1996). Crow et al. (1999) investigated the main factors affecting spatial heterogeneity in forest landscapes of Northern Wisconsin, USA. Forest ownership and physical environment explained significant portions of the spatial variation in the structure of forest patches (e.g., mean patch size).

Different types of forest characterize forest tenure in southern Italy: privately owned forests are mostly covered by broad-leaved species – such as *Quercus ilex*, *Q. trojana*, and *Q. pubescens* – that are managed through coppicing. Under public ownership rights, high forest systems are more common (INFC 2007). Over the past two-decades, trends towards coppice conversion into high forest have substantially increased especially in situation of large publicly-owned lands. On privately-owned forests, conversion is still limited because the small size of forest patches makes high forest systems not economic feasible (INFC 2007). Conversely, coppicing still represents a source of timber for many rural communities with an important role for the local economy. A number of regulations affect the forest tenure regime in this region, especially in situations where large and continuous forest patches characterise the landscape. In particular, new regulations have been recently proposed to guide the management and conversion of coppice woodlands thus integrating ecological functions (e.g., maintenance of forest ecosystem health, biodiversity conservation), with economic and social purposes (e.g., timber and recreation).

Forest tenure regime is affected in southern Ontario by management by-laws and forest agreements (short-term adjustments to ownership rights in exchange for technical assistance and low-cost reforestation of degraded agricultural lands). Local municipal by-law regulations limit tree cutting on private property. The limits are generally broad, but clear-cutting forest land without a permit is an actionable offense. Selective harvests of forest land and minor changes to size and shape of forest patches can occur incrementally without requiring a permit. Forest by-laws are enacted and enforced at a municipal (local) level and apply generally only to forests larger than two or four hectares (depending on jurisdiction) (Fitzgibbon and Summers 2002). Forest management agreements result in "agreement forests" – patches of forest where reforestation of abandoned, marginal agricultural land was made possible by technical and financial support from provincial government. In exchange, an agreement between land owner and provincial government was signed to protect and manage the forest land. These forests are sized and shaped based on the former field size and in recent history they are dominated by mono- or bi- or tri-culture stands of coniferous trees – such as *Picea glauca*, *Pinus strobus*, and *Pinus resinosa* – in landscapes that are typically deciduous-forest dominated.

Silvicultural traditions are the typical methods employed by local communities and forest managers for harvesting and regenerating forest stands to achieve production of fuelwood, fiber, and other types of products or commodities. Silvicultural traditions are generally based on an array of methods of carrying out the harvesting, regeneration, and stewardship and these methods vary according to the particular species, site conditions, cultural conventions, silvicultural system, and the type of forest in a given geographical region. Traditional methods for managing forest stands and woodlands generally incorporate a sustainable view of the system being managed, because these methods often support the idea of self-regenerating the forest through time with limited use of technologies and external inputs. In addition, these methods are often rooted in the traditional knowledge that communities have of forest and land-use management, thus representing a cultural heritage that is worth preserving (Rotherham 2007). Traditional methods are important cultural determinants of forest landscapes as they represent the inherent capacity of humans to benefit from forest resources without losing the linkage between ecological patterns and processes at landscape level.

In southern Italy a good example of traditional silvicultural method is "coppicing". This method involves cutting back trees periodically at the base of the stumps to produce new shoots regenerating the forest. If a woodland is managed in adequate blocks of rotation coppice, the structure that results could provide desirable ecological conditions for many species adapted to open woodland (e.g., woodland-floor vegetation). Various studies demonstrated how the lack of traditional management, such as overstood coppicing, could negatively affect the response of species to forest patterns, thus limiting the capacity of forest landscapes to support ecological processes (Sanesi et al. 2004; Sanesi et al. 2005). However, the increase in harvesting costs and the high level of parcelization of property is causing the abandonment of many coppice woodlands. Silvicultural techniques are also being drastically simplified in order to reduce costs, thus causing the loss of the wealth of traditional knowledge that has developed around this way of managing forest landscapes (Parrotta and Agnoletti 2007).

In southern Ontario forests, silvicultural traditions range from husbandry of sugar maple trees (*Acer saccharum*) for maple-syrup production, to promoting hardwood species for lumber and softwood species for timber, poles, and pulp. Small, privately-owned remnant woodlots on farms are commonly deciduous species valued for lumber, maple-syrup, or fuel-wood. Agreement forests are commonly held by farm or small private landowners and primarily provide timber, poles, or pulpwood. Selective harvesting is commonly used to favour sugar maples and for fuelwood harvest – beyond that there is not a traditional way of managing the remnant woodlot for products (either timber or others).

Management plans and practices are those strategies and activities commonly employed by forest professionals, forest-land owners, timber industries, or forest authorities to achieve timber production, forest conservation and recreation, or any combination of these (Kangas et al. 2000). Forest management plans are operational plans that provide landscape-level analysis and directions to enable tactical decisions for management of forest patches within a landscape context (Heiligmann 2002). Plans allow guiding the management of forest resources for long-term stewardship beyond the tenure of current ownership. Management plans could regulate various activities and management practices, such as clear-cut logging, seed-tree, shelterwood, selection harvesting and so on. In clear cutting, all trees are cleared from a forest site or patch and a new, even-aged stand of timber is grown naturally from seeds from the surrounding trees, or artificially from sown seeds or planted seedlings. This system generates even-aged forest patches. With the seed-tree system, an area is generally clearcut, except that a few seed-producing trees are left to naturally regenerate the area and the seed trees are removed after the seedling stand is established. In the shelterwood system, trees are removed in a series of cuts; some trees are left for several years to provide seeds and to protect the seedlings before being removed. The selection harvesting is an uneven-aged management system, resulting in stands with intermingled trees of many ages and a variety of sizes. If not well guided by large-scale ecological plans, management practices may drive severe changes in the spatial heterogeneity of forest landscapes, thus affecting a number of processes, e.g. flora/fauna dispersal, which are sensitive to forest ecological patterns. Using a combination of data from 34 studies, Dunn (2004) analysed the impacts of logging on species diversity of multiple taxa. Overall, logging did not decrease species diversity, however selective harvesting appeared to have much less impact on species than higher intensity and larger-scale management practices (e.g., clear-cut logging).

In southern Italy, management plans and practices follow a number of general rules and regional ordinances which apply mostly to publicly-owned lands. Within this type of ownership, shelterwood cutting is commonly applied on large extension of high forests. In situation of high heterogeneity of environmental variables (e.g., climate, soil and elevation) shelterwood cutting is normally preferred on small extension as this could allow creating mixed patterns of forest patches which facilitate a number of ecological processes related to species occurrence and dispersal. An important aspect of management in this region is the conversion of forest plantations – *Pinus halepensis* and *Pinus pinea* into patches of indigenous forest vegetation. Selective thinning is currently adopted to promote the establishment of native species through secondary successions. Indeed, in managed forests natural regeneration largely depends on the reduction of canopy cover after thinning that increases light availability in the understorey, allowing efficient resource exploitation by seedlings (Malcolm et al. 2001).

In southern Ontario most timber harvest is mechanized and requires access by wheeled or tracked harvest machines, whereas maple-syrup production varies from very fine-scale and horse-drawn sap collection to broad-scale and vacuum-pipeline sap collection. For maple syrup production most forest management activity occurs during a time of frozen soil which lessens the disturbance of forest soils and herbs. Forest harvest occurs throughout the year and can lead to soil compaction and damage to herb and shrub layers. Softwood harvests, especially of the regularly-spaced agreement forest plantings, tend to have regular intervals and patterns associated, such as thinning of selected rows (see Fig. 2.4).

Fig. 2.4 Silvicultural thinning in a spruce plantation, southern Ontario. Photograph shows thinning of every fourth row of plantation and inset shows machine technique for harvest (photograph by Robert Corry)

2.2.2 Cultural Patterns and Ecological Processes

The pressures generated by the driving forces may lead forest landscapes changing their state (*S*). Changes at landscape level are critical as they allow the creation of land mosaics that can be more or less heterogeneous depending on the type and magnitude of the changing (Forman 1995). If we consider forest landscapes as dynamic systems, it is important to evaluate the current status and past changes in terms of cultural patterns and ecological processes (Sanesi et al. 2006; Parrotta and Agnoletti 2007).

Cultural patterns in forest landscapes can be defined as a combination of different forest patches and other land units, whose arrangement follow the way humans have employed land resources within environmental constraints and in relationship with the cultural drivers of spatial heterogeneity and socio-economic context. Cultural patterns are consequences of human management and constitute the underlying structures of forest landscapes. The pervasive effects of land tenure systems, management practices and other cultural drivers gradually convert the intact forest ecosystems to smaller, fewer fragments having more geometrized shapes (mounting fragmentation and change in shape complexity: Impacts, Fig. 2.3). Forest fragments would not be simplified only under circumstances of unusually-high required effort, exorbitant costs, or technical limits (Corry and Nassauer 2002). A study in Wisconsin (USA) for example noted that forest patch shapes were simpler in disturbed (i.e., managed) forest landscapes containing scattered old-growth fragments and early successional hardwood and conifer forests (Mladenoff et al. 1993).

Ecological processes in forest landscapes are those functions supported or facilitated by cultural patterns and include things like movement and persistence of particular species (e.g., disturbance-specialist species, see: Dunn 2004; Otto 1996), susceptibility and spread of fires (Franklin and Forman 1987) or pest outbreaks (Coulson et al. 1999), and redistribution of matter and nutrients (see: Chao et al., Chapter 16). Haveri and Carey (2000) demonstrated how variable-density thinnings can enhance abundance and diversity of winter birds in second-growth Douglas-fir. Patterns of interspersed forest patches could increase understory and herb layer (i.e., niche diversification) and enhance populations of species associated with shrubs and herbaceous vegetation (see also: Carey 2001). However, the persistence of processes within forest landscapes is a function of the degree of patchiness of cultural patterns (i.e., fragmentation) and of patch-level attributes, such as: size, shape, and core area, which can influence the interaction of forest patches with adjacent patches, corridors, or matrix (see: Saura et al., Chapter 10). Large forest patches with irregular shapes are considered to provide more interior locations for specialist species (Bogaert et al. 2001) and to encourage more boundary-crossing animals through coves and lobes (Dramstad et al. 1996) than small and regular patches (typical of cultural driven landscapes). In addition, complex shapes lead to microclimatic variability and greater plant diversity along edges (Forman 1995). In highly-fragmented forest landscapes, patch sizes and shapes have been shown to have a relationship under particular management conditions (Corry et al. 2006). In southern Italy, for example, a recent study tested a number of models to predict the variation of forest patch shape and other landscapes metrics in relation to forest vegetation type, terrain slope, and distance from other land-cover types (Lafortezza et al. in prep.): indigenous forest patches (i.e., sclerophyllous forests) showed more irregularly shapes than coniferous forest or broad-leaved coppiced woodlands, especially in areas characterized by steep slopes or located at furthers distance from agricultural and urbanized areas.

2.3 Enhancing Cultural and Ecological Attributes

Cultural patterns and ecological processes are critical to form and function of forest landscapes because of the resilience, longevity and dominance associated with these factors. In many forest landscapes a delicate ecological balance has been maintained over centuries despite human exploitation and disturbance (Naveh 1995; Vos and Meekes 1999). Continuing changes to the socio-economic template may drive the transition of cultural patterns in terms of composition and configuration and this could limit a number of processes that depend on disturbances.

Human-induced changes could be beneficial or detrimental to forest landscapes, depending on the ecological consequences of management plans and interventions. Specific concerns, like maintenance of critical key wildlife habitat patches need to be addressed in current strategies of sustainable forest landscape management (Dunn 2004).

The goal of promoting a multiple use of forests cannot be achieved without considering forests within a landscape ecological context. Following this perspective, landscape ecology may be assumed to be a fundamental approach for integrating cultural patterns and ecological processes in a unique framework. Forest patches as any other landscape components need to be understood as spatial units having a combination of "vertical" natural elements modified by human interaction. Such units also have "horizontal" interactions with their surrounding habitats that contribute to the biodiversity and the ecological functionality of the rural landscape at large. Forest patches need to be analysed at the regional scale in order to understand their spatial arrangement and the spatial placement of neighbouring patches and corridors such as, wetlands, hedgerows, and other woodlands. Forest patch management needs to consider the temporal and spatial character of managed disturbance for multi-phase forest conditions. This may aid understanding of spatially-explicit processes like fragmentation and loss of species diversity that have reached substantial levels of concern in many forest landscapes, such as those in southern Italy and southern Ontario. In addition, forests have to be considered as part of the "cultural" heritage of the specific region while helping to sustain the production of multiple goods and services that enhance the livelihood security, quality of life and wellbeing (Parrotta and Agnoletti 2007). The use of and interaction with forests is part of reciprocal human-landscape interactions. Therefore forest management and conservation should be undertaken by local and regional authorities through the implementation of suitable policies and planning strategies (see: Azevedo, Chapter 14).

The cultural pedigrees of Italian and Ontario landscapes are markedly distinct and of very different ages. Though the resulting patterns differ, these examples of forest landscapes share similar constraints and potential: (1) complex land use patterns made by forest patches of relatively small dimension and regular shape that are controlled by forest and non-forest land uses; (2) high variability of landform units; (3) persistence of vernacular identities and uses in local rural-communities; (4) prospects of new "ancillary" functions of woodlands (e.g., greening the surrounding urban areas; carbon sequestration). Considering these factors, it is critical that management strategies are based on the characteristics of these cultural landscapes, thus avoiding uniform intervention over large surface areas. Spatial units of woodlands have to be differentiated in both their vertical and horizontal structures by allowing the presence, within the same patch, of trees that vary in species, age, height, stem diameter, crown development, and ecological niche. The collective integration of these principles may help to achieve a more sustainable mosaic of forest patches within cultural landscapes that is highly conducive to promoting biodiversity and other important ecological processes.

References

Baker WL (1993) Spatially heterogeneous multi-scale response of landscapes to fire suppression. Oikos 66:66–71

Bogaert J, Salvador-Van Eysenrode DS, Impens I, Van Hecke P (2001) The interior-to-edge breakpoint distance as a guideline for nature conservation policy. Environ Manage 27:493–500

Boose ER, Serrano MI, Foster DR (2004) Landscape and regional impacts of hurricanes in Puerto Rico. Ecol Monogr 74(2):335–352

Brown RD, Lafortezza R, Corry RC, Leal DB, Sanesi G (2007) Cultural patterns as a component of environmental planning and design. In Hong SK, Nakagoshi N, Fu B, Morimoto Y, Wu J (eds) Landscape ecological applications in man-influenced Areas – linking man and nature systems, Springer, pp 395–415

Bruce JW (1998) Review of tenure terminology. Tenure Briefs 1:1–8

Bürgi M, Hersperger AM, Schneeberger N (2004) Driving forces of landscape change – current and new directions. Landsc Ecol 19:857–868

Carey AB (2001) Induced spatial heterogeneity in forest canopies: response of small mammals. J Wildlife Manage 65(4):1014–1027

Chidumayo EN (2002) Changes in miombo woodland structure under different land tenure and use systems in central Zambia. J Biogeogr 29(12):1619–1626

Corry RC, Lafortezza R, Sanesi G, Brown RD (2006) The relationship of forest patch shapes and sizes in highly-fragmented landscapes of southern Ontario and southern Italy. In Lafortezza R, Sanesi G (eds) Proceeding of the IV IUFRO-8.01.02 (Landscape ecology) Conference, Locorotondo, Bari, Italy, Sept 26–29 2006. pp 319–326

Corry RC, Nassauer JI (2002). Managing for small patch patterns in human-dominated landscapes: cultural factors and corn belt agriculture. In Liu J, Taylor W (eds) Integrating landscape ecology into natural resource management, Cambridge University Press, Cambridge, UK. pp 92–113

Coulson RN, McFadden BA, Pulley PE, Lovelady CN, Fitzgerald JW, Jack SB (1999). Heterogeneity of forest landscapes and the distribution and abundance of the southern pine beetle. Forest Ecol Manag 114:471–485

Crow TR, Host GE, Mladenoff DJ (1999) Ownership and ecosystem as sources of heterogeneity in a forested landscape, Winsconsin, USA. Landsc Ecol 14:449–463

Dramstad WE, Olson JD, Forman RTT (1996) Landscape ecology principles in landscape architecture and land-use planning. Island Press, Washington, DC

Dunn RR (2004) Managing the tropical landscape: a comparison of the effects of logging and forest conversion to agriculture on ants, birds, and Lepidoptera. Forest Ecol Manag 191:215–224

FAO (2006) Understanding forest tenure: What rights and for whom? Secure forest tenure for sustainable forest management and poverty alleviation: the case of South and Southeast Asia. Forest Policy and Institutions Working Paper 14

FAO (2007) State of the World's Forests. Food and Agriculture Organization of the United Nations Rome, 157pp

Fitzgibbon and Summers (2002) Report on tree conservation by-laws in Southern Ontario. Unpublished technical report.

Forman RTT (1995) Land mosaics – Ecology of landscapes and regions. Cambridge University Press, Cambridge, UK

Forman RT, Godron M (1986) Landscape ecology. John Wiley, New York, 619pp

Franklin JF, Forman RTT (1987) Creating landscape patterns by forest cutting: ecological consequences and principles. Landscape Ecol 1(1):5–18

Gustafson EJ, Crow TR (1994) Modeling the effects of forest harvesting on landscape structure and the spatial distribution of cowbird brood parasitism. Landsc Ecol 9:237–248

Gustafson EJ, Crow TR (1996). Simulating the effects of alternative forest management strategies on landscape structure. J Environ Manage 46:77–94

Gustafson EJ, Loehle C (2006) Effects of parcelization and land divestiture on forest sustainability in simulated forest landscapes. Forest Ecol Manag 236:305–314

Hart JF (1998) The rural landscape. Johns Hopkins University Press, Baltimore, Maryland
Haveri BA, Carey AB (2000) Forest management strategies, spatial heterogeneity and winter birds in Washington. Wildlife Soc B 28(3):643–652
Heiligmann RB (2002) Forest management developing a plan to care for your forest. Extension Factsheet, Ohio State University 2:1–2
INFC (2007) Inventario Nazionale delle Foreste e dei Serbatoi Forestali di Carbonio. MiPAF – Corpo Forestale dello Stato – Ispettorato Generale, CRA, ISAFA, Trento, Italy [on line] URL: http://www.infc.it (last access 20/01/2008)
Kangas J, Store R, Leskinen P, Mehtatalo L (2000) Improving the quality of landscape ecological forest planning by utilising advanced decision-support tools. Forest Ecol Manag 132:157–171
LeMaster DC, Owubah CE (2000) Nation states and forest tenures An assessment of forest policy tools in Eastern European countries. In Schmithuesen F, Herbst P, Le Master DC (eds). Forging a new framework for sustainable forestry: Recent developments in European forest law. IUFRO World Series Volume 10:28–38
Lafortezza R, Corry RC (in prep) Investigating the origin and intensity of human control on forest patches in highly-fragmented landscapes. Landscape Ecol
Li H, Reynolds JR (1995) On definition and quantification of heterogeneity. Oikos 73(2):280–284
Lindenmayer DB, Fischer J (2006) Habitat fragmentation and landscape change: an ecological and conservation synthesis. Island Press, Washington DC
Malcolm DC, Mason WL, Clarke GC (2001) The transformation of conifer forests in Britain, regeneration, gap size and silvicultural systems. For Ecol Manage 151:7–23
Michael JL, Hermy M (2002) Synthesis: Ecological impacts of forest vegetation management fourth international conference on forest vegetation management Nancy, France, June 17–22, 2002, pp 7
Mladenoff DJ, White MA, Pastor J, Crow TR (1993) Comparing spatial pattern in unaltered old-growth and disturbed forest landscapes. Ecol Appl 3:294–306
Moss MR, Strath Davis L (1994) Measurement of spatial change in the forest component of the rural landscape of southern Ontario. ApplGeogr 14:214–231
Nassauer JI (1995) Culture and changing landscape structure. Landscape Ecol 10(4):229–237
Nassauer JI (1997) Placing nature: Culture and landscape ecology. Washington, DC. Island Press
Nassauer JI, Westmacott R (1987) Progressiveness among farmers as a factor in heterogeneity of farmed landscapes, Landscape Heterogeneity and Disturbance. In Goigel Turner M (ed) Ecological studies 64, Springer-Verlag, New York. pp 200–210
Naveh Z (1995) Interactions of landscapes and cultures. Landscape Urban Plan 32:43–54
Otto RD (1996) An evaluation of forest landscape spatial pattern and wildlife community structure. Forest Ecol Manag 89:139–147
Parrotta J, Agnoletti M (2007) Traditional forest knowledge: challenges and opportunities. Forest Ecol Manag 249:1–4
Pearce C M (1992) Pattern analysis of forest cover in southwestern Ontario. The East Lakes Geogr 27:65–76
Perera AH, Baldwin DLB (2000) Spatial patterns in the managed forest landscape of Ontario. Ecology of a managed terrestrial landscape. Perera AH, Euler D, Thompson ID and Ontario Ministry of Natural Resources (eds). Patterns and processes of forest landscapes in Ontario, UBC Press, Vancouver, British Columbia, pp 74–99
Risser PG, Karr JT, Forman RTT (1984) Landscape ecology: Directions and approaches. Special Publication, Illinois Natural History Survey, pp 22
Romano F (2007) Forest tenure changes in Africa: making locally based forest management work. Unasylva 228(58):11–17
Rotherham ID (2007) The implications of perceptions and cultural knowledge loss for the management of wooded landscapes: a UK case-study. Forest Ecol Manag 249:100–115
Runesson UF (2004) Glossary of forestry terms. Northwestern Ontario Forest. http://www.borealforest.org/nwgloss.htm (last access 23/01/2008)

Saastamoinen O, Matero J (2004) The Forest Policy and Economics Education and Research (FOPER) project. United Nations University and University of Joensuu. http://foper.unu.edu/ (last access 15/12/2007)

Sanesi G, Bottalico F, Lafortezza R (2005) Historical evolution of the Mediterranean forest landscape with prevailing Quercus trojana Webb. in the Putignano municipality, Southern Italy. In Agnoletti M, Armiero M, Barca S, Corona G (eds) History and sustainability, University of Florence Press, pp 109–114

Sanesi G, Lafortezza R, Yokohari M, Ciancio O, Nocentini S, Amati M (2004) Managing coppice woodlands in traditional rural landscapes: an international perspective. In Sano M, Miyamoto A, Sugimura K (eds) Proceedings of the III IUFRO-8.01.03 (Landscape ecology) Conference, October 25–28, 2004, Tsukuba, Japan, pp 1–4

Schooley RL, Wiens JA (2004) Movements of cactus bugs: patch transfers, matrix resistance, and edge permeability. Landscape Ecol 19(7):801–810

Smeets E, Weterings R (1999) Environmental indicators: typology and overview. Technical report n. 25. European Environmental Agency, Copenhagen, pp 19

Spies TA, Ripple WJ, Bradshaw GA (1994) Dynamics and pattern of a managed coniferous forest

Turner MG (1989) Landscape ecology: the effect of pattern on process. Annu Rev Ecol Evol S 20:171–197

Turner MG, Wear DN, Flamm RO (1996) Land ownership and land-cover change in the Southern Appalachian Highlands and the Olympic Peninsula. Ecol Appl 6(4):1150–1172

Vos W, Meekes H (1999) Trends in European cultural landscape development: perspective for a sustainable future. Landscape Urban Plan 46:3–14

Wear D, Turner MG, Flamm RO (1996) Ecosystem management with multiple owners: landscape dynamics in a Southern Appalachian Watershed. Ecol Appl 6(4):1173–1188

White A, Martin A (2002) Who owns the world's forest? Forest tenure and public forests in transition. Center for International Environmental Law, Washington DC, pp 32

Chapter 3

Managing Forest Landscapes for Climate Change*

Thomas R. Crow

Abstract Climate change is the defining issue of the day and probably for many subsequent generations of resource managers. Although the public and therefore the policymakers have been slow in grasping the far-reaching consequences of climate change on our social and economic institutions, they are now desperately seeking options for dealing with novel climates, ecological uncertainties, and potential social and economic dislocations. The challenges cannot be overstated, but neither can the role of scientists in helping provide the knowledge for making informed decisions. The science of landscape ecology, with its emphasis on integration and holism, has an important role to play in informing decision makers. In this paper, I explore this role in the context of managing forest landscapes.

3.1 Introduction

Climate change is the most challenging issue confronting contemporary and, for that matter, future resource managers. It affects every aspect of resource management, be it commodity production, biological diversity, land use, forest health, fire, water, invasive species, resource sustainability, and ecosystem services. For some interactions between climate change and resource management, a substantial body of knowledge has accumulated. Stocks et al. (1998), for example, used four current General Circulation Models (GCMs) to project forest fire danger levels in the Canadian and Russian boreal forests. The forecasts for seasonal fire weather severity were similar among all four GCMs – each projected a longer fire season and a large increase in the areal extent of extreme fire danger for both countries under a 2 X

T.R. Crow
Research and Development, Environmental Sciences Staff, USDA Forest Service, Stop Code 1113
1400 Independence Avenue, SW Washington DC 20250-1113 USA
e-mail: tcrow@fs.fed.us

R. Lafortezza et al. (eds.), *Patterns and Processes in Forest Landscapes*,

CO_2 climate scenario. Likewise, climate is a fundamental driver of the water cycle and so significant changes in precipitation patterns, storm intensity, evaporation, transpiration, timing of snowmelt, snow vs. rain events, and groundwater level can be expected with climate change (Houghton et al. 2001). With water, it is not only quantity but also quality and securing sources of safe drinking water that are of concern. The vulnerabilities of supply are linked not only to climate change, but also changes in demand related to a growing human population and economic development (Vörösmarty et al. 2000).

For other interactions, little is known. Although the potential for impacts of climate change on fire, water, forest health and productivity are widely acknowledged, how these impacts will play out in time and space is not well understood. Because of its emphasis on understanding ecological processes in time and space, the science of landscape ecology can play an important role in improving our understanding about the possible impacts of climate change on natural and human capital.

The movement toward investigating large systems reflects, in part, new technologies that enable scientists to study them. It also reflects an appreciation in the scientific world that phenomena observed and studied at small scales are part of larger systems. A corollary to this in the realm of resource management is the recognition that informed decisions can not be based solely on a single species or local site, but they require the context provided by larger spatial areas that encompass many species and multiple ecosystems (Turner et al. 2002).

Although the terms "managing landscapes" and "landscape management" are a common part of the lexicon for resource managers (Crow 2005; Rolstad 2005), the meaning of these terms is not always clear. For purposes of this paper, a landscape is defined as consisting of a mosaic of interacting ecosystems, both aquatic and terrestrial, that occupy a geographic space somewhere between the local and regional levels. While this may seem rather imprecise, the defining characteristic is clear – two or more interacting ecosystems constitute a landscape. Thus ecosystems are the fundamental unit defining landscapes. The ecosystem concept has deep roots in both science (Tansley 1935; Odum 1969; Golley 1996) and management (Schultz 1967; Crow and Gustafson 1997). It is a spatially-explicit unit for which the standards for identifying and classifying are in place (e.g., Rowe 1962; Bailey 1983; Host et al. 1996; Cleland et al. 1997), and for which boundaries can be identified and mapped (e.g., Spies and Barnes 1985; Albert 1995; Keys et al. 1995). By integrating the biota or biosphere with landform, soil, and atmosphere, the physical environment or geosphere, the ecosystem concept provides a useful and powerful tool for understanding and managing complex natural systems (Rowe and Barnes 1994).

In their August 2007 report, the Government Accountability Office (GAO) identified the challenges facing federal resource management agencies such as the USDA Forest Service, National Park Service, Fish and Wildlife Service, Bureau of Land Management, and National Oceanic and Atmospheric Administration in addressing climate change (GAO 2007). Specifically the GAO states that "resource managers have limited guidance about whether or how to address climate change and therefore, are uncertain about what actions, if any, they should take." The GAO further asserts that “resource managers do not have sufficient site-specific information to plan for and manage the effects of climate change on the federal resources they oversee.”

The purpose of this paper is to explore the value of a landscape perspective when addressing climate change. My intent is to provide general principles for managers as they consider strategies for adapting to a changing climate. Although these principles are not the specific prescriptions needed by managers, they are an important first step toward developing them.

3.2 Including People

This general principle is listed first because it is the most important. Although it seems rather obvious, never-the-less including people when managing landscapes for climate change needs to be stated explicitly.

People are part of the landscape and they are being affected by climate change. As stated by Allen (2004), landscape management “will not work unless it involves explicitly giving people options that they are prepared to use, and a sound basis to select among them.” In other words, we as scientists and managers need to understand what motivates people to act and how they make choices and assess risk while conducting their lives in the biological and physical worlds that we study and manage. The key is being responsive to society’s changing and diverse values and expectations (Bengston 1994).

3.3 Looking Beyond the Forest

The second principal deals with the scale at which resource management occurs. Scale is a characteristic dimension for both space and time which deals with the relative size, extent, or level at which an object, process, or action occurs (Lovell et al. 2002). The fundamental issue is this: focusing on a single scale in time or space obscures important processes or properties that are obvious at larger or smaller scales. In many respects, climate change is the perfect example for illustrating the need to consider multiple scales. Although climate change is a global phenomenon, it affects processes occurring at all spatial scales.

From a management perspective, this need to consider multiple scales in time and space complicates the already difficult task of managing natural resources. But there are some practical approaches that can be used when a landscape perspective of the forest is employed. The first is to make local management decision within the context of landscape, regional, national, and even global information. The USDA Forest Service and many other federal and state agencies have conducted numerous broad-scale biophysical and social assessments in response to a variety of issues and needs (Johnson et al. 1999; Jensen and Bourgeron 2001). Perhaps the best know of these assessments is the Northwest Forest Plan which is a bold attempt to implement adaptive resource management on the west side of the Cascades in the states of Washington and Oregon. Assessments for other regions of the United States exist as well – e.g., Southern Appalachian Assessment, Great Lake Ecological Assessment, Sierra Nevada Ecosystems Project, and Southern Forest Resource Assessment (Crow 2006). Although few of these sources currently deal explicitly with global change, you can expect much more emphasis given to this topic in the near future.

Within the United States, the Forest and Rangeland Renewable Resources Planning Act (RPA) of 1974 requires the USDA Forest Service to conduct periodic reviews of the nation's renewable natural resources. These reports, another good source of contextual information, include analyses of present and projected trends in the resource base at the national level. For example, in their RPA publication entitled *Land Use Changes Involving Forestry in the United States: 1952–1997, with Projections to 2050*, Alig et al. (2003) project a 3% or 23 million acre reduction in total forest area in the United States by 2050 compared to 1997. Most of these changes are due to projected increases in population and income that will likely result in demands for converting forests into urban and other developed landscapes. Other assessments such as the Heinz Center's *The State of the Nation's Ecosystems* or the United Nations' *Millennium Ecosystem Report* involve the efforts of many experts.

Collectively, these reports provide a state-of-the-knowledge about the conditions and trends of the world's ecosystems and the services they provide. When available, these assessments provide an excellent source of contextual information for making local management decisions.

3.4 Reading the Landscape

Interactions among soil, landform, climate (i.e., the physical environment) and human land use profoundly affect the composition and structure of the landscape (Crow et al. 1999). Landscape composition can be obtained using remote sensing, ecosystem classification, and mapping of natural features such as vegetative cover types or lakes and rivers as well as features resulting from human land use such as agricultural lands, urban and industrial lands, reservoirs, and transportation networks. Landscape structure is a measure of spatial heterogeneity. The spatial configuration as defined by the size, shape, and arrangement of cover types or patches create structure on the landscape. For managers, the size-class distribution of patches is a useful metric for characterizing landscape structure. Typically a landscape contains many small patches and a few large patches. The large patches serve an important "connecting" function that facilitates the flow of materials and organisms through a landscape.

Because of the physical linkage between the earth's surface and the atmosphere, changes in the composition and structure of the landscape can have pronounced effects on weather and climate (Pielke and Avissar 1990). Both observation and modeling demonstrate the influence of landscape characteristics on atmospheric conditions. Numerous studies document changes in near surface microclimate conditions with changes in surface conditions (see, e.g., Chen et al. 1999), so it is reasonable to conclude that changes in surface properties can affect the regional atmosphere (Pielke and Avissar 1990; Pielke 2005). The physical linkage between the surface and the atmosphere has to do fundamentally with albedo and the fractional partitioning of atmospheric turbulent heat flux into sensible and latent fluxes. Parameters in these budgets are directly dependent on the surface properties (e.g., an asphalt surface compared to forest cover) and the vegetation characteristics such as height, aerodynamic roughness, displacement height, percentage cover, and photometric properties. Both empirical and modeling studies suggest that even small changes in land use and changes in their spatial distribution can cause major alterations in local and regional atmospheric conditions. Much less is known, however, about the cumulative effects of the present surface properties (both amount and distribution) on global climate (Pielke and Avissar 1990).

Just as foresters can walk through a forest and make judgments about its suitability for a commercial harvest or for providing suitable habitat for deer or grouse, managers also need to read the landscape in which the forest exists in order to make judgments about the quality and quantity of goods and services that the landscape is capable of providing. Given the number of road crossing over streams, is it likely the landscape will produce clean water? Are critical habitats present in the landscape for threatened and endangered species? Are old as well as young age-classes for trees included in the landscape? Are undeveloped lakeshores present? What is the density of roads and trails in the landscape? Do the ephemeral ponds exist within the landscape and, if so, where? Given the soil types and landforms, what is the productive potential of the land? What forests types are present? What proportion of the landscape is forested compared to agriculture or urban and how are these different land covers arranged? By addressing these and similar questions, the resource manager and planner gains an understanding of the ecosystem services that the landscape can provide. By answering these and similar questions, the resource manager and planner are reading the landscape.

3.5 Managing Composition and Structure

Principle three, reading the landscape, is the antecedent to managing composition and structure. Desired changes in goods and services based on reading the landscape are derived from manipulating landscape composition and structure. Because of the demands humans are placing on the lands and waters, greater emphasis will be placed on designing landscapes to meet specific management objectives that, in combination, provide the array of goods and services desired by society. One example of manipulating structure relates to providing conductivity within the landscape. The migration of species during climate change is critical to their survival. The combination of land use and climate change place many species at risk unless migration is possible. Current land use patterns (especially habitat fragmentation) create barriers to migration and increase the isolation of some populations. As the landscape becomes dissected into smaller parcels of habitat, landscape connectivity (the functional linkage among habitat patches) may become disrupted. This limits the distribution and persistence of populations.

Corridors are one of many structural features that have been being studied within the landscape. Within the scientific community, a robust discussion is ongoing about the effectiveness of corridors to facilitate the movement of species and their populations in response to habitat fragmentation (e.g., Simberloff and 1987; Simberloff et al. 1992; Beier and Noss 1998). Corridors have been shown to maintain and enhance biological diversity, but corridors also facilitate the movement of invasive species, pests, diseases, as well as increase the risk to predation and the spread of disturbance. The effectiveness of corridors depends on many factors, including a species' habitat requirements, the dispersal ability of a species, and local environmental conditions. Habitat specialists with limited dispersal capabilities will likely have a much lower tolerance to habitat fragmentation (and climate change) compared to highly mobile, generalist species.

With the goal of creating forests that are resilient to a changing climate, greater emphasis is needed on the relation between landscape structure and the distribution and persistence of plant and animal populations. Spatially-explicit population and habitat models are useful tools for understanding species movement, their patterns of extirpation, and the processes that promote recolonization within a landscape. Gustafson and Gardner (1996) used an individual-based dispersal model to measure immigration and emigration rates between habitat patches within heterogeneous landscapes. The model was used to estimate the probabilities of disperser transfer between patches by varying the *a priori* probabilities of movement into each habitat type in order to estimate the effect of changing landscape heterogeneity on the transfer probabilities, and to visualize dispersal corridors and barriers as perceived by model organisms operating by specified rules and at specific scales. The results show that 89% of the variability in dispersal success can be accounted for by differences in the size and isolation of forest patches, with closer and larger patches having significantly greater exchanges of dispersing organisms (Gustafson and Gardner 1996). Further, their results suggest that corridors are often diffuse and difficult to identify from structural features in the landscape. Understanding how landscape structure, including well-defined corridors and other structural features, affects the movement of species will be critical to devising strategies that mitigate the effects of climate change (Taylor et al. 1993).

3.6 Designing the Landscape

This principle is an extension of the previous two principles. Designing landscapes deals fundamentally with managing the composition and structure of the landscape in order to meet some socially-defined outcomes and benefits. Most design principles utilized in land planning and management are directed at gaining economic efficiency or for providing aesthetic appeal rather than sustainability. Although there is increased recognition that more sustainable approaches are needed for planning and managing landscapes, much of the available information is conceptual and theoretical, and little of it deals directly with adapting to climate change. Leitão and Ahern (2002) present a conceptual framework for sustainable landscape planning through the application of ecological knowledge in the land planning process. Designs for local adaptation in which the behavior of resident individuals is guided by feedbacks from their immediate environment (e.g., Levinthal and Warglien 1999), based on organizational theory and behavior of complexity systems, certainly have implications for adapting to climate change. In the realm of urban planning, concepts such as "smart growth" to reduce urban sprawl can also reduce the greenhouse gases that are, in large part, responsible for climate warming.

The challenge is to relate the composition and structure of landscapes to the ecosystem services they provide. Ecosystem services are defined as those benefits people obtain from ecosystems (Mooney and Ehrlich 1997). Undeveloped forestlands provide clean drinking water, wood products, wildlife habitat, and recreational opportunities that may not be available in the developed landscape. In order to maximize these ecosystem services, Forman (1995) suggests arranging land uses based on what he calls the aggregate-with-outliers principle in which aggregates – contiguous areas of natural vegetation – are maintained along with smaller outliers distributed throughout the developed (i.e., agriculture, urban, industrial, infrastructure) landscape. The undeveloped forestland or aggregate provides unique ecosystem services and these services are supplemented by the dispersed outliers embedded in the developed landscape. Unfortunately, as lands become subdivided and developed, the likelihood for providing high quality ecosystem services such as clean water and air or the production of wood declines. Clearly, no one ownership class – public or private – can provide all the desired goods and services that can be derived from the forest. New mechanisms for coordination and collaboration are needed in order to make stewardship across boundaries possible (Yaffee 1998). Further, no one collaborative model will fit all cases; instead, many different partnerships, each tailored to meet local and regional needs, are likely.

3.7 Embracing Complexity

Landscapes are comprised of discrete, bounded patches that can be differentiated using either biotic or abiotic features (Pickett and Cadenasso 1995). The origins of this landscape mosaic relate to the interaction between the physical environment and human land use. Understanding how this spatial variation and changes in these patterns affect movement of organisms and fluxes of materials and energy is a centerpiece of landscape ecology. Because ecological phenomena respond to spatial heterogeneity, the response of ecosystems to climate change is likely to differ depending on the type of ecosystem and where that ecosystem occurs in the landscape.

Just as the responses to climate change will vary in space, ecosystem responses to climate change will vary with time. Some changes are likely to occur quickly and will become readily apparent, while other changes appear to occur more gradually and uniformly through time and thus will be less apparent in the short term. Some events such as an acute regional expose to ozone due to atmospheric conditions are episodic in their occurrence, while others such as seasonal monsoon rains are more cyclical in their temporal pattern. A 100-year drought may be stochastic when viewed in the short term, but are periodic when viewed in the long term. Other patterns are ephemeral (e.g., an intermittent stream) or continuous (e.g., groundwater movement) through time. Each of these varied responses adds temporal heterogeneity that necessitates considering many different time scales in order to understand the responses of ecosystems to climate change or any other disturbance.

Throughout much of its history, ecologists sought to minimize the variation in the systems they studied. Perhaps more insights would have been gained if just the opposite strategy had been employed. Likewise, resource managers have attempted to reduce the variation in the structure and composition of managed forests due largely to operational and economic considerations – creating a forest that is homogeneous in its composition and structure reduces the costs associated with extracting and processing wood. The cumulative effects of forest management at the landscape and regional scales tend to simplify the compositional and structural complexity of forest ecosystems (Schulte et al. 2007). Managing for complexity is likely to result in greater resilience and to promote adaptability of forest ecosystems to climate change.

3.8 Confronting Uncertainty

Resource managers are always making decisions under circumstances with high degrees of uncertainty. It is simply the nature of the job when we lack understanding about complex systems and given our inability to predict the future. This is the case with climate change. As previous mentioned, there is a high degree of uncertainly associated with how climate change will play out in both time and space. In their short but remarkable paper *Uncertainty, Resource Exploitation, and Conservation: Lessons from History*, Ludwig et al. (1993) provide useful suggestions for taking uncertainty into account. They remind us that although there are well-developed theories supporting decision-making under uncertainty, most times it is more about using common sense.

In their words:

> We must consider a variety of plausible hypotheses about the world; consider a variety of possible strategies; favor actions that are robust to uncertainties; hedge; favor actions that are informative; probe and experiment; monitor results; update assessments and modify policy accordingly; and favor actions that are reversible.

3.9 Conclusion

Adapting to climate warming will be the defining challenge for forest managers now and into the future. The challenges are enormous given the complexity of the underlying biological, physical, and social systems. Our science and therefore our management historically have featured reductionism in which the pieces are featured at the expense of the whole. While reductionism has its place, it alone is not sufficient to address the many complex, large-scale problems facing society. A landscape perspective is helpful for resource managers when developing adaptive strategies for climate change because it: (1) places humans at the center because they effect and are affected by climate change; (2) provides social, economic, and ecologic context for local decisions; (3) improves understanding of the cumulative impacts of multiple treatments and multiple stresses at large scales; (4) supports adaptive management when combined with monitoring; and (5) encourages a more holistic and integrated approach to science and resource management.

Climate change is but one compelling reason for conducting landscape-scale research. Success depends on our ability to understand interactions among biophysical components and between biophysical and social systems (Mills 2004). Landscape ecology provides a formal framework in time and space as we hedge, experiment, probe, monitor, assess, and finally adjust our actions as we learn more about the changing world in which we live.

References

Albert DA (1995) Regional landscape ecosystems of Michigan, Minnesota and Wisconsin: A working map and classification. USDA Forest Service, General Technical Report NC-178. St. Paul.

Alig RJ, Plantinga AJ, Ahn S, Kline JD (2003) Land use changes involving forestry in the United States: 1952 to 1997, with projections to 2050. USDA Forest Service, General Technical Report PNW-GTR-587. Portland

Allen TFH (2004) Summary remarks on views from the ridge. In Gucinski H, Miner C, Bittner B (eds) Proceedings: Views from the ridge considerations for planning at the landscape scale. USDA Forest Service, General Technical Report PNW-GTR-596, Portland, OR, pp 127–129

Bailey RG (1983) Delineation of ecosystem regions. Environ Manage 7:365–373

Beier P, Noss RF (1998) Do habitat corridors provide connectivity? Conserv Biol 12:1241–1252

Bengston DN (1994) Changing forest values and ecosystem management. Soc Natur Resour 7:515:533

Chen J, Saunders SC, Crow TR, Naiman RJ, Brosofske KD, Mroz GD, Brookshire BL, Franklin JF (1999) Microclimate in forest ecosystem and landscape ecology. BioScience 49:288–297

Cleland DT, Avers PE, McNab WH, Jensen ME, Bailey RG, King T, Russell WE (1997) National hierarchical framework of ecological units. In Boyce MS, Haney A (eds) Ecosystem management applications for sustainable forest and wildlife resources, Yale University Press, New Haven, pp 181–200

Crow TR (2005) Landscape ecology and forest management. In Wiens J, Moss M (eds) Issues and perspectives in landscape ecology, Cambridge University Press, Cambridge, pp 201–207

Crow TR (2006) Moving to the big picture: Applying knowledge from landscape ecology to managing U.S. national forests. In Perera AH, Buse LJ, Crow TR (eds) Forest landscape ecology, transferring knowledge to practice, Springer, New York, pp 157–180
Crow TR, Gustafson EJ (1997) Ecosystem management: Managing natural resources in time and space. In Kohm K, Franklin JF (eds) Creating a forestry for the 21st century, the science of ecosystem management, Island Press, Washington DC, pp 215–228
Crow TR, Host GE, Mladenoff DJ (1999) Ownership and ecosystem as sources of spatial heterogeneity in a forested landscape, Wisconsin, USA. Landscape Ecol 14:449–463
Forman RTT (1995) Land mosaics: The ecology of landscapes and regions. Cambridge University Press, Cambridge
GAO (2007) Climate change, agencies should develop guidance for addressing the effects on federal land and water resources. GAO-07-863. Washington, DC
Golley F (1996) A history of the ecosystem concept in ecology. Yale University Press, New Haven
Gustafson EF, Gardner RH (1996) The effect of landscape heterogeneity on the probability of patch colonization. Ecology 77:94–107
Host GE, Polzer PL, Mladenoff DJ, White MA, Crow TR (1996) A quantitative approach to developing regional ecosystem classifications. Ecol Appl 6:608–618
Houghton JT, Ding YU, Griggs DJ, Noguer M, van der Linden PJ, Dai X, Maskell K, Johnson CA (2001) Climate change 2001: The scientific basis. Published for the Intergovernmental Panel on Climate Change, Cambridge University Press, Cambridge
Jensen ME, Bourgeron PS (eds) (2001) A guidebook for integrated ecological assessments. Springer-Verlag, New York
Johnson KN, Swanson F, Herring M, Greene S (eds) (1999) Bioregional assessments: science at the crossroads of management and policy. Island Press, Washington DC
Keys J Jr, Carpenter C, Hooks S, Koenig F, McNab WH, Russell W, Smith ML (1995) Ecological units of the eastern United States – first approximation (map and booklet of map unit tables). USDA Forest Service, Atlanta
Leitão AB, Ahern J (2002) Applying landscape ecological concepts and metrics in sustainable landscape planning. Landscape Urban Plan 59:65–93
Levinthal DA, Warglien M (1999) Designing for local action in complex worlds. Organ Sci 10:342–357
Lovell C, Mandondo A, Moriarty P (2002) The question of scale in integrated natural resource management. Conserv Ecol 5:25 [published online at http://www.consecol.org/vol5/iss2/art25/]
Ludwig D, Hilborn R, Walters C (1993) Uncertainty, resource exploitation, and conservation: Lessons from history. Science 260:17 and 36
Mills TJ (2004) The future of landscape-level research at the Pacific Northwest Research Station. In Gucinski H, Miner C, Bittner B (eds) Proceedings: Views from the ridge – considerations for planning at the landscape scale. USDA Forest Service, General Technical Report PNW-GTR-596, Portland, pp 130–133
Mooney HA, Ehrlich PR (1997) Ecosystem services: a fragmentary history. In Daily GC (ed) Nature's services: Societal dependence on natural ecosystems, Island Press, Washington, DC
Odum EP (1969) The strategy of ecosystem development. Science 164:262–270
Pickett STA, Cadenasso (1995) Landscape ecology: Spatial heterogeneity in ecological systems. Science 269:331–334.
Pielke RA (2005) Land use and climate change. Science 310:1625–1226
Pielke RA, Avissar R (1990) Influence of landscape structure on local and regional climate. Landscape Ecol 4:133–155
Rolstad J (2005) Landscape ecology and wildlife management. In Wiens J, Moss M (eds) Issues and perspectives in landscape ecology, Cambridge University Press, Cambridge. pp 208–216
Rowe JS (1962) Soil, site and land classification. Forest Chron 38:420–432
Rowe JS, Barnes BV (1994) Geo-ecosystems and bio-ecosystems. Bull Ecol Soc Am 75:40–41
Schulte LA, Mladenoff DJ, Crow TR, Merrick LC, Cleland DT (2007) Homogenization of northern U.S. Great Lakes forests due to land use. Landscape Ecol 22:1089–1103

Schultz AM (1967) The ecosystem as a conceptual tool in the management of natural resources. In Ciriacy-Wantrup SV, Parsons JJ (eds) Natural resources: Quality and quantity, University of California Press, Berkeley, pp 139–161
Simberloff D, Cox J (1987) Consequences and costs of conservation corridors. Conserv Biol 1:63–71
Simberloff D, Farr JA, Cox J, Mehiman DW (1992) Movement corridors: Conservation bargains or poor investments? Conserv Biol 6:493–504
Spies TA, Barnes BV (1985) A multifactor ecological classification of the northern hardwood and conifer ecosystems of Sylvania Recreation Area, Upper Peninsula, Michigan. Can J For Res 15:949–960
Stocks BJ, Fosberg MA, Lynham TJ, Mearns L, Wotton BM, Yang Q, Jin J-Z, Lawrence K, Hartley GR, Mason JA, McKenney DW (1998) Climate change and forest fire potential in Russian and Canadian boreal forests. Clim Change 38:1–13
Tansley AG (1935) The use and abuse of vegetational concepts and terms. Ecology 16:284–307
Taylor PD, Fahrig L, Henein K, Merriam G (1993) Connectivity is a vital element of landscape structure. Oikos 68:571–573
Turner MG, Crow TR, Liu J, Rabe D, Rabeni CF, Soranno PA, Taylor WW, Vogt KA, Wiens JA (2002) Bridging the gap between landscape ecology and natural resource management. In Liu J, Taylor WW (eds) Integrating landscape ecology into natural resource management, Cambridge University Press, Cambridge, pp 433–460
Vörösmarty CJ, Green P, Salisbuy J, Lammers RB (2000) Global water resources: Vulnerability from climate change and population growth. Science 289:284–288
Yaffee SL (1998) Cooperation: A strategy for achieving stewardship across boundaries. In Knight RL, Landres PB (eds) Stewardship across boundaries, Island Press, Washington, DC, pp 299–324

Part II
Consequences of Management across Regions and Scales

Chapter 4

The Great Siberian Forest: Challenges and Opportunities of Scale

Igor M. Danilin and Thomas R. Crow

Abstract The vastness or scale of the Siberian forest presents both an opportunity and a challenge. It is a major source of softwood fiber in a world in which softwood fiber is in great demand. Its vastness and isolation from markets make it more difficult to regulate harvesting and to get both raw material and processed wood to consumers. Both natural and anthropogenic disturbances (e.g., fire, climate change) greatly alter forest landscapes and complicate the management of the resource for sustainability. We characterize the current condition of the Siberian forest in Russia and recommend future directions for this globally-important resource. The future is promising because Siberia has a relatively well-developed forest infrastructure, along with highly-trained scientists, an existing structure of forest enterprises, and some protective and regulatory measures that serve as a basis for developing and sustaining the resource. However, investments directed at modernization, especially technological, are needed to enhance the country's capacity to promote sustainable development in the forestry sector.

4.1 Forest Ecosystems and Forestry in Siberia

The vast Siberia is considered the Asian part of Russia. It encompasses area from the Urals to the Pacific Coast (nearly 8000 km) and from the Chinese and Mongolian borders to the Arctic islands ($\approx$ 3500 km). The total area is 1276.6 million ha, which is about 30% larger than the continental United States. Approximately 48% of Siberia (605 million ha) is forested, with about 450 million ha coniferous forests. The forested area of Siberia constitutes appropriately 20% of the world's total forested area and 50% of the world's total coniferous forest. Nearly 65% of Siberian forests are located in permafrost areas and more than 60% is classified as mountain forests (Pozdnyakov 1986; Siberian expectations 2003) (Fig. 4.1).

I.M. Danilin
V.N. Sukachev Institute of Forest, Russian Academy of Sciences, Siberian Branch, Akademgorodok, 50/28, Krasnoyarsk, 660036, Russia
e-mail: danilin@ksc.krasn.ru

R. Lafortezza et al. (eds.), *Patterns and Processes in Forest Landscapes*,

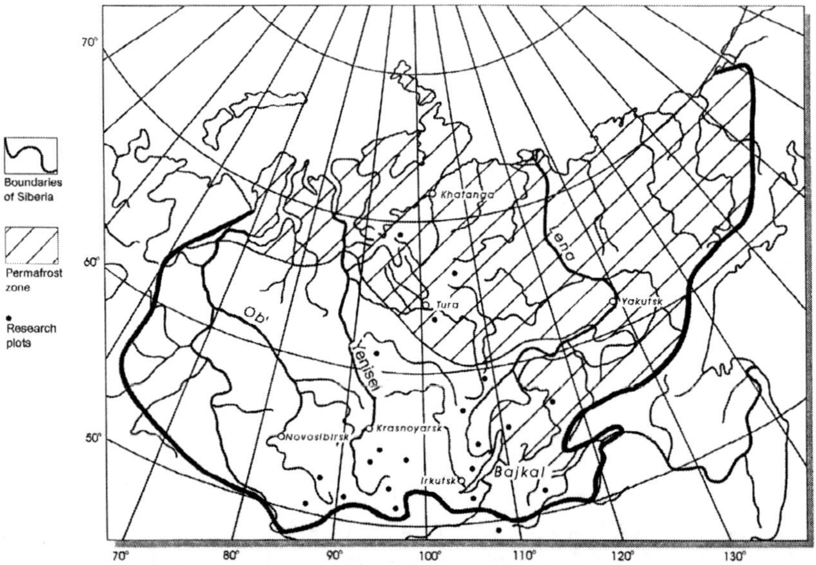

Fig. 4.1 Boundaries of the permafrost areas and geographic locations of the research plots in Siberia

Because of its vastness, the Siberian forest is well suited for applying a landscape and regional perspective to its management. Our purpose is to characterize the current condition of the Siberian forest and to consider the future for this forest given its management, its utilization, as well as related emerging issues such as forest health and climate change. Lastly, we make several recommendations aimed at promoting the sustainability of this globally-important natural resource.

4.1.1 Forest Resources

Siberia is divided into three major ecological and economic regions: West Siberia, East Siberia and the Far East. The percentage of forest cover (Forest Fund) is 53%, 57% and 45% of the total area, respectively, in these regions (Fig. 4.2).

In this chapter, we focus largely but not exclusively on East Siberia.

Forest resources are classified into broad categories that reflect their current status and their potential for growing trees. The Forest Fund consists of those areas currently covered by forests, along with areas with the potential for forest production but currently not in forest cover, i.e.,

- forest land – either covered by closed forests (i.e., forested areas), or areas temporarily not covered (i.e., unforested areas), including harvested areas and burned areas.
- non-forest land – areas which are not suitable for forest production under current conditions and areas with other land-use functions such as agriculture (Tables 4.1 and 4.2).

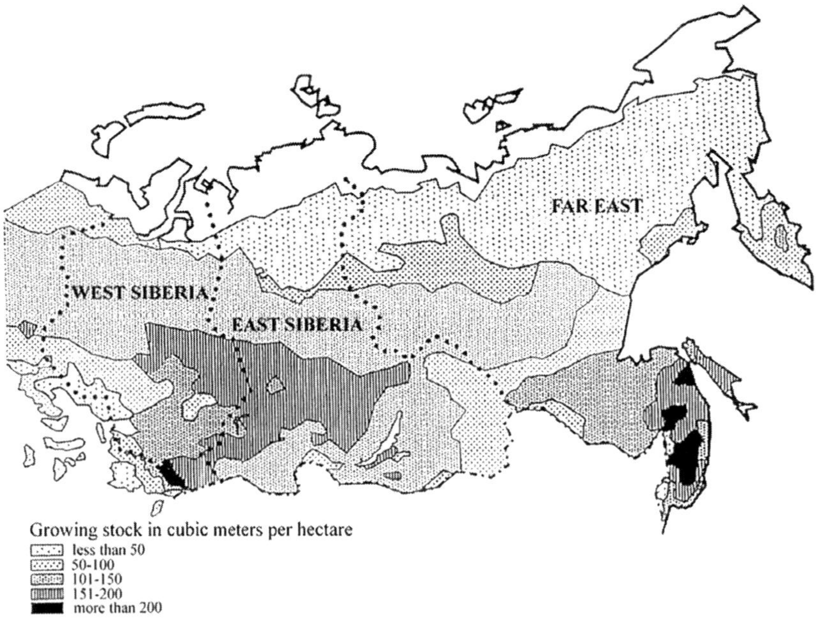

Fig. 4.2 Ecological and economic regions and growing stock of Siberian forests

Shrublands, another broad classification, are those landscapes where closed forests are unable to grow due to climatic conditions or other ecological factors.

Conifers are the dominant species group throughout Siberia (Table 4.2). Pine is the main species in West Siberian forest landscapes, but larch (*Larix sibirica*) dominates in other regions. Overall, larch is the most common species in Siberia. Additionally, soft deciduous species, mainly birch and aspen (*Betula pendula*, *Populus tremula*), are frequently present throughout Siberia. Hard deciduous species such as elm (*Ulmus japonica*) and oak (*Quercus mongolica*) are present represented in the Far East region. Siberian forests grow under severe climatic conditions and are often poorly stocked. More than 30% of the forested area have low stocking levels the majority of which are located in the East Siberia and the Far East. More than 40% of the forests grow on poor sites, predominately in the Far East region (Pozdnyakov 1986; Problemy 1998; Danilin 2003; Siberian Forests 2004) (Fig. 4.2).

Table 4.1 Extent of the Siberian forest resources, expressed in million hectares and growing stock (GS) in billion cubic meters

	Russia	Siberia total	West Siberia	East Siberia	Far East
Total Area	1707.5	1276.6	242.7	412.3	621.6
Forest Fund	1182.6	973.2	150.6	315.4	507.2
Forest Land	884.4	710.6	95.5	255.2	359.9
Forested Area	771.4	605.1	90.1	234.4	280.6
Growing Stock	81.6	61.4	10.8	29.3	21.3

Table 4.2 Distribution of forested area and growing stock of major forest types. Forested Area (FA) is expressed in million hectares and growing stock (GS) in billion cubic meters. Note that only major species are included and I do not take into account shrubs and other coppice, which are accounted for in Table 4.1

Species	West Siberia		East Siberia		Far East		Total	
	FA	GS	FA	GS	FA	GS	FA	GS
Coniferous	56.3	6.8	180.2	24.9	199.7	17.6	436.2	49.3
Pine (*Pinus sylvestris*)	28.7	3.0	32.1	5.5	12.0	1.2	72.8	9.7
Spruce (*Picea obovata*)	5.4	0.6	12.4	1.8	13.7	2.4	31.5	4.8
Fir (*Abies sibirica*)	3.8	0.5	9.4	1.6	1.8	0.3	15.0	2.4
Larch (*Larix sibirica and L. gmelinii*)	5.9	0.6	102.8	11.6	168.8	12.9	277.5	25.1
Cedar (*Pinus sibirica* and *P. koraiensis*)	12.5	2.1	23.5	4.4	3.4	0.8	39.4	7.3
Hard deciduous	–	–	–	–	10.6	0.9	10.6	0.9
Beech (*Ulmus japonica*)	–	–	–	–	6.6	0.6	6.6	0.6
Oak (*Quercus mongolica*)	–	–	–	–	4.0	0.3	4.0	0.3
Soft deciduous	21.7	2.8	31.2	2.8	12.7	0.8	65.6	6.4
Birch (*Betula pendula*)	17.0	2.0	26.4	2.1	11.6	0.7	55.0	4.8
Aspen (*Populus tremula*)	4.7	0.8	4.8	0.7	1.1	0.1	10.6	1.6
TOTAL	**78.0**	**9.6**	**211.4**	**27.7**	**223.0**	**19.3**	**512.4**	**56.6**

The total growing stock of stem wood is 61.4 billion m^3/ha with 51 billion m^3/ha consisting of coniferous species. Nearly 63% of the growing stock is classified as mature and over mature forests. A majority of the indigenous Russian people, some 40 different tribal groups, live in the Siberian forests, and depend on these natural resources for their livelihood. In terms of carbon sequestration, it has been estimated that nearly 94 billion tons are accumulated in Siberian forests with over 170 million tons of annual increment for this region. A substantial increase in carbon sequestration could be obtained through more intensive management practices (Isaev et al. 1995; Shvidenko et al. 2004).

4.1.2 Harvesting

The Annual Allowable Cut (AAC) includes both final felling and commercial wood for industrial and fuel uses. Although the level of harvesting during recent years (2000–2006) has increased, it still represents only 39–47% of the AAC. In 2006 the AAC was 382 million m^3 from forests managed by the forest authority with the allocation as follows: coniferous 261 million m^3, hard deciduous 6 million m^3 and soft deciduous 115 million m^3. The actual harvest in recent years ranged from 150 to 180 million m^3 but it is increasing and expected to exceed 200 million m^3 in 2007. In Siberia and the Far East, a 10% reduction in the AAC ($\approx$ 20% for hardwood forests) is expected due to excessive logging. In the Far East, the AAC for coniferous forests is estimated to be maintained at the present level and it is projected to increase by 29% for soft deciduous forests. The calculations concerning AAC only employ the commercial forests, hence the reason for low AACs in relation to the existing growing stock. Non-commercial forests and reserves (179 million ha) will not likely be harvested in the next 20 years, but will instead continue to produce non-timber benefits.

Significant problems exist with current forest harvest methods. First, the areas harvested are concentrated along existing transportation networks. For example, conifer stands along the Trans-Siberian Railroad are systematically being over harvested. Second, there were and still are few incentives or penalties in place to promote improved forest utilization. The stumpage fee is extremely low because per unit area volumes are low and penalties for poor utilization are minimal. And third, labor costs and forestry investments are increasing which results in "high-grading" of the forest for its best timber resources in order to increase profits.

Approximately 1 million hectares of forest are harvested annually in Siberia. Ninety-five percent of harvests occur via large-scale clear cuts, many of which are located in populated areas of the south and Far East where timber resources are overexploited. In some districts the AAC is substantially exceeded. In particular, pine forests are significantly affected by over harvesting. In contrast, larch and deciduous forests are underutilized. The result is a steady increase in the deciduous composition of the forest. Forests also experience high-grading, which has a negative influence on the future species composition of the forest. Problems relating to inefficient harvest and transport are also significant (Problemy 1998; Siberian expectations 2003). Large amounts of waste on the felling areas result in significant increases in insect and diseases infestation, and increases the risk to fire damage. The average residual following the harvest of a conifer forest ranges from 30 to 60 m^3/ha. For the East Siberia and the Far East, for every 3 m^3 of wood felled, 1 m^3 is left on the harvest site. Further losses from transportation can reach up to 60%. In East Siberia, an average of more than 70 m^3/ha of wood remains on site following clear cutting. An additional 20 m^3/ha of other biomass is left on the sites. The use of heavy harvesting equipment in Siberia causes serious damage to the understory, forest floor, and mineral soil. This damage includes changes to the soil moisture regime, increases in surface water run off, and increases in soil compaction as well as impacts to other ecological processes. Heavy harvesting equipment has a particularly negative impact in mountain landscape and permafrost regions. Skidding trails may not regenerate for at least 10 years or more and frequently skidding causes significant erosion problems.

Over harvesting in the southern region of the East Siberia has created serious ecological and social impacts. Subsequent decreases in harvesting have produced a large stock of logging equipment that is now underutilized and many manufacturing plants are no longer operating at full capacity. The equipment and plants are not transferable to other regions. Unemployment is therefore on the rise and industrial towns are experiencing social and economic disruption because they are heavily dependent on the timber industry. With large integrated manufacturing plants, problems are just as acute. When the local timber supply is exhausted, timber must be hauled for longer and longer distances or plants face shutdowns. Unemployed workers have limited options for relocation. Siberia is heavily dependent on the timber economy and therefore unsustainable practices are important for ecological, social, and economic reasons.

4.2 Forest Management

East Siberia, comprised of the Krasnoyarsk territory, the Khakass Republic, the Irkutsk region, the Republic of Buryatia, and the Chita region, represents one-third of all Russian coniferous forest area and 40% of the concomitant growing stock. Additionally, highly productive, high quality pine forests grow in the Yenisei-Angara River basins and nearly 20% of the national deciduous forests also are located within these basins (Osnovy 2000).

Growth potential – The total growth potential of the East Siberia is estimated at 361 million m^3 (Siberian Forests 2004). However, not all of this growth is available for use or development given factors such as access and remoteness. Approximately 95 million m^3 could be utilized in the short to medium term with the existing infrastructure. Of this amount, 70% is coniferous forest.

Roundwood harvest – In 2006, round wood harvest provided about 45 million m^3 in the East Siberia. Conifers accounted for over 90% of that harvest. The increase in harvest levels over time is representative of increasing developmental pressures. As a whole, forest resources in eastern Siberian are not being rapidly depleted, but there may be regional imbalances taking place. Most of these imbalances occur because of concentrated harvesting in response to increasing economic pressures in more heavily populated areas.

Intermediate stand treatments – Intermediate treatments include age-related thinning, pre-commercial thinning in young stands, commercial thinning, and selective sanitation harvest. These treatments are used for selecting preferred species for further growth, improving wood quality, providing wood for consumption, and reducing risk of loss due to fire, insects, and diseases. Over 11 million ha are in need of intermediate treatment and 30% of this area is in need of pre-commercial thinning. Commercial wood from sanitation harvest could yield 0.6 billion m^3. However, it is not economically feasible to treat all of these areas. The most significant factors preventing treatment include lack of a transportation network, lack of markets for small diameter trees, high transportation costs, dispersed treatment areas, and long haul distances to manufacturing plants. Thus, from a possible annual thinning volume of 96 million m^3, less than 10% is feasible under present economic conditions. Of these accessible areas, about 50% are in need of pre-commercial thinning and 5% are available for selection sanitation harvest.

4.2.1 Forest Regeneration

Approximately 800,000 ha are clear-cut annually in the East Siberia. The majority of harvested stands are suitable for natural regeneration by conifers. However, some areas are regenerated as plantations. As a general rule, the ratios between natural regeneration and plantations are as follows:

- northern and middle taiga 70:30
- southern taiga 50:50
- mixed forests 30:70
- forest steppe 5:95
- steppe 0:100

An exception is the Novosibirsk and Omsk regions (West Siberia), where plantations cover 60–70% of reforested areas. Natural regeneration is often insufficient due to destruction of undergrowth by use of inappropriate logging methods, inadequate assistance for natural regeneration, and inefficient forest fire protection. The forest regeneration system in Siberia includes:

(1) establishing forest plantations where natural regeneration is not expected
(2) assisting natural regeneration in the forest understory
(3) exposing mineral soils to promote natural regeneration
(4) encouraging natural regeneration of commercially valuable tree species
(5) converting soft deciduous young forests to coniferous or hard deciduous forests.

In reforested areas, survival rates tend to be low due to the low quality of planting stock as well as frequent forest fires. During the last 3 years (2004–2006) over 300,000 ha of reforested areas were destroyed by fire, roughly 10% of the accumulated total in the East Siberia. In the Far East region, only about 50% of the planted areas have survived. Research suggests that an increase in survival rate of 2–3 times is necessary for adequate reforestation. The Siberian forest has a huge potential for large-scale reforestation with carbon sequestration as the primary objective. Based on realistic projections, 50–80 million ha could be reforested during the next 50 years, resulting in an annual carbon sequestration of nearly 2.5 tons C ha^{-1} $year^{-1}$ (Problemy 1998; Siberian expectations 2003). At present, net primary production (NPP) for the Siberian forest averages 285 g C m^{-2} and NPP could increase 30–40% with global warming (Russian Forests 2007). On an ecosystem basis, however, the vast Siberian forest may well switch from a carbon sink to a carbon source with global warming due to increases in soil and forest floor respiration and tree mortality. Such a change would have significant implications on a global scale. Further, Schulze et al. (2000) argue that maintaining mature forests may have a greater positive effect on the carbon cycle than promoting reforestation and afforestation. That is to say, maintaining the forest is a better strategy than regrowth for carbon sequestration.

Restoration of forestlands destroyed or damaged by energy and mineral development, including coal, ore, peat, oil and gas exploitation, is an important reforestation issue in Siberia. Total areas of such lands are estimated to be nearly 10 million ha. During the last 2 years, planting and seeding on these lands occurred on less than 1000 ha in the entire territory of Siberia. There are large, low density forest stands of limited market value in Siberia that are in need of forest improvement (Danilin 1995, 2003). According to inventory data, areas requiring reconstruction

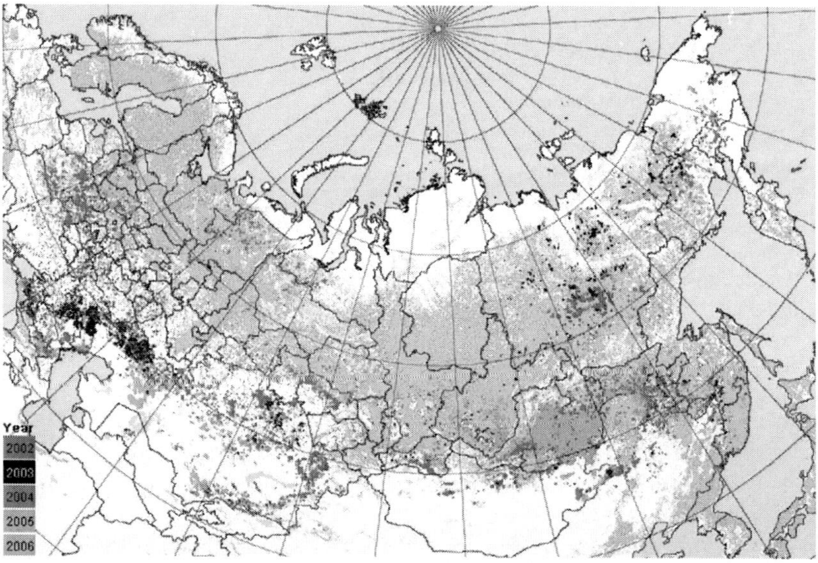

Fig. 4.3 Fire damage of forest and steppe vegetation at Northern Eurasia in 2002–2006 by SPOT-Vegetation and TERRA/MODIS data (Terra Norte 2007)

ranges from 238,000 ha in the Far East, 107,000 ha in East Siberia, 76,000 ha in West Siberia (Siberian Forests 2004). The current area of non-regenerated burns, cuts, and dead forests – nearly 15 million ha in Central and East Siberia – represents a serious environmental and economic problem (Figs. 4.3, 4.4, 4.5, 4.6).

The forested areas where plantations had been created recently in the East Siberia are highly productive pine and larch stands, commercially valuable, and accessible by vehicles (Figs. 4.7, 4.8, 4.9, 4.10). More than 1 million ha in the East Siberia are currently in forest plantations but only half of these are in good or satisfactory

Fig. 4.4 Fir (*Abies sibirica*) forest after two catastrophic fire in Krasnoyarsk territory in 2006. More than 50,000 ha were burned in 2006 fire

Fig. 4.5 A view of the 3-year old clearcut area with herb and shrubby dominating the stands in Krasnoyarsk territory

Fig. 4.6 Larch (*Larix sibirica*) forest severely damaged by Gypsy moth (*Lymantria dispar* L.) in Tyva Republic (Southern Siberia)

Fig. 4.7 Burned area replanted by pine (*Pinus sylvestris*) seedlings in Krasnoyarsk territory

Fig. 4.8 A 15-year old burned area of the mixed larch (*Larix sibirica*), pine (*Pinus sylvestris*), spruce (*Picea obovata*) and fir (*Abies sibirica*) forest naturally regenerated by birch (*Betula pendula*) and aspen (*Populus tremula*) in Southern Siberia (Krasnoyarsk territory)

Fig. 4.9 A 50-year old pine (*Pinus sylvestris*) plantation at Krasnoyarsk Research Center regenerated after logging

Fig. 4.10 A 118-year old larch (*Larix sibirica*) plantation at Krasnoyarsk territory

condition. Forest fires, insect outbreaks, and industrial pollution have caused extensive mortality in many plantations. During the period from 1996 to 2006, 700,000 ha of forest plantations were established in East Siberia, resulting in an increase in the area in forest cover by 160,000 ha. Sixty tree nurseries, including 15 temporary nurseries, are available in Siberia for growing planting stock. In 2006, these nurseries produced 33.5 million pine (*Pinus sylvestris*), cedar (*Pinus sibirica*), larch (*Larix sibirica*) and fir (*Abies sibirica*) seedlings.

4.3 Non-Timber Products

Many non-timber products are harvested from the Siberian forest, including fruits, berries, mushrooms, nuts, tree sap, and medicinal plants. Other ecosystem services include production of herbs, grazing, beekeeping, hunting, fishing, and recreation. About 45% of all medicines in Russia are produced from plants and more than 700 plant species with medicinal value are found in the forests and bogs of East Siberia. On average, 2 million tons of fruits and berries, 1.5–4.0 million tons of mushrooms, and 0.8–1.2 million tons of nuts are gathered each year from the forest (Encyclopedia 2006). This represents a small fraction of the total production of these forest products. About 80% of the total nut harvest in Russia was collected in Siberian pine forests (*Pinus sibirica* and *Pinus coraiensis*), although the actual harvest of cedar nuts does not exceed 2% of the production and reaches 18 to 20 thousand tons annually. The West and East Siberia have 37 species of fur animals. During the last 10 years these two regions supplied nearly 90% of the total fur in Russia. The area of cedar forests is less than 6% of all Siberian forests, yet they provide half of the total harvest of sable and gray squirrel fur, which are very important species to the fur trade. On average, 1000 ha of cedar taiga produces 17 times more fur than on a similar area of larch forest. In addition, there is a substantial population of wild animals that are important food supplies for people in Siberia. Experts estimate the populations for the last 10 years to be the following for West and East Siberia: 168,000 moose, 554,000 reindeer, and 241,000 roe deer (Encyclopedia 2006). These populations are increasing. For example, the moose population was estimated to have increased by 260,000–300,000 animals in spite of an annual harvest of about 180,000 animals over the past 5 years. There is a great potential for increasing the production of non-timber products in the Siberian forest. More research is needed to develop forest management techniques to enhance these ecosystem services. In addition, improved storage and manufacturing facilities are needed as well as more efficient collecting and improved harvesting techniques.

4.4 Disturbances

During the past 50 years, wide-scale forestland cultivation has created significant changes in forest development. Increased size of forest fires, decreased forest resistance to insects (worsened by pollution), and increased wood extraction have resulted in modifications of ecosystem structure and forest cover. National forest inventories from 1986 to 2006 show significant changes in cover over the East Siberia. Over this time period, fir stand area decreased by 23%, pine (*Pinus sylvestris*) stand area decreased by 14%, and deciduous cover increased by 60%. Additionally, young and middle-aged stands tripled, which is an indication of degradation of mature and overmature stands. After a disturbance event, such as fire or harvest, fir (*Abies sibirica*) stands are generally replaced by secondary deciduous species such as birch (*Betula pendula*) and aspen (*Populus tremula*). Pine forests occupy poorer sites, where stand replacement occurs much more slowly. Development of an understory tree layer composed of coniferous species is currently observed for 57% of the total forest area in the East Siberia.

4.4.1 Forest Fires

Forest fires are still the predominant forest disturbance in the Siberian forest and fires largely determine long-term forest dynamics at the landscape and regional scales (Figs. 4.3, 4.4, 4.8). In the 973 million ha Forest Fund area, 590 million ha (61%) are under some form of fire protection; 78% of West Siberia, 66% of East Siberia, and 52% of the Far East are under some form of fire protection. On non-protected areas, active fire fighting occurs only in exceptional cases, such as impending danger to residential and commercial property. On protected areas, between 15,000 and 35,000 fires occur annually, resulting in the destruction of 0.5–1.0 million ha of forested area. In these areas, the average fire size varies from 15 to 50 ha (Siberian Forests 2004). Based on current statistics and remote sensing data, the average individual fire across protected and unprotected areas causes a loss of about 100 ha of Forested Area and 2,000–5,000 $m^3$of timber. Large forest fires, which account for 10–15% of the total number of fires, are responsible for 80–85% of the area burned, and so these few large fires have a disproportional impact on the Siberian forest.

Global warming has serious implications for forest fire management in Siberia. Using four current General Circulation Models (GCMs), Stocks et al. (1998) projected large increases in the area extent of extreme fire danger in the Russian and Canadian boreal forests. These changes include both an extended fire season as well as increases in the area experiencing high to extreme fire danger. Changes in disturbance patterns associated with changes in climate in combination with anthropogenic disturbance are drastically “redesigning” the landscape in the boreal forest (Crow 2005, 2008).

Fire protection is the responsibility of the Federal Forest Protection Service and the Forest Fire Service. Satellite and other aerial monitoring techniques are the primary control mechanism, with most monitoring conducted through regional Aerial Forest Control Bases. The current system for monitoring and control does not provide effective forest fire protection for Siberia. Detection systems are inadequate to locate fires early before they spread and less than half of all forest fires in Siberia are discovered and suppressed before they cause extensive damage. The primary reasons for the low level of forest fire protection are lack of sufficient funding, scarce and poor technical equipment for both aerial and ground forest fire protection, and imperfect organizational structure, management, and administration for fire suppression.

4.4.2 Other Disturbances

After forest fires, the predominate cause of forest stand deaths are insect outbreaks, industrial pollution, and wind. The rapid change in forest ecosystems related to human activities is often associated with the loss of biological stability. In Siberia, outbreaks of insects and diseases have increased in recent years in apparent response to anthropogenic impacts such as intensive harvesting, pollution, as well as changes in hydrologic patterns. Insects and diseases have the ability to quickly multiply and spread over large forest areas. For example, residual trees and slash remaining after final felling provides ideal conditions for insect and disease outbreaks. Episodes of mortality in spruce-fir forests have been observed in the southern part of the Far East since 1926, with estimates of their extent ranging from hundreds to several million hectares. About 1 million ha of forested area are affected annually in the East Siberia by Gypsy moth (*Lymantria dispar*) and the Siberian silkworm (*Dendrolimus sibiricus*) outbreaks.

Vegetation cover of forest-tundra and northern taiga zones is polluted by a complex of smelters in Norilsk in north central Siberia. This complex pumps over 2 million tons of sulfur dioxide, heavy metals, and other pollutants into the air each year. The polluted zone extends for near 300 km from Norilsk in the south-eastern direction. Forest cover is completely destroyed or heavily disturbed in the surrounding area of 500,000 ha with measurable impacts occurring in an area several times as great. Strong winds (more than 17 m/s) are common in the West and East Siberia and the Far East and they produce large wind-falls. Other weather factors that create forest mortality include extended droughts and as well as flooding. The former can cause extensive mortality, while the latter is more localized in its impact.

During the past 10 years, sanitation treatments to reduce mortality from diseases and insects have been implemented on about 60,000 ha annually in Siberia. Most are biological control methods. Forest pathological surveys have been conducted on 1 million ha each in West Siberia and East Siberia, and on about 285,000 ha in the Far East.

4.5 Ecosystem Structure and Function

The dynamics of organic matter and biodiversity can serve as useful indicators for stability in forest ecosystems. Examination of production dynamics related to stand structure and growth makes it possible to define permissible limits for affecting ecosystem and the extent and nature of their effect on the environment. Studies dealing with ecosystem structure and function have been conducted in forests established on burned and commercial logging sites in the East Siberia (Krasnoyarsk territory, the Khakass Republic, Republic of Tuva, Republic of Buryatia, Irkutsk region and Chita regions) has been studied and analyzed (Fig. 4.1). The characteristics of permanent sample plots are described in Table 4.3.

The sample plots were established using conventional forest inventory and survey techniques (Danilin 1995, 2003). All trees at the sample plots were measured and mapped to establish their size-dependent position in the phytocenosis and to determine the structure of the aboveground biomass and diversity of forest vegetation. The forest composition in the experimental area was even-aged pine, larch and birch-aspen stands, both pure and mixed, belonging to different forest types and natural formation patterns/series (see Table 4.3).

Pure young pine stands occurred mostly on a river terraces and belong to the pine-bearberry-lichen (*Pinus sylvestris – Arctostaphylos uva-ursi, Cladina sylvatica*) and pine-red bilberry-green moss (*P. sylvestris – Vaccinium vitis idaea, Pleurozium schreberii*) forest types. The tree density in these stands varies from 1.6 to 94.6 thousand trees per ha. Tree mortality is moderate. Mixed pine, aspen and aspen-birch stands were common on flat interfluves with loamy soils. These belong to the pine-aspen, birch-red bilberry-herbaceous (*Pinus sylvestris, Populus tremula, Betula pendula, Vaccinium vitis idaea, Calamagrostis arundinacea, Carex macroura, Pulsatilla patens, Trifolium lupinaster, Buplerum aureum, Geranium sylvaticum*) forest type and the density of these stands varied from 6.4 to 11.2 thousand trees per ha.

The study was also conducted in larch stands occurring across an altitudinal range. The sites, located approximately at 51–56°N, 95–115°E, are dominated by *Larix sibirica* and *Larix gmelinii*, and are commonly located in the lower portions of mountain slopes of the East Siberia region. Because these are highly productive forests that are commercially valuable and with good access, they have been subjected to intensive harvesting during the past 30 years. Common cuts (i.e., commercial clear-cuts) and pseudo-clear-cuts are the norm. The harvested wood is usually skidded using tractors to a landing, often resulting in considerable disturbance of the understory vegetation, forest floor, and mineral soils. After harvesting, logging residues are evenly distributed over the harvest area and left untreated. In unlogged stands, densities of young larch in the understory generally number between 2400 and 5000 stems per ha; their quality and spatial distribution depend on their age, basal area, and the area occupied by the parent stand. Following harvest, 400–600 stems per ha typically survive in the understory. Both parent larch stands and cut areas suffer from frequent fires that enhance natural regeneration. Natural regeneration is often abundant in small clear cuts (up to 5 ha). Larch seedlings germinate annually and their resistance to environmental stresses is relatively high; in the first 5–6 years after cutting the number of seedlings can exceed 50,000 per ha, thus providing adequate regeneration in the cut area. By the time the regenerating canopy closes (in about 15 years), the herbaceous cover and litter layer characteristic of the previous stands are present.

Table 4.3 Structure and aboveground biomass of forest plantations

Plot number	Location of sample plot	Species composition, in % of timber stock	Dominant understorey	Tree species	Average					Above-ground biomass, ton/ha 100% dry matter	Annual biomass increment, ton/ha 100% dry matter
					Stand age, years	DBH, cm	Height, m	Density, thousand trees/ha	Growing stock, m^3/ha		
1	61°N 93°E	100% Pine	Bearberry lichen	Pine	12	1.1	1.2	94.6	23.1	10.9	0.9
2	58°N 104°E	100% Pine	Red bilberry green moss	Pine	15	2.4	2.6	6.8	7.6	5.0	0.3
		62.8% Pine		Pine	15	3.6	5.5	3.8	13.0	6.3	0.4
3	59°N 97°E	33.8% Aspen	Herbaceous	Aspen	15	3.2	6.2	2.3	7.0	3.4	0.2
		3.4% Birch		Birch	15	2.8	5.7	1.0	0.7	0.9	0.06
		61.1% Pine		Pine	28	6.5	8.4	4.8	69.0	42.3	1.5
4	59°N 97°E	23.9% Pine	Herbaceous	Aspen	30	4.7	8.8	3.3	27.0	16.5	0.6
		15.0% Birch		Birch	25	6.9	11.2	0.8	17.0	10.2	0.4
5	27°N 97°E	80.4% Pine	Herbaceous	Pine	36	5.3	6.2	10.4	86.0	40.4	1.1
		19.6% Birch		Birch	30	8.2	9.0	0.8	21.0	13.5	0.5
6	56°N 105°E	100% Pine	Red bilberry green moss	Pine	36	5.8	8.8	10.3	126.0	69.8	1.9
		95.8% Aspen		Aspen	15	4.4	8.5	4.7	30.0	13.8	0.9

Table 4.3 (continued)

Plot number	Location of sample plot	Species composition, in % of timber stock	Dominant understorey	Tree species	Average					Above-ground biomass, ton/ha 100% dry matter	Annual biomass increment, ton/ha 100% dry matter
					Stand age, years	DBH, cm	Height, m	Density, thousand trees/ha	Growing stock, m³/ha		
7	57°N 98°E	2.6% Pine	Herbaceous	Pine	15	5.3	6.0	3.2	0.8	0.39	0.03
		1.6% Birch		Birch	15	2.8	5.0	0.2	0.5	0.27	0.02
		77.3% Aspen		Aspen	15	3.6	6.9	8.1	34.0	19.9	1.3
8	55°N 95°E	20.4% Birch	Herbaceous	Birch	15	5.0	7.6	1.1	9.0	5.4	0.4
		2.3% Birch		Pine	15	2.3	2.9	0.8	1.0	0.27	0.02
9	52°N 95°E	100% Larch	Herbaceous	Larch	16	1.6	3.8	54.5	39.0	33.9	2.1
10	56°N 111°E	100% Larch	Herbaceous	Larch	28	5.9	7.9	5.2	74.0	55.2	2.0
11	49°N 109°E	100% Larch	Herbaceous	Larch	30	7.5	9.5	9.1	212.0	117.9	3.9
12	51°N 110°E	100% Larch	Herbaceous	Larch	37	18.4	15.4	1.2	305.0	163.7	4.4
13	51°N 114°E	100% Larch	Herbaceous	Larch	70	18.9	18.1	1.5	397.0	189.1	2.7
14	64°N 103°E	100% Larch	Read bilberry green moss	Larch	30	3.5	6.7	5.2	20	11.6	0.4
15	50°N 109°E	93.2% Birch	Herbaceous	Birch	60	13.6	14.0	1.6	159.9	123.2	2.1
		6.8% Aspen		Aspen	70	22.6	14.7	0.04	11.6	6.1	0.1

The structural relationships between biomass fractions change with increasing average stand age and density. In dense stands, tree crowns were best developed. Consequently, the total crown biomass of dense, young stands was greater than in older stands with lower tree density. This may be due to the fact that trees, at early development stages, make maximum use of their assimilation apparatus and branches, and there is coenopopulation competition for light, nutrients, and water. Biomass increment is a more objective index for the production process and ecosystem stability than total biomass. Over the last five years, the current annual increment of diameter and height increased consistently with increasing tree size. Trees with the best developed crowns and stems showed the greatest increment, but in underdeveloped trees, these values were the lowest. The maximum average annual biomass increment was found to be 4.4 and 3.9 tons ha^{-1} (dry matter) at sample plot 12 and 11 (Table 4.3).

If the process of succession follows its natural course and the existing rate of biomass increment is maintained, these forests could be expected to regain their original state (i.e., a relatively stable climax forest ecosystem whose components are in balance with the environment) in about 100 years for the pure larch and pine and mixed stands, and about 50 years for birch and aspen phytocenosis. However, this time period would become much longer if disturbances occurred, such as fire or insect outbreaks. In this case, if a phytocenosis is partly or completely destroyed, succession would take the form of replacement of coenopopulations: secondary aspen and birch stands on flat interfluves with loamy soils, or the native edificator tree species on river terraces with sand and loamy sand (Danilin 1995; Danilin et al. 1996).

With no regeneration of birch, larch or pine due to poor silvicultural activities, harvest of seeds, droughts, forest fires or after logging, successional development may take the course of forming a meadow or steppe ecosystem. Such a development decreases forested area, reduces protective and environment-forming functions of the ecosystem, its tolerance to environmental effects, which make it undesirable. Specific forest management activities, such as thinning, can considerably reduce the time necessary for a native phytocenosis to fully recover. Thinning, however, is a complex biotechnical treatment that needs planning. Before thinning, account should be taken of the biological characteristics of the stand (i.e., its composition, age, density and productivity) and site characteristics (e.g., topography, soils, climate, the rate and character of anthropogenic disturbance, and wildfire dynamics). When controlling phytocenosis succession through intervention (anthropogenic succession), it is preferable for the forest ecosystem to regenerate naturally to form uneven-aged (of several generations) pine or larch forests with birch and aspen admixtures (and to create mosaics on the landscape) that are more resistible to disturbance and more productive than monocultures (Danilin 2003).

In conclusion, human influences are profoundly affecting the composition, structure, and health of forest landscapes in Siberia. Some of these are direct influences such as forest harvesting, while others are indirect such as pollution or altered disturbance patterns. In Siberia, research dealing with the establishment and culture of forest plantations, especially at burned and logged areas, is a high priority. Other critical needs include the application of well-designed resource assessments, fire protection, and the control of disease and insect outbreaks. Applications of remote sensing and GIS technologies are needed to create up-to-date forest database. And finally, more suitable machinery for conducing forest operations are needed (Osnovy 2000; Organizaciya 2002; Danilin et al. 2005).

4.6 Status Summary

Forest ecosystems and forest landscapes in Siberia have the following features:

- Relatively low productivity – about 50% of the region is occupied by stands of low productivity with a timber stock below 100 m^3 ha^{-1}.
- Fire losses – annually nearly 500,000 ha of forested areas are affected by wildfires.
- Harvesting pattern – overcutting of timber has occurred along the main railroad transportation routes and close to the manufacturing centers.
- Harvesting area – there is a significant increase in the rate of harvest in remote regions.
- Utilization – the high grading of timber resources is widespread. There are serious losses of wood during transport from harvest site to consumer.
- Species change – the combination of clear cut harvesting and fire are reducing the conifers and increasing the soft deciduous species.
- Forest health – large territories have damaged forest health due to attacks by insects and diseases, unsound final harvesting methods, pollution, and other factors. The area of non-regenerated cuts, burns and dead stands is nearly 16 million ha in East Siberia alone.
- Silviculture – commonly used silvicultural methods are not necessarily creating an efficient forest renewal program. Reforestation is inadequate relative to the actual need.
- Research – there is limited application of academic research findings in commercial practical forestry and landscape management.
- Forest dynamics – the forest resources are deteriorating slowly but significantly in Siberia. Generally, the development of the Siberian forestry cannot be considered sustainable. The key issue in Siberian forestry is to establish a sustainable form of landscape management and development of forest resources from an ecological, economic and social point of view.
- Opportunities – the vast Siberian forest is an important natural resource at a global scale.

4.7 Conclusions

From a resource perspective, opportunities exist to seek new markets for the deciduous fiber supply, and to better manage and control utilization of the softwood coniferous forest. From an industry perspective, opportunities are more likely to be found in meeting the rising demand for wood in Pacific Rim countries. Higher domestic demands for wood products are also boosting harvest activity. To offset longer transport distances to markets in the west, higher prices are needed for both raw materials and processed wood. Siberia has a relatively well-developed forest infrastructure, highly-trained scientists, a structure of forest enterprises, and some protective and regulatory measures that serve as a skeleton for rapid and productive development of the sector. What is needed are essential investments directed at modernization, technical support and basic materials (especially technological), to enhance the country's capacity to promote sustainable development of the forest sector. During the past five years, most international assistance (USAID, USDA Forest Service, World Bank, WWF and the others) in the forestry sector has been directed toward increasing forest productivity and sustainability, as well as improving forest resource and landscape management. However, much can be accomplished by reducing the perceived risk attached with investing in forest sector in Siberia. This includes improving financial and transportation infrastructures, information technology, and improved training for the labor force.

References

Crow TR (2005) Landscape ecology and forest management. In: Wiens J, Moss M (eds) Issues and perspectives in landscape ecology, Cambridge University Press, Cambridge, pp 201–207

Crow TR (2008) Managing forest landscapes for climate change. In: Lafortezza R, Chen J, Sanesi G, Crow TR (eds) Patterns and processes in forest landscapes – Multiple use and sustainable management, Springer

Danilin IM (1995) Structure and biomass of larch stands regenerating naturally after clear cut logging. Water Air Soil Pollut, 82(12):125–131

Danilin IM (2003) Morfologicheskaya struktura, produktivnost' i distancionnye metody taksacii drevostoev Sibiri (Morphological structure, productivity and remote sensing methods for forest inventory of the Siberian tree stands). Dissertation, V.N. Sukachev Institute of Forest, Russian Academy of Sciences, Siberian Branch, Krasnoyarsk, Russia, 537p (in Russian)

Danilin I, Kobayashi S, Abe N (1996) Growth dynamics and biodiversity of larch forest after wildfire at the north of Central Siberia. World Res Rev 8(2):215–230

Danilin IM, Medvedev EM, Abe NI et al. (2005) Lesoustroistvo v zapovednikah – principy i metody. Podkhody k obobscheniyu opyta (Forest planning at natural reserves – principles and methods. Approaches for summarizing the experience). V.N. Sukachev Institute of Forest, Russian Academy of Sciences, Siberian Branch Publ., Krasnoyarsk, Russia, 118p (in Russian)

Enciklopedia lesnogo khozyaistva (Encyclopedia of Forestry) (2006) VNIILM, Moscow V1 p 424 (in Russian)

Isaev AS, Korovin GN, Zamolodchikov DG et al (1995) Carbon stock and deposition in phytomass of the Russian forests. Water Air Soil Pollut 82(1t2):247–256

Organizaciya osobo okhranyaemykh prirodnykh territorii (Organization of especially protected natural areas) (2002) Sokolov VA, Farber SK, Danilin IM et al. Russian Academy of Sciences, Siberian Branch, Publ., Novosibirsk, Russia, p 264 (in Russian with abstract in English)

Osnovy lesnoi politiki v Krasnoyarskom krae (Bases of forest policy at Krasnoyarsk territory) (2000) Sokolov VA, Semetchkin IV, Danilin IM et al. Russian Academy of Sciences, Siberian Branch, Publ., Novosibirsk, Russia, p 246 (in Russian with abstract in English)

Pozdnyakov LK (1986) Les na vechnoi merzlote (Forest at permafrost). Nauka Publ., Siberian Branch, Novosibirsk, Russia, p 192 (in Russian)

Problemy ustoichivogo lesopol'zovaniya (Problems of sustainable forest use) (1998) Sokolov VA, Danilin IM, Farber SK et al. Russian Academy of Sciences, Siberian Branch, Press, Krasnoyarsk, Russia, 225p (in Russian with abstract in English)

Russian Forests and Forestry. Data and Analysis. Version 1.0. IIASA, 2007 http://www.iiasa.ac.at/Research/FOR/forest_cdrom/home_en.html

Schulze E-D, Wirth C, Heimann M (2000) Managing forests after Kyoto. Science 289(5487): 2058–2059

Shvidenko AZ, Schepaschenko DG, Nilsson S et al. (2004) Sistema modelei rosta i dinamiki produktivnosti lesov Rossii (tablicy biologicheskoi produktivnosti) (System of models for growth and productivity dynamics of Russian forests (tables of biological productivity)). Lesnoe khozyaistvo (Forest management) 2:40–44 (in Russian)

Siberian forests: approach from Russian and Japanese researchers (2004) Abe N, Sokolov VÀ, Danilin IM. Japan Forestry Investigation Committee, Tokyo, Japan, 243p (in Japanese)

Sokolov VA, Danilin IM, Semetchkin IV et al. (2003) Siberian expectations: an overview of regional forest policy and sustainable forest management, World Forest Institute Publ., Portland, Oregon, USA, 67p, http://www.worldforestry.org/wfi/SiberianExpecations.pdf. Cited 15 Jan 2003

Stocks BJ, Fosberg MA, Lynham TJ, Mearns L, Wotton BM, Yang Q, Jin J-Z, Lawrence K, Hartley GR, Mason JA, McKenney DW (1998) Climate change and forest fire potential in Russian and Canadian boreal forests. Climatic Change 38(1):1–13

Terra Norte (2007) http://terranorte.iki.rssi.ru/

Chapter 5

Fragmentation of Forest Landscapes in Central Africa: Causes, Consequences and Management

Jan Bogaert, Issouf Bamba, Kouao J. Koffi, Serge Sibomana, Jean-Pierre Kabulu Djibu, Dominique Champluvier, Elmar Robbrecht, Charles De Cannière and Marjolein N. Visser

Abstract Forest fragmentation has a paramount impact on landscape pattern and has therefore been a key focus of landscape ecology. Trends and causes of deforestation are analysed for the Democratic Republic of the Congo, Rwanda and Burundi, and are put in a regional, continental and global perspective. In order to investigate the role of shifting cultivation as a driver of fragmentation, the dynamics of a forest landscape between 1970 and 2005 for a study area in the Bas-Congo province of the Democratic Republic of the Congo were analysed. Using a transition matrix and the identification of the spatial land transformation processes involved, historical data are compared with the current situation based upon field visits and remote sensing imagery. As a consequence of non sustainable shifting agriculture, forest fragmentation is observed, leading to an expansion of savannah, fallow lands and fields which replace secondary forest vegetation and limit forest succession towards primary forest. Since forest ecosystems are known to be the habitat of indicator species only observed in one specific phytogeographic territory, the potential impact of habitat preservation for these species is investigated. A dataset of 310 Acanthaceae species containing 6362 herbarium samples for the Democratic Republic of the Congo, Rwanda and Burundi is analysed and species presence is compared with the phytogeographic theories of Robyns (1948), White (1979, 1983) and Ndjele (1988). Study of the spatial distribution and analyses of species habitats reveal the importance of forest preservation to protect these indicator species. Conservation of these habitats should therefore be given priority to avoid loss of genetic resources for future generations. Implications for the management of forested landscapes are discussed, regarding the role of local populations, the application of ecological principles, the conservation of virgin forests, the potential role of forest plantations, and the importance of landscape pattern analysis.

J. Bogaert
Université libre de Bruxelles, Laboratoire d'Ecologie du Paysage, 50 Avenue F.D. Roosevelt, CP169, B-1050 Bruxelles, Belgium
e-mail: jan.bogaert@ulb.ac.be

R. Lafortezza et al. (eds.), *Patterns and Processes in Forest Landscapes*,

5.1 Fragmentation of Forests in Central Africa: Facts, Figures, and Possible Causes

One of the earliest and still continuing human impacts on the biosphere is the removal of the original vegetation cover and its replacement by either another or by man-made structures. At the global scale, the ecologically most significant impact of this kind is deforestation. It has been estimated that about half of the forest or woodland that originally covered two-thirds of the earth's surface has already been removed and reduction is being maintained at a rate with outstrips replacement (Tivy 1993).

Forest area decrease is the main parameter to describe forest fragmentation (Gascon et al. 2003); the loss of primary forest results in the creation of a new matrix habitat. Matrix habitat will be important in the evolution of ecosystem dynamics in forest patches because (1) it will act as a filter for movement between landscape features; (2) disturbed area-associated species will be present and may invade forest patches and edge habitat; (3) depending on land-use, the matrix habitat will take on a different form, such as pasture, degraded pasture, or second growth forest, and the nature of the matrix habitat will influence the severity of the edge effect in patches (Williamson et al. 1997 in Gascon et al. 2003). Extent of forest resources is also a main element characterizing sustainable forest management; it is an easily understood baseline variable, which provides a first indication of the relative importance of forest in a country or region (FAO 2005). Estimates of its change over time provide an indication of the demand for land for forestry and other land uses, as well as of the impact of significant environmental disasters and disturbances on forest ecosystems (FAO 2005). Fragmentation is considered a main indicator of landscape degradation, next to an increased rate of movement of surface soil particles, a change of the phenology of the vegetation (perennial towards annual), and a change in the hydrologic regime (Groves 1998). Forest extent is relatively easy to measure, and this variable has therefore been selected as one of the 48 indicators for monitoring progress towards the Millennium Development Goals agreed by the United Nations. Unless mentioned otherwise, all data and metadata cited further on in this section are based upon the Global Forest Resources Assessment of 2005 by the Food and Agriculture Organisation (FAO 2005), which constitutes, to our knowledge, the most recent and reliable source on global forest extent.

Total forest area in 2005 is estimated to be 3952 million ha or 30.3% of total land area. This corresponds to an average of 0.62 ha per capita, for a world population equal to 6.3 billion people (2004 data). Forest area in Africa was found equal to 635 million ha or 16.1% of global forest area, 21.4% of the continents land area, and 0.73 ha per capita for a population of 868 million people. Only Asia has a lower proportional forest cover of 18.5%.

Deforestation, mainly due to conversion of forest to agricultural land, continues at an alarming rate, some 13 million ha per year globally. At the same time, forest planting, landscape restoration and natural expansion of forests have significantly reduced the net loss of forest area; net global change in forest area in the period 2000–2005 is estimated at –7.3 million ha per year (–0.18%), down from –8.9

million ha per year (–0.22%) for 1990–2000. From 2000 to 2005, Africa lost 4.0 million ha annually (–0.62%), against 4.4 million ha annually for the 1990–2000 period (–0.64%).

In Africa, a majority of countries have a negative change rate. Among the 10 countries with the largest annual net negative change rates for 2000–2005, the following African countries are found: Comoros (–7.4%), Burundi (–5.2%), Togo (–4.5%), Mauritania (–3.4%), Nigeria (–3.3%), Benin (–2.5%) and Uganda (–2.2%). Eighteen countries are characterized by an estimated annual positive change of 1% or more due to natural expansion of forests and to reforestation, among which four countries of the African continent: Rwanda (+6.9%), Lesotho (+2.7%), Egypt (+2.6%) and Tunisia (+1.9%). Among the 10 countries with largest annual net loss in forest area 2000–2005, six African countries are found: Sudan (-589×10^3 ha/yr), Zambia (-445×10^3 ha/yr), the United Republic of Tanzania (-412×10^3 ha/yr), Nigeria (-410×10^3 ha/yr), the Democratic Republic of the Congo (-319×10^3 ha/yr) and Zimbabwe (-313×10^3 ha/yr). No African country is found among the 10 countries with largest annual net gain in forest area over the same period.

For Central Africa, defined here as the region composed of Burundi, Cameroon, the Central African Republic, Congo, the Democratic Republic of the Congo, Equatorial Guinea, Gabon and Rwanda (FAO 2001a), forest cover equals 224 million ha in 2005, which is a reduction of 1.3% compared to 2000 and 4.8% compared to 1990. Firstly, it can be observed that the larger the country, the larger the forest extent in 2005 of that country. The largest country in the region, i.e. the Democratic Republic of the Congo, is characterized by the largest 2005 forest extent (134 million ha). Although the trend curve suggests that Central Africa should have a forest cover of ~57% (Fig. 5.1), considerable differences are observed when forest proportion is considered.

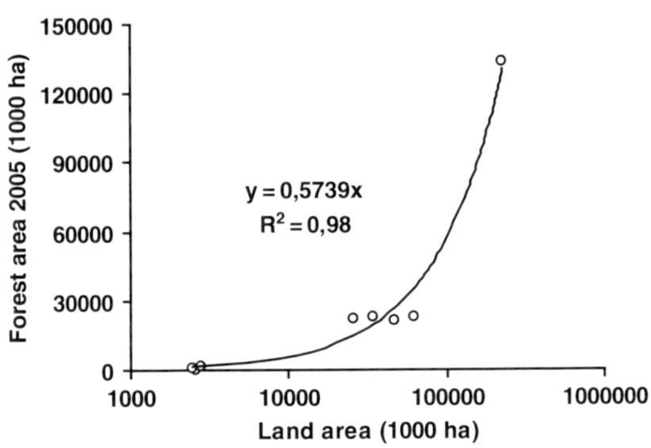

Fig. 5.1 Forest area as a function of land area for Central African countries Burundi, Cameroon, the Central African Republic, Congo, the Democratic Republic of the Congo, Equatorial Guinea, Gabon and Rwanda. Larger countries are characterized by larger extents of forest. A trend towards a forest proportion of 57% is observed, but large differences exist between the countries considered (see text for explanation)
Data source: Global Forest Resources Assessment of 2005 of the Food and Agriculture Organisation (FAO 2005).

While Congo (65.8%), the Democratic Republic of the Congo (58.9%), Equatorial Guinea (58.2%) and Gabon (84.5%) are dominated by forest cover, Rwanda and Burundi are characterized by 19.5%, respectively 5.9% of forest cover. Forest area has decreased annually since 1990 except in Rwanda, where it increased from 318×10^3 ha (1990) to 344×10^3 ha (2000) and 480×10^3 ha (2005). Between 1990 and 2005, the largest loss was observed in the Democratic Republic of the Congo (6.9 million ha). Expressed as a fraction of the 1990 forest cover Rwanda shows the largest increase (+50.9%) and Burundi the largest decrease (–47.9%), the latter being an alarming observation. For Gabon, no significant trends were observed. Population density (Fig. 5.2) clearly forces forest area decline. Rwanda (0.29 ha per capita) and Burundi (0.35 ha per capita) are more densely populated than, for example, the Democratic Republic of the Congo (4.13 ha per capita). Their 2005 forest covers are clearly at the lower end, with 19.5% (Rwanda) and 5.9% (Burundi) against, for example, 58.9% for the Democratic Republic of the Congo or even 84.5% for Gabon. The trend curve shown in Fig. 5.2 illustrates this causality with anthropogenic pressure.

The aforementioned and for the region exceptionally forest area increase in Rwanda is due to the change in extent of forest plantations (productive and protective forest plantations combined). Their extent increased from 78.0% of the total forest area in 1990 (248×10^3 ha) to 87.2% (419×10^3 ha). Next to these plantations, no primary forest is found, only a small fraction of modified natural forest (62×10^3 ha). This trend observed in Rwanda has not been found elsewhere in Central Africa. It should be noted, however, that also forests in Burundi are dominated by plantations (2000: 43.2%; 2005: 56.2%). This turning point from net deforestation to net reforestation, as observed for Rwanda, is defined in literature as a forest transition (Rudel et al. 2005; Kauppi et al. 2006). Forest transitions have been described to

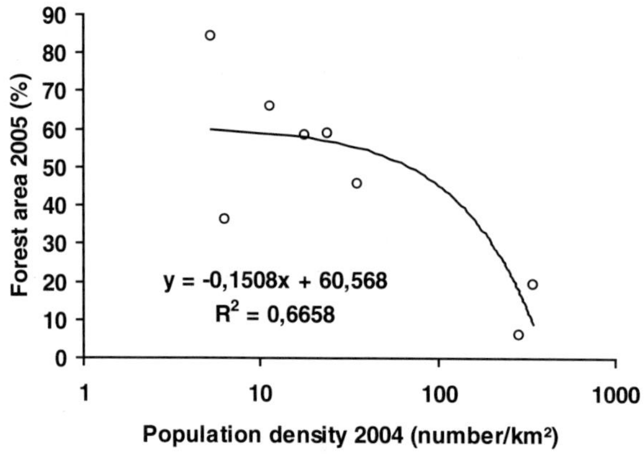

Fig. 5.2 Proportional forest cover as a function of population density for Central African countries Burundi, Cameroon, the Central African Republic, Congo, the Democratic Republic of the Congo, Equatorial Guinea, Gabon and Rwanda. The decreasing trend suggests a negative causality between demographic pressure and forest cover
Data source: Global Forest Resources Assessment of 2005 of the Food and Agriculture Organisation (FAO 2005).

occur in two, sometimes overlapping circumstances (Rudel et al. 2005). In some places, economic development has created enough non-farm jobs to pull farmers off the land, thereby inducing the spontaneous regeneration of forests in old fields. In other places, a scarcity of forest products has prompted governments and landowners to plant trees (Verheyen et al. 2006). Both circumstances initiate only an increase of the forest area, and therefore no deforestation (Kauppi et al. 2006). The transitions do little to conserve biodiversity, but they do sequester carbon and conserve soil (Rudel et al. 2005).

Forests may be fragmented by a number of activities or events, such as road construction, logging, conversion to agriculture, or wildfire, but ultimately, the fragmenting cause is either anthropogenic or natural in origin (Wade et al. 2003). Intuitively, forest fragmentation by anthropogenic sources is at higher risk of further fragmentation or removal than forest fragmented by natural causes; identifying only human-caused forest fragmentation may be a useful tool for policy and decision makers allowing for improved risk assessments and better targeting of areas for protection and remediation (Wade et al. 2003). Based upon local-case studies, Geist and Lambin (2002) concluded that too much emphasis has been given to population growth and shifting cultivation as primary causes of deforestation. Considering proximate causes (human activities or actions at the local level such as agricultural expansion) and underlying driving forces (fundamental social processes such as human demographic evolution), no universal link between cause and effect has been established and tropical forest decline was found caused by different combinations of various proximate causes and underlying driving forces in varying historical and geographical contexts (Geist and Lambin 2002; Lambin and Geist 2003; Lambin et al. 2003). At the underlying level, public or individual decisions largely corresponded to changing national- to global-scale economic opportunities and policies, and at the proximate level, regionally distinct modes of agricultural expansion, wood extraction, and infrastructure extension prevailed in causing deforestation (Geist and Lambin 2002). Nevertheless, traditional shifting cultivation for subsistence has been cited, next to timber logging by private – often foreign – companies, as dominant proximate cause of deforestation for West and Central Africa (Lambin and Geist 2003), as confirmed *in situ* by Bamba (2006).

It should be emphasized that fragmentation constitutes more than area decline only: spatial pattern change is a main characteristic of fragmentation. The magnitude of the ecological impacts of habitat loss can be exacerbated by the spatial arrangement of remaining habitat (Ewers and Didham 2006). Ecologists agree that fragmentation changes the landscape regarding interior-to-edge ratios, patch shape, total patch boundary length, connectivity and patch number (e.g., Collinge 1998; Davidson 1998). Nevertheless, detailed data of pattern such as the number of patches or patch perimeter at local to regional scales are generally lacking and difficult to obtain via remote sensing, which hampers a true assessment of fragmentation impact. This link between pattern and ecological function is yet central to landscape ecology (Turner 1989; D'Eon 2002) and is known as the pattern-process paradigm (Coulson et al. 1999; Gustafson and Diaz 2002). A triangular relationship describing the interdependence of configuration (spatial arrangement and geometry of the system elements), composition (types of elements present) and processes (fluxes; spatial, biological and ecological processes) of every ecological system (Noon and Dale 2002) forms the baseline of landscape ecological research.

Other spatial processes exist that alter the pattern of land cover, but in a distinctive way. Fragmentation is usually considered a phase in the broader sequence of transforming land by natural or human causes from one type to another (Forman 1995). Regardless of the type of land conversion, there appears to be a limited number of common spatial configurations that result from such land transformation processes (Franklin and Forman 1987; Collinge and Forman 1998). Often the term fragmentation is used to denote all these types of pattern changes (e.g., Knight and Landres 2002), although their ecological impact will be different. To determine fast and objectively the processes involved in landscape transformation, a decision tree model was conceived based upon the change of the area, the perimeter and number of patches of the class of interest (Bogaert et al. 2004; Koffi et al. 2007). Figure 5.3 shows a stripped version of the original model with only those five processes causing area decrease of the class of interest. These processes can be divided in two groups based on the change of the number of patches. Fragmentation and dissection lead to an increase; while perforation, attrition and shrinkage do not cause patch density increase. To separate fragmentation from dissection (the carving up or subdividing of an area or patch using equal width lines), it was accepted

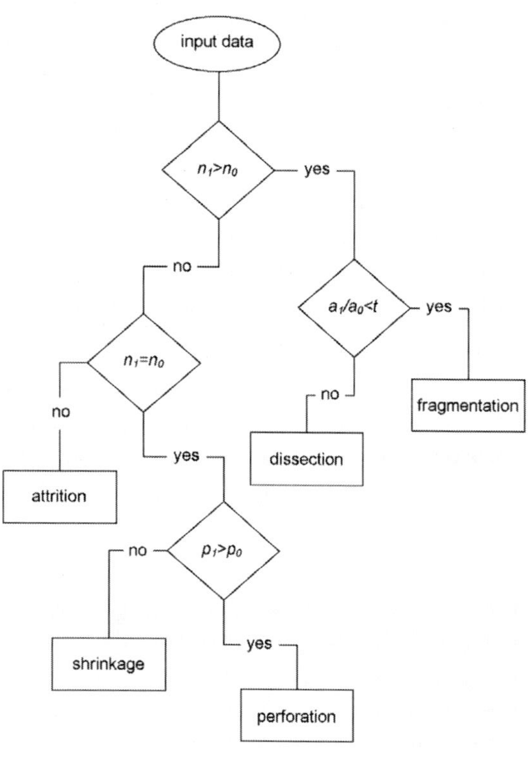

Fig. 5.3 Spatial processes in landscape transformation characterized by area loss ($a_1/a_0 < 1$) of the class of interest. All decision steps in the flow chart, represented by the diamond-shaped components, are based on a comparison of either the area (a), the perimeter (p) or the number of patches (n) before (a_0, p_0, n_0) and after (a_1, p_1, n_1 the transformation of the landscape, which constitute the input data. Comparison of a_1/a_0 with a – by the user – predefined area loss ratio (t) enables distinction between fragmentation and dissection, which generate similar patterns. Fragmentation is accepted to cause smaller a_1/a_0 ratios

that fragmentation is associated with considerable area loss, while in the case of dissection, area loss was limited. To distinguish "considerable area loss" from "limited area loss", a predefined threshold value has to be used. The inclusion of this criterion was essential since both processes cause similar pattern changes. Considering the aforementioned processes of landscape transformation, every observation of attrition, dissection, fragmentation, perforation or shrinkage followed by aggregation, creation or enlargement could be denoted as a forest transition (Rudel et al. 2005; Kauppi et al. 2006).

5.2 The Impact of Shifting Agriculture on Forest Succession and Land Cover Dynamics in the Bas-Congo Province (Democratic Republic of the Congo)

Until the turn of the 20th century, human impact on the tropical rain forests in Africa was limited to shifting cultivation on a long-term rotation by indigenous peoples (Tivy 1993). Fire always has been an essential tool of the peasant rain forest farmer: the above-ground biomass of a forest contains plant mineral nutrients and these are mobilized when it is burned (Whitmore 1998). Patches of forest felled and burned preparatory to crop planting were initially small – not much larger than the natural gaps created by the death fall of primary trees – and widely scattered. A cover of a diverse mixture of perennial crops of varying height replicated consequently morphologically the structure of the forest in miniature and protected the soil from accelerated erosion (Ruthenberg 1976 in Tivy 1993). After a few years cropping yields generally declined – because the soil becomes exhausted and also because of a build-up of pests, diseases and weeds (Whitmore 1998) – and the cultivated patch was abandoned and would not be re-cultivated for at least 30 years, by which time a secondary forest would have re-established itself and its associated soil fertility (Tivy 1993; Whitmore 1998). Rapid increase in native populations in Africa has resulted in the shortening of this tree fallow period to as little as 3 years, insufficient time for other than a poor grass shrub vegetation to regenerate, which can easily be further degraded by overgrazing of domestic livestock and wild herbivores (Tivy 1993). Removal of the original forest cover over increasingly larger areas exacerbates the erosivity of the torrential rainfall while increased surface evaporation causes drying and hardening of the exposed mineral soil. In addition, a rapidly growing demand for firewood from both urban and rural settlements in the tropics has resulted in the selective cutting of secondary and primary forest instead of wood culling for this purpose (Tivy 1993). In many areas of Africa, "savannah-isation" of the tropical rain forest has occurred with a concomitant decline in soil fertility, an increase in soil erosion, and exposure of hard indurated sterile crusts near or at the surface (Tivy 1993; Bamba 2006).

In order to investigate the role of shifting cultivation as a driver of landscape dynamics, land cover changes over a 35 year period were analysed for a test zone of about 410 km^2 situated in the Bas-Congo province of the Democratic Republic of the Congo. The study area (15°23'E–15°38'E; 4°52'S–5°00'S) is situated at less than 100 km from the country's capital Kinshasa (15°24'E; 4°24'S), and at less than 50 km from the main road "National 1", which leads to the port of Matadi (13°27'E; 5°50'S).

Two data sets were used in this study: (1) a digitized version of the land cover map of Compère (1970) at a scale of 1:250000 based on aerial photos of 1959 and field visits; (2) a map at scale 1:75000 based on a mosaic of ASTER remote sensing imagery (dd. 18.07.2003) supported by field surveys (Wolff 2005). Since both data sets were conceived differently, map homogenization was effectuated. Firstly, the ten original classes of the Compère (1970) map were reduced to the same four classes as present on the Wolff (2005) map: savannah, fallow lands and fields, secondary forest, primary forest. Secondly, the technique of the Minimum Mapping Unit (Saura 2002) was applied using the "Dissolve by area" function of ArcView 3.3 to homogenize the precision of both maps; the Wolff (2005) map was modified using fusion of patches in order to obtain a smallest patch of approximately the same size as the smallest patch on the Compère (1970) map. In this way, the precision on both maps was equalized.

Two types of analysis were applied to assess landscape dynamics over the 35 year period. Firstly, a transition matrix was composed to interpret land cover changes among the classes considered. Secondly, the landscape transformation processes involved in the spatial dynamics of the classes were determined using the decision tree model (Bogaert et al. 2004; Koffi et al. 2007).

Several tendencies could be observed for the study area. Firstly, a "savannah-isation" had taken place, since savannah area rise from 19.80% (1970) to 29.61% (2005). Secondly, the area of the fallow lands and fields increased from 22.72% to 54.61%; this class forms actually the new landscape matrix. In 1970, the landscape matrix was still formed by the secondary forest (49.95%), of which the extent was reduced to 5.67% in 2005. Thirdly, primary forest increased slightly from 7.52% to 10.10% in the same time period. Savannah increase mainly originated from secondary forest (14.23%) (Figure 5.4). This was also observed for the fallow land and field class (27.32%). The secondary forest class also contributed to the formation of primary forest (5.62%) via spontaneous or natural succession (Figure 5.5). About 41.44% of the landscape occupied in 1970 by the secondary forest was degraded in savannah, fallow lands and fields. Analysis of the dynamics of class area hence signals the importance of "savannah-isation" in landscape dynamics, the increase of the agricultural activities as a consequence of demographic pressure, and the presence of natural succession of the secondary forest remnants.

Fragmentation of the secondary forest was evidenced by analysis of the landscape transformation processes. All classes showed an increase of their number of patches, which suggested an overall fragmentation of the landscape, but which could also be influenced by the different mapping procedures. It could be considered a refinement of pattern texture. The increase of the number of patches was characterized by different magnitudes; while the number of patches of secondary (primary) forest increased with a factor of ~100 (~50), much smaller increases were found for savannah (multiplication factor ~3) and fallow lands and fields (multiplication

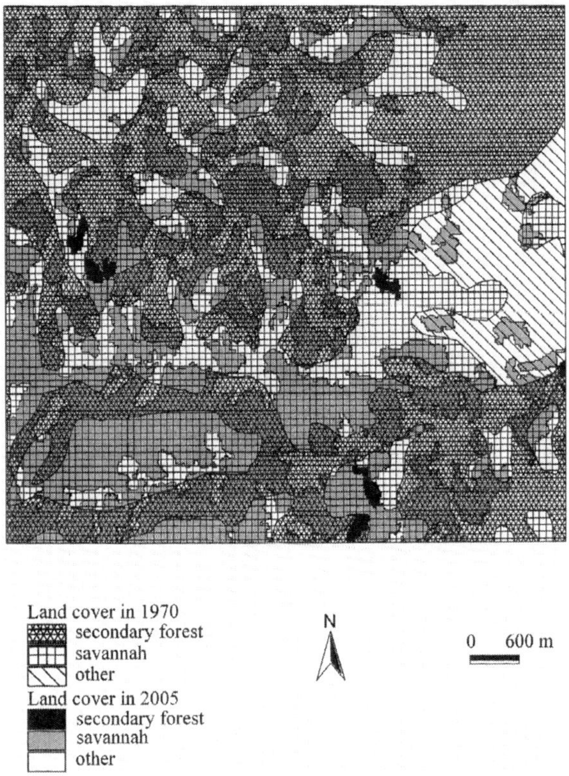

Fig. 5.4 Landscape transformations observed in the Bas-Congo province between 1970 and 2005: "savannah-isation". As a consequence of shifting agriculture with too short fallow periods due to demographic pressure, savannah vegetation systematically replaces the secondary forest. The fragmentation of the secondary forest is evidenced by the profound reduction of its extent. Only a section of the study area is shown for clarity

factor ~10). These observations limited the number of potential spatial landscape transformation processes to three: creation, dissection and fragmentation (Bogaert et al. 2004; Koffi et al. 2007). The differences in magnitude in patch number rise suggested the impact of distinct spatial transformation processes between the forest classes on one hand, and the savannah and fallow land and field classes on the other hand. This observation was partly confirmed by the evolution of the class area values, which decreased for secondary forest only. Strong increases of perimeter length were likewise observed for savannah (+235.8%), fallow land and field (+971.4%) and primary forest (+126.8%). When compared to these values, the total perimeter length of the secondary forests hardly changed (+0.2%). Implementation of the decision tree model showed that the spatial patterns of classes savannah, fallow land and field, and primary forest have been altered between 1970 and 2005 by creation, while the extent and spatial configuration of the secondary forest has been changed profoundly by fragmentation. No threshold value has been applied to distinguish between fragmentation and dissection in this case study, since the value of the area decrease (–88.7%) left no discussion or doubt on this point.

Shifting agriculture has the limitation that it can usually only support 10–20 persons per km^2, though occasionally more because at any one time only ~10% of the area is under cultivation (Whitmore 1998). It breaks down if either the bush

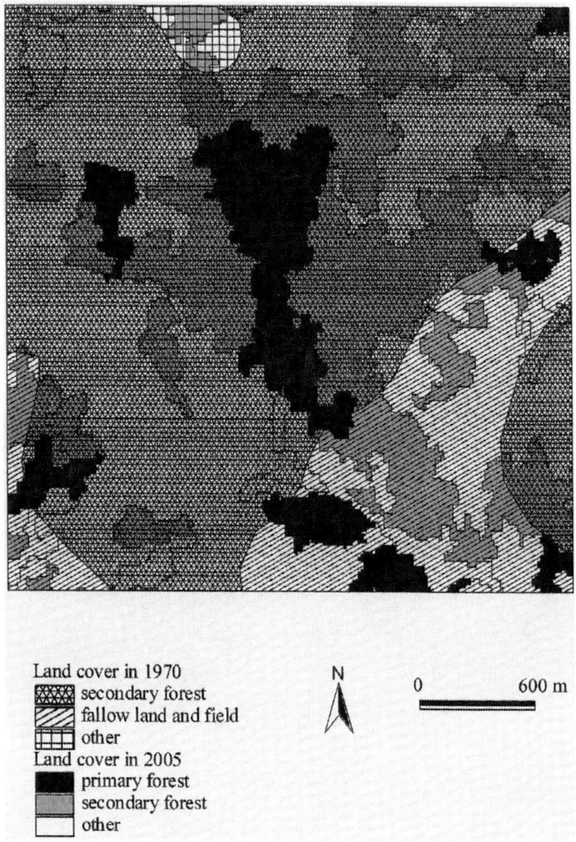

Fig. 5.5 Landscape transformations observed in the Bas-Congo province between 1970 and 2005: natural or spontaneous succession. Due to natural vegetation dynamics, secondary forest evolves into primary forest. The development of secondary forest vegetation on formerly cultivated land is also observed, which illustrates the recovery potential of the forest. Primary forest patches superimposed on land cover types other than secondary forest have probably passed by this stage during the considered 35-year period. Only a section of the study area is shown for clarity

fallow period is excessively shortened or if the period of cultivation is extended for too long, either of which is likely to occur if population increases and a land shortage develops (Whitmore 1998). This phenomenon was observed in our case study, (1) with about 54.61% of the landscape classified as fallow land and fields in 2005 against 22.72% in 1970, and (2) with no recovery of the natural vegetation (reduction of the secondary forest to 5.67%). The slight increase of the primary forest should not be overestimated. Since its formation relies on the presence of secondary forest, a fragmentation of this latter class will be the bottleneck for future primary forest development in this region.

Being located in the hinterland of Kinshasa, landscapes suffered from an increased demographic pressure and demands for higher production. In the Democratic Republic of the Congo, about 60% of the population lives in rural areas (FAO 2001b). After Kinshasa and the Nord Kivu province, the Bas-Congo province is the third with regard to population density, which equals ~52 inhabitants per km^2, mostly concentrated in the cities of Matadi and Boma (Tshibangu 2001). Consequently, a overexploitation of the natural resources has taken place with systematic deforestation along the main roads (corridor type mosaic sequence, Forman 1995) and bad agricultural practices (Binzangi 2004). The Kikongo people living in this region apply customarily the technique of shifting agriculture.

An overall tendency towards fragmentation of the landscape could be suggested, primarily evidenced by an increase of the number of patches for every class. Nevertheless, analysis of the total area per class showed that two antagonistic processes dominated landscape dynamics: creation and fragmentation. It could be suggested that this overall trend towards more patches was a methodological artefact due to the comparison of two data sets relying on different technologies. The aforementioned technique of the Minimum Mapping Unit was applied to counter this deficiency. However, it remains possible that cartographic precision has not been completely equalized between the maps, which could have given pattern differences between 1970 and 2005 that were not a consequence of real landscape dynamics. Raster maps, such as the Wolff (2005) map based on satellite imagery, can show a tendency for upward bias of perimeter lengths because of the stair stepping pattern of the line segments, and the magnitude of bias will vary in relation to the spatial resolution of the image (McGarigal and Marks 1995; Hargis et al. 1997); the degree of curve roughness is also influenced by pixel resolution. This strong increase of perimeter lengths has also been observed in the current study. Comparison between images with different resolutions should therefore be handled with caution (Bogaert and Hong 2003). There seems to be a strong relationship between the degree of detail (spatial resolution) used and the information present on land cover maps (Farina 1998). Rare land cover types are lost when resolution becomes coarser, and patchy arrangements disappear more rapidly with increasing resolution than contagious ones (Turner 1989; Haines-Young and Chopping 1996). Nevertheless, the differences in magnitudes between the classes (for the number of patches), and the opposite tendencies (for the class area) leave no doubt in interpreting the landscape transformation of our study area and exclude the possibility that the observed dynamics are only the result of different mapping procedures. These conclusions were confirmed by an analysis of landscape metrics (data not shown) for fragmentation on the same data set (Bamba 2006).

5.3 Forest Conservation to Preserve Indicator Species in Central Africa

The humid rain forest is above all characterized by the richness and diversity of its fauna and flora. It contains the largest known assemblage of plants and animals in the world (Tivy 1993). Therefore, forest fragmentation is a major cause of loss of biodiversity, in particular in the species-rich wet tropics, where landscape transformation is an ongoing process (Kattan and Murcia 2003). A large body of literature gives evidence of the negative effects of fragmentation, which include changes in the physical environment, and regional and local extirpation of populations and many species of plants and animals. Obviously, the loss of primary habitat will lead to the disappearance of many forest-associated species. Moreover, the appearance of barriers in the modified landscape can significantly alter the metapopulation dynamics of the surviving species (Gascon et al. 2003).

Deforestation is likewise rarely spatially random; instead it may be concentrated on certain areas depending on factors such as topography and soil types. This may result in the elimination of entire habitats and their associated species assemblages, as well as species that depend on these habitats for some stages of their life cycle (Kattan and Murcia 2003). Consequently, forest conservation is crucial to preserve these species that depend on these forest habitats.

Phytogeographic data enable testing of hypotheses regarding the geographic origin of a species, its speed of evolution, and its migration pathways (Koffi 2005). A phytogeographic analysis is often executed to delimit smaller (homogeneous) entities such as regions, districts, and sectors in vast geographic zones. Three major phytogeographic theories have been proposed for Central Africa, defined here as the geographic zone covered by the Democratic Republic of the Congo, Burundi and Rwanda. Robyns (1948) subdivided Central Africa in 11 districts. White (1979, 1983) subdivided Africa and Madagascar in 20 regional entities. The Guineo-Congolian regional centre of endemism, the Zambezian regional centre of endemism, the Afromontane archipelago-like regional centre of endemism, the Guineo-Congolian/Zambezian regional transition zone and the Guineo-Congolian/Sudanian regional transition zone cover Central Africa. Ndjele (1988) developed a phytogeographic system subdividing the Democratic Republic of the Congo in 13 sectors.

Phytogeographic data will reflect the spatial variation of plant and community diversity, and can constitute a useful tool in conservation policy development (Koffi et al. 2007). The aforementioned theories rely on many parameters, such as plant physiognomy, bioclimatic data (precipitation, dry season length) and on the concept of endemism, a notion central to the study of biogeography (Crisp et al. 2001). A taxon is considered endemic to a particular area if it occurs only in that area (Anderson 1994). Conservationists are strongly interested in areas of endemism because narrowly endemic species are by definition rare, and therefore potentially threatened (Crisp et al. 2001). Due to this large number of parameters, these theories are less practical for direct use in conservation policy development. Therefore, it was investigated if indicator species could be found of which the spatial distribution is bound to one single phytogeographic zone and which, by means of their presence or absence, proxy the phytogeographic subdivisions proposed by Robyns (1948), White (1979, 1983) and Ndjele (1988). This research for functional relationships between species richness and the occurrence of indicator species – defined generally as a small set of species with presence or absence patterns that are correlated functionally with species richness of a larger group of organisms – is a common practice in conservation biology (Fleishman et al. 2005). In this way, the consideration of phytogeographic theories could be useful to determine the conservation value (*sensu* Menon et al. 2001) of a region. The question remains, however, whether species from one taxonomic group might serve as indicators of the species richness of other taxonomic groups. As indicator species, those species found in one single phytogeographic entity only were chosen; they were denoted as "unique species". We investigated whether unique species of the Acanthaceae family existed for the aforementioned phytogeographic theories. Consequently, we investigated whether forest conservation could have a positive impact on the preservation of these species.

A data base composed of 9181 herbarium samples of the Acanthaceae family has been used in this study. The use of the Acanthaceae family is justified because (1) this family has been submitted to a profound taxonomic revision which forms a guarantee for data quality (Champluvier 1991, 1997, 1998), (2) no phytogeographic research has been done up to today for this family and for Central Africa, (3) this family contains species that colonize a variety of biotopes which cover our entire study area, (4) the family is dominated by herbaceous species that are easily collected and identified by means of their inflorescence (Cronquist 1981), and (5) the current study forms a part of a larger project on the realization of a flora of the Acanthaceae for Central Africa. Each herbarium sample contained, next to the species name, its taxonomic classification and a plant specimen, the geographical coordinates of the observation. Using these coordinates, species distribution maps have been made using ArcView 3.3. The data have been collected by 417 botanists between 1888 and 2001. The herbarium samples represented 48 genus, 310 species, and 6362 different geographical sites. The number of samples per species is quite variable. Nineteen species were represented by more than 100 samples, 35 species by a number of samples between 99 and 50, 141 species by a number of samples between 49 and 10, and 114 species by less than 10 samples. The data set was made available by the National Botanical Garden of Belgium. Since ruderal, aquatic and cultivated species do not show natural spatial distribution patterns, they were excluded from the data set (64 species).

Remarkable differences in the spatial distribution of the species of the Acanthaceae family have been observed throughout the study area. Figure 5.6 gives some examples of unique species associated with the equatorial forest. Visual inspection of the distribution patterns of the unique species suggested that forest fragmentation would have a direct impact on their preservation.

Analysis of the habitat type(s) reflects and confirms the importance of forest conservation. For the system of White (1979, 1983), 117 unique species have been found, of which 19 (16.2%) could be associated with forest habitat, but also with other types. Thirty species (25.6%) have been found uniquely in forest habitat. For the phytogeographic system of Robyns (1948), 79 unique species have been identified, with 18 (22.8%) only found in forest habitat and four (5.1%) occasionally collected in forest vegetation. When the Acanthaceae species distributions were compared with the Ndjele (1988) system, 84 unique species have been found, of which ten (11.9%) could be found in forest habitat but not exclusively; 21 (25.0%) were bound uniquely to the presence of forest habitat. On average for the three phytogeographic systems, 11.1% of the unique species have been found in forest habitat, next to other habitat types. For these species, forest fragmentation is less threatening, since their ecological amplitude them enables to develop also in other types of habitat. Nevertheless, forest degradation will certainly have a negative impact. The unique species inextricably bound to forest habitat and not found elsewhere,

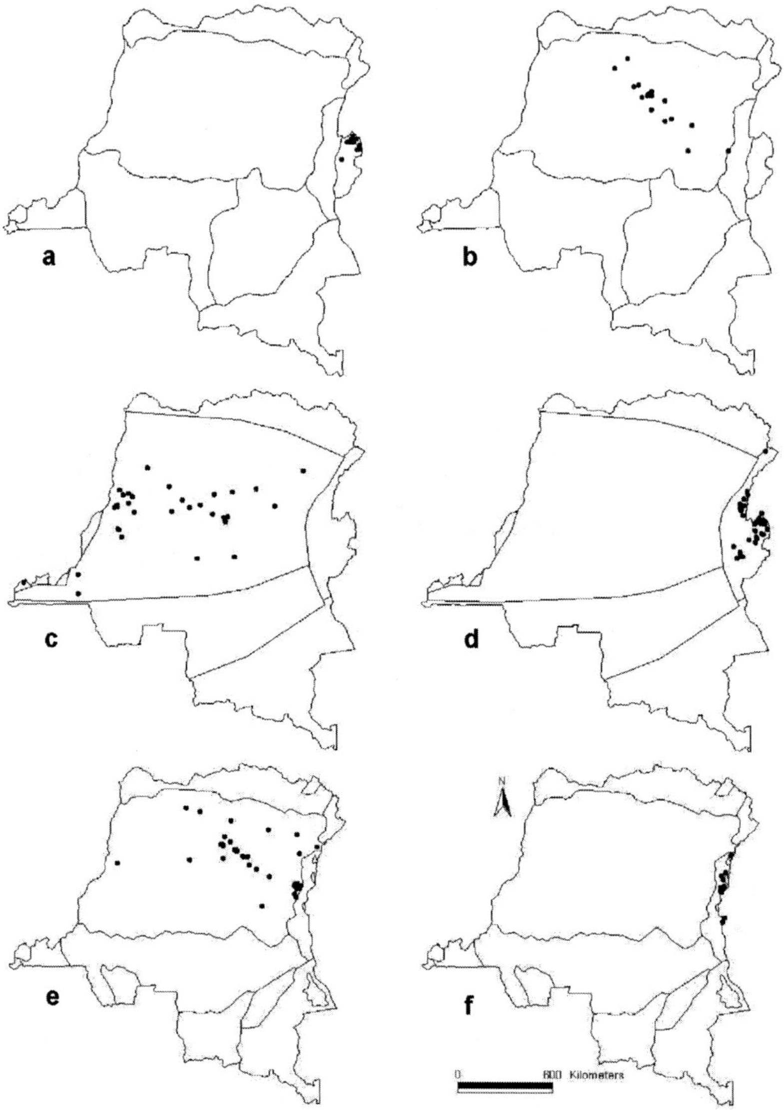

Fig. 5.6 Examples of unique species and their number of samples (n): (**a**) *Blepharis cristate* ($n = 26$), only observed in the Ruanda-Urundi district of Robyns (1948); (**b**) *Justicia pynaertii* ($n = 30$), only observed in the Central Forest district of Robyns (1948); (**c**) *Physacanthus batanganus* ($n = 43$), only observed in the Guineo-Congolian regional centre of endemism of White (1979, 1983); (**d**) *Dyschoriste radicans* ($n = 59$), only observed in the Afromontane Archipelago-like regional centre of endemism of White (1979, 1983); (**e**) *Whitfieldia arnoldiana* ($n = 60$), only observed in the Central Forest sector of Ndjele (1988); (**f**) *Phaulopsis imbricate* subsp. *imbricate* ($n = 13$), only observed in the Mountainous sector of Ndjele (1988)

on average 24.5% for the three systems considered, will be influenced directly and irreversible by disappearance of forest cover by fragmentation, which underlines the importance of counteracting this latter process of landscape degradation.

It is appealing to link conservation priorities to the vulnerability concept. A wild plant species is considered vulnerable when it shows an increased extinction risk (Koffi et al. 2007). According to Cunningham (1994), vulnerable species are highly demanded by man, are characterized by slow growth rates, difficult reproduction methods and limited spatial distribution, and can be associated with fragile or endangered habitats. In Betti (2001), a quantitative method is proposed to evaluate the vulnerability status of a species, based upon six parameters: zonation or altitude range, biotope, morphology, geography, diaspore type, and use by man. For each parameter, a score is assigned between one and three which increases with the risk of extinction. With regard to biotope, a score of two will be given to those species associated with secondary forests and a score of three to species of the primary, undisturbed forest. The reader is referred to Betti (2001) for the exact quantification of the vulnerability parameters. Finally, an average score (V) is calculated which reflects the overall extinction risk or vulnerability of the species. If $V \geq 1.5$, the species is considered vulnerable; for $V \geq 2.5$, the species is considered highly vulnerable. This latter case will be observed when it is bound to particular altitudinal limits, when it is associated with undisturbed of primary forests, when being a tree, shrub or liana species, when it is an endemic or Afromontane species, when disseminating by sarcochory or desmochory, and when it is used by man for construction or in traditional medicine practices (Koffi et al. 2007). Fragmentation of primary and secondary forests will consequently menace directly species that are already disfavoured by other parameters of vulnerability. Fragmentation will therefore aggravate the vulnerability status of these species, and future forest fragmentation should therefore be avoided to preserve their role as sanctuary for endangered species. The comprehensive index V should be interpreted with caution since it equals the mean value of ordinal data, which is strictly spoken mathematically incorrect. It is therefore advised to consider V only as a proxy of species vulnerability and to base the final assessment of vulnerability on the interpretation of the underlying ordinal data.

5.4 Implications for Management

If the forest frontier is to be stabilized, rural people must live in balance with the landscape they inhabit. Their exclusion from the forest – known as "protection conservation" (Blom 1998) – is unrealistic and may involve annulling traditional rights. The difficulty is the exorable, and in many places rapid, increase of the human population (Whitmore 1998). The forest and the people who depend on it need to be considered as a single ecosystem, managed to maintain a continual but changing stream of goods and services (Sayer 1995 in Whitmore 1998). Broad participation in management and conservation has become a standard element of good practice: effective planning, implementation, and monitoring require an institutional framework with elements that include broad participation in planning, administrative capacity, field presence, and effective knowledge management (Sheil et al. 2004). Funding is another key factor in determining institutional and administrative capacity (Sheil et al. 2004).

It is hard to know where to start or stop in discussing ecological principles that are relevant to the management of protected areas. For example, edge effects and other factors related to disturbance can favour some species at the same time as they hurt others. For managers of protected areas, the question must not be what will maximize species numbers, but rather how to preserve the target species or communities (May 1994). For many rain forest species, it is difficult to make meaningful statements about rarity and hence about conservation priority (Whitmore 1998). Knowledge of species richness and diversity is biased to the places that have been studied (Koffi 2005; Koffi et al. 2007). The only sound scientific basis is to conserve adequate habitat, spread across its geographical extent, and to sample all biogeographical regions (Whitmore 1998).

There is an indisputable case for retaining parts of the rain forest inviolate, as natural reserves, kept intact for species to continue to interact between themselves an with the environment; these reserves of natural forest can act as benchmarks against which change elsewhere can be monitored. Their usefulness is even increased if they are surrounded by production forests, not cultivated land (Whitmore 1998). Land use patterns and other activities outside protected areas are considered crucial, both in maintaining general landscape connectivity as well as minimizing any direct effects of edges on forest biota (Opdam and Wiens 2002 in Githiru and Lens 2004). Introducing and fostering activities such as agro-forestry will simultaneously address socioeconomic and ecological problems, by providing alternative fuel wood and fodder while reducing edge effects and promoting dispersal (Githiru and Lens 2004). This is the integrated landscape management notion that unites the principles of metapopulation theory, landscape ecology, corridors and buffer areas (Saunders et al. 1991; Githiru and Lens 2004). The conservation focus could be diverted from the forest reserves themselves to include activities outside the reserves, particularly different land uses. Implementation of these principles would stimulate a more balanced and self-sustaining landscape mosaic, proving basic goods and services to the rural human population, which is key to poverty alleviation, while maintaining the habitats and ecosystems functions that are required to provide services to the people (Githiru and Lens 2004).

The possibility remains to manage rain forests for multiple purposes, in order to meet the needs of conservation as well as to produce useful products. But to retain the long term benefits implied by conservation it is necessary to forgo some immediate cash profit. Multiple uses involve compromises (Whitmore 1998). There is also a case for plantation (Evans 1984 in Whitmore 1998). Timber sold from plantations takes pressure off natural forests as a source of foreign exchange. They should only be established on already degraded sites, never at the expense of good natural forest; restoration of forest via plantations should consequently be considered an important tool for the land manager of tomorrow (Whitmore 1998).

When developing management plans for forested landscapes, the potential of landscape pattern analysis should not be neglected. Many studies that address landscape monitoring emphasize the calculation of numerous indices using remote sensing and geographic information systems (Sheil et al. 2004). Landscape metrics are important for what they may reflect about the disruptive effects of forest fragmentation on ecological processes and species viability. The information they provide is very useful for planners wishing to detect, evaluate, and monitor threats to biodiversity. The measures provided should be considered as a first step in the development of an effective management strategy for a region once conservation goals have been determined and more information is available on the relationship between ecologic processes, species characteristics, and their interactions (Tole 2006). A landscape level conservation strategy is necessary, for example to maximize area to perimeter relationships across reserves, to protect reserve edges using buffer zones (managed ecotones) or to minimize matrix harshness (Hill and Curran 2005). The spatial structure of populations persisting in fragmented landscapes governs their response to habitat fragmentation and hence dictates the remedial actions that will be most effective for species and habitat conservation (Githiru and Lens 2004). Ecological processes that operate over large distances must not be ignored (Hill and Curran 2005). Biologists need to be pressed to research and monitor the effectiveness of different land use patterns and to develop active management tools for biodiversity conservation in cultural landscapes of which rain forest is only a part (Whitmore 1998).

In addition to well balanced and realistic management plans, there must be a focus on defence: what precautions are to be taken against threats such as agricultural encroachment or fires; sustainable monitoring systems are required (Sheil et al. 2004). It is essential to enforce laws to minimize the damage caused by logging and to prevent hunters, collectors or farmers from entering along roads and causing damage, depletion or destruction (Whitmore 1998). Management of protected areas will in many cases involve weaving ecology together with social and economic considerations. Given the inherently dynamic and non-linear character of the biological and other processes involved, it is often hard to foresee what the outcome of the well intentioned policies may be (May 1994). The problems are often complex, and while there are few simple solutions, one point is clear: without greater commitment from wealthy and developing nations alike, most of the world's tropical forests will disappear within our lifetime (Laurance 1999); humankind must bear in mind the ultimate constant: the scarcest conservation resource is time (Myers 2003).

5.5 Conclusions

Forest fragmentation is one of the most important conservation issues of recent times (D'Eon 2002). Forest area harbours biodiversity, beautifies landscape and bestows solitude. Forest area also anchors soil, slows erosion, and tempers stream flow (Kauppi et al. 2006). The tropical rain forest is the most massive, diverse and productive of the earth's ecosystems. The environmental significance of deforestation is related to the particular attributes of the forest ecosystems of the world, the relative size of the forest biomass, which accounts for 75% of the total global plant biomass, and its carbon-storage capacity (Tivy 1993). Degradation of this ecosystem type is taking place at an alarming rate: about five per cent of the forest area in Central Africa has disappeared since 1990; only Rwanda showed a forest transition, due to active plantation of forests.

Traditional shifting agriculture and demographic pressure are often cited as the main cause of forest fragmentation, although various drivers and local factors can overrule the dominant impact of these traditional practices, depending on the historical and geographical context (Geist and Lambin 2002; Lambin and Geist 2003; Lambin et al. 2003). Nevertheless, shifting agriculture was found the driver of landscape change in the Bas-Congo province of the Democratic Republic of the Congo, where a matrix of secondary forest was converted into a mosaic of fallow lands, fields and savannah in 35 years time. The fragmentation of the secondary forest was also assumed the limiting factor for the future creation of primary forests.

Forests are, next to a natural resource of timber or non timber forest products, also habitats and therefore a resource of natural biodiversity. Many species depend on the specific microclimatic and ecological conditions of forests to complete their life cycle or a part of it, and cannot survive without our outside these particular environments. A study of the Acanthaceae family in Central Africa showed that about 25% of the unique species which presence or absence are characteristic for a given phytogeographic entity, have only been observed in forest habitat. Since they can be useful to assist in conservation policy development based upon phytogeographic concepts, their habitat should be preserved without any delay.

Habitat fragmentation has become a worldwide environmental issue (Forman 1995). Because landscape ecology concerns the study of the reciprocal effects between ecological pattern and ecological function (Gustafson and Diaz 2002), and since fragmentation implies, next to area decline, also pattern change, landscape ecologists have contributed significantly to the scientific debate regarding the definition of fragmentation, the identification of its drivers, and its ecological impact. Field and satellite observations, ecological modelling and the use of geographic information systems have enabled scientists to detect and map fragmentation, and to assess its impact on ecological parameters such as landscape connectivity, microclimate, species dispersal, and biodiversity. It is now up to the scientific community to translate these academic concepts and findings into clear, applicable and realistic objectives useful to be incorporated in conservation policy initiatives in order to counteract or prevent further fragmentation and degradation of the world's tropical forests. Nevertheless, empirical data from well-designed fragmentation studies is still needed to validate theoretical predictions stemming from the fragmentation paradigm (D'Eon 2002). It should be emphasized that anthropogenic fragmentation is a recent phenomenon in evolutionary time and the final, long-term impacts of habitat fragmentation may not yet have shown themselves (Ewers and Didham 2006).

Acknowledgments The authors acknowledge the Government of Ivory Coast for the fellowships of I. Bamba and K.J. Koffi. The research of S. Sibomana and J.-P. Kabulu Djibu is supported by CTB fellowships. The authors acknowledge the FWO – Vlaanderen (G.0019.04), the FNRS

(1.5.028.05), the CIUF and the ULB (CER) for financial support. A. Mama and the SLCD are acknowledged for their assistance regarding the research in the Bas-Congo Province.

References

Anderson S (1994) Area and endemism. Q Rev Biol 69:451–471

Bamba I (2006) Etude de la structure spatiale et de la dynamique spatio-temporelle d'un paysage dans l'ex-Bas-Congo (R.D. Congo). Dissertation, Université libre de Bruxelles

Betti JL (2001) Vulnerabilité des plantes utilisées comme antipaludiques dans l'arrondissement de Mintom au Sud de la Réserve du Dja (Cameroun). Syst Geogr Pl 71:661–678

Binzangi K (2004) Impact de la production des combustibles ligneux en RDC. Cas du Katanga, de Kinshasa et du Bas Congo. In: Actes des séminaires de formation et ateliers de haut niveau en évaluation Environnementale, Association Nationale pour l'Evaluation environnementale, Kinshasa, 12–17 Janvier 2004

Blom A (1998) A critical analysis of three approaches to tropical forest conservation based on experiences in the Sangha region. Yale FES Bull 102:208–215

Bogaert J, Ceulemans R, Salvador-Van Eysenrode D (2004) A decision tree algorithm for detection of spatial processes in landscape transformation. Environ Manag 33:62–73

Bogaert J, Hong SK (2003) Landscape ecology: monitoring landscape dynamics using spatial pattern metrics. In: Hong SK, Lee JA, Ihm BS et al (eds) Ecological issues in a changing world, Kluwer Academic Publishers, Dordrecht, pp 109–131

Champluvier D (1991) Révision des genres *Staurogyne* Wall., *Anisosepalum* E. Hossain et *Saintpauliopsis* Staner (Acanthaceae) en Afrique tropicale. Bull Jard Bot Nat Belg 61:93

Champluvier D (1997) *Brachystephanus glaberrimus* (Acanthaceae), espèce nouvelle gynomonoïque de la dorsale Congo-Nil (Congo, Rwanda, Uganda). Bull Jard Bot Nat Belg 66:187–200

Champluvier D (1998) *Justicia violaceotincta* (Acanthaceae), espèce nouvelle d'Afrique Centrale (Congo, Rwanda, Burundi). Observations palynologiques et morphologiques. Belg J Bot 130:221–235

Collinge SK (1998) Spatial arrangement of habitat patches and corridors: clues from ecological field experiments. Landscape Urban Plan 42:157–168

Collinge SK, Forman RTT (1998) A conceptual model of land conversion processes: predictions and evidence from a microlandscape experiment with grassland insects. Oikos 82:66–84

Compère P (1970) Carte des sols et de la végétation du Congo, du Rwanda et du Burundi, INEAC, Bruxelles

Coulson RN, Saarenmaa H, Daugherty WC et al (1999) A knowledge system environment for ecosystem management. In: Klopatek JM, Gardner RH (eds) Landscape ecological analysis. Issues and applications, Springer, Berlin Heidelberg New York, pp 57–79

Crisp MD, Laffan S, Linder HP et al (2001) Endemism in the Australian flora. J Biogeogr 28:183–198

Cronquist A (1981) An integrated system of classification of flowering plants, Columbia University Press, New York

Cunningham AB (1994) Management of medicinal plant resources: an Africa-wide overview. In: Seyani JH, Chikuni AC (eds) Proceedings of the XIIIth Plenary meeting of AETFAT, National Herbarium and Botanic Gardens of Malawi, Zomba, 2–11 April 1991

Davidson C (1998) Issues in measuring landscape fragmentation. Wildlife Soc B 26:32–37

D'Eon RG (2002) Forest fragmentation and forest management: a plea for empirical data. Forest Chron 78:686–689

Evans J (1984) Plantation forestry. Clarendon Press, Oxford

Ewers RM, Didham RK (2006) Confounding factors in the detection of species responses to habitat fragmentation. Biol Conserv 81:117–142

Farina A (1998) Principles and methods in landscape ecology. Chapman & Hall, London
FAO (2001a) Global forest resources assessment of 2000 – main report. FAO Forestry Paper 140. FAO, Rome
FAO (2001b) FOSA document national de prospective – République Démocratique du Congo. FAO, Rome
FAO (2005) Global forest resources assessment of 2005. Progress towards sustainable forest management. FAO Forestry Paper 147. FAO, Rome
Fleishman E, Thomson JR, Mac Nally R et al (2005) Using indicator species to predict species richness of multiple taxonomic groups. Conserv Biol 19:1125–1137
Forman RTT (1995) Land mosaics. The ecology of landscapes and regions. Cambridge University Press, Cambridge
Franklin JF, Forman RTT (1987) Creating landscape patterns by forest cutting: ecological consequences and principles. Landscape Ecol 1:5–18
Gascon C, Laurance WF, Lovejoy TE (2003) Forest fragmentation and biodiversity in Central Amazonia. In: Bradshaw GA, Marquet PA (eds) How landscapes change. Human disturbance and ecosystem fragmentation in the Americas, Springer, Berlin Heidelberg New York, pp 33–48
Geist HJ, Lambin EF (2002) Proximate causes and underlying driving forces of tropical deforestation. BioScience 52:143–150
Githiru M, Lens L (2004) Using scientific evidence to guide the conservation of a highly fragmented and threatened Afrotropical forest. Oryx 38:404–409
Groves RH (1998) Ecological indicators of landscape degradation. In: Rundel PW, Montenegro G, Jaksic FM (eds) Landscape disturbance and biodiversity in Mediterranean-type ecosystems. Springer, Berlin Heidelberg New York, pp 55–62
Gustafson EJ, Diaz N (2002) Broad-scale ecological science and its application. In: Gutzwiller KJ (ed) Applying landscape ecology in biological conservation. Springer, Berlin Heidelberg New York, pp 244–265
Haines-Young R, Chopping M (1996) Quantifying landscape structure: a review of landscape indices and their application to forested landscapes. Prog Phys Geog 20:418–445
Hargis CD, Bissonette JA, David JL (1997) Understanding measures of landscape pattern. In: Bissonette JA (ed) Wildlife and landscape ecology, Springer, Berlin Heidelberg New York, pp 231–261
Hill JL, Curran PJ (2005) Fragment shape and tree species composition in tropical forests: a landscape level investigation. Afr J Ecol 43:35–43
Kattan GH, Murcia C (2003) Forest fragmentation and biodiversity in Central Amazonia. In: Bradshaw GA, Marquet PA (eds) How landscapes change. Human disturbance and ecosystem fragmentation in the Americas. Springer, Berlin Heidelberg New York, pp 183–200
Kauppi PE, Ausubel JH, Fang J et al (2006) Returning forests analyzed with the forest identity. P Natl Acad Sci USA 103:17574–17579
Knight RL, Landres PB (2002) Broad-scale ecological science and its application. In: Gutzwiller KJ (ed) Applying landscape ecology in biological conservation. Springer, Berlin Heidelberg New York, pp 22–33
Koffi KJ (2005) Analyse des structures spatiales des données de distribution phytogéographique en Afrique centrale. Dissertation, Université libre de Bruxelles
Koffi KJ, Deblauwe V, Sibomana S et al (2007) Spatial pattern analysis as a focus of landscape ecology to support evaluation of human impact on landscapes and diversity. In: Hong SK, Nakagoshi N, Fu B et al (eds) Landscape ecological applications in man-influenced areas: linking man and nature systems, Springer Berlin Heidelberg New York, pp 7–31
Lambin EF, Geist HJ (2003) Regional differences in tropical deforestation. Environment 45:22–36
Lambin EF, Geist HJ, Lepers E (2003) Dynamics of land-use and land-cover change in the tropical regions. Annu Rev Env Resour 28:205–241
Laurence WF (1999) Reflections on the tropical deforestation crisis. Biol Conserv 91:109–117
May RM (1994) Ecological science and the management of protected areas. Biodivers Conserv 3:437–448

McGarigal K, Marks BJ (1995) FRAGSTATS. Spatial pattern analysis program for quantifying landscape structure (PNW-GTR-351). Pacific Northwest Research Station, Oregon.
Menon S, Gil Pontius R, Rose J et al (2001) Identifying conservation-priority areas in the tropics: a land-use change modeling approach. Conserv Biol 15:501–512
Myers N (2003) Conservation of biodiversity: how are we doing? Environmentalist 23:9–15
Ndjele MB (1988) Les éléments phytogéographiques endémiques dans la flore vasculaire du Zaïre. Dissertation, Université libre de Bruxelles
Noon BR, Dale VH (2002) Broad-scale ecological science and its application. In: Gutzwiller KJ (ed) Applying landscape ecology in biological conservation. Springer, Berlin Heidelberg New York, pp 34–52
Opdam P, Wiens JA (2002) Fragmentation, habitat loss and landscape management. In: Norris K, Pain DJ (eds) Conserving bird biodiversity: general principles and their application, Cambridge University Press, Cambridge, pp 202–223
Robyns W (1948) Les territoires phytogéographiques du Congo Belge et du Ruanda-Urundi. Atlas général du Congo Belge. Académie Royale des Sciences Coloniales, Bruxelles
Rudel TK, Coomes OT, Moran E et al (2005) Forest transition: towards a global understanding of land use change. Global Environ Chang 15:23–31
Ruthenberg H (1976) Farming systems in the tropics. Oxford University Press, Oxford
Saunders DA, Hobbs RJ, Margules CR (1991) Biological consequences of ecosystem fragmentation: a review. Conserv Biol 5:18–32
Saura S (2002) Effects of minimum mapping unit on land cover data spatial configuration and composition. Int J Remote Sens 23:4853–4880
Sayer J (1995) Forestry research, a way forward to sustainable development. CIFOR News 6:1–2
Sheil D, Nasi R, Johnson B (2004) Ecological criteria and indicators for tropical forest landscapes: challenges in the search for progress. Ecol Soc 9:7 [online] http://www.ecologyandsociety.org/vol19/iss1/art7
Tivy J (1993) Biography. A study of plants in the ecosphere. Longman Scientific & Technical, Essex
Tole L (2006) Measurement and management of human-induced patterns of forest fragmentation: a case study. Environ Manage 37:788–801
Tshibangu KWT (2001) Etude du déboisement et de la crise de combustibles ligneux en tant que source d'énergie domestique à Kinshasa (République Démocratique du Congo). Dissertation, Université libre de Bruxelles
Turner MG (1989) Landscape ecology: the effect of pattern on process. Annu Rev Ecol Syst 20:171–197
Verheyen K, Fastenaekels I, Vellend M et al (2006) Landscape factors and regional differences in recovery rates of herb layer richness in Flanders (Belgium). Landscape Ecol 21:1109–1118
Wade TG, Riiters KH, Wickham JD et al (2003) Distribution and causes of global forest fragmentation. Conserv Ecol 7:7 [online] http://www.consecol.org/vol7/iss2/art7
White F (1979) The Guineo-Congolian region an its relationships to other phytochoria. Bull Jard Bot Nat Belg 49:11–55
White F (1983) The vegetation of Africa. A descriptive memoir to accompany the UNESCO/AETFAT/UNSO vegetation map of Africa. Natural Resources Research 20. UNESCO, Paris
Whitmore TC (1998) An introduction to tropical rain forests. Oxford University Press, Oxford
Williamson GB, Mesquita R de CG, Ickes K et al (1997) Estratégias de àrvores pioneiras nos Neotrópicos. In: Gascon C, Moutinho P (eds) Floresta Amazonica: dinâmica, regeneração e manejo. Instituto Nacional de Pesquisas da Amazônia, Manaus, pp 131–144
Wolff E (2005) Rapport de mission cartographique à Kisantu (R.D. Congo) du 15 au 30 Septembre 2005. SLCD, Bruxelles

Chapter 6

Human-Induced Alterations in Native Forests of Patagonia, Argentina

Francisco Carabelli and Roberto Scoz

Abstract Since less than a century ago, human activity in the Patagonian Andes native forests has played a major role in landscape modification. *Austrocedrus chilensis* is one of the indigenous tree species in Patagonia that most clearly shows human influence. Thus, we have been concerned with the magnitude of those changes arising from the substitution of this species by exotic ones, from fires – due to the non-planned development of human settlements – and timber exploitation. The comprised *Austrocedrus* area represented 10% of the total area covered by this species in Argentina. Two main locations, named "Epuyén" and "Trevelin", both placed in the northwest of Central Patagonian Chubut province, were selected for this Argentinean-German cooperative research. We based our work mainly on remote sensing material that was orthorectified and classified recognizing *Austrocedrus* patches according to three classes of density. In the "Epuyén area", Austrocedrus–dominated stands composed, in 1970, a predominantly continuous or interconnected area of 3400 ha. In 2001, 14% of these forests had been substituted by Pinus plantations and 10% were removed by fires. In addition, a strong change in the forest landscape heterogeneity was verified, due to *Austrocedrus* fragmentation – 34% in the considered time period. In the "Trevelin" area, the changes in the 30-year period affected 4400 ha of *Austrocedrus*. Thirty four percent – 1500 ha – were differentially altered by forest fires or flooded by a hydroelectric dam. Nevertheless, 2900 ha are now discernible by growth of young trees or due to regeneration recruitment. Most of these assessments are worrisome because *Austrocedrus* forests cover the smallest area among those indigenous – and also singular – forest species in Patagonia under a fast-increasing human pressure.

F. Carabelli
Patagonia National University, Esquel, Chubut Province, Argentina
e-mail: francisco.carabelli@gmail.com

R. Lafortezza et al. (eds.), *Patterns and Processes in Forest Landscapes*,

6.1 Introduction

The ecological processes and mainly the human settlement and the development of these activities cause spatial and temporal changes on composition and structure at a landscape level. Therefore, the analysis of landscape alterations and their causal factors related to human activities are central issues of landscape ecology and land-use planning. This topic has certainly been well-addressed because as the UN's Food and Agriculture Organisation (FAO) had pointed out in 1993, degradation of land resources may be attributed to greed, ignorance, uncertainty or lack of an alternative but, essentially, it is a consequence of using land today without investing in tomorrow. Even though forest landscapes have natural levels of spatial and temporal heterogeneity, human-induced alterations tend to change their natural heterogeneity and spatial patterning. This way, most forest landscapes exist in various states of structural modifications (Loyn and McAlpine 2001). Thus, the knowledge of landscape dynamics acquires great significance for the evaluation and assignment of rational and sustainable management alternatives of land resources. The expression *landscape* used here refers to the systemic conception, resulting from combining different features of the land research which constitute a convenient mapping unit. Usually, the resulting combination is a geographical region in which "horizontal" aspects are analysed from a spatial point of view in reciprocity with natural phenomena or ecological "vertical" approach (Forman and Godron 1986).

Several studies assessing human-induced changes of spatial and temporal heterogeneity regarding tree species in forest landscapes have been developed worldwide (Lida and Nakashizuka 1995; Rescia et al. 1997; Silbernagel et al. 1997; Grez et al. 1998; Roth 1999; Ripple et al. 2000; Jenkins and Parker 2000; Puric-Mladenovic et al. 2000; Hessburg et al. 2000; Löfman and Kouki 2001; Fukamachi et al. 2001; Lawes et al. 2004; Tucker et al. 2005) and even at a global scale (Riitters et al. 2000; Wade et al. 2003). It is worrisome that such studies are still scarce in the distant and vast Patagonia, where mostly non-planned human activities concerning changes on landscape heterogeneity in indigenous forests have an increasing influence on environmental degradation and biodiversity reduction. Within the group of indigenous species traditionally used during the last century, *Austrocedrus chilensis* (ciprés de la cordillera) has been a tree species under permanent and intense pressure, due to its most favourable microclimatic characteristics and very accessible locations (Carabelli et al. 2002).

Austrocedrus is an endemic forest species in the cold temperate forests of the Patagonian Andes region in Argentina and Chile. In Argentina, it forms relatively dense pure stands in a west-east precipitation gradient between 500 and 1600 mm/year, being the conifer with the largest geographical distribution, from 37°08' up to 43°43'S (Bran et al. 2002). It is preferably located at altitudes that range between 300 and 1000 m, depending on the latitude, in a 60–80 km-wide strip, representing the forest boundary between the Patagonian steppe to the east and the humid forests of *Nothofagus* to the west (Dezzoti and Sancholuz 1991), where it usually develops dense mixed forests along with the evergreen *Nothofagus dombeyi*

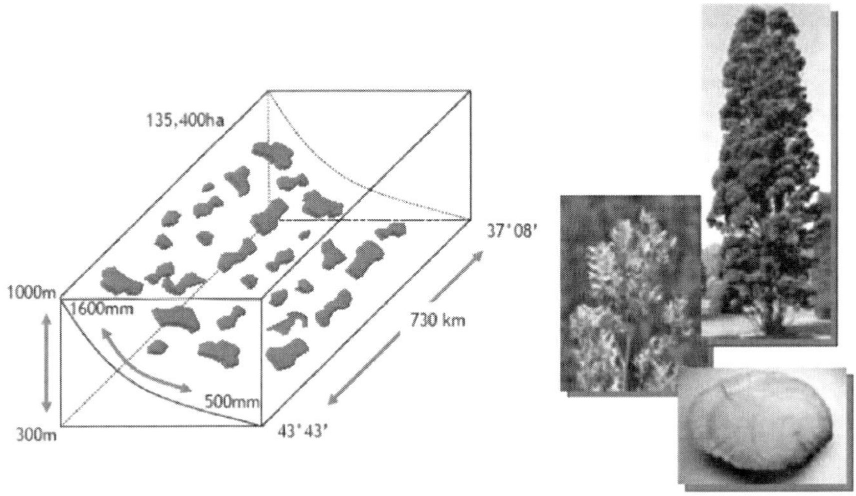

Fig. 6.1 Schematic representation of the *Austrocedrus* geographical distribution and main botanical features

(Fig. 6.1). At present, the pure *Austrocedrus* forests and the mixed *Austrocedrus-Nothofagus* areas cover 135400 ha (Bran et al. 2002).

Forest fires have traditionally represented a great problem for *Austrocedrus* forests. Between 1890 and 1940, a close correspondence of the European colonization and the installation of sawmills with the increase of fires in *Austrocedrus* forests was verified (Bondel and de Almeida 1996). In earlier times, these accidents were attributed to the aboriginals that used them as a hunting strategy of *Lama guanicoe* (guanaco) (Muster 1971; Fonck 1900). Many of the present *Austrocedrus* forests have developed on areas affected by great fires (Rothkugel 1916; Veblen and Lorenz 1987). Still, at present, fire is the most important disturbance affecting these forests in northwest Patagonia (Veblen et al. 1992) and shaping north Patagonian landscapes (Kitzberger and Gowda 2004). Reforestation with exotic species in pure or mixed *Austrocedrus* forests has also contributed to the reduction of the original area (Loguercio et al. 1999).

In such a context, at least two issues concerning forest environment changes related to human-induced alterations are nowadays of great relevance: the alterations at a landscape level through time and a better comprehension of the fragmentation process at a landownership level (Carabelli et al. 2003). An improved knowledge of the first matter would allow an overview of the direction of human-related current forces threatening native forests, to propose scientific-based conservation measures. In addition, a reconstruction with historical perspective of land uses affecting forest areas on representative selected small ownerships would offer a bridge between landscape and estate levels to develop defined strategies in order to support a more realistic land-use planning towards the sustainability of natural resources in Patagonia. During the last five years, we have been dealing with the quantitative analysis of changes of the forest heterogeneity at a landscape level to determine the rates of decrease and fragmentation of native forest areas and the causes behind them. Thus, the aim of this chapter is to present and discuss some aspects of forest fragmentation in northwest Patagonia, emphasizing the fact that the underlying unplanned growing development of several human activities is putting at risk not only the *Austrocedrus* forests, but also the whole biological richness –including the fresh water – of these unique ecosystems.

6.2 Two Case Studies in Northwest-Patagonia

In the Northwest Andes cordillera in Chubut Province, the forest management has proved to be unsustainable and it has been responsible for the resource degradation (Loguercio 2005). For a long-term development, able to guarantee an environmental sustainability, it is necessary to begin studies that can provide a quantitative treatment of changes on landscape patterns and its dynamic toward a comprehensive understanding of the influence of such changes on biodiversity. On the basis of two case studies, we attempt to characterize the landscape changes mainly related with the identification of causes forcing them and the quantification for a time period of 30 years.

The first considered area, named "**Epuyén**", covers 6000 ha and it is placed NW of Chubut Province, Argentina (Fig. 6.2) The area center coordinates are:

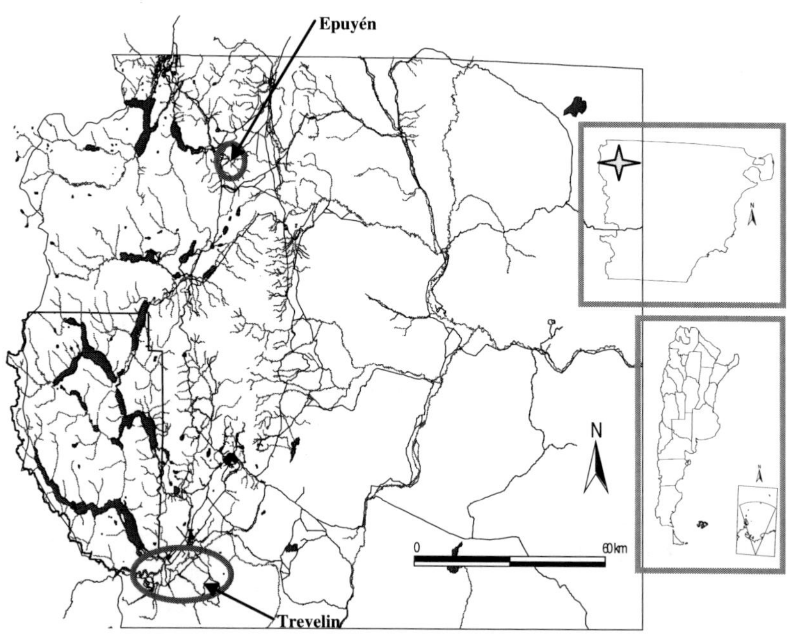

Fig. 6.2 Location of "Epuyén" and "Trevelin" areas in Province Chubut, Patagonia, Argentina

42°09'06"S 71°22'48"W. This area is characterized by mountainous topography with altitudes between 300 m.a.s.l. at the valley bottoms and 2000 m.a.s.l. It is dissected by the valley of the Epuyén river, where human presence is very noticeable with numerous settlements, infrastructure and crop areas. On the slopes, cattle-raising and wood extraction are the main activities. Vegetative cover, predominantly forests, corresponds biogeographically to the Andean Region, Sub Antarctic Sub region (Morrone 2001). *Austrocedrus* and *Nothofagus dombeyi* (coihue) are the dominant tree species, which form pure and mixed forests. The average annual temperature is 9.6°C and the annual precipitation is 1375 mm. The region presents a conspicuous dry season coinciding with the summer period, when temperatures are higher. This conjunction of high temperatures and scarce precipitation grants other environmental factors related to the hydric regime (e.g., soil depth, slope steepness, aspect), a supreme importance in the genesis of vegetation patches and their dynamics. This issue favours the occurrence and spread of fires and it is a hindrance for the subsequent recovery of the affected areas.

In the second selected area, called "**Trevelin**", covering 30000 ha (area coordinates: 42°38'–43°34'S 71°22'–71°51'W) (Fig. 6.2), the climate is temperate with an annual average temperature of 10°C. Precipitation oscillates between 600 and 1200 mm/year in a west-east gradient of only 30 km. Prevailing winds, originating in the west, alternate with southwest-oriented winds (Córdoba and González Capdevilla 1999). The herbaceous vegetation is characterized by different species of *Poa sp.* and *Mulinum spinosum* (neneo). In the shrub layer predominate *Nothofagus antarctica* (ñire), *Lomatia hirsuta* (radal), *Schinus patagonicus* (laura) and *Maitenus boaria* (maitén). Representative tree species are *Austrocedrus*, *N. pumilio* (lenga) and *N. dombeyi* (Dimitri 1974). From the beginning of the 20th century, the main land-use in the valley was agriculture; however, during the 1960s, cattle-raising almost replaced this former land-use. This situation increased the pressure on forest areas, some of which were converted into grazing sectors consequently damaging forest regeneration due to browsing and trampling.

In both areas, a detailed analysis of landscape elements on photo mosaics (infrared aerial photographs 1:20000) taken in 1970 was carried out. For the "Epuyén" area, the identification of burnt forest sectors and plantations with exotic species was performed on an IKONOS multispectral satellite image from January, 2001. For the "Trevelin" area, the detection of burnt forest sites and timber exploitations was performed on a SPOT XS-PAN satellite image from March, 2001.

On the "Epuyén" photo mosaic of the year 1970, the defined landscape elements were delimited as forest patches according to the following classification: (1) pure *Austrocedrus*, (2) *Austrocedrus-Nothofagus dombeyi* (*Austrocedrus* cover > 50%), (3) *N. dombeyi-Austrocedrus* (*N. dombeyi* cover > 50%), (4) *Austrocedrus*-shrubs with the following tree or shrubby species: *N. antarctica*, *Schinus patagonicus* and *Lomatia hirsuta* and (5) Matrix: all other landscape elements not belonging to the above mentioned forest types. On the "Trevelin" photo mosaic, the only class considered was pure *Austrocedrus*. When comparing the different classifications of *Austrocedrus* forests in this area, it was necessary to define if the variations were owed to problems arising from the material quality or if they must be assessed to a real landscape change. These verifiable changes were classified into the following categories: (1) Young trees, (2) Interesting elements masked in the matrix, (3) Human-induced alteration deteriorating or removing the *Austrocedrus* forests and (4) "Natural" diminishing of *Austrocedrus* on very unfavourable sites.

On both areas, the forest patches were distinguished in three classes according to the canopy density. A minimum cartographic unit of 2500 m^2 was set, thus individualizing only elements of an equal or bigger area. The same landscape elements and the indicated alterations were identified on the 2001 IKONOS and SPOT images. Comprehensive field controls helped adjust the initial classification of forest types. Due to the small size of the patches covered by the *Austrocedrus-N. dombeyi* and *Austrocedrus*-shrubs categories in the "Epuyén" area, they were excluded from the posterior analysis.

The area calculation for the different classes and the analysis of the considered landscape changes were carried out with Xtools and Patch Analyst programs, both working as extensions of the ArcView software. The following indices were used: patch number per class (N), class area (A) (area of all polygons belonging to the same class, expressed in hectares), percentage of class area ($A_{\%}$) regarding the total landscape area, mean patch size (MPS), representing the arithmetic average size of every patch of a given class type and area-weighted mean patch size (MPS^2). Under these conditions, a simple arithmetic average does not reflect the expected patch size that could be found through a simple location of random points on the map (Turner et al. 2001). The last selected metric was the maximum patch size (MaxPS), used to get a representation of the connectivity degree of the class of interest.

6.3 Results And Discussion

6.3.1 Fragmentation of Austrocedrus Forests in "Epuyén"

In 1970, the zones dominated by *Austrocedrus* forests constituted a predominantly continuous or interconnected area and occupied almost 3400 ha (Fig. 6.3a,b). This situation was drastically modified being 360 ha affected by forest fires, exclusively in the western sector (Fig. 6.3c), and 475 ha replaced by plantations, mainly on the eastern side of the study area (Fig. 6.3d).

The net 24%-decrease in area – 835 ha (Table 6.1) – was accompanied by a strong negative change in the heterogeneity of the forest landscape due to the fragmentation of the *Austrocedrus* area – almost 34% in the considered time period. Fragmentation of *Austrocedrus* forests was more intense in semi-dense and sparse classes on the eastern side (Fig. 6.4a), whereas the decline of the *Austrocedrus* area was variable within these two sectors, being superior in dense and sparse classes in the western sector (Fig. 6.4b). The *N. dombeyi* area came down in the western sector, whereas this class area increased in the eastern sector. The matrix enlarged its area in the burnt sector and decreased in the afforested one. In both cases, there was a reduction in the number of polygons.

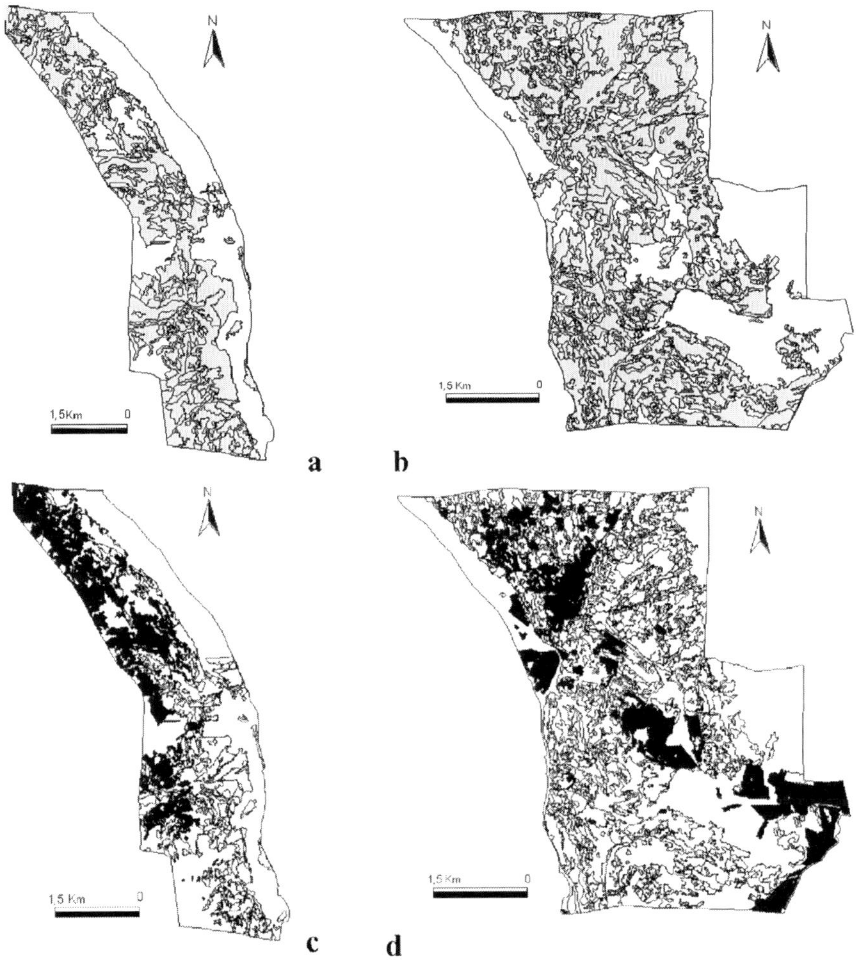

Fig. 6.3 (**a–b**) Eastern sector with *Austrocedrus* and *N. dombeyi* forest types (*grey*) and other landscape elements included as matrix (*white*) in 1970, (**c**) Burnt areas in western sector (*black*) and other landscape elements including *Austrocedrus* and *N. dombeyi* forest types (*white*) in 2001, (**d**) Plantations of exotic species in eastern sector (*black*) and other landscape elements including *Austrocedrus* and *N. dombeyi* forest types (*white*) in 2001

In 1970, the total number of polygons of *Austrocedrus* forest was 551, whereas in 2001, this number reached 831 (Table 6.1). In the analyzed landscape, the *Austrocedrus* forests occupied 56% of the study area in 1970 and 42% in 2001.

Relevant reductions were also checked in the mean patch size and the area-weighted mean patch size. In 1970, the value of the latter was 35 ha (Table 6.1), whereas in 2001, it had been reduced to only 10 ha. We also noticed that in 1970, the biggest patch occupied an area of almost 500 ha, whereas in 2001, it had reduced its size almost 60% (210 ha) as compared to the original condition.

Table 6.1 Quantification of changes on the heterogeneity of *Austrocedrus* forests due to substitution by afforestations and forest fires in "Epuyén"

	Landscape element	N	A	$A_{\%}$	MSP	MSP^2	MaxSP
1970	*A. chilensis*	551	3375	56	19.5	35.5	486
	N. dombeyi	164	484	8	5.2	0.8	51
	Plantations	0	0	0	0	0	0
	Matrix	176	2153	36	12.2	200	801
	Total	891	6013	100	–	–	–
2001	*A. chilensis*	831	2540	42	9.8	9.9	209
	N. dombeyi	160	517	9	5.1	1.5	63
	Plantations	86	677	11	7.9	5.1	101
	Matrix	158	2280	38	14.4	332.9	1212
	Total	1235	6013	100	–	–	–

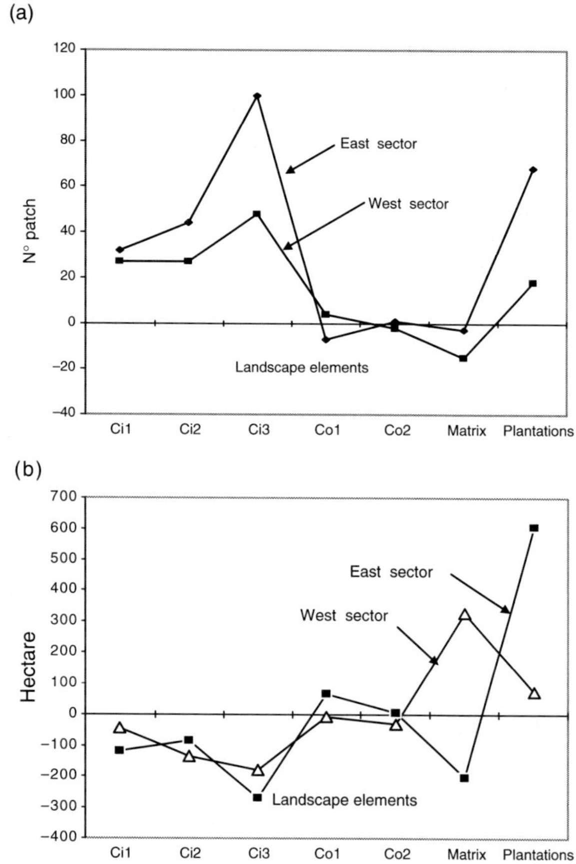

Fig. 6.4 (**a**) Net patch number variation for the considered landscape elements between 1970 and 2001 in western and eastern sectors, (**b**) Net area variation for the considered landscape elements between 1970 and 2001 in western and eastern sectors

During the field survey, monospecific plantation blocks of different ages were identified. The most common species were *Pinus radiata*, *Pinus murrayana*, *Pinus ponderosa* and *Pseudotsuga menziesii* at a smaller extent. Plantation density was 1100 plants per hectare or higher. None of the surveyed plantations showed any kind of forest management, so the older stands presented an excessive density. The inexistence of prevention schemes that would allow a reduction of the risk of fires, naturally high due to the presence of human settlements as well as to the state of the plantations (high density, dense undergrowth and abundant dry material, low living or dead branches, continuous plantation blocks of considerable area) is quite troublesome. In the summer of 2001, a fire focus was intentionally lit in a *Pinus murrayana* plantation situated near a national road, which affected 120 ha that surpassed the boundaries of this plantation spreading to a considerable area of indigenous forest.

Plantations were carried out on different substrata and vegetation types (Fig. 6.5). In some cases, this activity was oriented toward the substitution of indigenous forest

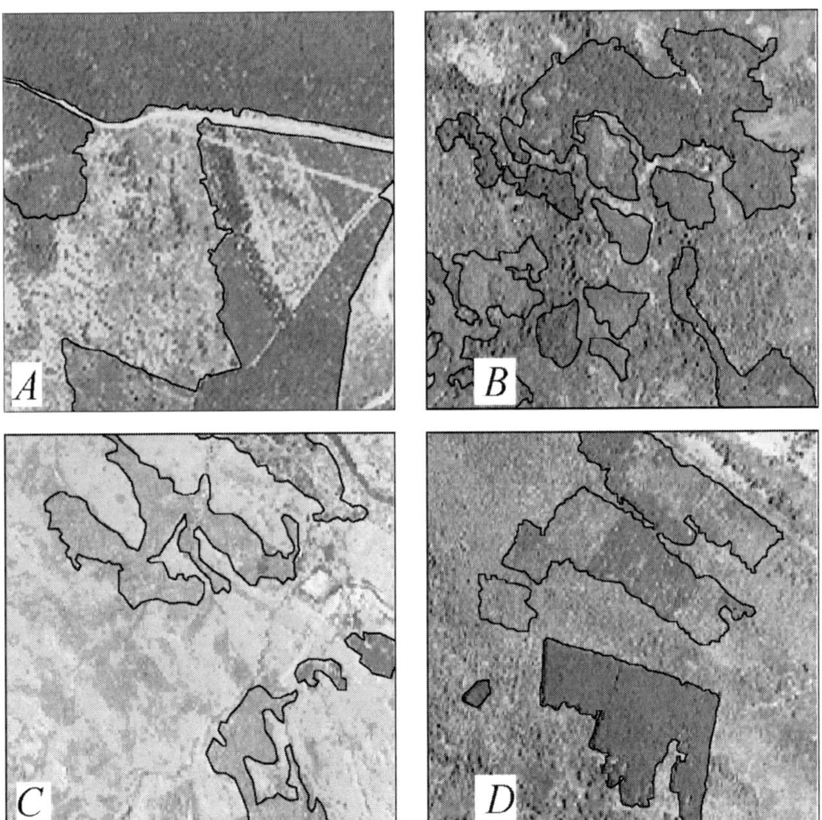

Fig. 6.5 Plantation patches established on different substrata and vegetation types (see explanatory text above)

communities by exotic species of a faster growth rate, which was legally allowed. Paradoxically, the goal of the provincial government was "the increase of the existing wooded stocks without harming the natural species". Plantations of *Pinus ponderosa*, *P. radiata* and *Pseudotsuga mensiezii* were carried out on fluvioglacial deposits in sectors dominated by shrubs of *Diostea juncea*, *Lomatia hirsuta* and *Schinus patagonicus* (Fig. 6.5a). Plantations were also established on areas of complex topography with rocky outcrops, slopes of different aspects and small valleys, dominated by *Austrocedrus*, *Austrocedrus-N. dombeyi*, or *N. dombeyi-Austrocedrus* (Fig. 6.5b), where we could still observe some remnants of these indigenous species. In some other cases, plantations intended to recover burnt forest areas. In Fig. 6.5c, plantations, generally of *P. ponderosa*, were established in the most humid sectors of a northeast slope burnt in 1987. Prior to this disturbance, vegetation was dominated by semi-dense or sparse *Austrocedrus* forests. A mixed situation showing plantation patches on slopes with a similar aspect as the previous one with *Austrocedrus* forest of different densities, sometimes accompanied by shrubby vegetation, is presented in Fig. 6.5d. Formerly, pure *Austrocedrus* forests of different densities dominated this sector with understory strata of *Lomatia hirsuta* and *Schinus patagonicus.*

The substitution with exotic species is currently a practice not legally allowed nor so extended, yet it is still carried out on burnt *Austrocedrus* stands as well as on sectors affected by "mal del ciprés" disease. In fact, this sanitary problem is the main factor influencing the forest management performed by the provincial forest services (Rajchenberg and Gomez 2005), supported on the existence of relevant affected areas (La Manna and Carabelli 2005). Furthermore, the habitat alterations that these procedures have brought along and indeed keep acting differentially in distinct spatial and temporal scales must not be underestimated (Carabelli 2004). We agree with Haila (2002), when indicating that as different organisms and ecological systems "experience" the fragmentation degree of a particular environment in variable forms, even contradictory, it is necessary to consider multiple spatial and temporal scales, taking into account that the relevant scales probably vary across species, geographical regions and types of environments.

On the other hand, there still remains the question of the incidence of forest fires (Fig. 6.6). Recent statistics of the Chubut forest service (Dirección General de Bosques y Parques 2002) reveals that between May, 2001 and March, 2002, about 700 ha of *Austrocedrus* forests in this area were damaged by fire (25% of the total *Austrocedrus* forests affected by fires in the distribution range of the species in the provincial territory that season) and 630 ha of *N. dombeyi* forests (28% of the total *N. dombeyi* forests affected by fire in Chubut in this time period).

A policy of indigenous forest replacement by plantations had been not sustained by a program capable to guarantee the monitoring of management activities in the new plantations. Currently, this situation has produced a chain of legal and jurisdictional problems, where land is illegally occupied and confrontations are permanently present. A good-quality forest has been substituted by an unmanaged one that will provide raw material of bad quality, laying the basis for a wide discussion on even the financial convenience of the actions carried out. Lack of management in the plantations threatens their persistence because the risk of fires increases (Fig. 6.6).

Fig. 6.6 Plantations replacing *Austrocedrus* (**a**) and burnt areas of *Austrocedrus* forests (**b**) in "Epuyén" area

Huge timber masses representing a high quantity of fuel with vertical and horizontal continuity could be consumed in a single event without any possibility of controlling it, risking the adjacent indigenous forests. Finally, these areas with high proportions of weakened and diseased trees constitute a focus for the development and propagation of plagues, as it had been recently demonstrated (Gomez et al. 2006).

6.3.2 Changes on Austrocedrus Forest Landscapes in "Trevelin"

We considered two subjective kinds of changes over the *Austrocedrus* forests: "positive" changes and "negative" changes. The first one encompasses the expansion of forest areas owed to *Austrocedrus* regeneration or preexistent young trees that could be detected in the SPOT image from 2001. On this remote-sensed material, *Austrocedrus* distribution areas were delimited (Fig. 6.7) but not the precise *Austrocedrus* patches configuration, since the access to all forested sectors was very complicated due to the extreme irregular topography and the scarce net of roads. On the other hand, negative changes are those showing a decrease or loss on original *Austrocedrus* forests areas (Fig. 6.7). Causes vary depending on intensity and circumstances but predominantly, it was an area reduction due to a total loss or a strong density decrease (for example, a low-intensity forest fire not burning all of the trees). Moreover, the combination of different human-induced alterations also produced changes when the fire-affected area disappeared by the harvesting of the few still-standing trees.

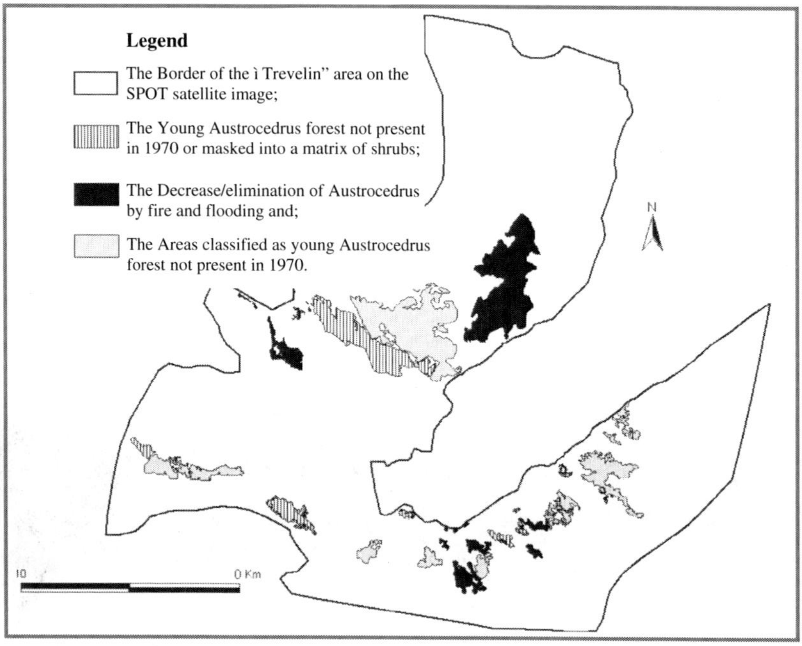

Fig. 6.7 Positive and negative changes on the heterogeneity of *Austrocedrus* forest landscape patterns in "Trevelin" between 1970 and 2001

The changes in the considered time period affected 4400 ha of *Austrocedrus* forests, being 1500 ha (34%) flooded by the construction of a dam with hydroelectric purposes (Fig. 6.8a), affected by fire (Fig. 6.8b) or by timber harvesting (Fig. 6.8c). In other sectors, 2900 ha were computed as "new" *Austrocedrus* forests by the growth of young trees not visible in the photo mosaic from 1970 or by the regeneration recruitment.

The three classes of *Austrocedrus* density assessed in the photo mosaic from 1970 were grouped in only one class to conduct the quantitative analysis of changes with the landscape indices (Table 6.2), since the classification of the SPOT image from 2001 was grouped in only one class (*Austrocedrus* and non-*Austrocedrus*). A strong enlargement of the total edge and edge density in the classification of 2001 can be attributed to the increase of the *Austrocedrus* forest area and the number of patches between 1970 and 2001. The mean patch size in 2001 decreased 25% compared to 1970, with more patches between 0.25 and 5 ha. Differences on the standard deviation of the mean patch size (four times bigger in 2001) can be explained by the wide range between the extreme values (0.25–1187 ha) as compared to the range in 1970 (0.25–154 ha). The values for the mean patch edge, mean shape index and perimeter-mean area relationship are relatively close between the compared years.

The quality of the aerial photographs represented a noticeable constraint for the identification of *Austrocedrus* areas, thus limiting the potential of the comparative analysis related to data in 2001. As it has been referred, the quality of

Fig. 6.8 (**a**) Flooded *Austrocedrus* forests by the "Futaleufú" hydroelectric dam, (**b**) Burnt *Austrocedrus* forests, (**c**) Cutting of *Austrocedrus* forests for timber, (**d**) Young *Austrocedrus* forests not visible in 1970

this remote-sensed material usually restricts the differentiation of cover types or land-use types in order to analyze landscape changes (Dunn et al. 1991; Chuvieco 2000, pp. 471–482; Scoz et al. 2005). The photo-interpretation process is also high time-demanding and more expensive than the satellite image classification. In our

Table 6.2 Class indices for *Austrocedrus* forests in "Trevelin" in 1970 and 2001

Metric	1970	2001
Area (ha)	3945	7200
Number of Patches	913	1949
Mean Patch Size -MPS- (ha)	4.87	3.69
MPS Standard Deviation (ha)	9.36	41.36
Total Edge (km)	1857	5047
Edge Density (m/ha)	157	701
Mean Patch Edge (m/patch)	2257	2589
Mean Shape Index (>1)	3.12	2.98
Perimeter-mean area relationship (m/ha)	839	1204

experience, the classification costs per hectare were almost 20 times higher (i.e., \$1US/ha vs. \$0.005US/ha) (Carabelli and Claverie 2005).

A central assessment in the "Trevelin" area was the enlargement of the *Austrocedrus* area due to the growth of young trees that, in 1970, were not detectable on the aerial photos or not present (Fig. 6.8d). These changes represent 66% of the global area of *Austrocedrus* forests effectively modified during the considered time period. This area of 2900 ha represented 73% of the 1970-identified *Austrocedrus* area and 95% of the incremented forest area between 1970 and 2001.

Also remarkable, due to the consequences, is one of the so-considered "negative" changes. Forest fires affected 1500 ha but the forest was not completely ruined, even when it was harvested after that. During the field work, we observed that forest stands strongly damaged by fire had often good and extended *Austrocedrus* regeneration. Taking this situation into account, the burnt *Austrocedrus* areas did not disappear, even though a decrease of both former forest area and connectivity was settled. The classification on the SPOT image detected the surviving *Austrocedrus* trees and the further regeneration. This circumstance was considered when quantifying the *Austrocedrus* areas in 2001.

An increase of the connectivity between *Austrocedrus* patches in 2001 was assessed. Two reasons are plausible to explain this fact. The first one refers to the classification algorithm for the satellite data, which is more sensitive to detect the spectral signatures of *Austrocedrus* compared to the skills of the photo interpreter to distinguish the landscape features that more precisely match *Austrocedrus* forest type on the aerial photos. The second reason is the verifiable increase of the *Austrocedrus* forest area due to the growth of the shrubby *Austrocedrus* regeneration in 1970. In addition, very isolated trees not registered in the class of density "sparse" on the photo mosaic of 1970 (because they did not fulfill the distance requirement) allowed the regeneration that reconnected areas formerly sheltering dense *Austrocedrus* forests.

This assessment supports a hypothesis suggested by Kitzberger and Gowda (2004) who analyzed the Patagonian-Andean landscape to define structures and potential range models for different tree species in the Cholila-Lake basin over about 174000 ha. These authors found out that *Austrocedrus* forests are sparsely connected with patches of different tree species. They pointed out that even though a high *Austrocedrus* fragmentation was verified, some indicators showed a coalescence process mainly on shrubby areas, induced by a lower frequency of forest fires. Although a long-time fire record for the area does not exist, the recuperation and regeneration of burned *Austrocedrus* forests had been very significant for the last 30 years.

This research also allowed an improvement in the classification details related to area and distribution of *Austrocedrus* forests for the considered areas. The selected scale was 1:50000 corresponding to the recommendations of the Cartographic International Association for maps derived from SPOT data (Chuvieco 2000, p. 152). Up to now, the scale of the best available maps for this area and forests was 1:250000 (Bran et al. 2002).

The analyzed *Austrocedrus* areas represent 10% of the total region (135400 ha) covered by this species in the Patagonian Andes. Thus, these assessments are particularly worrisome because *Austrocedrus* occupies the smallest area among those indigenous forest species that have traditionally and currently been involved into a wide range of human uses.

Furthermore, other processes are affecting the integrity of this species, such as the relatively recent but intense landownership subdivision, the unplanned use for grazing, the timber exploitation on sectors affected by "mal del ciprés" disease and intensive farming (Carabelli et al. 2006). This context plays its own role by intensifying a complexity of alterations that notoriously has more negative than positive effects on indigenous forests. Such circumstances highlight the need for an integral insight of landscapes and for developing management actions with sound technical and scientific bases and acceptance in the different community sectors, so that land-use practices on forest environments in our region contribute to mitigating the deterioration processes of our indigenous ecosystems (at least those that have visible manifestations and are quantifiable in the short term).

Acknowledgments During these 5 years, a great team of colleagues gave efficient but, most important, enthusiastic and faithful support to this research. We know we are just at the beginning because hard work must still be done – but at least it had been done. It was possible due to Marcelo Jaramillo, Mariano Gómez, Silvio Antequera, Ivonne Orellana, Horacio Claverie, Matías Acetti and Lilian Soto. The research was financially supported by the German Volkswagen Foundation and the National Council for Scientific and Technical Research of Argentina.

References

Bondel CS, de Almeida A (1996) El Bolsón en la cuenca del lago Puelo. Descripción inicial de la relación hombre – medioambiente. Anales de la Sociedad Argentina de Estudios Geográficos, Tomo 20, p 69

Bran D, Pérez A, Barrios D, Pastorino M, Ayesa J (2002) Eco-región valdiviana: distribución actual de los bosques de "cipres de la cordillera" (*Austrocedrus chilensis*) – escala 1:250.000. Informe Preliminar. INTA-APN-FVSA, p 12

Carabelli FA (2004) Quantitative analysis of forest fragmentation in Patagonia, Argentina. In Sano M, Miyamoto A, Sugimura K (eds) Proceedings of IUFRO International Workshop on landscape ecology 2004: conservation and management of fragmented forest landscapes, Tsukuba, Ibaraki, Japan, pp 83–87

Carabelli F, Claverie H (2005) Análisis de cambios en los paisajes de bosques de ciprés de la cordillera en el noroeste de la Provincia de Chubut. In Libro de Actas de la I Reunión sobre ecología, conservación y uso de los bosques de ciprés de la cordillera, Esquel, Chubut, Argentina, pp 61–66

Carabelli F, Jaramillo M, Antequera S (2003) Cambios en la heterogeneidad del bosque nativo en la Patagonia Andina de Argentina y su impacto sobre la biodiversidad en los sectores de borde. Cuadernos de Biodiversidad (14):10–15

Carabelli F, Scoz R, Claverie H, Jaramillo M, Gómez M (2002) Detection and assessment of positive and negative changes on the heterogeneity of forest landscapes: fragmentation analysis of "ciprés de la cordillera" forests in Patagonia Andina, Argentina. In: Abstracts of the international conference "Disturbed landscapes: Analysis, Modeling and Valuation", Brandenburg University, Cottbus, Germany, 24–27 Sep 2002

Carabelli F, Scoz R, Claverie H, Jaramillo M, Gómez M (2006) Changes on landscape heterogeneity and spatial patterning of native forests in Patagonia, Argentina. Investigación Agraria. Sistemas y Recursos Forestales (Madrid) 15(2):160–170

Chuvieco E (2000) Fundamentos de Teledetección espacial. 3ra. edición revisada. Ed. RIALP. Madrid

Córdoba ML, González Capdevilla F (1999) Comarca de los Alerces, Planificación Estratégica y Participativa. Organismo Provincial de Turismo, Provincia de Chubut, 163pp

Dezzoti A, Sancholuz L (1991) Los bosques de *Austrocedrus chilensis* en Argentina: Ubicación, estructura y crecimiento. Bosque 12(2):43–52

Dimitri MJ (1974) Pequeña flora ilustrada de los Parques Nacionales Andino– Patagónicos. Anales de Parques Nacionales Tomo XIII, 122pp

Dirección General de Bosques y Parques (2002) Informe de la campaña de incendios temporada 2001–2002. Informe Técnico, 17pp

Dunn CP, Sharpe DM, Guntenspergen GR, Stearns F, Yang Z (1991) Methods for analyzing temporal changes in landscape pattern. In: Turner MG, Gardner RH (eds) Quantitative methods in landscape ecology, ecological studies, vol 82. pp 173–198, Springer

Fonck F (1900) Viajes de Fray Francisco Menéndez a Nahuel Huapi. Valparaíso: C. F. Niemeyer

Food and Agriculture Organisation of the United Nations (1993) Guidelines for land-use planning. FAO Development Series 1. Soil resources, management and conservation service, interdepartmental working group on land use planning

Forman R, Godron M (1986) Landscape ecology. John Wiley & Sons, New York

Fukamachi K, Oku H, Nakashizuka T (2001) The change of a satoyama landscape and its causality in Kamiseya, Kyoto Prefecture, Japan between 1970 and 1995. Landscape Ecol 16:703–717

Gomez C, Greslebin A, Jovanosvski A, Rajchenberg M (2006) Enfermedades y plagas en plantaciones de coníferas exóticas en Patagonia. Informe Final, Proyecto PIA 1104. Secretaría de Agricultura, Ganadería, pesca y Alimentación, p 71

Grez AA, Bustamante RO, Simonetti JA, Fahrig L (1998) Landscape ecology, deforestation and forest fragmentation: the case of the Ruil Forest in Chile. In: Salinas Chávez E, Middleton J (eds) Landscape ecology as a tool for sustainable development in Latin America. http://www.brocku.ca/epi/lebk/lebk.html Cited 12 Aug 1998

Hessburg PF, Smith BG, Salter RB, Ottmar RD, Alvarado E (2000) Recent changes (1930s–1990s) in spatial patterns of interior northwest forests, USA. Forest Ecol Manage 136:53–83

Jenkins MA, Parker GP (2000) Changes in the forest landscape of the Charles C. Deam Wilderness, Southern Indiana, 1939–1990. Nat Area J 20:46–55

Kitzberger T, Gowda J (2004) Análisis del paisaje andino patagónico. Estructura y modelos de rango potencial. Workshop on fragmentation of native forests in Argentina. In: Mitteilungen der Abteilung für Forstliche Biometrie 3, Freiburg University, Germany, pp 13–17

La Manna LF, Carabelli F (2005) Distribución espacial de la declinación de los bosques de *Austrocedrus chilensis* en el Valle 16 de Octubre (Chubut, Argentina). Primeras Jornadas Argentinas de Ecología del Paisaje. Grupo de Ecología de Paisajes y Medio Ambiente (GEPAMA), Facultad de Arquitectura y Urbanismo de la Universidad de Buenos Aires, pp 1–11

Lawes MJ, MacFarlane DM, Eeley HAC (2004) Forest landscape pattern in the KwaZulu–Natal midlands, South Africa: 50 years of change or stasis? Austral Ecol 29(6):613–623

Lida S, Nakashizuka T (1995) Forest fragmentation and its effect on species diversity in sub-urban coppice forests in Japan. Forest Ecol Manag 73:197–210

Löfman S, Kouki J (2001) Fifty years of landscape transformation in managed forests of Southern Finland Scand. J For Res 16:44–53

Loguercio G (2005) Posibilidades y limitaciones del manejo forestal de los bosques de ciprés de la cordillera. In: Libro de Actas de la I Reunión sobre ecología, conservación y uso de los bosques de ciprés de la cordillera, Esquel, Chubut, Argentina, pp 21–28

Loguercio G, Burschel P, Rey M (1999) El Bosque de Ciprés de la Cordillera: su conservación y uso. Guía Práctica. Folleto de Divulgación N° 14. CIEFAP. 21pp

Loyn RH, McAlpine C (2001) Spatial patterns and fragmentation: indicators for conserving biodiversity in forest landscapes. In: Raison RJ, Brown AG, Flinn DN (eds) Criteria and indicators for sustainable forest management. CAB International, Chapter 19, pp 391–422

Morrone JJ (2001) Biogeografía de América Latina y el Caribe. M&T–Manuales & Tesis SEA, vol. 3. Zaragoza

Muster GC (1971) At home with the Patagoniens: a years wanderings over untrodden ground from the straits of Magellan to the Río Negro. London, John Murray
Puric-Mladenovic D, Kenney WA, Csillag F (2000) Land development pressure on peri-urban forests: a case study in the regional municipality of York. Forest Chron 76:247–250
Rajchenberg M, Gomez C (2005) Enfermedades y plagas en *Austrocedrus chilensis*. In: Libro de Actas de la I Reunión sobre ecología, conservación y uso de los bosques de ciprés de la cordillera, Esquel, Chubut, Argentina, pp 35–39
Rescia AJ, Schmitz MF, Martín de Agar P, de Pablo CL, Pineda FD (1997) A fragmented landscape in northern Spain analyzed at different espatial scales: implications for management. J Vegetation Sci 8:343–352
Riitters K, Wickham J, O'Neill R, Jones B, Smith E (2000) Global-scale patterns of forest fragmentation. Conserv Ecol 4, 2:3, http://www.consecol.org/vol4/iss2/art3 Cited 13 Jun 2000
Ripple WJ, Hershey KT, Anthony RG (2000) Historical forest patterns of Oregon's central Coast Range. Biol Conserv 93:127–133
Roth LC (1999) Anthropogenic change in subtropical dry forest during a century of settlement in Jaiqui Picado, santiago Province, Dominican Republic. J Biogeogr 26:739–759
Rothkugel M (1916) Los bosques patagónicos. Oficina de Bosques y Yerbales. Dirección de Agricultura y Defensa Agrícola. Boletín N° 3. Buenos Aires
Scoz R, Carabelli F, Soto Meza L (2005) Quantifizierung der Waldstrukturen in Nord-West-Patagonien mit Hilfe von Fragmentierungsindizes. Grüne Reihe, Sektion Forstliche Biometrie und Informatik. Zeitschrift des Deutschen Verbands Forstlicher Forschungsanstalten (J German Union Forest Res Organ), pp 120–131
Silbernagel J, Chen J, Gale MR, Pregitzer KS, Probst J (1997) An interpretation of landscape structure from historic and present land cover data in the Eastern Upper Peninsula of Michigan. United States Department of Agriculture, Forest Service, North Central Forest Experiment Station, General Technical Report, NC, 192, p 30
Tucker CM, Munroe DK, Nagendra H, Southworth J (2005) Comparative spatial analyses of forest conservation and change in Honduras and Guatemala. Conserv Soc 3(1):174–200
Turner MG, Gardner RH, O'Neill RV (2001) Landscape ecology in theory and practice. Pattern and process. Springer
Veblen TT, Kitzberger T, Lara A (1992) Disturbance and forest dynamics along a transect from Andean rain forest to Patagonian shrubland. J Veg Sci 3:1–14
Veblen TT, Lorenz D (1987) Post-fire stand development of *Austrocedrus-Nothofagus* forests in northern Patagonia. Vegetatio 71:113–126
Wade TG, Riitters KH, Wickham JD, Jones KB (2003) Distribution and causes of global forest fragmentation. Conserv Ecol 7(2):7. http://www.consecol.org/vol7/iss2/art7. Cited 19 Sept 2003

Chapter 7

Landscape-Scale Factors Influencing Forest Dynamics in Northern Australia

Daniel S. Banfai and David M.J.S. Bowman

Abstract Understanding the extent and causes of savannah-forest dynamics in tropical regions is vital as small but widespread changes to tropical forests can have a major impact on global climate, biodiversity and human well-being. There is emerging evidence from aerial photography that an overall expansion of monsoon rainforests has occurred in northern Australia over the last few decades. Factors that may have driven the observed rainforest dynamics include the management of local scale disturbance events such as fire and buffalo numbers, as well as regional scale factors such as increases in rainfall and atmospheric CO_2. Landscape ecology studies conducted in Kakadu National Park are provided as a case study as they together provide a cohesive methodology for investigating the consequences of management. The extent of boundary change at individual rainforest patches supported an effect of fire on the rainforest dynamics. The effect of historical buffalo impact was also supported by modelling analyses. However disturbance factors were unable to account for the overall expansion of rainforest. We conclude that fire and buffalo management have mediated the boundary dynamics. However, the overall boundary expansion is likely to have been primarily driven by factors that have shown similar increases during the study period, such as annual rainfall and atmospheric CO_2. The methodology presented can be applied to forests in other regions and will contribute to 'adaptive management' programs, particularly with respect to fire management.

7.1 Introduction

The monsoon rainforest patches of northern Australia are integral to the maintenance of both the natural and cultural values of the region. In terms of biodiversity, the rainforest patches support a high number of plant species and are important refugia for animals such as mammals from climatic extremes and disturbance events

---

D.S. Banfai
School for Environmental Research, Institute of Advanced Studies, Charles Darwin University, Darwin, Australia
e-mail: heals11@yahoo.com

R. Lafortezza et al. (eds.), *Patterns and Processes in Forest Landscapes*,

such as fire (Russell-Smith 1991; Woinarski et al. 2001). They are also critical to the conservation of frugivorous birds which move between patches of rainforest and thus depend on the existence of a network of patches. A spatial analysis of six bird species by Price et al. (Price et al. 1999) demonstrated that the destruction of individual patches can have far-reaching regional effects by reducing the amount of connectedness between patches. The rainforest patches are also of great cultural value, being harvested by local Aboriginal people for traditional foods such as yams (Lucas et al. 1997).

The rainforests in northern Australia are a microcosm of the ecology and conservation challenges facing tropical rainforests globally. Tropical forests house more than half of the Earth's species, and are integral to climate change as they cycle huge amounts of carbon each year (Malhi and Grace 2000). Understanding the factors that control the boundaries of tropical forests and the consequences of management is therefore critical as small changes in the extent of tropical forests can have major impacts on climate, biodiversity and human well-being. Understanding the mechanisms of changes to tropical forests is particularly important given the risk of many forests drying out over the next century and burning, which may lead to a rapid acceleration of climate changes (Lewis 2006).

While the rainforests in northern Australia are clearly of great natural and cultural value, a number of processes are thought to be threatening the integrity of rainforest boundaries. For example, surveys by Russell-Smith (1992) indicated that monsoon rainforests are contracting at the regional scale due to the combined effects of an increase in late dry season fires, feral animal damage and weed invasion. These authors found that one-third of rainforest sites surveyed had boundaries severely degraded by fire.

Understanding the drivers of rainforest boundary change and the consequences of human management of these factors is therefore critical to the future conservation of these systems. In this chapter we aim to (1) provide a summary of the extent and causes of rainforest boundary dynamics in northern Australia, and (2) provide a methodology for assessing the consequences of localised management of rainforest boundaries and the relative importance of regional factors such as climate changes. Research conducted in Kakadu National Park is used as a case study. The methodology presented is a synthesis of a number of studies undertaken in Kakadu National Park which together provide a cohesive approach which can be applied to other regions.

7.2 Australian Rainforests

There is no consensus as to the precise definition of rainforest in Australia. The term 'rainforest' is used to define a broad variety of atypical Australian forest types (Bowman 2000). These include a variety of structural and floristic types and are broadly classified by the climatic regime in which they occur as either tropical, subtropical, monsoonal and both cool and warm temperate rainforest types. The current distribution of rainforests in Australia occurs as an arc of rainforest fragments along the eastern and northern coastlines (Fig. 7.1). The monsoonal rainforests, which are

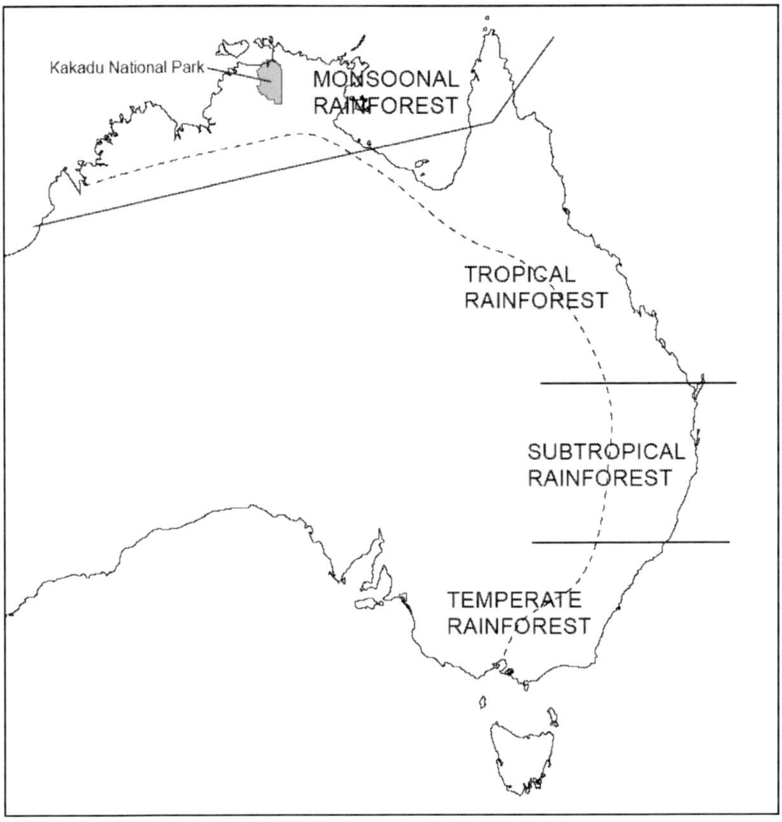

Fig. 7.1 The distribution of the four major types of rainforest in Australia (adapted from Bowman 2000). The dashed line indicates the approximate inland extent of rainforest (includes the whole of Tasmania). The location of Kakadu National Park is also indicated

the subject of this chapter, are adapted to a hot climate which is characterized by seasonally wet and dry conditions (Bowman 2000). The northern Australian region has a monsoonal climate with over 90% of the c.1.5 m annual rainfall occurring in the wet season (October to March). Day time air temperatures remain high throughout the year (~30°C) with cooler nocturnal temperatures (~25°C) occurring during the dry season (Bureau of Meteorology 2003).

The pioneering work on describing and classifying Australian rainforests was carried out by Webb and Tracy (1981). More recently, extensive surveys of 1219 monsoon rainforest patches by Russell-Smith (1991) allowed the classification of rainforests of northern Australia into 16 floristic types. Monsoon rainforests occur as vegetation with a closed canopy that is not dominated by species such as eucalypts (*Eucalyptus* and *Corymbia* spp.) or paperbarks (*Melaleuca* spp.). The two basic types of rainforest that occur are 'wet monsoon rainforest' which occurs on sites with perennial moisture supplies such as springs, and 'dry monsoon rainforest' which occurs on freely drained sites that are often associated with topographic fire protection such as rocky areas, cliff lines and hill tops (Bowman et al. 1991;

Russell-Smith 1991). For simplicity these vegetation types are referred to here as 'wet rainforest' and 'dry rainforest'. Additionally, *Allosyncarpia ternata* S.T.Blake dominated rainforest is a unique rainforest type which occurs on the Arnhem Land plateau region (Bowman and Dingle 2006).

7.3 Rainforest Boundary Dynamics in Northern Australia

Rainforest boundaries are known to be highly dynamic on both long and short time scales, and these changes appear to have been driven by shifts in climatic variables as well as disturbance events (Bowman 2000). Information on early rainforests in Australia comes from a wide variety of fossil types. Although there are many gaps and uncertainties, there is sufficient evidence to show that a large part of the continent was covered with rainforest for much of the late Cretaceous and Tertiary period (Truswell 1990). Greater uncertainty exists regarding the history of monsoonal rainforests, due to the absence of relevant paleorecords in the region where this rainforest type occurs. However, the current distribution of monsoon rainforests as an archipelago of islands separated by drier vegetation such as savannah is consistent with the view that this forest type was once much more widespread (Russell-Smith and Dunlop 1987).

The fossil record for eastern Australia suggests that a dramatic reduction in rainforest area occurred in the late Tertiary period, and was most likely primarily driven by climatic cooling and drying (Truswell 1990). There is debate as to the relative contribution of Aboriginal burning to the retreat of rainforest vegetation in Australia, however it is almost certain that such fire practices would have influenced the process of forest fragmentation over the last c. 50,000 years (Bowman 2000).

A high degree of dynamism is also apparent in the rainforest boundaries in northern Australia at the decadal scale. A field based assessment of the integrity of 1,219 rainforest patches by Russell-Smith and Bowman (1992) suggested monsoon rainforests had contracted at a regional scale as a result of unfavorable fire regimes and feral animal disturbance. However, this evidence has been overturned by the large number of more recent localized studies using historical sequences of aerial photography which have revealed rainforest expansion at the expense of more open vegetation types. These include rainforests in Litchfield National Park near Darwin (Bowman et al. 2001), on the Arnhem Land Plateau (Bowman and Dingle 2006), in the Gulf of Carpentaria (Brook and Bowman 2006) and in Kakadu National Park (Banfai and Bowman 2006).

7.4 Causes of Boundary Dynamics

The observed changes in rainforest boundaries in northern Australia raises the question of what has driven the changes, and what are the consequences of human management of rainforest boundaries. Plausible drivers of the changes that have occurred at the decadal scale include the effects of fire, feral animal impact, rainfall and atmospheric CO_2 (Banfai and Bowman 2006). The past research on each of the main drivers of change is reviewed below.

7.4.1 Fire Management

Previous research throughout the tropics has highlighted that fire regimes can have a major impact on rainforest boundary dynamics. Fire can lead to rainforest contraction by killing seedlings and consuming live foliage, thus reducing tree growth and survival on rainforest boundaries (Bowman 2000). High intensity fires in the latter part of the dry season are often thought to be particularly threatening to the integrity of rainforest boundaries (Russell-Smith and Bowman 1992). However, because rainforest seedlings are usually able to survive the effects of at least a single fire (Bowman and Panton 1993; Russell-Smith and Dunlop 1987) fire frequency is of greater importance in determining the rate of boundary dynamics. The combination of regular late dry season fires and exotic flammable weeds can lead to the rapid contraction of rainforest boundaries in northern Australia, while fire protection can promote rainforest expansion. The importance of fire in driving the boundary dynamics is consistent with studies of other tropical forests, where the local dynamics of the forest-savannah boundary zone were inferred to be related to fire incidence (Favier et al. 2004; Furley et al. 1992; King et al. 1997).Globally, a reduction in fire frequency has often been attributed as the cause of rapid expansion of tropical forests (Hopkins 1992; Swaine et al. 1992).

Little is known of the frequency and extent of burning in northern Australia prior to the record of fire scars from satellite imagery, which are available from 1980 onwards (Bowman et al. 2007). The available ethnographic, historical and contemporary data about Aboriginal burning in northern Australia suggests that, prior to European settlement, Aboriginal people used fire in a skilful manner for a diverse range of both cultural and ecological (i.e. resource management) applications (Bowman 1998; Preece 2002; Russell-Smith 2001). Burning tended to be highly patchy, creating a fine-grained habitat, and was concentrated in the second half of the dry season (Bowman et al. 2004). Under traditional Aboriginal management, rainforest patches in northern Australia were commonly afforded habitat-specific fire management. Examples include the careful burning of rainforest boundaries to protect food resources such as yams (Russell-Smith et al. 1997). Such practices were widespread in northern Australia at least until the end of the 19th century (Preece 2002). While traditional fire management practices continue in many localized regions (Bowman et al. 2004; Press and Lawrence 1995) there has been a substantial disruption of traditional fire regimes over the last century due to a dramatic reduction in the Aboriginal population in many areas and the establishment of European management practices. Current European management in northern Australia is primarily focused on reducing the incidence of destructive late season fires, with a focus on burning early in the year to reduce fuel loads and create fire breaks (Edwards et al. 2003).

This transition from Aboriginal to European management is a possible driver of rainforest expansion through causing a reduction in the frequency of fire (Bowman et al. 1990; Bowman et al. 2001; Crowley and Garnett 1998), however this is uncertain given that in many areas of northern Australia fire frequencies remain very high. For example, Kakadu National Park has a detailed fire history based on interpretation of fire scars from satellite imagery which shows that an average of 46% of Kakadu National Park was burnt each year between 1980 and 1995 (Russell-Smith et al. 1997).

7.4.2 Feral Animal Management

The Asian water buffalo (*Bubalus bubalis*) was introduced into northern Australia from South-east Asia in the 1820s (Letts 1962). Large populations became established in the Kakadu region in the late 1800s. Buffalo were hunted for their hides from the 1880s until 1956 when the industry failed. Their population began to increase dramatically after this time. In the absence of reliable historical estimates of population size, it is not possible to determine precisely when the population reached peak levels (Skeat et al. 1996). However evidence from aerial photography and anecdotal evidence from local residents (e.g. David Lindner *pers. comm.*) suggests that peak levels were reached in the 1970s.

The Brucellosis and Tuberculosis Eradication Campaign (BTEC) commenced in the early 1980s which had a dramatic impact in reducing Buffalo numbers in northern Australia. For example, between 1983 and 1988 buffalo densities in Kakadu National Park reduced from 5.6 to 1.2 animals km^{-2}(Skeat et al. 1996). However higher numbers may have occurred in rainforest patches as buffalo preferentially used rainforest habitat. For example, Ridpath et al. (1983) estimated the density in forest vegetation along floodplain margins as 34 animals km^{-2}. Buffalo are still present in northern Australia, at low density. The populations of other feral animals such as pigs and horses were also dramatically reduced by the BTEC campaign.

Buffalo are known to have dramatic impacts on rainforests. A study in Kakadu National Park showed that rainforest patches that were more intensively used by buffalo had a lower density of vegetation <3 m during the dry season and a higher foliage height diversity because buffalo knocked down many young trees (Braithwaite et al. 1984). Soil compaction was also thought to have led to the death of large trees due to poor recharge of groundwater. Regional surveys of monsoon forest patches by Russell-Smith and Bowman (1992) also found that buffalo had caused widespread damage to rainforest boundaries.

The feral pig (*Sus scrofa*) can also cause damage to rainforest by upturning soil while foraging and by rubbing trees, wallowing and trampling. Regional surveys by Russell-Smith and Bowman (1992) indicated that extensive rooting activity was present in 10.3% of rainforests, with wet rainforests being particularly susceptible. In the short term, pigs do far less structural damage to rainforest than buffalo, however, pig disturbance could have a major impact in the long term as they limit recruitment of rainforest seedlings. The feral pig has been implicated in contributing to the decline of the palm *Ptychosperma bleeseri* in the Darwin area (Barrow et al. 1993).

7.4.3 Rainfall and CO_2

Climate is a primary factor determining the location of forest boundaries, and shifts in rainfall patterns can alter the balance between forests and more open vegetation types (Cramer et al. 2004; Vanacker et al. 2005). An increase in rainfall has occurred in the Northern Territory over the last century (Fig. 7.2b). Between 1910 and 1995 the total annual rainfall increased by 15–18% (Hennessy et al. 1999).

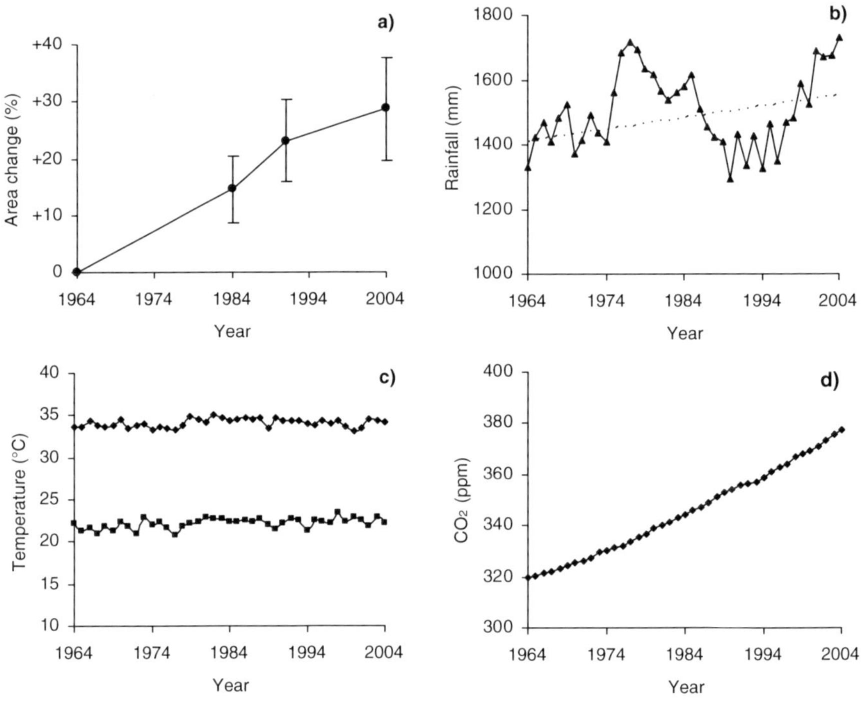

Fig. 7.2 Changes in area of rainforest in Kakadu National Park compared to trends of various climate variables (Banfai and Bowman 2007). Percentage change (± SE) in rainforest area over study period relative to 1964 (**a**). Five-year running average rainfall for Oenpelli, the closest rainfall station to Kakadu National Park with a complete record (Bureau of Meteorology 2003). The least squares regression line is also shown (**b**). Annual mean maximum and minimum temperatures recorded at Oenpelli (Bureau of Meteorology 2003) (**c**). Atmospheric CO_2 concentrations collected at Mauna Loa Observatory, Hawaii (Brook and Bowman 2006) (**d**)

The increasing trend has been considerably steeper for the second half of the 20th century (Smith 2004). Increased rainfall may have facilitated expansion of rainforest by increasing supply of water to tree roots through influencing catchment scale soil moisture patterns. Soil moisture has been shown to be important for establishment of rainforest seedlings in north Australian savannahs (Bowman and Panton 1993). In addition to the increase in rainfall, there was an increase of almost 20% in the number of rain days in the last 100 years (Hennessy et al. 1999), which may have extended the growing season for rainforest trees limited by water availability.

Similarly to rainfall, atmospheric CO_2 has shown a steady rise over the last few decades, increasing from 320 ppm in 1964 to 377 ppm in 2004 (Keeling and Whorf 2004, Fig. 7.2d). Increased levels of atmospheric CO_2 may facilitate the expansion of forest into more open vegetation types, as it favors the growth of trees (predominantly C_3 photosynthetic pathway) over grasses (predominantly C_4) (Berry and Roderick 2006; Bond et al. 2003).

Controlled experiments have consistently shown an increase in plant growth rates under elevated CO_2, known as the 'CO_2 fertilization effect'. For example, seedlings of *Maranthes corymbosa* Blume, a rainforest species that occurs in northern Australia, showed a marked increase in growth in a doubled CO_2 environment with total shoot dry weight increasing by 163% (Berryman et al. 1993). Faster growth rates may therefore have allowed rainforest trees to more readily escape the 'fire trap' posed by regular fires (Bond et al. 2003). Increased atmospheric CO_2 also increases the water use efficiency of trees, which may have allowed rainforest to establish in areas that were previously water limited.

However, increases in atmospheric CO_2 may not have contributed substantially to the expansion of rainforest where other factors are limiting. Such factors may include mycorrhizas and soil fertility, which have been shown to be important determinants of rainforest seedling establishment into savannahs (Bowman and Panton 1993). Considerable uncertainty remains as to the effect of elevated CO_2 on vegetation change in Australia (Hughes 2003).

7.5 Case Study of Kakadu National Park

The biological and cultural importance of rainforest, and the large variety of processes that may be threatening the systems, means that research is urgently needed to assess the consequences of management of rainforest boundaries, and the relative importance of regional scale factors such as climate changes. These issues have been investigated through a number of studies conducted in Kakadu National Park (Banfai and Bowman 2006, 2007; Banfai et al. 2007). Together these studies provide a cohesive methodology which can be applied to other regions and forest types. The methodology applied to Kakadu National Park is summarised below and the core results obtained in relation to each step of the methodology are then presented.

7.5.1 Methodology

Step 1: Extent of Change

Assessment of the extent of change of rainforest boundaries is the critical first step in assessing the consequences of management, and provides an invaluable historical context of vegetation change. Monsoon rainforest exists in Kakadu National Park as an archipelago of mostly small (less than 5 ha) patches within savannah matrix. The extent of change of these rainforest patches was assessed using aerial photography. High resolution satellite imagery could also have been used, although the time period to assess change would have been greatly reduced given that comparable spatial data has only been available since the 1990s. Previous mapping of rainforest patches was used to select 50 patches for analysis across the full geographic range of Kakadu National Park. It was ensured that both wet and dry rainforest types were sampled adequately, and a variety of management regimes were represented. Aerial photography was sourced for each rainforest patch as contact prints. The years chosen for analysis were those that had a complete photographic coverage of Kakadu National Park within two years at a scale suitable for the analysis; 1964/5, 1983/4, 1991 and 2004.

The 1991 photographs were georectified to 1:50,000 topographic maps. All other photographs were registered to the 1991 images using image-to-image registration. A 20 m × 20 m point lattice with fixed geo-coordinates was overlaid on each aerial photograph. The extent of the dot grid was proportional to the size of the closed rainforest patch allowing for a 100 m buffer around the edge. Each dot grid point was manually classified as rainforest or non-rainforest for each year at a common scale of 1:3000 with reference to the area within 10 m radius of the point. The rule-based dot grid method to classify vegetation was used to minimize the error associated with defining boundaries across ecotones. Transition matrices, size class distributions and fragmentation indices were calculated. Field samples were also taken of a subset of 30 rainforest patches to assess the accuracy of the boundary mapping.

A set of plausible hypotheses (reviewed in the previous section) for the causes of the observe changes was then developed based on existing literature, and extensive consultation with Park Rangers and Traditional Owners. A preliminary assessment was conducted of each of these hypotheses based on the observed extent of change.

Step 2: Field Assessment of Causes of Change

One approach which has been shown to help narrow down the hypotheses for the causes of vegetation dynamics is to assess the ecological 'fingerprint' of the changes based on their field attributes. In Kakadu National Park floristic, structural, environmental and disturbance attributes were investigated by sampling 588 plots across 30 rainforest patches. Field survey plots were 20 × 20 m and were centred on grid points used to classify the vegetation. Grid points which had different histories of change were sampled to compare the attributes of these areas.

Data analysis was conducted on wet and dry rainforest types separately to allow comparison. The proportion of rainforest species in plots with different histories of change was compared. Other variables investigated were grass cover, flammable weed cover, pig impact, buffalo impact, time since fire, soil type and slope. All variables were contrasted between plots with different histories of change using ANOVA analyses. Ordination analyses were conducted to compare the floristics of newly established rainforest with longer established rainforest. Generalised linear models were also used to predict the vegetation changes based on disturbance and environmental variables.

Step 3: Modelling of Causes of Change

Model selection techniques, such as those based on information theory, provide a robust methodology to investigate the relative importance of factors affecting the rate of rainforest boundary change. Various hypotheses can be represented by models, and the relative strength of evidence of each model can be evaluated. For Kakadu National Park linear and mixed effects models were constructed to assess the role of fire, buffalo impact and patch characteristics in determining the rate of boundary change. The analysis was conducted at both the patch scale, and within-patch (plot) scale, to capture the different processes operating at different spatial scales. They examined: (i) what determines the variation in the rate of change in total patch size? and (ii) what determines the probability of change for areas on the boundary within a patch?

Table 7.1 Hypotheses for the causes of boundary dynamics for patch scale modelling analyses, with the corresponding variables included in the generalised linear mixed effects models. Patch was included as a random effect in all models (Banfai et al. 2007)

Hypothesis	Model
Direct effect of fire on tree mortality	Fire frequency
Rainforest type is primary mediating factor due to water source and/or landscape setting	Rainforest type
Buffalo grazing and trampling limits tree recruitment and causes direct mortality	Buffalo impact
Landscape setting mediates changes due to topographic fire protection	Rainforest type X fire frequency
Buffalo impact varies between rainforest types due to hydrology and landscape setting affecting buffalo density and habitat preferences	Rainforest type X buffalo impact
Smaller patches more vulnerable to fire due to higher edge/core ratio	Patch size X fire frequency
More fragmented patches more vulnerable to fire due to higher edge/core ratio	Fragmentation X fire frequency
Smaller patches more vulnerable to buffalo impact due to higher edge/core ratio	Patch size X buffalo impact
More fragmented patches more vulnerable to buffalo impact due to higher edge/core ratio	Fragmentation X buffalo impact
Buffalo interacts with the impact of fire due to altering fuel loads from browsing and transport of flammable weeds	Fire frequency X buffalo impact
Variation in rate of boundary change is due to factors other than those captured by the models	Null, with random effect for patch

The rate of change of the patch boundaries was based on the mapping from Step 1. The model sets were developed *a priori* based on hypotheses arising from the previous steps in the analysis. Interactions between variables were included in the models where these interactions were biologically defensible. The effect of spatial autocorrelation was minimised in the plot scale analyses by sub-sampling lattice points. Model variables in the plot scale analyses were (i) the distance from the rainforest boundary, (ii) the number of rainforest neighbours and (iii) the aspect of the rainforest boundary. The candidate set of models for the patch scale analysis is provided in Table 7.1. The relative strength of evidence for each of the models and the amount of deviance explained was then assessed. Relationships between variables were also explored graphically to identify patterns.

Step 4: Synthesis
The inferences gained from each of the previous steps were then synthesised to assess the relative support for each of the hypotheses. Where the results were not consistent, the strengths and weaknesses of each of the approaches was considered. Regional scale factors such as trends in rainfall and atmospheric CO_2 were compared to the observed rates of change in rainforest boundaries, and the strength of these factors in driving the observe changes was assessed relative to that of other factors such as fire management. Interactions between factors were also carefully considered.

7.5.2 Results

7.5.2.1 Extent of Change

The comparison of aerial photographs between 1964 and 2004 revealed an overall expansion of rainforest patches by an average of 28.8% (Fig. 7.2). Expansion was observed in both wet and dry rainforest types, although localised contraction of rainforest boundaries also occurred (Fig. 7.3). The changes observed in these studies parallel rainforest expansion observed on the east coast of Australia (Harrington and Sanderson 1994; Russell-Smith et al. 2004) as well as global trends in vegetation thickening in savannah landscapes (Archer et al. 1995; Lewis et al. 2004).

Assessment of the extent of change in the rainforest boundaries also provided insight into the causes of the observed changes and the consequences of human management. The effect of fire was supported by the large differences in the rate of boundary change between rainforest patches that were situated nearby geographically. Case studies of individual rainforest patches also supported an effect of fire: The dry rainforest site that experienced one of the greatest reductions in percentage area (–22% from 1964 to 2004) is claimed by Park staff to have received regular late dry season fires. This patch was also one of the only patches to have a large abundance of the flammable exotic weed *Pennisetum polystachion* (L.) Schult.

The three sites situated on coastal islands, locations largely protected from fire, all experienced an increase in rainforest area. Additionally, the two rainforest

a)

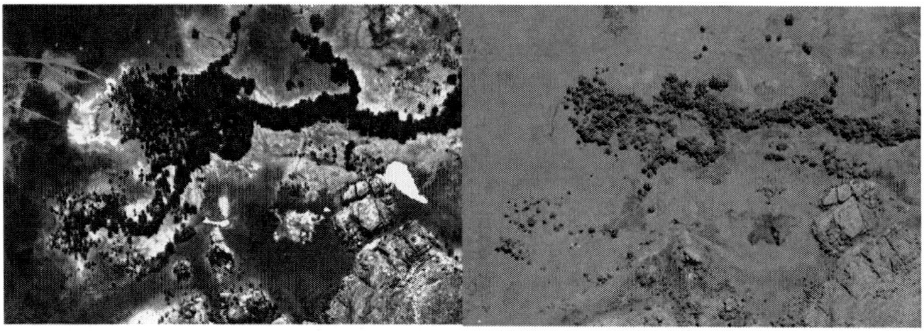

b)

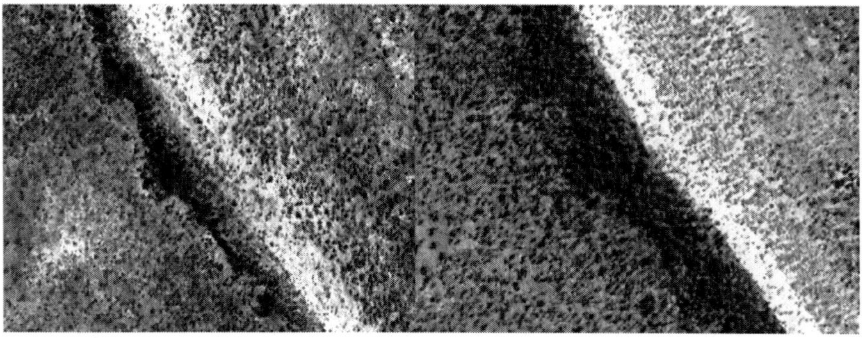

Fig. 7.3 An example of rainforest boundary contraction (**a**) and expansion (**b**) of a patch in Kakadu National Park between 1964 and 2004

patches that had the highest rates of expansion (+352% and +256% from 1964 to 2004) have experienced a substantial reduction in the frequency of fire following a lessening in the intensity of Aboriginal management. Both these sites were claimed to be of great cultural significance by traditional owners of the area as they are adjacent to historical Aboriginal camps and were important sites for traditional food resources (Lucas et al. 1997).

7.5.2.2 Field Assessment of Causes of Change

The trend of rainforest expansion in Kakadu National Park was also supported by the field surveys. These found that those areas in which the aerial photography indicated that rainforest expansion had occurred were associated with a significantly higher abundance of rainforest trees and less grasses, relative to stable savannah areas. Additionally, the field surveys suggested that the floristic composition was similar between newly established rainforest and longer established rainforest, supporting the view that the rainforest boundaries had been highly dynamic at a decadal scale. The changes were found to be not strongly related to disturbance variables recorded in the field. However, the contemporary field surveys were unable to capture the historical impacts of these factors.

7.5.2.3 Boundary Modelling of Causes of Change

At the broad patch scale, the rate of change was best explained by rainforest type and historical buffalo impact, although a substantial amount of deviance remained unexplained. An example of the modelling results for the period 1984–2004 is provided in Table 7.2. This shows the model containing an interaction between rainforest type and buffalo impact was the best selected model for this period, explaining 17% of the deviance. Fire frequency, patch size and fragmentation were not important predictors of the rate of change. The lack of support for an effect of fire frequency was inconsistent with the case studies of individual patches based on the extent of change (Step 1). However, the limited explanatory power of fire in this study may have been due to a mismatch between the temporal scales of the frequent fires and the 20 year photographic intervals used.

At the plot scale, distance from rainforest edge was the most important predictor of the probability of change, while fragmentation and aspect of the boundary were unimportant. Rainforest expansion has occurred through a process of margin extension rather than eruption of new patches.

Table 7.2 Fire: results of AIC analyses of generalised linear mixed effects models for period 1984–2004, with fire variable included. RT = Rainforest type; BI = Buffalo impact; FG = Fragmentation; FI = Fire frequency; PS = patch size (Banfai et al. 2007)

Model	log(L)	K	AICc	Δ AICc	w_i	Deviance (%)
RT X BI	−120.022	4	249.019	0.000	0.401	17.041
Null	−124.319	1	250.729	1.709	0.171	–
FG X BI	−120.966	4	250.908	1.889	0.156	13.564
RT	−124.134	2	252.548	3.529	0.069	0.799
BI	−124.211	2	252.702	3.683	0.064	0.466
FI	−124.276	2	252.831	3.812	0.060	0.186
FG X FI	−122.410	4	253.796	4.777	0.037	7.963
RT X FI	−123.457	4	255.890	6.871	0.013	3.676
PS X BI	−123.503	4	255.981	6.961	0.012	3.487
PS X FI	−123.609	4	256.193	7.173	0.011	3.041
FI X BI	−124.030	4	257.035	8.016	0.007	1.249

7.5.2.4 Synthesis

The modelling results provided moderate support for an effect of buffalo impact. Case studies of the extent of boundary dynamics at individual rainforrest patches also supported an effect of fire managment. Buffalo and fire managment appear to have mediated the observed boundary dynamics at the decadal scale. However, they are unable to account for the overall expansion of rainforest that has occurred in Kakadu National Park. The expansion of rainforest boundaries continued throughout the period of peak buffalo numbers around the 1970s when their impact on rainforest boundaries is likely to have been greatest. A reduction in fire frequency is also unlikely to explain the observed rainforest expansion, as fires continue to occur at very high frequencies and intensities at the rainforest boundaries.

The overall expansion of rainforest is therefore likely to have been primarily driven by factors that promote rainforest expansion at a regional scale. Prime candidates for this are increases in rainfall, atmospheric CO_2 or both. These factors have shown increasing trends over the study period (Fig. 7.2), and would have promoted the establishment of rainforest species in the surrounding savannah. Interactions between fire and these factors may have been critical. For example, the increased growth rates caused by enhanced atmospheric CO_2 may have allowed the newly established rainforest trees to escape the 'fire trap' posed by regular fires.

Fire management will be an important determinant of the extent and direction of future changes in rainforest boundaries in northern Australia. A continuation of the current fire regime is likely to allow the rainforest patches to continue to expand. Fire frequency in itself is not an important determinant of the rate of boundary dynamics, even if these fires occur in the late dry season. However a shift in the fire regime to regular late dry season fires will result in contraction of rainforest where high fuel loads are available on the boundary, such as when exotic grasses become established. Thus flammable weed management of introduced grass species at a regional scale remains a priority for land managers.

The rainforest boundary dynamics have also been mediated by buffalo management as buffalo have acted to limit the expansion of rainforest and appear to have caused boundary contraction in extreme cases. The impact of buffalo on rainforest boundaries is likely to be low in the contemporary environment, as the numbers of buffalo in northern Australia was dramatically reduced between 1980 and 1991 during the Brucellosis and Tuberculosis Eradication Campaign (Skeat et al. 1996). However control of buffalo numbers remains a priority for land managers, as high numbers are still present in some areas (Robinson and Whitehead 2003).

The successful application of 'adaptive management' programs relies on the attainment of new information about the system being managed which can help to guide management actions. In the case of the rainforest patches in Kakadu National Park, this new information produced a variety of changes to the contemporary management of these systems. For example, identification of patches where rapid contraction of rainforest boundaries had occurred resulted in additional management effort being applied to these areas to help stop the rainforest patches from being completely eliminated. This was usually focused on reducing the frequency of high intensity fires at the rainforest boundary, through control of flammable grasses. Management of rainforest patches was also impacted by the raised awareness among land managers about the sensitivity of these systems to management actions, and the interactions between the various drivers of change.

7.5.3 Avenues for Further Research

The network of aerial photographic and survey plots established in studies of rainforest dynamics in Kakadu National Park provides a baseline for future monitoring of rainforest. Given the biological and cultural importance of these systems and their susceptibility to rapid rates of change, an adaptive management approach is needed based on continued monitoring and evaluation of the impact of factors such as fire. Linking aerial photography to high resolution satellite imagery (Salami et al. 1999) may be a more economically viable way for this monitoring to continue in the future.

The analysis of the drivers of the boundary dynamics in northern Australia has been limited by a lack of detailed historical information on disturbance regimes. Useful historical data would include field records of fire and buffalo impacts on the rainforest boundary. Future collection of this data is critical to increasing the statistical power of future analyses, and would allow for the accurate prediction of the boundary dynamics under various potential management regimes. Small autonomous aircraft (drones) may become powerful method to monitor specific locations and assess fire and feral animal damage. An alternative approach to investigate the effect of fire would be to simulate fire occurrence on the rainforest boundary using cellular automata modelling (Favier et al. 2004). This would allow various hypotheses regarding the potential impact of various future fire frequencies and intensities to be investigated.

7.6 Conclusions

The rainforest boundaries in northern Australia have been highly dynamic over the last few decades. Evidence from aerial photography has suggested that a rapid overall expansion of rainforest has occurred, however localised boundary contraction has also been common. The case study of Kakadu National Park has provided an example of a successful landscape-scale methodology for assessing the extent and causes of changes to rainforest boundaries, and thus the consequences of human management. In Kakadu National Park the rainforest boundaries have also been strongly influenced by global change phenomena such as increases in rainfall and atmospheric carbon dioxide. It has been shown that at a regional scale these factors may be overwhelming the impacts of human management of rainforest boundaries. Nonetheless, at a local scale, the integrity of rainforest boundaries is still largely dependent on human management, such as control of high intensity fires. The structural and floristic similarity of recently established rainforest to older more established rainforest, and the lack of clear environmental limits to change, suggest that the current trend of rainforest expansion will continue if climatic trends and disturbance regimes persist.

However, reaching any firm conclusions as to causality is made difficult by (i) the multiple simultaneous shifts in environmental variables and disturbance regimes, (ii) the many complex interactions between the possible drivers of change, and (iii) the lack of detailed historical information. Nonetheless, the methodology outlined has been shown to provide information which is useful for adaptive management programs.

This study adds to the emerging evidence for an effect of global change phenomena on tropical forests and the importance of human management (Lewis et al. 2004), however, substantial uncertainty remains as to the extent and dynamics of tropical forests both within northern Australia and globally (Lewis 2006). The methodology outlined in this chapter could be applied to forests in other regions to help assess, on a global scale, the consequences of human management of forests. The historical dynamics of rainforest boundaries observed in northern Australia and the potential for future changes driven by shifts in global change phenomena, reinforces the view that stability cannot be assumed and conservation values cannot be maintained passively in these environments (Edwards et al. 2003).

Acknowledgments This work was funded by an Australian Research Council Linkage Grant (No. LP0346929) in conjunction with Kakadu National Park. We thank the Traditional Owners and staff of Kakadu National Park for their input and support. Thanks also to Rebecca Mitchell for her assistance with the preparation of the manuscript.

References

Archer S, Schimel DS, Holland EA (1995) Mechanisms of shrubland expansion: landuse, climate or CO_2? Clim Chang 29:91–99

Banfai DS, Bowman DMJS (2006) Forty years of lowland monsoon rainforest expansion in Kakadu National Park, northern Australia. Biol Conserv 131:553–565

Banfai DS, Bowman DMJS (2007) Drivers of rain-forest boundary dynamics in Kakadu National Park, northern Australia: a field assessment. J Trop Ecol 23:73–86

Banfai DS, Brook BW, Bowman DMJS (2007) Multi-scale modelling of the drivers of rainforest boundary dynamics in Kakadu National Park, northern Australia. Divers Distrib 13:680–691

Barrow P, Duff G, Liddle D et al (1993) Threats to monsoon rainforest habitat in Northern Australia: the case of Ptychosperma bleeseri Burret (Arecaceae). Aust J Ecol 18:463–471

Berry SL, Roderick ML (2006) Changing Australian vegetation from 1788 to 1988: effects of CO2 and land-use change. Aust J Bot 54:325–338

Berryman CA, Eamus D, Duff GA (1993) The influence of CO_2 enrichment on growth, nutrient content and biomass allocation of *Maranthes Corymbosa*. Aust J Bot 41:195–209

Bond WJ, Midgley GF, Woodward FI (2003) The importance of low atmospheric CO_2 and fire in promoting the spread of grasslands and savannahs. Global Chang Biol 9:973–982

Bowman DMJS (1998) Tansley review No. 101 – The impact of Aboriginal landscape burning on the Australian biota. New Phytol 140:385–410

Bowman DMJS (2000) Australian rainforests: Islands of green in the land of fire. Cambridge University Press, Cambridge

Bowman DMJS, Dingle JK (2006) Late 20th-century landscape-wide expansion of Allosyncarpia ternata (Myrtaceae) forests in Kakadu National Park, northern Australia. Aust J Bot 54:1–9

Bowman DMJS, Dingle JK, Johnston FH, Parry D, Foley M (2007) Seasonal patterns in biomass smoke pollution and the mid 20th century transition from Aboriginal to European fire management in northern Australia. Global Ecol Biogeogr 16:246–256

Bowman DMJS, Panton WJ (1993) Factors that control monsoon-rainforest seedling establishment and growth in north Australian *Eucalyptus* savannah. J Ecol 81:297–304

Bowman DMJS, Panton WJ, McDonough L (1990) Dynamics of forest clumps on Chenier Plains, Cobourg Peninsula, Northern Territory. Aust J Bot 38:593–601

Bowman DMJS, Walsh A, Milne DJ (2001) Forest expansion and grassland contraction within a *Eucalyptus* savannah matrix between 1941 and 1994 at Litchfield National Park in the Australian monsoon tropics. Global Ecol Biogeogr 10:535–548

Bowman DMJS, Walsh A, Prior LD (2004) Landscape analysis of Aboriginal fire management in Central Arnhem Land, north Australia. J Biogeogr 31:207–223

Bowman DMJS, Wilson BA, McDonough L (1991) Monsoon forests in North-western Australia. I. vegetation classification and the environmental-control of tree species. J Biogeogr 18: 679–686

Braithwaite RW, Dudzinski ML, Ridpath MG et al (1984) The impact of water buffalo on the monsoon forest ecosystem in Kakadu National Park. Aust J Ecol 9:309–322

Brook BW, Bowman DMJS (2006) Postcards from the past: charting the landscape-scale conversion of tropical Australian savannah to closed forest during the 20th century. Landsc Ecol 21

Bureau of Meteorology (2003) Climate data: Australia. Climate Services, Kent Town, SA

Cramer W, Bondeau A, Schaphoff S et al (2004) Tropical forests and the global carbon cycle: impacts of atmospheric carbon dioxide, climate change and rate of deforestation. Philos Trans R Soc London, Ser B 359:331–343

Crowley GM, Garnett ST (1998) Vegetation change in the grasslands of east-central Cape York Peninsula, Australia. Pac Conserv Biol 4:132–148

Edwards A, Kennett R, Price O et al (2003) Monitoring the impacts of fire regimes on vegetation in northern Australia: an example from Kakadu National Park. Int J Wildland Fire 12:427–440

Favier C, Chave J, Fabing A et al (2004) Modelling forest-savannah mosaic dynamics in man-influenced environments: effects of fire, climate and soil heterogeneity. Ecol Model 171:85–102

Furley P, Proctor J, Ratter J eds (1992) Nature and dynamics of the Forest-Savannah boundaries. Chapman and Hall, London

Harrington GN, Sanderson KD (1994) Recent contraction of wet sclerophyll forest in the wet tropics of Queensland due to invasion by rainforest. Pac Conserv Biol 1:319–327

Hennessy KJ, Suppiah R, Page CM (1999) Australian rainfall changes, 1910–1995. Aust Meteorol Mag 48:1–13

Hopkins B (1992) Ecological processes at the forest-savannah boundary. In: Furley P, Proctor J and Ratter J (eds) Nature and dynamics of the Forest-Savannah boundaries, Chapman and Hall, London, pp 21–33

Hughes L (2003) Climate change and Australia: trends, projections and impacts. Aust Ecol 28: 423–443

Keeling CD, Whorf TP (2004) Atmospheric CO_2 records from sites in the SIO air sampling network. In: Trends: a compendium of data on global change. Carbon Dioxide Information Analysis Center, U.S. Department of Energy, Oak Ridge

King J, Moutsinga JB, Doufoulon G (1997) Conversion of anthropogenic savannah to production forest through fire-protection of the forest-savannah edge in Gabon, Central Africa. Forest Ecol Manag 94:233–247

Letts G (1962) Feral animals in the Northern Territory. Aust Vet J 40:84–88

Lewis SL (2006) Tropical forests and the changing earth system. Philos Trans R Soc London, Ser B 361:195–210

Lewis SL, Malhi Y, Phillips OL (2004) Fingerprinting the impacts of global change on tropical forests. Philos Trans R Soc London, Ser B 359:437–462

Lucas D, Gapindi M, Russell-Smith J (1997) Cultural perspectives of the South Alligator River floodplain: continuity and change. In: Rose D and Clarke A (eds) Tracking knowledge in North Australian landscapes, North Australian Research Unit, Australian National University, Darwin, pp 120–135

Malhi Y, Grace J (2000) Tropical forests and atmospheric carbon dioxide. Trends Ecol Evol 15:332–337

Preece N (2002) Aboriginal fires in monsoonal Australia from historical accounts. J Biogeogr 29:321–336

Press T, Lawrence D (1995) Kakadu National Park: reconciling competing interests. In: Press T, Lea D, Webb A and Graham A (eds) Kakadu: natural and cultural heritage and management, Australian Nature Conservation Agency, Darwin

Price OF, Woinarski JCZ, Robinson D (1999) Very large area requirements for frugivorous birds in monsoon rainforests of the Northern Territory, Australia. Biol Conserv 91:169–180

Ridpath MG, Begg R, Dudzinski ML et al (1983) Counting the same populations of large tropical mammals from the ground and from the air. Aust Wildl Reserves 10:487–498

Robinson CJ, Whitehead P (2003) Cross-cultural management of pest animal damage: a case study of feral Buffalo control in Australia's Kakadu National Park. Environ Manage 32:445–458

Russell-Smith J (1991) Classification, species richness, and environmental relations of monsoon rain forest in northern Australia. J Veg Sci 2:259–278

Russell-Smith J (2001) Pre-contact Aboriginal, and contemporary fire regimes of the savannah landscapes of Northern Australia: patterns, changes and ecological responses. Ngoonjook 20

Russell-Smith J, Bowman DMJS (1992) Conservation of monsoon rainforest isolates in the Northern Territory, Australia. Biol Conserv 59:51–63

Russell-Smith J, Dunlop C (1987) The status of monsoon vine forests in the Northern Territory: a perspective. In: World Wildlife Fund (ed) The rainforest legacy: Australian National Rainforest Study, Australian Government Public Service, Canberra, pp 227–288

Russell-Smith J, Ryan PG, Durieu R (1997) A LANDSAT MSS-derived fire history of Kakadu National park, monsoonal northern Australia, 1980–94: seasonal extent, frequency and patchiness. J Appl Ecol 34:748–766

Russell-Smith J, Stanton PJ, Whitehead PJ et al (2004) Rain forest invasion of eucalypt-dominated woodland savannah, iron range, north-eastern Australia: I. Successional processes. J Biogeogr 31:1293–1303

Salami AT, Ekanade O, Oyinloye RO (1999) Detection of forest reserve incursion in south-western Nigeria from a combination of multi-date aerial photographs and high resolution satellite imagery. Int J Remote Sens 20:1487–1497

Skeat AJ, East TJ, Corbett LK (1996) Impact of feral water buffalo. In: Finlayson CM and von Oertzen I (eds) Landscape and vegetation ecology of the Kakadu Region, Northern Australia, Kluwer Academic Publishers, Netherlands, pp 155–177

Smith I (2004) An assessment of recent trends in Australian rainfall. Aust Meteorol Mag 53: 163–173

Swaine MD, Hawthorne WD, Orgle TK (1992) The effects of fire exclusion on Savannah Vegetation at Kpong, Ghana. Biotropica 24:166–172

Truswell E (1990) Australian rainforests: the 100 million year record. In: Webb L and Kikkawa J (eds) Australian tropical rainforests: science – values – meaning, CSIRO, Australia

Vanacker V, Linderman M, Lupo F et al (2005) Impact of short-term rainfall fluctuation on interannual land cover change in sub-Saharan Africa. Global Ecol Biogeogr 14:123–135

Webb L, Tracy J (1981) Australian rainforests: patterns and change. In: Keast A (ed) Ecological biogeography of Australia, Dr W Junk, The Hague

Woinarski JCZ, Milne DJ, Wanganeen G (2001) Changes in mammal populations in relatively intact landscapes of Kakadu National Park, Northern Territory, Australia. Aust Ecol 26: 360–370

Chapter 8

Spatial Patterns and Ecology of Shifting Forest Landscapes in Garo Hills, India*

Ashish Kumar, Bruce G. Marcot and P.S. Roy

Abstract In many parts of the world, increasing rates of shifting cultivation – also called slash-and-burn cultivation, swidden, and (in India) *jhum* – has compromised native forest biodiversity. We explore this relationship with a case study from North East India where much of the remaining, intact, old tropical forest is found in the few protected areas and reserved forests (collectively PAs) of the region, and where *jhum* has largely permeated much of the rest of the landscape. Our analysis and mapping of land use and cover types, levels of forest fragmentation, and occurrence of *jhum* lands suggests that: buffer zones around PAs could contain additional, intact forest; incursion into PAs can reduce their effective interior core forest area; and forest wildlife habitat, particularly for Asian elephant, can be delineated among PAs in corridors consisting of low-fragmented, native forest cover. As human population density and concomitant anthropogenic stressors increase, however, more severe effects of increased rates of *jhum* on forest biodiversity will be felt. Offsetting such effects will entail not just redirecting *jhum* activities but also addressing the full cultural, social, economic, and even religious context in which shifting cultivation is pursued. Solutions must consider effects on nutrition, health, education, economic trade, and traditional lifestyles.

8.1 Introduction

The history of land use and agriculture in the forest landscapes of the greater Indian subcontinent and south Asia is strikingly diverse. Over the centuries, multiple overlapping cultures have occupied and tilled the forests of sal (*Shorea robusta*), teak (*Tectona grandis*), and hundreds of other evergreen, semi-evergreen, and deciduous tree species. In North East India, the dense tropical forests of the Himalayan hill

A. Kumar
Gujarat Institute of Desert Ecology, Bhuj, India
e-mail: ashishkumarphd@gmail.com

R. Lafortezza et al. (eds.), *Patterns and Processes in Forest Landscapes*,

regions and the vast river plains of the Brahmaputra and Ganges in the areas of Bengal, Assam, and adjacent regions were originally viewed as major obstacles to the expansion of rice paddies and other agricultural cultivation. Over time, the land was tamed and soil seen as having immense fertility. Today, however, it is human population density and its toll on soils and native forests of the region that have become an impediment to prosperity (Ludden 1999). Increasing rates of shifting cultivation have led to increased fragmentation of intact, native forests, and the implications of such changes in forest landscape patterns on native biota, particularly Asian elephants (*Elephas maximus*).

The hill country of this region has always had distinctive, indigenous tribal farming societies dating back at least six millennia. These societies have had complex relations with the lowland agrarians and with the 19th century British settlers. Today, the sustainability of these societies and their forest resources are facing the greatest challenges as markets for forest products and other natural resources are being more fully opened to pressures and demands of the outside world.

8.1.1 The Role of Shifting Cultivation in Cultural and Landscape Ecology

Jhum – also known elsewhere as shifting cultivation, swidden, and other terms – is a primitive but sometimes complicated form of forest agriculture practiced mostly in tropical countries world-wide, for example in Honduras (House 1997), Indonesia (Sunderlin 1997), Brazil (Metzger 2002), and Mexico (Pulido and Caballero 2006) as well as India (Momin 1995; Sachchidananda 1989). Under low human population density, fallow periods can extent to 20–30 years or more and much of the forest landscape can escape the slash and burn cycle in any given year, and thus there is minor influence on overall forest biodiversity and soil productivity. But high human density in many tropical areas of the world, including North East India, has caused fallow periods to drop to just a few years and large portions of forest landscapes to be converted, resulting in major losses of old, native forest cover, soil fertility, biodiversity, and crop health (e.g., Raman 2001).

From 1980 to 1990, $> 6\%$ of worldwide tropical forests and 10% of Asian tropical forests were converted to shifting cultivation (WRI 1996). As per the 1979 report of the North Eastern Council, in the Indian state of Meghalaya a total of 4116 km^2 was placed under *jhumming*, of which 760 $km^2$was used at one point of time every year by 68000 *jhummias*, i.e., families involved in *jhumming* (DSWC 1995).

Such forest perturbations in western Meghalaya included the Garo Hills, which are one of the richest botanical regions of India (Awasthi 1999). The Garo Hills represent the remnant of an ancient plateau of the pre-Cambrian peninsular shield (Momin 1984) and are prominently inhabited by the native Garo tribes. The major stressor to native forest biodiversity in the Garo Hills is the increasing rate of anthropogenic conversion of mature and primary forests to *jhum* land. Apart from *jhum,* other major land uses are the habitation and practice of permanent agriculture in valley plains.

8.1.2 Shifting Cultivation in Garo Hills, Meghalaya

Our study focused on the South Garo Hills district, which includes much community *jhum* land as well as several protected areas, notably Balpakram National Park and the adjoining Nokrek Biosphere Reserve and National Park (Fig. 8.1). Our study area (hereafter, also "landscape") represents the western-most hill ranges of Meghalaya state in North East India. The landscape contains four protected areas (PAs) and four reserved forests (RFs) (Fig. 8.1) which collectively comprise 15% of the area and which offer excellent prospects of conserving native forest and the associated biodiversity of the region. The PAs include Balpakram National Park (BNP; 220 km^2), Nokrek National Park (NNP; 47.48 km^2), Siju Wildlife Sanctuary (SWS; 5.18 sq km^2), and Baghmara Pitcher Plant Sanctuary (BPPS; 2.7 ha). The four RFs are Baghamara Reserved Forest (BRF; 44.29 km^2), Rewak Reserved Forest (RRF; 6.48 km^2), Emangiri Reserved Forest (ERF; 8.29 km^2), and Angratoli Reserved Forest (ARF; 30.11 km^2). Like the PAs, these RFs have been considered by Kumar et al. (2002) as elements of a Protected Area Network (PAN) because forests in the RFs are not being actively harvested and are not occupied by native Garo communities. Little work has been done to evaluate the landscape, PAN, and plant and animal communities of the Garo Hills except for a few studies (Kumar and Rao 1985; Haridasan and Rao 1985; Khan et al. 1997; Sudhakar and Singh 1993; Kumar and Singh 1997; Roy and Tomar 2001; Talukdar 2004).

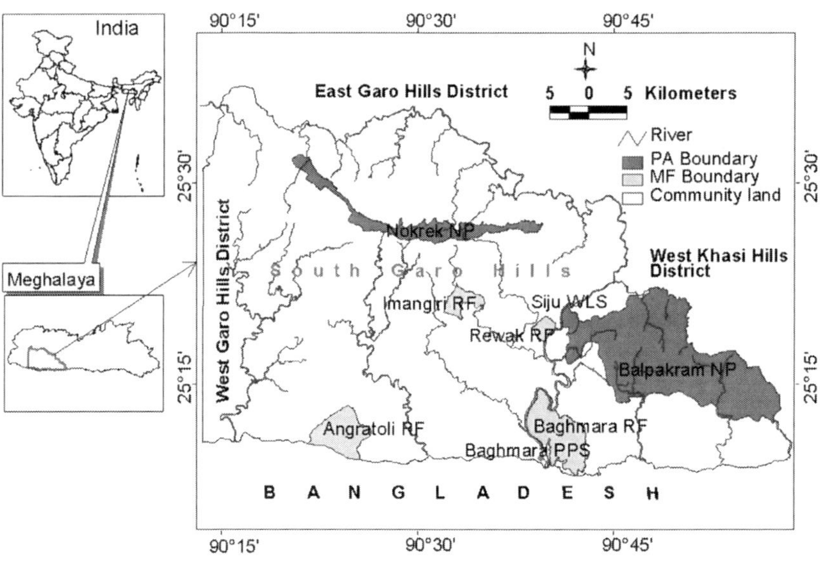

Fig. 8.1 Study area of South Garo Hills and geographical location within India and Meghalaya PA = Protected Areas, NP = National Park, WLS = Wildlife Sanctuary, MF = Managed Forests, RF = Reserved Forests and PPS = Pitcher Plant Sanctuary

In general, human activities can influence patterns and processes of forests in diverse ways. In many parts of the world, for want of food and shelter, agrarian societies have impinged upon the distribution and amount of native forests across landscapes. In this chapter, we address a particular kind of rural agricultural activity – shifting ("slash and burn") cultivation or *jhum* – and its influence on conservation of native forest biodiversity at the landscape scale, using our research in India as a case study. We offer some guidelines for conservation or restoration of native forest biodiversity. Most North East Indian forests are under tremendous pressure of exploitation from unplanned traditional forestry practices, especially the widespread use of *jhum*. *Jhum* entails native people clearing and burning the old forest growth over a piece of land to get fertile land for raising agricultural crops. A given plot of land is used for crops typically for only one or two years, and then it is left fallow for several years before being cleared and used again. In this paper we explore the effects of *jhum* on forest diversity and conditions in Garo Hills as a case study. We generalize results as lessons to learn for other tropical forests of the world undergoing accelerated shifting cultivation with associated loss of old, native forests and their attendant biodiversity. Our present study examines the spatial patterns and processes of the *jhum*-influenced landscape to identify and prioritise the wildlife habitat areas for conserving native biodiversity.

8.2 Methods

We first prepared a base map of the study area showing boundaries and locations of all PAs and RFs at 1:50 000 scale with use of Survey of India topographic data and other maps available from the State Forest Department of Meghalaya (SFDM). We also prepared a land use and land cover (LULC) map using remotely sensed satellite data (IRS-1D LISS III, 23.5 m resolution) of February 1999. "Land use" reflects categories of human activities including industrial, residential, agricultural, and other uses. "Land cover" refers to 9 categories of vegetation: active *jhum* (0 to approximately 3 years old) and grassland; scrub and abandoned *jhum* (3–6 years old) on degraded sites; bamboo brakes and secondary forest (6–10 years old); deciduous forest; semi-evergreen forest (approximately 15–30+ years old); evergreen forest; permanent agriculture; water bodies; and shadows. LULC mapping was done using guided classification of satellite data at a 1: 50 000 scale, with additional attributes of old-forest cover (viz., "dense" and "open" forest conditions) being mapped at 1: 250 000 (FSI 2001). Details of methods were presented in Kumar et al. (2000). We could not differentiate permanent agricultural fields from sandy river banks in valley plains because they have similar spectral characteristics.

Land use and land cover categories were identified using unsupervised and supervised classification techniques. The unsupervised classification consisted of a remote sensing image with 50 distinct spectral classes, which more or less represented the natural features of the landscape. This image was taken to the field and verified on 80 ground control points. Half of the ground control points easily identifiable on the image were used as training areas to perform the supervised classification, while the remainder of the points was used to evaluate the classification accuracy of the land use and land cover categories. The training areas provided a numerical description of the spatial attributes of each class and the basis for merging similar spectral classes into the more meaningful and identifiable land use and land cover classes.

We then computed selected patch indices from the LULC map in a geographic information system (GIS). We computed and mapped various indices of forest landscape pattern for each LULC category, and mapped forest patch core areas at two distances of 250 m and 500 m from the forest patch edge, by using Bio_CAP, a GIS-based programme developed by the Indian Institute of Remote Sensing, Dehradun. The landscape pattern indices included average, minimum, and maximum map polygon (patch) area; and indices of terrain complexity (topographic relief variation), patchiness, porosity, interspersion, fragmentation, polygon (patch) edge length, and disturbance (see Kumar et al. 2002; Marcot et al. 2002 for definitions and details of analyses). These indices mainly represented the degree to which forest and non-forest patches were intermixed. Core area calculations were based on 250 m and 500 m distances because these distances were suggested by a general review of the literature on "depth of edge" influences and boundary effects within protected areas (Kumar et al. 2002).

We next overlaid the map of forest fragmentation index results with the base map of boundaries, and delineated areas of low levels of fragmentation that span adjacent PAs and RFs as potential wildlife forest-habitat corridors, particularly for Asian elephant. We considered the Asian elephant an "umbrella species" so that habitat corridors identified for elephants might also benefit a wide diversity of other wildlife species.

We calculated forest fragmentation as the normalized number of forest and non-forest polygons found within a 6.25 ha area (in a roving map window of 250 m $\times$ 250 m); the lower the number of such polygons, the lower the degree of fragmentation. We defined low fragmentation as $\leq$ 30% of the normalized number of polygons, medium fragmentation as $\leq$ 80%, and high fragmentation as >80%. Corridors were thus mapped as polygons (1) that linked adjacent or nearest PA and RF boundaries, and (2) that consistently contained low fragmentation index values.

In Garo Hills, most villagers tend to restrict their movements inside forests up to two km and five km from forest edges, for collecting non-timber forest products and for *jhumming*, respectively. Hence we calculated forest area within buffers of two km and five km extending out beyond the boundaries of the PAs and RFs, and we termed these buffer areas "zones of influence" (ZIs). ZIs represent areas of potential conservation value for biotic communities of the PAN that could be affected by human activities of land use.

We then overlaid the ZI map onto the LULC and core area maps, and used Chi-square analysis to determine significant differences of forest cover and core areas among ZIs, the PAs and RFs, the wildlife habitat corridors, and other ZIs within community lands. We also used secondary information on elephant census records of SFDM for the years 1993 and 1998 and spatial information of Garo Hills (Talukdar 2004) to analyze elephant habitat relationship at the landscape level. We also used information on *jhum* families and other socio-cultural and economic factors to help suggest conservation strategies for biodiversity, productivity and sustainability of the ecosystem.

8.3 Results

8.3.1 Mapping Land Use and Land Cover

Results of the LULC analysis included mapping of the following forest and cover classes: tropical moist evergreen forest (TMEF), tropical semi-evergreen forest (TSEF), tropical moist deciduous forest (TMDF), bamboo growth & young secondary forests up to six years old, shola type forests along with associated grasslands at Balpakram Plateau, habitation and abandoned *jhum*, agriculture and sand, shifting cultivation and grasslands and water bodies, especially rivers (Fig. 8.2). Six of these nine land cover classes including various land uses and forest type were grouped into two broad categories, i.e., (i) land uses, including land area under habitation, permanent agriculture and *jhumming*; and (ii) old forest growth including TMEF, TSEF, and TMDF. The classification accuracy for these two broad categories combined was 100% and 83% when assessed for each land cover class separately.

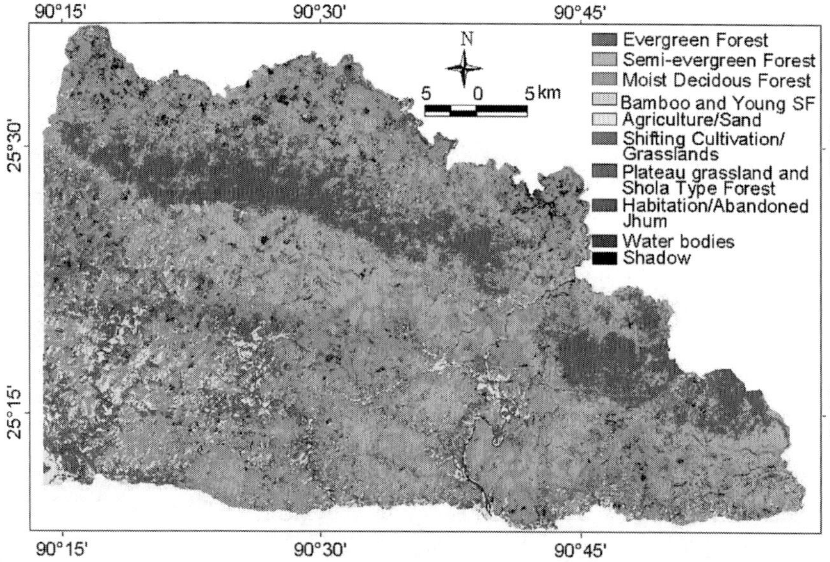

Fig. 8.2 Land use and land cover map of the study area

8.3.2 Landscape Patterns of Forest and Jhum Patches

The Garo Hills area, totaling 2459 km^2,was 75% forested (Table 8.1). The single largest forest patch occupied nearly 5% of the entire landscape area and was represented by old forest growth, specifically, the TSEF. *Jhum* patches covered only 4% of the landscape area, but were well dispersed. In general, forest patches averaged more than five times as large as *jhum* patches with three times the average patch edge length (Table 8.2). As the larger, native forest patches have become carved into smaller patches of *jhum* fields, the densities of individual patches and patch edges have increased (Table 8.2).

One-third of all native Garo families inhabiting the landscape were engaged in *jhumming* (DSWC 1995) over 47 km^2 of the total landscape area per year (Table 8.3). The land consumption per family for *jhumming* varied across the landscape from 0.28 ha for Rongra Community Development Block to 0.80 ha for Chokpot Community Development Block.

Table 8.1 Patterns of the Garo Hills study area, forests and core areas

Component	Number of patches	Area (km^2)
Garo Hills	227 977	2459
Forest	8 921	1844
Core areas >250 m from edge	2 236	561
Core areas >500 m from edge	644	291

Table 8.2 Patterns of all, forest, and *jhum* patches in Garo Hills, Meghalaya

Component	Patch size (km^2), mean ± 1SD	Patch edge length (km), mean ± 1SD	Patch density (n/ km^2)	Patch edge density (km/km^2)
All patches	0.10 ± 1.23	3 ± 18	10	28
Forest patches	0.17 ± 1.86	4 ± 27	6	24
Jhum patches	0.03 ± 0.04	1.3 ± 1.3	37	49

Table 8.3 Families engaged in shifting cultivation (*jhumming*) in selected locations within Garo Hills, Meghalaya

Community Development Block	Total no. of *Jhumia* families	*Jhum* area (mean ha/family)	Total area under *Jhumming* (km^2)	Total Households
Chokpot	2991	0.80	24	5519
Baghmara	763	0.70	5	6175
Rongra	989	0.28	3	2698
Samanda	1960	0.75	15	5782
Total	6703	0.70	47	20174

Source: Directorate of Soil and Water Conservation, Meghalaya (1995)

8.3.3 Forest Cover and Core Areas

The forest types TMEF, TSEF and TMDF together constituted 68% of the landscape area. TMEF represented old primary forest growth and occupied about 14% of the landscape area. TSEF occurred mostly as a buffer to TMEF and occupied 26% of the landscape area. TMDF usually occurred along fringes of human settlements or habitations and other land use subjected to frequent anthropogenic disturbances, and occupied 29% of the landscape area.

The core area analysis revealed that the total area within PAs and RFs >250 m from the edge was nearly twice the total area >500 m from the edge, with nearly 3.5 times as many patches (Table 8.1). This means that greater depths of incursion into PAs and RFs, such as from *jhum* or human habitation, may leave lesser total areas and smaller patch sizes undisturbed. This could have ecological ramifications for protecting plants and animals that require or select for undisturbed forest interior conditions.

8.3.4 Forest Fragmentation and Wildlife Forest Corridors

Most of the existing old forest area, occurring in TMEF, was intact or subjected to very low levels of fragmentation or anthropogenic disturbance. Only 1% of the landscape was under high levels of forest fragmentation, and 21% of the landscape was under medium levels of forest fragmentation (Fig. 8.3).

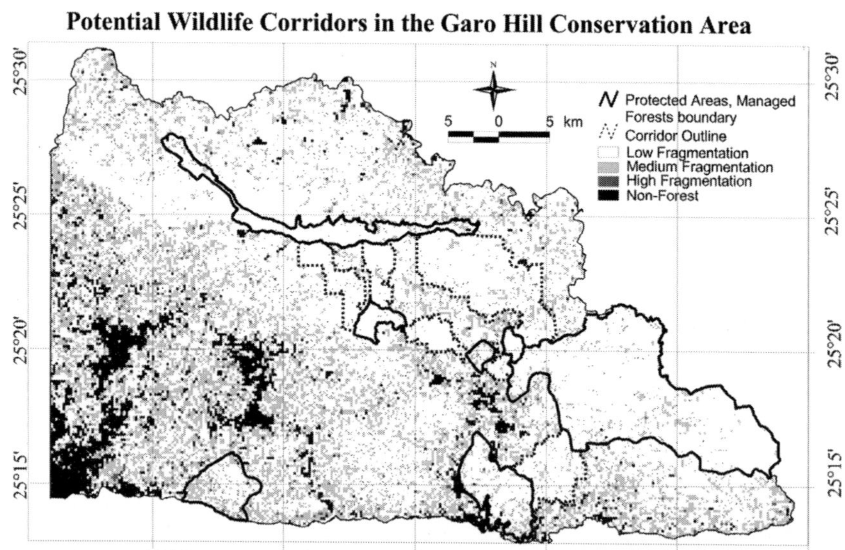

Fig. 8.3 Forest fragmentation within the study area. Dotted lines denote potential wildlife habitat corridors

We mapped seven potential wildlife forest corridors which totalled 14,340 patches of all cover types including forest and non-forest (Fig. 8.3). The corridors consisted mostly of native forests with low levels of fragmentation. The total corridor area included 6944 forest patches of TMEF, TSEF and TMDF and constituted 92% of the total corridor area.

8.3.5 Zones of Influence

The 2-km ZIs had low proportions of agricultural, *jhum* and scrubland areas, suggesting a low degree of stress from human use and occupation. The 2-km ZIs of BNP, SWS, and RRF overlapped (with a total non-overlapping area of 135 km^2). Therefore, we calculated a combined ZI for these areas.

The 5-km ZIs had much overlap among the zones around BNP, NNP, SWS, BRF, RRF and ERF. The total area under the 5-km ZIs was seven times greater than that under the 2-km ZIs, although the proportion of forest and non-forest was the same, i.e., almost 80% under forest growth (TMEF, TSEF, TMDF). The overall land uses in this ZI comprised about 13% of land area of 5 km ZI.

8.3.6 Comparison of Landscape Segments

Our findings revealed that larger and more intact (less fragmented) patches of native forest occur within the PAs and RFs, the wildlife habitat corridors, and the ZIs buffering the PAs and RFs, as compared with the rest of the community forest landscape of Garo Hills (Fig. 8.4). More specifically, the area of the three main forest

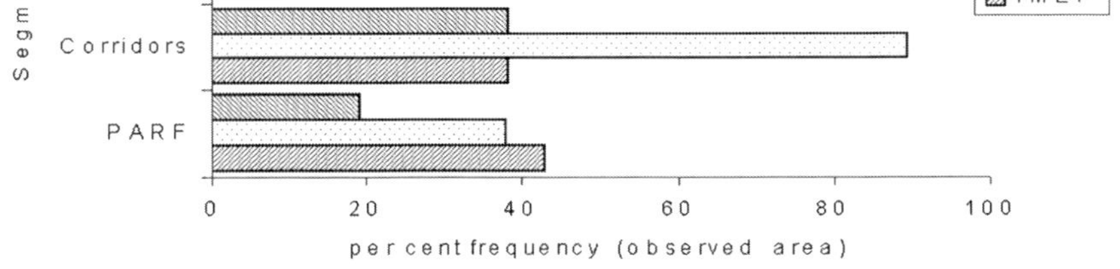

Fig. 8.4 Observed per cent area of forest types among landscape segments
TMEF = Tropical Moist Evergreen Forests, TSEF = Tropical Semi-evergreen Forests; and TMDF = Tropical Moist Deciduous Forests. ZI1 = zone of influence out to 2 km from boundaries of protected areas and reserved forests; ZI2 = out to 5 km; ZI3 = land area beyond 5 km

types (TMEF, TSEF, TMDF) differed significantly among PAs and RFs combined, wildlife habitat corridors, 2-km and 5-km ZIs, and all land area outside the 5-km ZI (likelihood Chi square ratio = 411.472, df = 8, P< 0.001).

8.4 Discussion

8.4.1 Landscape Patterns and Trends in Garo Hills

Our analysis suggests that, although most of the land area in Garo Hills is forested, residential and agricultural use by the Garo community is widely dispersed throughout the area. Also, most of the area around human settlements is extensively used for *jhumming* until the recent past. Past *jhumming* has left degraded scrub areas concentrated around villages or settlements. Most arable land is being used either for settled permanent agriculture or *jhumming*, and other forest resource uses such as firewood gathering (Bhatt and Sachan 2004) are also having a toll on native forests of the region.

Lower mean patch size and smaller edge length of patches with intense human use, as compared to those of forest cover patches, suggested that forest cover is being fragmented by human use. This is also suggested by patches with intense human use having a higher patch density and edge density as compared to those of forest cover patches.

However, the landscape still holds larger tracts of old forest cover. Several large forest patches in the PAs and in BRF and ARF are the best examples, as these forest patches provide promising habitat for hoolock gibbons (*Bunopithecus hoolock*), which have gradually disappeared during the past two decades and have become locally extinct from these areas mainly due to increasing human disturbances in their habitats. Forest managers may consider protecting large patches of native forests that occur outside but adjacent to the PAs and RFs as part of restoration programs for locally extirpated wildlife species.

Jhum has had an obvious impact on reducing native forest cover of the area. During 2000, a total of 7900 families (39 500 people) used 68 km^2 land for *jhumming*, at an annual rate of *jhumming* of 3.67% in South Garo Hills (DSWC 2001). Such rates of converting forest to *jhum* likely have adversely impacted some habitats and populations of wildlife species of the area.

The impact of *jhum* can be described by identifying the levels of fragmentation of native forests and the dispersion of *jhum* patches over the landscape. Fortunately, 71% of the landscape area was at a low level of fragmentation, whereas most of the medium or high fragmentation areas were concentrated in the south-west corner of the landscape. This portion lies on the flat land south and away from Nokrek Ridge and far from Balpakram National Park. Nokrek and Balpakram are important protected areas of the region, and both seem have mostly retained extensive cover of native old forests.

Unfortunately, such intact forest cover in *Garo* Hills is suffering an increased rate of fragmentation from *jhum* and other human use. The seven wildlife habitat corridors we identified encompass three corridors identified previously by Williams and Johnsingh (1996). We observed that ARF is the most isolated of all PAs and RFs in the landscape, and has lacked forest connectivity with any other such elements, although evidence suggests that it was historically connected with ERF by a corridor of old native forest cover. This historic corridor once facilitated movement of elephants across NNP in the north and the plains of Bangladesh in the south. Such migratory routes could be restored with timely management interventions.

It is unclear the degree to which the existing PAs, which constitute over 15% of the landscape, will conserve the rich biodiversity of these old forests. Our finding of higher mean forest patch size and lower mean forest patch density within the PAs as compared with outside the PAs reflects the lower degree of forest fragmentation within PAs as compared to RFs and community land, and we speculate that fragmentation might sacrifice some biodiversity elements. However, most forest cover (60% of landscape) is found in community land. However, in a companion analysis, Kumar (2005) reported that most community forests had high tree species diversity, but some tree species were found only with, or at least more dominant within, the PAs.

Our findings included that the 2-km ZIs had lower proportions of TMEF forest and with negligible area under various land use activities, whereas the 5-km ZIs contained a higher proportion of forest and a moderate proportion of land use activities. The community forest land area beyond the 5-km ZIs contained the lowest proportion of TMEF but most of TMDF and the highest proportion of residential and agricultural (settled or *jhumming*) areas. The area of TMDF represented more or less open or disturbed forest growth.

Thus, efforts to conserve wildlife habitat and tropical evergreen forest in the 2-km ZI could focus on preservation and perhaps enlargement of PA boundaries or designation of intact forests as parts of corridors. Forest conservation in the 5-km ZI could focus on additional protection measures because most of this ZI contains TMEF and TSEF native forest with only low levels of land use activities. And forest conservation in the land area beyond the 5-km ZI could focus more on restoration (rather than preservation) activities, which may be coupled with additional protection measures to help protect at least some of the remaining, larger forest tracts within community land.

8.4.2 Elephant Habitat

As reported by Marcot et al. (2002 and in press), elephant populations were most dense in Balpakram, Mahadeo, Chimitab, Siju, Baghmara, Nokrek and Samanda areas, but were also dispersed in lower densities in other parts of Garo Hills, for example, Dambu, Dagal, Kherapara, Adugre, Ranggira and other areas (Marak 1998). Therefore, examining the requirements of such a widely distributed species as elephant within the South Garo Hills study area should also entail understanding its distribution and habitat associations in a broader geographic context (Fig. 8.5). For example, using spatial information from Talukdar (2004), Marcot et al. (2002 and

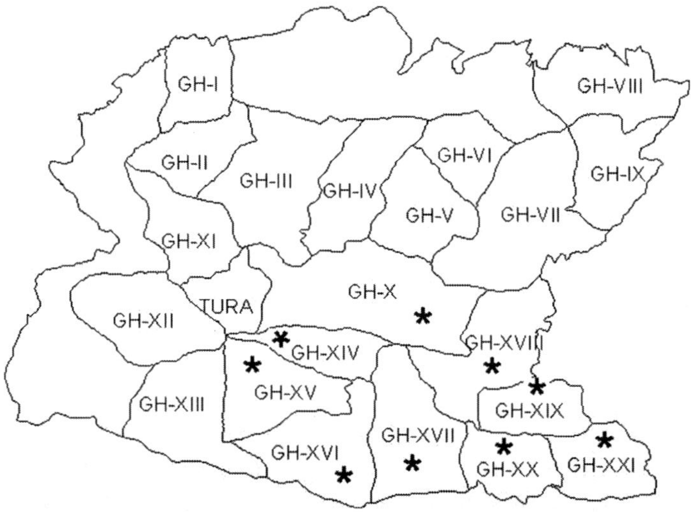

Fig. 8.5 Elephant census zones in the Garo Hills (GH), western Meghalaya. As used by Marcot et al. (2002 and in press), the All Garo Hills area included all labeled census zones, and the South Garo Hills area included zones marked with an asterisk (*). Map source: Marak (1998), as digitized by us into ArcInfo GIS

in press) analysed elephant-habitat relationships across all of Garo Hills, by using elephant censuses from 1993 and 1998. They presented a statistical model suggesting the following critical values of specific habitat variables significantly correlating with elephant density. Across the entire Garo Hills, elephant densities were reported to be greater in landscapes with:

- $< 30\%$ current and abandoned *jhum* (current *jhum* $< 5\%$, abandoned *jhum* $< 25\%$).
- $< 20\%$ in high forest patchiness (caused by *jhum*).
- Village density <about 0.4 villages/km^2.
- Annual *jhum* rates $< 2\%$ of the land *jhummed*/year.
- Evergreen, semi-evergreen, and mixed moist deciduous forest cover is $> 40\%$.

These correlations, along with the wildlife habitat corridors, could also be used to help guide conservation or restoration of elephant forest habitat in Garo Hills. In addition, activities for maintaining or restoring overall forest biodiversity could also use our findings for conserving intact blocks of native forest, especially of TSEF, in the 2- and 5-km ZIs, and encouraging rates of *jhum* in the land areas beyond the 5-km ZIs to allow for some degree of forest regrowth and restoration.

The 2002 amendments in the Wildlife Protection Act (1972) of Government of India bestowed the State Governments and Forest Departments with a strong tool through designating some forests on private non-government lands (Garo community land in present study) as "Community Reserves," whereas government lands may be designated as "Conservation Reserves." The landscape under investigation during present study offers excellent prospects for declaring both community reserves and conservation reserves for the purpose of conserving or restoring elephant populations and overall forest biodiversity.

8.5 Conclusion

Results of our study support similar findings of adverse effects on native forest cover and diversity from intensive and accelerated shifting cultivation in India and elsewhere. For example, studies in the Chittagong Hills of Bangladesh, which borders the Garo Hills to the south, have shown that shifting cultivation had little affect on forests until it accelerated at the beginning of the colonial period (Thapa and Rasul 2006). Together with dam construction, expansion of permanent plot agriculture, commercial and clandestine logging, and population migration and increase, shifting cultivation in Chittagong Hills has caused loss of native forests and reduction in soil productivity (Gafur et al. 2003). Changing the course of this tide has been impeded there by policy problems in land rights and trade, and lack of infrastructures and support services (Thapa and Rasul 2006). Agroforestry – the mixing of semi-permanent crops with different harvest cycles and life forms on the same plot of land – has been suggested as an alternative to shifting cultivation in this area (Rasul and Thapa 2006), and indeed in many areas of the world such as in Amazonia (Mcgrath et al. 2000).

In other examples, studies by Lawrence (2004, 2005) of 10–200 years of shifting cultivation in rainforests of Borneo suggested that, over many cultivation cycles, tree diversity and regeneration has shifted from seed-banking species to resprouting species, and that the overall carbon sequestration capacity of the secondary forests may become compromised. In Peru, Naughton-Treves et al. (2003) reported that hunting had an additional impact on persistence of mammals in shifting cultivation forest landscapes. In Arunachal Pradesh, India, Arunachalam and Arunachalam (2002) found that degraded soil of *jhum* fallows, particularly soil microbial carbon, nitrogen, and phosphorous, could be rehabilitated by planting of *Bambusa nutans* bamboo. Tawnenga et al. (1996) suggested that second-year cropping and use of fertilizers could maintain yields in *jhum* fields in Mizoram, North East India, and those shorter cycles (6 yrs vs. 20 yrs) of traditional *jhum* methods result in declines in primary productivity and economic yield of rice crops. However, as our case study also revealed, *jhum* cannot be singled out as the villain, nor easily altered or replaced wholesale with other less-extensive forms of agriculture and land use, without considering the tapestry of fuller cultural traditions, society norms, and even religious beliefs in which old forms of forest agriculture have arisen. Moreover, it is not *jhum* per se that is the concern, but rather the unimpeded expansion of human populations into forest landscapes with fragile or erodable tropical and subtropical soils. Strategies for changing such long-standing traditions as well as ameliorating adverse effects of human density and occupation of forest landscapes must integrate consideration for nutrition, health, education, access to economic trade, and effects on traditional lifestyles.

Acknowledgments We gratefully acknowledge project support from our colleagues at Wildlife Institute of India, Dehradun, Indian Institute of Remote Sensing, Dehradun, National Remote Sensing Agency, Hyderabad in India and US Forest Service in USA. The manuscript benefited from comments from an anonymous peer reviewer and from Dr. Raffaele Lafortezza.

References

Arunachalam A, Arunachalam K (2002) Evaluation of bamboos in eco-restoration of '*jhum*' fallows in Arunachal Pradesh: ground vegetation, soil and microbial biomass. For Ecol Manag 159(3):231–239

Awasthi A (1999) Plant diversity of Meghalaya and some issues for concern. Indian J For 22(1): 14–21

Bhatt BP, Sachan MS (2004) Firewood consumption pattern of different tribal communities in Northeast India. Energ Policy 32(1):1–6

DSWC (1995) A report on the survey of shifting cultivation in Meghalaya. Shillong. A report by the Directorate of Soil and Water Conservation, Government of Meghalaya.

FSI (2001) State of Forest Report. A report by the Forest Sruvey of India, Ministry of Environment and Forests, Government of India, Dehradun. 130pp.

Gafur A, Jensen JR, Borggaard OK, Petersen L (2003) Runoff and losses of soil and nutrients from small watersheds under shifting cultivation (*Jhum*) in the Chittagong Hill Tracts of Bangladesh. J Hydrol 279(1–4):293–309

Haridasan K, Rao RR (1985) Forest lora of Meghalaya Volumes I and II. Bishen Singh Mahendra Pal Singh, Dehradun, p 937

House, P (1997) Forest farmers: a case study of traditional shifting cultivation in Honduras. Rural development forestry network Paper 21a. Rural Development Forestry Network, Overseas Development Institute, London, UK, 24pp

Khan ML, Menon S, Bawa KS (1997) Effectiveness of the protected area network in biodiversity conservation: a case study of Meghalaya state. Biodivers Conserv 6:853–868

Kumar A (2005) Assessment of Tree Diversity, Successional Changes and Forest Fragmentation in *Jhum* Influenced Forest Ecosystem of South Garo Hills, Meghalaya. Ph.D. thesis. Forest Research Institute of India and Deemed University, Dehadun, 192pp

Kumar A, Gupta AK, Marcot BG, Saxena A, Singh SP, Marak TTC (2002) Management of Forests in India for Biological Diversity and Forests Productivity – A New Perspective, Volume IV: Garo Hills Conservation Area (GCA). WII-USDA Forest Service Collaborative Project Report, Wildlife Institute of India, Dehradun, 206pp

Kumar A, Saxena A, Marcot BG, Sawarkar VB, Roy PS, Mathur PK, Singh SP (2000) Forest fragmentation in the tropical forest ecosystem of Garo Hills, Meghalaya, northeast India. In Roy PS, Singh S, Toxopeus AG (eds) Proceedings of a workshop on Biodiversity and Environment: Remote Sensing and Geographic Information System Perspective. Indian Institute of Remote Sensing, National Remote Sensing Agency, Dehra Dun, India, pp 174–196

Kumar P, Singh TB (1997) Habitat Characterisation of Balpakram National Park (Meghalaya) using Remote Sensing and Geographic Information System. M.Sc. Dissertation. Indian Institute of Remote Sensing, Dehradun

Kumar Y, Rao RR (1985) Studies on Balpakram Wildlife Sanctuary in Meghalaya – 3: General account, forest types and fauna. Indian J For 8(4):300–309

Lawrence D (2004) Erosion of tree diversity during 200 years of shifting cultivation in Bornean rain forest. Ecol Appl 14(6):1855–1869

Lawrence D (2005) Biomass accumulation after 10–200 years of shifting cultivation in Bornean rain forest. Ecology 86(1):26–33

Ludden D (1999) The new Cambridge History of India IV. An agrarian history of south Asia .Cambridge University Press, Cambridge, UK, 261pp

Marak TTC (1998) The elephant census report for Meghalaya. Wildlife Wing, Forest & Environment Department, Shillong, Meghalaya, 22pp

Marcot BG, Kumar A, Roy PS, Sawarkar VB, Gupta AK, Sangama SN (2002) Towards a landscape conservation strategy: analysis of *jhum* landscape and proposed corridors for managing elephants in South Garo Hills District and Nokrek area, Meghalaya. (English with Hindi summary). The Indian Forester 2:207–216

Marcot BG, Kumar A, Roy PS, Talukdar G (In press) Elephants, people and landscapes in the Garo Hills of Meghalaya. In: Alfred JRB and Ahmad M (eds) Biodiversity conservation and utilisation, Jyoti Publisher, Dehra Dun

Mcgrath DA, Duryea ML, Comerford NB, Cropper WP (2000) Nitrogen and phosphorus cycling in an amazonian agroforest eight years following forest conversion. Ecol Appl 10(6):1633–1647

Metzger JP (2002) Landscape dynamics and equilibrium in areas of slash-and-burn agriculture with short and long fallow period (Bragantina region, NE Brazilian Amazon). Landsc Ecol 17(5):419–431

Momin PG (1984) Physical setting of Garo Hills. In: Gassah LS (ed) Garo Hills land and the people, Omsons publications, New Delhi, 235pp

Momin KC (1995) Traditional *jhum*-based economy of the Garos and its changing patterns. In Sangma MS (ed) Hill societies their modernisation: a study of North East with special reference to Garo Hills, Omsons Publications, New Delhi, India, pp 103–109

Naughton-Treves L, Mena JL, Treves A, Alvarez N, Radeloff VC (2003) Wildlife survival beyond park boundaries: the impact of slash-and-burn agriculture and hunting on mammals in Tambopata, Peru. Conserv Biol 17(4):1106–1117

Pulido MT, Caballero J (2006) The impact of shifting agriculture on the availability of non-timber forest products: the example of Sabal yapa in the Maya lowlands of Mexico. For Ecol Manag 222(1–3):399–409

Raman TRS (2001) Effect of slash-and-burn shifting cultivation on rainforest birds in Mizoram, northeast India. Conserv Biol 15(3):685–698

Rasul G, Thapa GB (2006) Financial and economic suitability of agroforestry as an alternative to shifting cultivation: the case of the Chittagong Hill Tracts, Bangladesh. Agr Syst 91(1–2):29–50

Roy PS, Tomar S (2001) Landscape cover dynamics pattern in Meghalaya. Int J Remote Sens 22(18):3813–3825

Sachchidananda (1989) Shifting cultivation in India. Concept Publishing Company, New Delhi, India, 181pp

Sudhakar S, Singh SB (1993) Forest type and density mapping in Meghalaya through digital image processing of Indian Remote Sensing Satellite data. Under the collaborative project of Forest Resource Division, Meghalaya and Regional Remote Sensing Service Centre, Kharagpur, 45pp

Sunderlin WD (1997) Shifting cultivation and deforestation in Indonesia: steps toward overcoming confusion in the debate. Rural development forestry network Paper 21b. Rural Development Forestry Network, Overseas Development Institute, London, UK, 18pp

Talukdar G (2004) Geospatial Modelling of Shifting Cultivation Induced Landscapes of Meghalaya. Ph.D. thesis. University of Pune, Pune

Tawnenga U, Shankar, Tripathi RS (1996) Evaluating second year cropping on *jhum* fallows in Mizoram, north-eastern India—Phytomass dynamics and primary productivity. J Biosci 21(4):563–575

Thapa GB, Rasul G (2006) Implications of changing national policies on land use in the Chittagong Hill Tracts of Bangladesh. J Environ Manage 81(4):441–453

Williams AC, Johnsingh AJT (1996) Threatened elephant corridors in Garo Hills, Northeast India. Gajah 16:61–68

World Resources Institute [WRI] (1996) World Resources, Oxford University Press, Oxford

Synthesis

Cultural Controls of Forest Patterns and Processes

Raffaele Lafortezza and Giovanni Sanesi

1 Forest Landscapes and Cultural Controls

Forest landscapes are one of the most tractable examples of the human influence on pristine ecosystems and habitats. A closer look at the various components charactering a given forest landscape could reveal traces of past and current management practices and, at some extent, the consequences of management or cultural control on the ecological patterns and processes (Rotherdam 2007). Current human ability to modify the forest environment at various scales is unprecedented (Franklin 2001). For example, changing historic disturbance regimes through fire suppression and reforestation has significantly modified the composition and structure of many forest landscapes throughout the globe (Crow 2002) – e.g., by altering species dynamics, abundance and age structure (Wang et al. 2007). The pervasive effects of non-sustainable management practices could limit forest successional patterns and species response to natural disturbances (Host and Pastor 1998). In their review, Guariguata and Ostertag (2001) analyzed tropical forest successions after complete clearance due to agricultural activities and pasture and concluded that patterns of species replacement is highly dependent by the interaction between site-specific factors and intensity of past and present land use activities (i.e., landscape-level factors).

One of the most evident consequences of these cultural controls on forest landscapes is the conversion of large forest patches into smaller fragments having more geometrized shapes and less interior locations for specialist species (see: Saura et al., Chapter 10): fragments that remain are associated with large amounts of habitat edge that is unsuitable for many species, and the remnants are isolated from one another, so individuals are, in fact, impeded to move across the landscape to forage and maintain gene flow (Ewers et al. 2007). Under this perspective, an important role is played by the intervening matrix which may include recently logged areas, agricultural fields, urban settlements, and other areas of human disturbance (Lindenmayer

R. Lafortezza
Department of Scienze delle Produzioni Vegetali, Università degli Studi di Bari, Italy
e-mail: r.lafortezza@agr.uniba.it

R. Lafortezza et al. (eds.), *Patterns and Processes in Forest Landscapes*,

and Fischer 2006). Disturbances that begin in the matrix can spread to influence processes in the forest remnants and this could ultimately affect the dynamics of species and the resulting biodiversity at ecosystem and landscape level. The persistence of ecological processes within forest landscapes is therefore a function of the type of cultural control which in turn depends on the type of benefits, services or commodities targeted into management plans and decisions (see: Lafortezza et al., Chapter 2).

Understanding the consequences of management on forest patterns and processes is a prerequisite to devise a more realistic and relevant foundation for landscape ecological studies. By focusing on these issues, landscape ecologists could gain new insights into some of the applicable ecological theory that underlies forest management with a specific focus on how human interventions affect forest landscapes and, in turn, how forest landscapes may influence humans and their culture or traditions (Rotherdam 2007). In addition, landscape ecologists could attain enlightenment by practicing their principles and models at forest landscape level, thus facilitating the transfer of knowledge to practioners and the dissemination of research findings to policy makers or even the general public (see: Chen et al., Chapter 1).

2 Human-Induced Alterations at Forest Landscape Level

Forest landscape management requires considering the multitude of cultural and natural forces that control for patterns and processes across dimensions of time and space (i.e., changes in patterns and processes at different scales). Following this *cause-effect* relationship, the chapters in this section focused on some of the most relevant mechanisms and factors regulating the ecological impacts of management on forest ecosystems and landscapes throughout the globe. Specifically, authors explained these mechanisms on the basis of applicative studies and research works conducted in five different regions, such as: eastern Siberia, central Africa, north-western Patagonia, northern Australia, and north-eastern India, with various degrees of spatial and temporal resolutions. Overall, these chapters provided insights into some of the applicable landscape ecological theories that underlies forest management, placing emphasis on the impact of humans in shaping forest landscape mosaics and on the role of cultural and environmental constrains.

In Chapter 4, Danilin and Crow described the great Siberian forest by putting current management practices into a broad socio-economic context that includes both a regional and global perspective. The vastness of this forested region presents serious challenges in managing for sustainability. Large-scale clear-cut logging is one of the main causes of soil compaction and alteration of forest floor properties that negatively affect natural regeneration of pine and larch stands. Forest fires, insect and disease outbreaks are described as concurrent disturbances that complicate natural succession and affect the health and productivity of forest landscape ecosystems in Siberia. The authors stressed the need for more applied research on forest landscape management in this boreal forest and the adoption of ecologically-sound harvesting practices that promote natural regeneration. New opportunities are envisaged for promoting ecosystem services such as carbon sequestration and biodiversity conservation.

In Chapter 5, Bogaert et al. analyzed the role of shifting cultivation as a driver of forest landscape dynamics in a province of the Democratic Republic of the Congo. Forest fragmentation of tropical rain forest is described in relation to the expansion of savannah, fallow lands and crop fields with the concomitant decline in soil fertility and increase in soil erosion. Various sources of spatial information and field data were used to assess the transition from secondary forests to savannah over 35 year time period. The development of secondary forest patches on formerly cultivated land is also observed and discussed as possible consequence of natural succession. Significant differences in the spatial distribution of some endangered species have been observed throughout the study area, suggesting that forest fragmentation would have a direct impact on the vulnerability status of species associated with undisturbed primary or secondary forests. From this study, an important lesson could be learned that is the importance of rural communities in fostering activities such as agro-forestry as alternative source of fiber and fuel wood. The application of landscape ecological principles could be an asset in creating a more balanced and self-sustaining landscape mosaic providing a wide range of services and goods to local communities and the global population at large.

Another relevant example of human-induced alterations at forest landscape level is brought by Carabelli and Scoz (Chapter 6), with a case-study in the Patagonian Andes. The authors discussed some of the main factors related with the mounting fragmentation of *Austrocedrus* forest landscapes in this region, such as: high incidence of forest fires, intensive timber harvesting, disease outbreaks, and widespread introduction of exotic species through plantations. Using a combination of remote sensing techniques and landscape pattern measures, the authors provided a quantitative estimation of the net balance between indigenous forest-cover reduction and reforestation/afforestation processes (i.e., fragmentation) and concluded with not encouraging news on the current status of forest landscapes in Patagonia. More research efforts should be placed in order to understand the factors limiting natural regeneration of indigenous species, e.g. as a consequence of after fires. Linking large-scale landscape assessments with additional field surveys could represent an important step forward the analysis of forest landscapes in this region, thus supporting management activities and actions mitigating disturbances.

In Chapter 7, Banfai and Bowman gave evidence of the main causes influencing the dynamics of monsoon rainforest landscapes in northern Australia. Factors of forest dynamics included unfavourable fire regimes, feral animal disturbance, rainfall and atmospheric CO_2. Linear and mixed effects models were used to assess the role these factors along with patch-level characteristics in determining the rate of rainforest boundary change in the Kakadu National Park. The authors explained how local-scale factors (e.g., feral animals and fire management) are unable to explain the overall expansion of rainforests in this region. Regional-scale factors, such increasing trends in rainfall and atmospheric CO_2 may have promoted the occurrence of new rainforest patches in the surrounding savannah. A clear message from the authors is the emerging evidence for a global-to-local interaction between factors controlling for patterns and processes in forest landscapes. Long-terms protocols are therefore required to make accurate predictions of rainforest dynamics under different scenario of climate change and/or modified disturbance regimes, such as those associated with forest fire and feral animals. Adaptive management programs are also needed as a way to gather new information on rainforest dynamics which can help to determine if desired conditions and management practices are resulting in expected outcomes at forest ecosystem and landscape-level and this could guide management actions and improved plans.

In Chapter 8, Kumar et al. explored the effects of shifting cultivations (i.e., *jhum*) on native forest biodiversity and the implications of such changes on forest landscape patterns and processes. Using various sources of spatial information and an array of landscape ecological measures, the authors quantified the process of fragmentation in *jhum*-modified forest landscapes and identified corridors potentially suitable for the Asian elephant. Such corridors, consisting of sequence of native forest patches, could be used to support the conservation or restoration of elephant forest habitats in this region. Results from this study could be generalized as lessons to learn for other forest landscapes of the globe, especially in situation of developing economies and escalating demand for fuelwood, fiber, and other types of products and commodities. Changing traditional ways of exploiting forest resources, such as *jhuming*, may appear impracticable because of the cultural values and religious beliefs associated with these practices. However, possible alternatives for using forests in this region should be sought and discussed with local communities, taking into account a number of cultural issues related with nutrition, health, education, economic trade, and traditional lifestyle.

3 Sustainable Use and Management of Forest Landscapes

The goal of promoting a sustainable use of forest resources could not be achieved without considering the socio-economic template of the region in which forest management is practiced. The authors of these chapters suggest that uncertainty is often associated with the effects of management on forest patterns and processes, thus envisaging a great deal of ecological research that should be consistent with existing knowledge and historical experience of the system being managed (Crow 2002). Cross-disciplinary and cultural backgrounds are therefore required for guiding forest landscape management towards the goal of multiple use and sustainability. In this direction, landscape ecology can be seen as the common language between the sciences of ecology, resource management, and land use planning. Through this language, forest managers and landscape planners could strength their synergy and devise farsighted strategies on how to use and manage forest resources at landscape level. Although challenging, the integration of landscape ecological principles into current management plans and practices could assist the conservation of natural and cultural values in many forested regions of the globe. Landscape ecology should be considered as a way to integrate human ecology and behaviour into a broader context, such as the landscape-scale patterns and processes, but also the global-scale conditions and influences. With the help of landscape ecology, the outcomes of multiple uses and non-sustainable management practices could be modelled and predicted in a much more integrative fashion, thus providing the mean for considering forest plans and management actions in a much larger cultural, economic, and ecological template.

References

Crow TR (2002) Putting multiple use and sustained yield into a landscape context. In: Liu J, Taylor WW (eds) Integrating landscape ecology into natural resource management, Cambridge University Press, Cambridge, UK, pp 349–365

Ewers RM, Thorpe S, Didham RK (2007) Synergistic interactions between edge and area effects in a heavily fragmented landscape. Ecology 88:96–106

Franklin SE (2001) Remote sensing for sustainable forest management. CRC Press LLC, Boca Raton, Florida, US, p 407

Guariguata MR, Ostertag R (2001) Neotropical secondary forest succession: changes in structural and functional characteristics. Forest Ecol Manag 148:185–206

Host G, Pastor J (1998) Modeling forest succession among ecological land units in northern Minnesota. Conserv Ecol [online] 2(2): http://www.consecol.org/vol2/iss2/art15/

Lindenmayer DB, Fischer J (2006) Habitat fragmentation and landscape change: an ecological and conservation synthesis. Island Press, Washington, DC

Rotherham ID (2007). The implications of perceptions and cultural knowledge loss for the management of wooded landscapes: a UK case-study. Forest Ecol Manag 249:100–115

Wang X, Hong H, Li X (2007) The long-term effects of fire suppression and reforestation on a forest landscape in Northeastern China after a catastrophic wildfire. Landsape Urban Plan 79(1):84–95

Part III
Landscape-Scale Indicators and Projection Models

Chapter 9

Tools for Understanding Landscapes: Combining Large-Scale Surveys to Characterize Change*

W. Keith Moser, Janine Bolliger, Don C. Bragg, Mark H. Hansen, Mark A. Hatfield, Timothy A. Nigh and Lisa A. Schulte

Abstract All landscapes change continuously. Since change is perceived and interpreted through measures of scale, any quantitative analysis of landscapes must identify and describe the spatiotemporal mosaics shaped by large-scale structures and processes. This process is controlled by core influences, or "drivers," that shape the change and affect the outcome depending on their magnitude and intensity. Our understanding of landscape change and its drivers depends upon many different sources of information of varying quality and breadth – some quantitative, some systematic, others anecdotal or qualitative. In this respect, large-scale surveys and inventories capable of documenting landscape composition, structure, and dynamics, both past and present, can prove to be vital tools for addressing contemporary resource issues. This chapter examines the role of large-scale inventories in identifying landscape change and developing hypotheses about the underlying drivers. Although a number of such sources exist, we shall focus on two from the United States: the Public Land Surveys (1785–1900), and the US Forest Service's Forest Inventory and Analysis program (1930s-present). After defining landscapes and providing definitions and examples of landscape change, we evaluate these surveys with respect to their potential use for ecological analysis, and present examples of their use for ecosystem reconstruction. These longitudinal comparisons are a good first step in understanding the biophysical processes that drive landscape change, but determining the influence of other drivers – social, cultural, or economic – requires other sources of information that are rarely systematic or conclusive. To this end, cautious analysis and conservative conclusions are essential when employing this mix of data sources.

W.K. Moser
U.S. Department of Agriculture, Forest Service, Northern Research Station, Forest Inventory and Analysis Program, St. Paul, Minnesota, USA
e-mail: wkmoser@fs.fed.us

R. Lafortezza et al. (eds.), *Patterns and Processes in Forest Landscapes*,

9.1 Landscapes and Landscape Change

Landscapes are the expression of the inherent productive capability of any given area as shaped by climate, parent materials, the biota, and environmental history and as influenced by continuums of endogenous and exogenous biophysical drivers (Bolliger 2005; Bolliger et al. 2003). As a result, landscapes continually change over time and space. While these changes may or may not be desired, particular outcomes are certainly preferred. To this end, understanding landscape change can help society mitigate the effects of change or at least identify undesirable patterns and processes to be avoided. This chapter examines the role of large-scale surveys in defining landscapes and, by inference, landscape change. After a brief introduction to the concept of driving forces, three examples of landscape change analysis are presented that compare landscapes separated by almost 2 centuries.

For recent changes, evidence tends to be well documented and relatively easy to investigate. This, however, is not the case with historical landscape change. Given fewer and often less accurate sources of information, discerning the mechanisms behind past landscape change becomes more challenging. Not that historical information is without value – prior events and observations can contribute towards the understanding of previous environmental conditions (Fei 2007; Goforth and Minnich 2007). Rather, more exacting research is needed to identify processes and consequences of prior land use to foster collaboration with fields other than ecology to ensure interdisciplinary science (Wu and Hobbs 2002; Bürgi et al. 2004; Bürgi et al. 2007). After all, the factors driving past environmental change, though often the same as those occurring nowadays, can have fundamentally different consequences on modern landscapes.

Furthermore, assessment of landscape change involves looking beyond the local landscape or research question to search for general properties that can be applied elsewhere (Bürgi et al. 2004). Are there common drivers that might explain landscape change across ecoregions or even climatic zones? If so, are these drivers temporally extensible? Assuming that at least some of the factors that shaped past landscapes still affect those observed today, what does this tell us of future conditions? For instance, natural disturbances continue to resonate across forests, with concurrent biotic responses to these alterations. Knowing how the environment responded to perturbations in the past should provide at least a hint about how a landscape may respond to similar disturbances.

Another challenge in understanding landscape change lies in the linkage of data of inherently different origins, structure, and quality. Bürgi et al. (2004) emphasized this point by comparing data from unique disciplines. However, even the comparison of information collected within a given field can prove challenging. As an example, natural resource surveys conducted a century or more ago are noticeably different from current ones. This difference arises from differences in the data being measured, the tools available to assess the resource, and our ability to understand the available information. While some attributes, such as species composition, are still utilized, many attributes now considered important were rarely incorporated in inventories even a few decades ago. For instance, measurements of large woody debris (Woodall and Williams 2005) or vertical forest structure (Ferris and Humphrey 1999) are now common in ecological surveys.

Finally, we must consider societal influences as an explicit and prominent portion of any model of landscape change (Bürgi et al. 2004). While environments can change dramatically under natural processes, few have proven to be more pervasive and intensive than human activities, which typically result in simpler conditions than those caused by natural disturbances (Skånes and Bunce 1997). For example, the globalization of the forest products industry has resulted in many natural forests being replaced by even-aged, short rotation monocultures. Hence, a purely economic driver (fiber production) has supplanted established patterns of natural disturbance, plant succession, soil development, and carbon accumulation, amongst others.

9.2 What Are Drivers of Landscape Change?

Even though our understanding of change in the face of uncertainty challenges any model we may wish to construct, the measured analysis of data in light of known landscape drivers has been remarkably successful in explaining large-scale pattern and processes. This understanding of pattern and process is possible because driving forces are considered to be the most "... influential processes in the evolutionary trajectory of the landscape..." (Bürgi et al. 2004, p. 858). Like the large- and small-scale disturbances impacting the dynamics of a forest stand, these forces shape and change landscapes over time (Oliver and Larson 1996). Driving forces may be natural or socioeconomic (including political, technological, and cultural factors (Brandt et al. 1999)), and are often exceedingly complex and inextricably intertwined, making it impossible to consider them as discrete phenomena.

Most ecologists are familiar with natural driving forces, which can be either directly observed or inferred from biotic responses to certain environmental conditions. The former is self-evident, while an example of the latter can be taken from the presettlement forests of the Ozark Plateau of Missouri and Arkansas (U.S.A.). These *Quercus*-dominated woodlands were primarily composed of low density stands or isolated denser groves in sheltered coves or narrow strips along riparian zones (Beilmann and Brenner 1951; Schroeder 1981; Foti 2004). The historically low forest density and species composition of Ozark landscapes are usually attributed to frequent fires (Batek et al. 1999; Guyette et al. 2002) and extensive areas of poorly suited soils (Schoolcraft 1821). Applying the landscape drivers model, we find that the historical driving forces of poor soils and fire imposed upon vegetative patterns, producing a feedback loop that helped sustain presettlement landscape patterns.

From the previous example, we can see how individual drivers can combine to effect landscape change. These drivers can also act in concert with each other over time. In eastern North America, for example, forested landscapes changed following the evolution of human economic activity from hunting and gathering to row-crop agriculture, government-promoted settlement of lands, the influence of the railroad in timber harvesting, and a trend towards maximizing economic productivity (Beilmann and Brenner 1951; Kersten 1958; Fitzgerald 1991; Benac and Flader 2004). This final industrialization driver is witnessed in the growing prominence of loblolly pine (*Pinus taeda* L.) plantation monocultures across the southern U.S.A. Over much of this region, the potential for increased financial returns has encouraged many landowners to significantly intensify their silvicultural practices (Stanturf et al. 2003; Rousseau et al. 2005). As a result, much of the region has been cleared of the existing timber and converted to short-rotation (15- to 30-year) loblolly pine plantations (Wear and Greis 2002; Allen et al. 2005). Older and larger tracts are most susceptible to conversion, greatly simplifying landscape composition and structure (Rogers and Munn 2003; Arano and Munn 2006). These alterations also affect other large-scale phenomena, such as variations in site quality or the frequency of damaging storms (Read 1952; Rebertus et al. 1997).

9.2.1 Using Large-Scale Data to Identify Landscape Change

Given the lasting legacy of past events and conditions on current systems (Bürgi and Turner 2002; Bürgi et al. 2007), an understanding of historical environments is a valuable asset in natural resource management (Landres et al. 1999). Knowledge of past conditions can provide a baseline for assessing change, help us understand important processes associated with ecosystem conditions, and provide potential targets for restoration activities (Bolliger et al. 2004). Fortunately, many types of information are available on past conditions and processes, including diaries, newspaper reports, official forest and agricultural statistics, maps, photos, and public and private archives (Russell 1997; Bürgi et al. 2007; Fei 2007; Goforth and Minnich 2007). Note that these data sources can be either quantitative or qualitative in character, represent different spatial or temporal extents, and vary in their accuracy regarding past conditions, so their interpretation must be carefully undertaken.

A critical prerequisite for the study of landscape change and the drivers propelling it is knowing how to acquire accurate baseline information. With this in mind, Antrop (1998) provided the following questions:

1) What is being changed?
2) How often does the change occur?
3) How significant is the change? and
4) What is the reference period the changed environment is compared to?

Large-scale inventories can provide answers to these questions. However, the longer the time between measurements, the more that differences in inventory design – such as the scale and resolution of sampling, individual performance in data collection and taxonomic identification, variation in the units of measurement, and lack of consistency in quality control – make the comparison complex and uncertain. Furthermore, some drivers such as natural ones (severe wildfires or landscape-level soil productivity) or political ones (government support of land settlement) are more easily documented than others (e.g., changing cultural attitudes towards land use or the rate of technological progress).

The appropriateness of large-scale data depends in part on the analytical method(s) employed. Whereas documentation of a particular landscape condition based on anecdotal descriptions may suffice for qualitative analysis, quantitative analyses of past conditions require spatially- and temporally-representative data. Surveys and inventories across multiple levels are used to inform this process by cataloging the current state of the landscape, flora, or fauna, and can be used to assess the likely consequences of environmental change. For instance, contemporary land-cover and land-use surveys usually employ remotely-sensed data from aerial photographs or satellites to develop geospatially and chronologically comparable datasets.

Examples of landscape change detected by land-cover and land-use surveys can be seen in the large-scale trends affecting agricultural regions. The primary drivers influencing these agricultural lands are associated with economic and technological changes in crop production. In many parts of the world, particularly in mountainous and other marginal areas, farmlands are being lost to other land uses, driven by declines in the economic significance of agriculture (Bolliger et al. 2007; Laiolo et al. 2004). Often, this results in the reforestation of formerly open land, which may lead to a short-term increase in species richness due to an increase in the variety in landscape structure (Söderström et al. 2001) and the offset of forestlands lost to urbanization (Wear 2002). However, there are instances of pastoral abandonment that result in significant habitat loss for open-land species (Dirnböck et al. 2003; Bolliger et al. 2007) and can potentially threaten species diversity (Tilman et al. 2001). The trend in North America has been toward simplified agricultural landscapes, with a diminishing number of cover types arranged in fewer and larger patches (Schulte et al. 2006). In central North America, this simplification has been linked to a decline in populations of grassland birds (Murphy 2003) and degradation of water quality (Turner and Rabalais 2003).

Models are also gaining importance in formulating spatiotemporal interactions within and between landscape elements. A range of quantitative model types can be distinguished based on various aspects of the modeling approach. Models differ in the way landscape heterogeneity is taken into account, based on the research focus and the availability of data on exogenous and endogenous factors and processes (for reviews see Guisan and Zimmermann 2000; Lischke et al. 2007). Yet, whether it is a stochastic Markov analysis of potential transitions between differing species mixtures (Moser et al. 2003), detailed modeling of individual driving forces, or qualitative Delphi-type techniques that incorporate all of the underlying driving forces into one category of change magnitude (Moser et al. 2006), each method describes the transition of a landscape from one state or condition to another. However, quantitative methods do not provide certitude by themselves. Ecologists increasingly need to incorporate ancillary data, circumstantial evidence, and inferential reasoning from other information for their analyses to avoid misinterpretations of landscape change (Bürgi and Russell 2000), or to combine data from different resources to optimize spatial information (Edwards et al. 2006).

The combination of such information from drastically different sources is fraught with challenges. For instance, taxonomic data (particularly for infrequent species) are often acquired via purposive sampling (Edwards et al. 2006; Lütolf et al. 2006). This type of sampling, which is generally not statistically or spatially representative, provides information on species presence. While the presence of a species may be easily determined in the field, absences are more difficult to confirm (Kéry 2002). A species may be absent for any number of reasons, but only unsuitable habitat is considered a real absence in habitat modeling (Lütolf et al. 2006). Thus, many species surveys include presence-only data (i.e., data with confirmed presences, but unconfirmed absences). Although there are ways to model species distributions with presence-only data, the generation of pseudo-absences should be made a priority in habitat distribution modeling, e.g., by using auxiliary species whose habitat(s) resembles that of the focus species (Lütolf et al. 2006). Another option would be to pool taxonomic information from other sampling strategies. However, it has been demonstrated that the overall sampling design has significant influence on the validity of the statistics (Edwards et al. 2006) and, hence, on the interpretability of the habitat distribution patterns. A comparison of purposive sampling and design-based strategies shows that the model performance from simulations originating from the former method is lower compared to those from the latter (Edwards et al. 2006).

The conflict between data types (whether sampled or modeled) and reliability shows that when they are integrated to address landscape-to-regional questions, close attention should be paid to their limitations. The data, analytical methods, and resultant interpretation must be carefully evaluated so that conclusions are not tied more to the inherent tendencies of the source than to the ecology of the system. Diary records, newspaper reports, or personal photos may provide details for a particular time and location, but are heavily influenced by the writer's perception of what conditions were noteworthy. Hence, this source of information is likely to over-represent sensational, large, or unique landscape features. Examples of potentially misleading ecological information in generally reputable outlets are historical photographs of old-growth timber or large "trophy" trees in lumber trade journals (Bragg 2004) and dramatized newspaper reports of large-scale fires in the California chaparral (Goforth and Minnich 2007). Official historical surveys, maps, or land statistics may be more representative over broader spatial scales (Manies and Mladenoff 2000), but should be carefully examined to minimize interpretation errors or spurious correlations. It is also critical to avoid observer biases made from contemporary experiences with modern-day landscapes. For example, the current distribution of species such as red maple (*Acer rubrum* L.) and loblolly pine has drastically increased from what existed in presettlement times as natural disturbance regimes and land use patterns have changed (Abrams 1998; Bragg 2002). On the other hand, some once dominant taxa have declined precipitously (e.g., American chestnut [*Castanea dentata* (Marsh.) Borkh.] or American elm [*Ulmus americana* L.]) because of introduced diseases.

9.3 Examples Using Historical Data and Current Large-Scale Surveys

Ecologists and other resource professionals in North America trying to establish criteria for sustainability have looked to pre-European settlement landscapes as a contrast to today's highly altered landscapes (Swetnam et al. 1999; Foti 2004). Although these early landscapes were known to be disturbed by indigenous peoples (Guyette et al. 2002) and biotic and abiotic forces (Schulte and Mladenoff 2005), many people believe that they represent examples of "natural variability" (Landres et al. 1999). Yet, serious questions remain. For instance, how does one define the nature of these presettlement landscapes? Given that historical surveys were rarely collected specifically for the study of the biota, how must the information contained within them be interpreted? What sources are best suited for this task?

Probably best known among the official historical resource surveys of the U.S.A. is the General Land Office's public land surveys (PLS). Implemented across most of the country during the 19th century, the PLS was a rectangular, rule-based system of land subdivision that opened the public domain to private ownership, provided a key source of revenue to a growing federal government, and brought development to heretofore "wild" landscapes (Linklater 2002). These north-south and east-west running demarcations divided the land into nominal 9,324 ha (36 mi^2) squares called "townships," which were then further subdivided into 259 ha (640 ac) "sections" (Stewart 1935; White 1983). At corners and selected points in between, the surveyors recorded information (e.g., species, estimated diameter, and distance) on two to four trees near the posts. In addition to these witness trees, the PLS field notes also usually recorded conspicuous features, such as stream and river crossings, the predominant trees, and obvious changes in forested condition or geology. Furthermore, the surveyors also drew geographic plat maps of many of the features (e.g., streams, lakes, springs, bluffs, prairies, early settler improvements) reported in the field notes.

Despite many shortcomings (e.g., Bourdo 1956; Manies and Mladenoff 2000; Schulte and Mladenoff 2001; Foti 2004), the PLS records provide useful large-scale information on vegetation composition and structure due to their resolution, extent, and detail. In part, this is because the PLS field instructions have been thoroughly documented in the literature (Stewart 1935; White 1983), allowing users to evaluate their applicability to the question at hand and interpret the surveys accordingly. Decades of experience have resulted in the PLS' being used to interpret (1) local and regional vegetation patterns using both descriptive and quantitative approaches (Batek et al. 1999; Schulte et al. 2002; Bolliger et al. 2004; Bolliger and Mladenoff 2005); (2) the characteristics of historical disturbance events (Zhang et al. 2000; Schulte and Mladenoff 2005); (3) landscape change (Radeloff et al. 2000); and (4) land-use change (Foster et al. 1998; Bürgi et al. 2000). They have also been used with spatially dynamic landscape models to evaluate relationships between pattern and process (Bolliger et al. 2003; Bolliger 2005) and have revealed early socioeconomic trends (Silbernagel et al. 1997).

The following examples present very different approaches to using historical and contemporary large-scale survey data to examine issues of landscape change and their drivers. Each uses PLS and Forest Inventory and Analysis (FIA)[1] data to address the topics. First, a series of resource inventories was used to reconstruct a shift in dominance between two native pine species in the southern portions of Arkansas. Here, the study specifically examined the drivers that propelled the landscape change. A second example addresses the problem of conforming two different inventories to a common metric capable of summarizing landscape change and guiding restoration priorities. Acknowledging the drivers that promoted landscape change, this study amalgamates the driving influences into a dimensionless restoration-suitability category.

9.3.1 Shifts in Pine Dominance Across the Gulf Coastal Plain of Arkansas

In the early decades of the 20th century, foresters were concerned about an apparent decline in pine abundance across the Upper West Gulf Coastal Plain (UWGCP) (Chapman 1913; Hall 1945). Bruner (1930) reported that the forested lands of Arkansas had dropped from almost 13 million ha before settlement to 8.9 million ha, and standing volume had declined from an estimated 0.9–1.4 billion m^3 in the original forests to about 0.2 billion m^3 in 1930. In addition, a variety of less valuable hardwood species had markedly increased their presence across the landscape (Reynolds 1956). Over the intervening decades, it became obvious that agricultural abandonment and the spread of scientific forestry had stemmed the loss of pine-dominated timberlands (Hall 1945). Indeed, as silviculture became increasingly lucrative following World War II, management of a greater proportion of the land was driven by the interest in a single species – loblolly pine. Over the years, structurally-complex, naturally-regenerated mixed pine-, pine-hardwood-, and hardwood-dominated stands have been replaced with increasingly loblolly-dominated, intensively-cultivated stands (Bragg et al. 2006).

Evidence suggests that shortleaf pine (*Pinus echinata* Mill.) was considerably more abundant in presettlement times over much of the UWGCP in southern Arkansas (Mohr 1897; Bragg 2002). Estimates of the shortleaf pine composition of the pine-dominated presettlement upland forests of the UWGCP in Arkansas ranged from approximately 25–50 percent shortleaf pine, with an increasing representation of shortleaf as one traveled from east to west and localized pockets of

[1] The national inventory conducted by the US Forest Service FIA program uses permanent sample plots located systematically across the U.S.A. at an intensity of approximately one plot every 2,400 ha to produce a random, equal-probability sample. Over the years, other types of environmental measurements, such as forest health monitoring or state-based assessments, have been tied to the FIA plot system, and therefore share considerable concordance in the statistical nature of the data collected (McRoberts 1999). Complete documentation of the plot design and all measurements can be found at http://socrates.lv-hrc.nevada.edu/fia/dab/databandindex.html.

"pure" (>80 percent) shortleaf pine across the UWGCP. Loblolly pine's abundance in historical upland forests also varied considerably, but in general the species was considered prominent only in smaller bottomland or on more mesic sites protected from frequent fires (Mohr 1897; Olmsted 1902; Chapman 1913; Bragg 2002).

Modern-day assessments of forest cover in the UWGCP of Arkansas reveal that loblolly pine is now the most dominant species (e.g., Rosson et al. 1995). Documenting the shift from shortleaf to loblolly pine, however, is not a simple matter. The PLS surveyors did not differentiate between the pine taxa of Arkansas, although other sources of historical information report considerable differences in pine abundance by species, geography, and landform (e.g., Bragg 2002). Furthermore, the PLS represents vegetation conditions at an instant in time, and thus does not reflect changes in pine dominance.

However, combining PLS data with the US Forest Service's FIA data can be used to derive long-term species dynamics across large geographic regions. Using periodic inventories of the 22 counties conducted since the late 1930s (Eldredge 1937; Duerr 1950; Sternitzke 1960; Hedlund and Earles 1970; Quick and Hedlund 1979; Hines 1988; Rosson et al. 1995), the long-term relative trends of shortleaf pine, loblolly pine, and hardwoods were determined over the last seven decades for southern Arkansas (Fig. 9.1). From its peak abundance during presettlement times, shortleaf pine abundance dropped following the historical logging, burning, and agricultural clearing of the forests of southern Arkansas. Up until 1970, however, shortleaf pine maintained a respectable presence in the overstory, comprising between 20 and 25 percent of the standing sawtimber. Over the last 35 years, shortleaf pine has declined dramatically, dropping to less than 7 percent of all sawtimber-sized trees in the latest FIA information available for this region (Moser et al. 2007b).

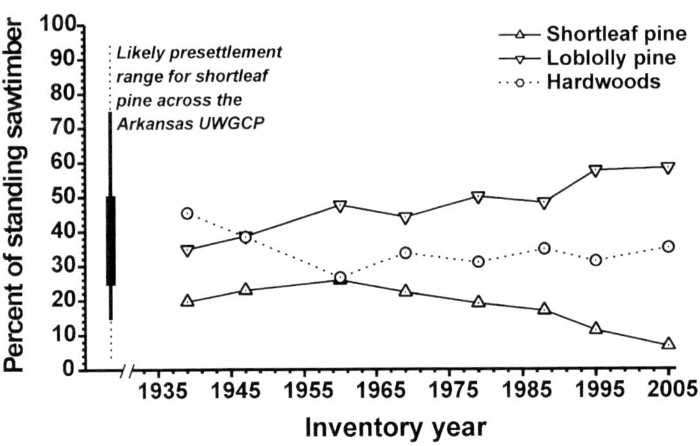

Fig. 9.1 Long-term trend in loblolly pine, shortleaf pine, and hardwood species in southern Arkansas compiled from multiple inventory reports. The presettlement abundance of shortleaf pine has been adapted from several historical references, and the thickness of the bar indicates the relative likelihood of that proportion of shortleaf in the Upper West Gulf Coastal Plain (UWGCP)

Loblolly pine, on the other hand, has steadily increased from about 35 percent of the overstory volume in 1937 to over 58 percent in 2005.

9.3.2 Comparing Current and Historical Resource Surveys as a Tool for Targeting Landscape Restoration in Missouri

In 2003, a team of natural resource professionals from the Missouri Department of Conservation and the University of Missouri developed a forest classification scheme based on current and potential forest-type groups with the suitability (and, by implication, the ease) of conversion from one type to another based on site index (Moser et al. 2006).[2] This system, excerpted in Table 9.1 (Nigh et al. 2006), is analogous to a "state and transition" approach to restoration (Fig. 9.2, Hobbs and Norton 1996).[3] Although the categories of suitability presented in Table 9.1 refer to all management activities (not just restoration), the overall concept applies.

To evaluate historical forest land structure, 1815–1855 PLS data from Missouri were used. Current data was obtained from the annualized inventory of Missouri's forest resources, collected by the FIA program between 2001 and 2005 (Moser et al. 2007a) to assess the present-day landscape. The study employed a "moving window" analysis – where each pixel was assigned a value based on a function of the ground observations within a particular radius (similar to what was employed in Moser et al. 2006). Because of the different sampling intensities of the two surveys, each analysis required different-sized windows: a 2000 m radius for the historic (PLS) data and a 4000 m radius for the current FIA data. The two datasets were then reduced to a common data structure to facilitate analysis (Table 9.1).

The output from the classification scheme was a conversion suitability map that estimated the effort required to restore the landscape of the 1820s (Fig. 9.3). Of the 1.4 million ha in the study area, 11 percent was classified as low-effort sites, 11 percent as medium-effort sites, 6 percent as high-effort sites, 2 percent as maximum-effort sites, 12 percent as non-forest and 45 percent as not possible (Table 9.2). The remaining 12 percent was classified as having no information. The large number of hectares considered unsuitable or for which there were no data resulted largely from an inability to delineate particular combinations of present-past forest types. Among these was savanna, for which there was no definition in the conversion matrix. As savanna represented a considerable portion of the historic landscape,

---

[2] The effort required to maintain a particular composition and structure depends upon many factors, including the dynamics of the current forest, the degree of difference between current and desired states, and site factors such as soil productivity and climate.

[3] In their article, Hobbs and Norton defined State 1 as a non-degraded ecological state, States 2 and 3 as partially degraded states, and State 4 as a highly degraded state. Stressors or some other debilitating agent caused the transition from State 1 to States 2, 3, and 4. Removing the stressor in States 2 and 3 can result in an unaided return to State 1, analogous to natural resiliency. However, additional management action beyond merely removing the stressor will be required in State 4, as the threshold represents a level of degradation that would preclude any unaided restoration.

Table 9.1 Classification system and management options for upland forest/woodland types in Missouri, excerpted from Nigh et al. 2006. Numbers associated with each site quality class indicate degree of suitability and effort from 1 = highly suitable and low effort to 4= low suitability and maximum effort. An "X" indicates a very unlikely occurrence

Present	Suited	Site Quality			
Forest Type	Forest Type	1	2	3	4
		12–16 m (40–54 ft)	16–19 m (55–64 ft)	19–22 m (65–74 ft)	22 m+ (75 ft+)
Post oak woodland	Post oak woodland	1	2	4	X
	Mixed oak woodland	2	1	1	X
	Mixed oak forest	X	2	1	X
Mixed oak woodland	Post oak woodland	1	2	4	X
	Mixed oak woodland	1	1	2	4
	Pine-oak woodland	1	2	3	X
	Mixed oak forest	X	3	1	2
	Pine-oak forest	X	2	1	3
	White oak forest	X	3	2	1
	Pine woodland	1	2	4	X
	Pine forest	2	1	3	X

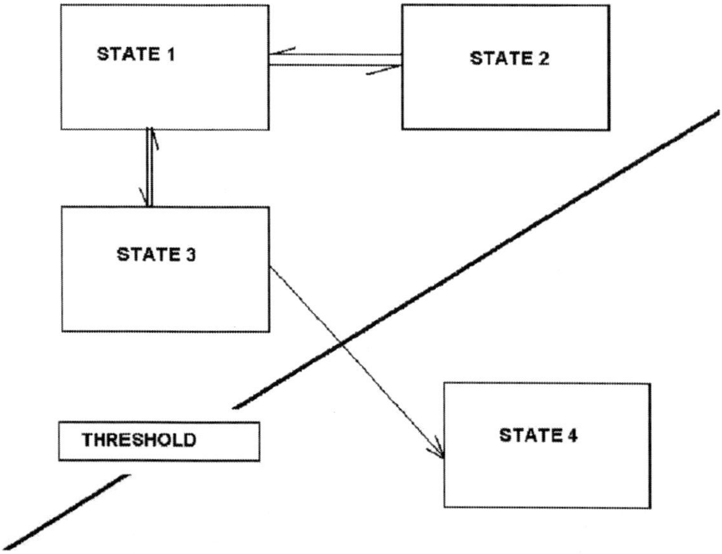

Fig. 9.2 A state and transition approach to restoration (Hobbs and Norton 1996). States 2 and 3 represent conditions that could naturally return to the predisturbance state 1 once the stressor is removed. State 4 is beyond the limit of natural resiliency and additional restoration efforts must occur for this state to be returned to State 1

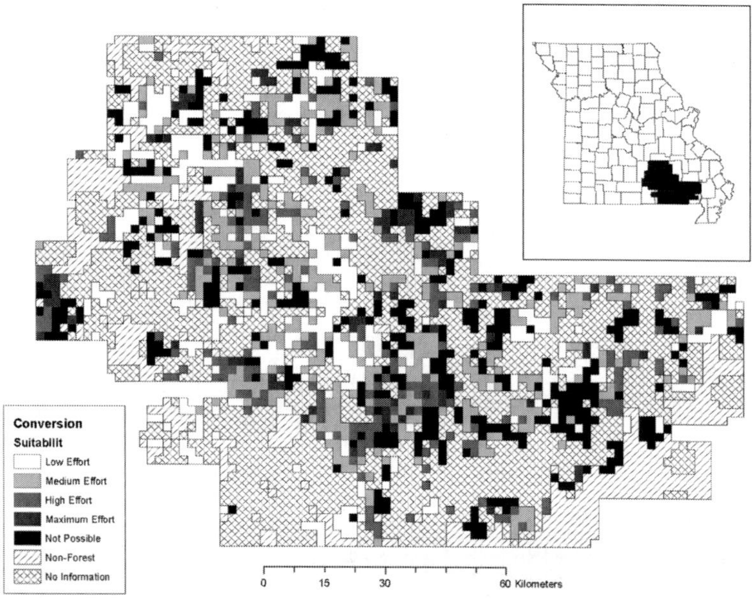

Fig. 9.3 Map of categories of conversion suitability and effort. The scale is from "low effort" (a "1" in the matrix in Table 9.1) to "maximum effort" (a "4" in the matrix in Table 9.1)

Table 9.2 Summary of categories of conversion suitability in the study area, Missouri Ozark region. Percentages do not add up to 100 due to rounding

Suitability	Hectares	Percentage of Total
Low Effort	158,800	11
Medium Effort	153,200	11
High Effort	86,400	6
Maximum Effort	34,800	2
No Information	165,200	12
Non-Forest	168,000	12
Not Possible	636,800	45
Total	1,403,200	

omitting it meant that a substantial segment went "unclassified" ("no information" in Table 9.2). Nevertheless, the results are consistent with other analyses that use a more disturbance-based protocol (e.g., Guyette et al. 2002). Hence, an understanding of landscape change via analysis using large-scale inventories can be used to develop drivers to predict potential future conditions.

9.4 Conclusions

Recognizing the presence and influence of drivers of landscape change improves the ability to predict outcomes from current and future resource management activities, especially large-scale restoration work. Practitioners documenting landscape change with an eye toward restoration should first determine the primary historical structures and functions, followed by a series of inquiries patterned after the questions posed by Antrop (1998) to identify and quantify landscape protection:

1) How often must the landowner invest in restoration? Is this a one-time effort, or will there need to be continued maintenance?
2) How much effort will it take to restore the landscape to the desired state? Is the restoration effort worth the perceived benefits? Will the investment in restoration be rewarding to the landowner, perhaps as a result of a subsidy?
3) What are the criteria for success?

In conjunction with these questions, landscape analyses can help identify practical constraints in restoration activities (Bell et al. 1997). For instance, environmental degradation can result from extensive and intensive causes (Hobbs and Norton 1996), so effective, sustainable restoration efforts should also be at a comparable scale.

Characterization of landscape attributes involves more than just comparing patterns over time and space. Rather, it involves explicitly connecting past environments with the underlying processes that drive them towards specific patterns. Not surprisingly, the more complex the processes influencing the landscape, the more important it is to understand them and their role in landscape change. However, models of landscape change should follow "Occam's Razor" and be only as sophisticated as needed to answer the question – if for no other reason than that simpler models will likely fit the available data better than more complex ones. After all, surveys such as the PLS of the 19th century or FIA in the 21st collect a limited set of information. The apparent changes noted between these surveys are the result not only of biophysical processes but also cultural, technological, social, and economic drivers that frequently go undocumented.

Effective analysis of these drivers requires that scientists move beyond mere comparisons of two inventories at different points in time to a more holistic analysis that incorporates different types of information reflecting the different influences upon landscape change. Balanced against this goal is the reality that our understanding of the influences – human and environmental – is limited not only by our personal understanding of the subject but also by the data available. Large-scale inventories are a good first step, but they are, by themselves, incomplete. The scientist gains understanding of the past as tidbits of information are revealed: a settler's account of the land he cleared, fire scars on those few surviving trees, commercial records two centuries old, remnants of an old cord road or a railroad line. In the end, scientific honesty demands that one be conservative in the analyses in order to take into account the fractured and incomplete nature of the evidence.

Despite the humbling reality of the available information, landscape ecologists are still able to discern interesting patterns of change that hold lessons beyond the region, watershed, or process in which they were found. As scientists become even more practiced in integrating qualitative and quantitative information, they will improve their ability to assemble mechanistic relationships from survey information and to incorporate knowledge from other ancillary sources as different as cultural surveys, historical accounts, and satellite imagery. The ultimate objective, to gain an understanding of agents of landscape change, is then within their grasp.

Acknowledgments The authors wish to thank Matthias Bürgi, Mike Shelton, William McWilliams, Cynthia Moser, and an unknown reviewer for their thoughtful comments and recommendations on earlier versions of this manuscript.

References

Abrams MD (1998) The red maple paradox. BioScience 48(5):355–364

Allen HL, Fox TR, Campbell RG. (2005) What is ahead for intensive pine plantation silviculture in the South? S J Appl For 29(2):62–69

Antrop M (1998) Landscape change: plan or chaos. Landsc Urban Plan 41:155–161

Arano KG, Munn IA (2006) Evaluating forest management intensity: a comparison among major forest landowner types. For Pol Econ 9: 237–248

Batek JJ, Rebertus WA, Schroeder TL, Haithcoat EC, Guyette RP (1999) Reconstruction of early nineteenth-century vegetation and fire regimes in the Missouri Ozarks. J Biogeogr 26:397–412

Beilmann AP, Brenner LG (1951) The recent intrusion of forests in the Ozarks. Ann Missouri Bot Gar 38:261–282

Bell SS, Fonseca MS, Motten LB (1997) Linking restoration and landscape ecology. Restor Ecol 5(4):318–323

Benac D, Flader S (2004) History of Missouri forests in the era of exploitation and conservation. Spetich, Martin A., ed. Upland oak ecology symposium: history, current conditions, and sustainability. Gen. Tech. Rep. SRS–73. Asheville, NC: U.S. Dep Agric, For Serv, Southern Research Station, pp 36–41
Bolliger J (2005) Simulating complex landscapes with a generic model: sensitivity to qualitative and quantitative classifications. Ecol Complex 2(2):131–149
Bolliger J, Mladenoff DJ (2005) Quantifying spatial classification uncertainties of the historical Wisconsin landscape (USA). Ecography 28:141–156
Bolliger J, Sprott JC, Mladenoff DJ (2003) Self-organization and complexity in historical landscape patterns. Oikos 100:541–553
Bolliger J, Schulte LA, Burrows SN, Sickley TA, Mladenoff DJ (2004) Assessing restoration potentials of Wisconsin (USA) using historical landscape reconstructions. Restor Ecol 12(1): 124–142.
Bolliger, J, Kienast F, Soliva R, Rutherford G (2007) Spatial sensitivity of species habitat patterns to scenarios of land-use change (Switzerland). Landsc Ecol 22(5):773–789.
Bourdo EA, Jr (1956) A review of the General Land Office survey and of its use in quantitative studies of former forests. Ecology 37:754–768
Bragg DC (2002) Reference conditions for old-growth pine forests in the Upper West Gulf Coastal Plain. J Torr Bot Soc 129:261–288
Bragg DC (2004) Dimensionality from obscurity: revisiting historical sources of big tree size. In Yaussy D, Hix DM, Goebel PC, Long RP (eds) Proceedings of the 14th central hardwood forest conference. Gen. Tech. Rep. NE-316. Newton Square, PA: U.S. Dept. Agriculture, Forest Service, Northeastern Research Station. pp 440–447
Bragg DC, Guldin JM, Shelton MG (2006) Stacking the log deck, or some fallacies about natural pine management. In Konnor KF (ed) Proceedings of the 13th biennial Southern Silvicultural Research Conference. Gen. Tech. Rep. SRS-92. Asheville, NC: U.S. Dept. Agriculture, Forest Service, Southern Research Station. pp 37–41
Brandt J, Primdahl J, Reenberg A (1999) Rural land-use and dynamic forces – analysis of 'driving forces' in space and time. In Krönert R, Baudry J, Bowler IR, Reenberg A (eds) Land-use changes and their environmental impact in rural areas in Europe. UNESCO, Paris, France, pp 81–102. Cited in Bürgi et al. 2004
Bruner EM (1930) Forestry and forest fires in Arkansas. University of Arkansas Agricultural Extension Service Extension Circular 281. p 30
Bürgi M, Russell EWB (2000) Evaluating accuracy of two types of early land survey records in the northeastern United States. J Torr Bot Soc 127:94–98
Bürgi M, Turner MG (2002) Factors and processes shaping land cover and land cover changes along the Wisconsin River. Ecosystems 5:184–201
Bürgi M, Russell EWB, Motzkin G (2000) Effects of postsettlement human activities on forest composition in the north-eastern United States: a comparative approach. J Biogeogr 27: 1123–1138.
Bürgi M, Hersperger AM, Schneeberger N (2004) Driving forces of landscape change – current and new directions. Landsc Ecol 19(8):857–868
Bürgi M, Hersperger A, Hall M, Russell EWB, Schneeberger N (2007) Using the past to understand the present land-use and land-cover.In Kienast F, Ghosh S, Wildi O (eds) A changing world: challenges for landscape research, Springer, pp 133–145
Chapman HH (1913) Prolonging the cut of southern pine. I. Possibilities of a second cut. Yale University Forestry School Bulletin 2:1–22
Dirnböck T, Dullinger S, Grabherr G (2003) A regional impact assessment of climate and land-use change on alpine vegetation. J Biogeogr 30(3):401–417
Duerr WA (1950) Forest statistics for southwest Arkansas. U.S. Dep Agric, For Serv Forest Survey Release 65, p 21
Edwards TC, Cutler DR, Zimmermann NE, Geiser L, Moisen GG (2006) Effects of sample survey design on the accuracy of classification tree models in species distribution models. Ecol Model 199:132–141

Eldredge IF (1937) Forest resources of southwest Arkansas. U.S. Dep Agric, For Serv Forest Survey Release 27, p 21
Fei S (2007) The geography of American tree species and associated place names. J Forestry 105:84–90
Ferris R, Humphrey JW (1999) A review of potential biodiversity indicators for application in British forests. Forestry 72(4):313–328
Fitzgerald D (1991) Beyond tractors: The history of technology in American agriculture. Technol Cult 32(1):114–126
Foster DR, Motzkin G, Slater B (1998) Land-use history as long-term broad-scale disturbance: regional forest dynamics in central New England. Ecosystems 1:96–119
Foti TL (2004) Upland hardwood forests and related communities of the Arkansas Ozarks in the early 19th Century. In: Spetich MA (ed) Upland oak ecology symposium: history, current conditions, and sustainability. Gen. Tech. Rep. SRS–73. Asheville, NC: U.S. Dept. Agriculture, Forest Service, Southern Research Station. pp 21–29
Goforth BR, Minnich RA (2007) Evidence, exaggeration, and error in historical accounts of chaparral wildfires in California. Ecol Appl 17:779–805
Guisan A, Zimmermann NE (2000) Predictive habitat distribution models in ecology. Ecol Mod 135:147–186.
Guyette RP, Muzika R-M, Dey DC (2002) Dynamics of an anthropogenic fire regime. Ecosystems 5(5):472–486
Hall WL (1945) Is pine coming or going in south Arkansas? J Forestry 43:634–637
Hedlund A, Earles JM (1970) Forest statistics for Arkansas counties. U.S. Dep Agric, For Serv Res Bull SO-22, 52pp
Hines FD (1988) Forest statistics for southwest Arkansas counties—1988. U.S. Dep Agric, For Serv Res Bull SO-140, 32pp
Hobbs RJ, Norton DA (1996) Toward a conceptual framework for restoration ecology. Rest Ecol 4(2):93–110
Kersten EW, Jr (1958) Changing economy and landscape in a Missouri Ozarks area. Ann Assoc Amer Geographers 48(4):398–418
Kéry M (2002) Inferring the absence of a species: a case study of snakes. J Wildl Manage 66: 330–338
Laiolo P, Dondero F, Ciliento E, Rolando A (2004) Consequences of pastoral abandonment for the structure and diversity of the alpine avifauna. J Appl Ecol 41:294–304
Landres PB, Morgan P, Swanson FJ (1999) Overview of the use of natural variability concepts in managing ecological systems. Ecol Appl 9(4):1179–1188
Linklater A (2002) Measuring America: how an untamed wilderness shaped the United States and fulfilled the promise of democracy. Walker Publ Company, New York, NY, 288pp
Lischke H, Bolliger J, Seppelt R (2007) Dynamic spatio-temporal landscape models. A changing world: challenges for landscape research (eds. Kienast F, Ghosh S, Wildi O), Springer, Dordrecht, pp 273–296
Lütolf M, Kienast F, Guisan A (2006) Strategies to improve species distribution model performance using occurrence data. J Ecology 43:802–815
Manies KL, Mladenoff DJ (2000) Testing methods used to produce landscape scale presettlement vegetation maps from the U.S. General Land Office (GLO) survey records. Land Ecol 15: 741–754
McRoberts RE (1999) Joint annual forest inventory and monitoring symposium: The North Central perspective. J Forestry 97(12):27–31
Mohr C (1897) The timber pines of the southern United States. USDA Forestry Division, Bulletin 13. 176pp
Moser WK, Treiman TB, Johnson EE (2003) Species choice and risk in forest restoration: A comparison of methodologies for evaluating forest management decisions. Forestry 76(2):137–147
Moser WK, Hansen MH, Hatfield MA, Nigh TA (2006) Evaluating sustainability: A method for assessing vegetation change in southern Missouri, U.S.A.: 1820–2003. In: Lafortezza R et al.

(eds) IUFRO landscape ecology conference, patterns and processes in forest landscapes: Consequences of human management, Sept. 26–29, 2006 – Locorotondo, Bari, Italy. pp 399–408
Moser WK, Hansen MH, Brand GJ, Treiman TB (2007a) Missouri's forest resources in 2005. USDA Forest Service, Resource Bulletin NRS-15. 25pp
Moser WK, Hansen MH, McWilliams W, Sheffield R (2007b) Shortleaf pine composition and structure in the United States. In: Kabrick JM, Dey DC, Gwaze D (eds) Shortleaf pine restoration and ecology in the Ozarks: proceedings of a symposium; 2006 November 7–9; Springfield, MO. Gen. Tech. Rep. NRS-P-15. Newtown Square, PA: U.S. Department of Agriculture, Forest Service, Northern Research Station: 19–27
Murphy MT (2003) Avian population trends with the evolving agricultural landscape of eastern and central United States. The Auk 120(1):20–34
Nigh TA, Larsen DR, Kabrick JM, Moser WK (2006) Forest types management matrix. Columbia, MO: Missouri Dept. Conservation. Unpublished table. 3p
Oliver CD, Larson BC (1996) Forest stand dynamics. Update Edition. John Wiley and Sons Inc., New York
Olmsted FE (1902) A working plan for forest lands near Pine Bluff, Arkansas. USDA Bureau of Forestry, Forest Bulletin 32. Washington, DC.
Quick RT, Hedlund MS (1979) Forest statistics for Arkansas' delta counties. USDA Forest Service, Resource Bulletin SO-71. 26 p
Radeloff VC, Mladenoff DJ, Boyce MS (2000) A historical perspective and future outlook on landscape scale restoration in the Northwest Wisconsin Pine Barrens. Restor Ecol 8:119–319
Read RA (1952) Tree species occurrence as influenced by geology and soil on an Ozark north slope. Ecology 33(2):239–246
Rebertus AJ, Shifley SR, Richards RH, Roovers LM (1997) Ice storm damage to an old-growth oak-hickory forest in Missouri. Amer Midl Nat 137:48–61
Reynolds RR (1956) Upland hardwoods in south Arkansas pine stands—aerial or underground fifth column? J Forestry 54:585–586
Rogers WR, Munn IA (2003) Forest management intensity: a comparison of timber investment management organizations and industrial landowners in Mississippi. S J Appl For 27(2):83–91
Rosson JF, Hartsell AJ, London JD (1995) Forest statistics for southwest Arkansas counties—1995. USDA Forest Servive, Resource Bulletin SO-194. 40pp
Rousseau R, Kaczmarek D, Martin J (2005) A review of the biological, social, and regulatory constraints to intensive plantation culture. S J Appl For 29:105–109
Russell EWB (1997) People and the land through time: Linking ecology and history. Yale Univ Press, New Haven
Schoolcraft HR (1821) Journal of a tour into the interior of Missouri and Arkansaw, from Potosi, or mine a Burton, in Missouri Territory, in a south-west direction, toward the Rocky Mountains, performed in the years 1818 and 1819. Richard Phillips & Co, London, 96pp
Schroeder WA (1981) Presettlement prairie of Missouri. Natural History Series, No. 2, Missouri Dept. Conservation, Jefferson City. 37pp
Schulte LA, Mladenoff DJ (2001) The original U.S. Public Land Survey records: Their use and limitations in reconstruction pre-European settlement vegetation. J Forestry 99:5–10
Schulte LA, Mladenoff DJ (2005) Severe wind and fire regimes in northern forests: Historical variability at the regional scale. Ecology 86(2):431–445
Schulte LA, Mladenoff DJ, Nordheim EV (2002) Quantitative classification of a historic northern Wisconsin (USA) landscape: mapping forests at regional scales. Can J For Res 32:1616–1638
Schulte LA, Liebman M, Asbjornsen H, Crow TR (2006) Agroecosystem restoration through strategic integration of perennials. J Soil Water Conserv 61:164A–169A
Silbernagel J, Martin SR, Gale MR, Chen J (1997) Prehistoric, historic, and present settlement patterns related to landscape structure in the Eastern Upper Peninsula of Michigan, USA. Land Ecol 12(4):223–240
Skånes HM, Bunce RGH (1997) Directions of landscape change (1741–1993) in Virestad, Sweden – characterised by multivariate analysis. Landsc Urban Plan 38(1):675

Söderström B, Svensson B, Vessby K, Glimskär A (2001) Plants, insects, and birds in semi-natural pastures in relation to local habitat and landscape factors. Biodiv and Conserv 10:1839–1863

Stanturf JA, Kellison RC, Broerman FS, Jones SB (2003) Productivity of southern pine plantations—where are we and how did we get here? J Forestry 101:26–31

Sternitzke HS (1960) Arkansas forests. USDA Forest Service Forest Survey Release 84, 58pp

Stewart LO (1935) Public land surveys: history, instructions, and methods. Collegiate Press, Inc., Ames, IA, 202pp

Swetnam TW, Allen CG, Betancourt JL (1999) Applied historical ecology: Using the past to manage for the future. Ecol Appl 9(4):1189–1206

Tilman D, Fargione J, Wolff B, D'Antonio C, Dobson A, Howarth R, Schindler D, Schlesinger WH, Simberloff D, Swackhamer D (2001) Forecasting agriculturally driven global environmental change. Science 292(5515):281–284

Turner RE, Rabalais NN (2003) Linking landscape and water quality in the Mississippi River Basin for 200 years. BioScience 56:563–572

Wear DN (2002) Land use. In: Wear DN, Greis JG (eds) Southern forest resource assessment. USDA Forest Service, General Technical Report SRS-53, pp 153–173

Wear DN, Greis JG (eds) (2002) Southern forest resource assessment. USDA Forest Service, General Technical Report SRS-53, 635pp

White CA (1983) A history of the rectangular survey system. Washington, DC: U.S. Dept. Interior, Bureau of Land Management

Woodall CW, Williams MS (2005) Sampling protocol, estimation, and analysis procedures for down woody materials indicator of the FIA program. USDA Forest Service, General Techical Report NC-256, 47pp

Wu JJ, Hobbs R (2002) Key issues and research priorities in landscape ecology: An idiosyncratic synthesis. Landsc Ecol 17:355–365

Zhang Q, Pregitzer KS, Reed DD (2000) Historical changes in the forests of the Luce District of the Upper Peninsula in Michigan. Amer Midl Nat 143:94–110

Chapter 10

Shape Irregularity as an Indicator of Forest Biodiversity and Guidelines for Metric Selection

Santiago Saura, Olga Torras, Assu Gil-Tena and Lucía Pascual-Hortal

Abstract The development of quantitative methods in landscape ecology has provided new perspectives for analysing the distribution of forest biodiversity. The shape of landscape patterns may be linked to the imprint of the factors that have configured the boundaries and affected the diversity of forest patches. There is now available a large number of spatial metrics for characterising the shape of landscape patterns. However, the properties, behaviour and adequacy of these shape metrics for landscape pattern analysis have not been sufficiently evaluated, and there is a risk of potential misuse and arbitrary metric selection. We review the main characteristics and limitations of existing landscape shape metrics, and explore the relationships between shape irregularity metrics and forest landscape biodiversity in the regions of Galicia and Asturias (NW Spain). We analysed data from the Spanish Forest Map, the Third Spanish National Forest Inventory and the Spanish Atlas of Vertebrates at two different levels: forest types with homogenous composition and different total areas, and equally-sized heterogeneous UTM 10 × 10 km cells. We found that shape irregularity metrics were significantly correlated with forest vegetation diversity and with the richness of forest birds, mammals and total vertebrate species. Shape metrics correlated more with forest biodiversity variables than fragmentation metrics. We conclude that shape irregularity metrics may serve as valuable spatial indicators of forest biodiversity at the landscape scale, and suggest that more attention should be paid to shape as a key characteristic of landscape patterns.

10.1 Introduction

10.1.1 Shape and the Imprint of Human and Natural Factors in the Forest Landscape

The landscape is a mosaic of patches with varying sizes and shapes resulting from the interaction of natural and human factors (e.g. Mladenoff et al. 1993; Forman

S. Saura
Department of Agroforestry Engineering, Higher Technical School of Agrarian Engineering, University of Lleida, Lleida, Spain; Forest Technology Centre of Catalonia, Solsona, Lleida, Spain
e-mail: ssaura@eagrof.udl.es

R. Lafortezza et al. (eds.), *Patterns and Processes in Forest Landscapes*,

1995; Hulshoff 1995). Shape refers to the form of an area (e.g. patch) as determined by variation in its margin or border, and is considered one of the most relevant properties of landscape patterns (Forman 1995). Many factors and ecological processes influence the shape of the forest landscape in different ways. Topography and geomorphology are major determinants of landscape shape, with plains usually having the smoothest and most-compact shaped patches, and slopes having the most elongated and convoluted patches (Forman 1995). Forest patches resulting from some natural disturbances, such as forest fires and blowdowns, present considerably complex and lobulated boundaries (Haydon et al. 2000; Lindemann and Baker 2001). Elongated patches of natural riparian forest vegetation are mostly influenced by hydrogeomorphic processes, such as floodings and sedimentation (Rex and Malanson 1990). Soil types and moisture patterns may also be key variables to understand the shape of forest landscapes (Saura and Carballal 2004).

Landscapes resulting from human activity tend to present simpler shapes than more natural landscapes. The imprint of human influences in the landscape is often reflected in simplified shapes, such as those resulting from cultivation, transportation lines, urbanisation, forest harvesting, etc. Differentiating natural from human-created patches may be in some cases simple because of distinctive boundary forms; boundaries of natural patches are curved, while human-created patches typically contain one or more straight lines (Forman 1995).

Therefore, shape features may be successfully related to the origin or degree of human alteration of the patches in the landscape (e.g. Moser et al. 2002; Saura and Carballal 2004). For example, Krummel et al. (1987) noted that the shape of the forest patches varied considerably between agricultural areas (in which human activities imposed regular boundaries to the remnant forest) and other more natural areas where the same class exhibited more complex shapes. Iverson (1988) found that deciduous forests presented more irregular boundaries than evergreen (plantation) forests in Illinois (USA). Mladenoff et al. (1993) concluded that an intact primary old-growth forest landscape in northern Wisconsin (USA) was significantly more complex in shape than a human-disturbed forest landscape in the same area. Lindemann and Baker (2001) showed that the shapes of forest patches resulting from blowdowns after windstorms were more complex than those resulting from silvicultural treatments. Saura and Carballal (2004) found that native forest presented both more complex and elongated boundaries than exotic forests in NW Spain.

However, Hulshoff (1995) concluded that there was no difference between the shape of natural and human-modified patches in the Netherlands, because the shape of natural patches was mostly fixed by human-modified neighbour patches. In that study Hulshoff (1995) considered a five-class classification with only one forest type. When a single forest class is discriminated, all forest patches will necessarily fall next to other dominant and possibly less natural cover types (e.g. agricultural lands), and hence their borders will be determined by adjacent land uses that may impose simpler shapes to the forest. However, if a relatively large number of forest classes is considered, a natural forest may be adjacent to other more or less natural forest types, and then the boundaries between these classes may be determined by physical and biological factors that produce more complex shapes. Therefore, and as noted by Saura and Carballal (2004), a relatively high thematic detail may be required in order to make evident the differences in shape between different land cover classes.

10.1.2 Landscape Pattern as an Indicator of Forest Biodiversity

Conserving biodiversity is currently an imperative issue for sustainable forestry (Hagan and Whitman 2006). Characterisation and monitoring of forest biodiversity has therefore received increasing attention in recent years (e.g. Noss 1990, 1999; Lindenmayer 1999; Lindenmayer et al. 2000; Newton and Kapos 2002; Allen et al. 2003; McAfee et al. 2006). However, progress has been made difficult by the complexity of all the aspects involved in the concept of biodiversity and the large areas to be studied. If we consider biodiversity as a metaconcept that includes all forms of life, in practice it is not possible to deal with its whole assessment.

There is an urgent need to develop cost-effective methods for biodiversity assessment, avoiding complex and time-consuming approaches that may not be able to address the need for continuous and operational information on forest biodiversity. In this context, it is recognised that in many circumstances it may not be possible to measure the target directly and hence it may be necessary to seek indirect or surrogate measures, which are called forest biodiversity indicators (Lindenmayer et al. 2002). These indicators should be proven linear correlates of the biodiversity aspect being evaluated (Duelli and Obrist 2003). The development of biodiversity indicators aimed at assessing general trends in the different components of forest ecosystems is a key research in applied forest ecology (Noss 1999) and a way to better understand biodiversity for its conservation.

Most of the forest biodiversity indicators used so far are based on field surveys (forest inventory plots), which in general are costly and can only be undertaken with low sampling intensities in large areas. However, the recent development of quantitative methods in landscape ecology offers new perspectives in this context (Forman 1995; Moser et al. 2002; Saura and Carballal 2004; Loehle et al. 2005). It is now recognised the need to consider different spatial and temporal scales in designing forest biodiversity indicators, in order to capture the complex dynamics and relationships existing in biodiversity management (Failing and Gregory 2003).

From this point of view, landscape pattern may be an important feature for inventorying, monitoring and assessing terrestrial biodiversity structure at the regional landscape level of organisation (Noss 1990). Some landscape pattern metrics may serve as spatial indicators for assessing whether critical components and functions of forests are being maintained (García-Gigorro and Saura 2005). In this sense, the Improved Pan-European Indicators for Sustainable Forest Management by the Ministerial Conference on the Protection of Forests in Europe (MCPFE) include the indicator 4.7 'Landscape pattern' (landscape-level spatial pattern of forest cover) within the criteria 'Maintenance, Conservation and Appropriate Enhancement of Biological Diversity in Forest Ecosystems' (MCPFE 2002). In the same way, Allen et al. (2003) suggested for New Zealand that forest biodiversity indicators should include, among others, the spatial arrangement of forests, and similar considerations were made by Newton and Kapos (2002) in the context of national forest inventories. Although many international processes and conventions as well as most national forest assessments do not yet concern for spatial pattern (Kupfer 2006), this situation is changing rapidly. For example the recent Third Spanish National Forest Inventory is now including several landscape pattern metrics for the assessment of forest habitat biodiversity (Ministerio de Medio Ambiente 2005).

Forest landscape pattern indicators are easy and cost-effective to monitor, requiring less intensive ground-truthing than monitoring landscape composition (Noss 1990), since they can usually be obtained through remote sensing systems and GIS at various spatial resolutions (Noss 1999; Kupfer 2006). This is beneficial for the development of forest biodiversity indicators for sustainable forestry (Hagan and Whitman 2006). Nonetheless, further validation of the use of landscape pattern as a biodiversity indicator is needed (Noss 1999), particularly because landscape metrics cannot unambiguously explain the response of an ecological process to landscape structure (Tischendorf 2001).

Forest fragmentation is one of the most commonly known features of landscape pattern, commonly regarded as a major determinant of biodiversity loss. Forest fragmentation can refer to the process of forest cover loss and isolation or more specifically to the shifts in the spatial configuration of forest patches (Fahrig 2003). Indeed, most of the relevance of landscape structure to biodiversity is due to the voluminous literature on habitat fragmentation (Noss 1990). Comparatively, the effects of landscape shape on biodiversity have been much less studied. However, recent research have shown the relevance of this aspect of landscape pattern and its influence on several components of biodiversity (Moser et al. 2002; Honnay et al. 2003; Saura and Carballal 2004; Barbaro et al. 2005; Brennan and Schnell 2005; Radford et al. 2005). Indeed, the relationships between landscape patterns and the interaction of natural and human factors that are and have been influencing the landscape may provide a basis to link those spatial patterns with biodiversity distribution (Moser et al. 2002; Saura and Carballal 2004). As noted by Moser et al. (2002), shape complexity may be a good predictor of species richness as it may serve as a measure of land use intensity. Potentially, shape metrics may be a valuable indicator of forest biodiversity at the landscape scale. However, this hypothesis has only been subjected to limited testing and we here wish to provide further insights in this respect.

10.2 Shape Metrics for Landscape Analysis

10.2.1 A Brief Review of Landscape Shape Metrics

The rapid development of quantitative methods in landscape ecology in the last two decades has made available a large number of spatial metrics for characterising the shape of landscape patterns (e.g. Lagro 1991; Gustafson and Parker 1992; Bogaert et al. 1999; McGarigal et al. 2002; Moser et al. 2002; Saura and Carballal 2004; Kojima et al. 2006). Some of these metrics have been specifically developed within the landscape ecology literature while others have been adapted from other fields such as geography, image processing, or physical sciences. As noted by Forman (1995) ‘patch shape is a much richer concept than size because it varies in so many ways’. Therefore no single metric is able to characterise all the aspects of shape, and it may be considered natural that different shape metrics are available and are being used for landscape ecological applications.

However, the properties, behaviour and adequacy for landscape pattern analysis of these shape metrics have not been sufficiently evaluated. There is a considerable risk of potential misuse, since for uninformed users almost any available metric may be considered suitable to measure ‘landscape shape’ without further concern. Indeed, many of these available metrics are often applied without interrogation about what are they really measuring, in a context where landscape ecologists usually lack of solid and objective guidelines for selecting an appropriate metric for their particular applications. As stated by Li and Wu (2004) ‘after two decades of extensive research, interpreting indices remains difficult because the merits and caveats of landscape metrics remain poorly understood. What an index really measures is uncertain even when the analytical aspects of most indices are quite clear’. They also note that ‘without theoretical guidance, landscape ecologists are often overwhelmed by numerous indices and spatial statistical methods, as well as by increasing volumes of GIS and remote sensing data’.

For these reasons, we here intend to briefly review some of the most common and useful metrics for characterising landscape shape, and to provide some guidelines on what each of these metrics is really quantifying (Table 10.1). This may be useful for selecting the most appropriate metrics for different landscape analysis applications, and particularly for their potential use as biodiversity indicators at the landscape scale.

Most of the available metrics can be computed either at the patch level (i.e. a metric value characterising the shape of a single patch), class level (i.e. a metric value summarising the shape characteristics of a certain land cover type) or landscape level (i.e. a metric value summarising the shape of all the patches in a certain landscape). This is the case of all the patch-level metrics described in Table 10.1, which can be easily extended to the class or landscape level. This is done for most of the metrics through an average or area-weighted average of the metric values for the individual patches in that class or landscape (e.g. McGarigal et al. 2002; Saura and Carballal 2004). For some other metrics, such as the number of shape characteristic points, the class or landscape-level values are obtained just as the sum of patches, values or, when units with very different areas are to be compared, as a density by dividing that sum by total area (Moser et al. 2002; Saura and Carballal 2004). Other two shape metrics that have been commonly used in landscape literature are the landscape shape index and the perimeter-area fractal dimension (e.g. McGarigal et al. 2002; Saura and Carballal 2004). These two metrics are only defined for sets of several patches (i.e. class or landscape level) and not at the patch-level, and therefore are not reported in Table 10.1. The landscape shape index is computed similarly to

Table 10.1 Description and main characteristics of different patch-level shape metrics, including their reaction to the four spatial changes illustrated in Figure 10.1, here indicating if the metric increases (+), decreases (–) or is not affected (0) by that spatial change. Patch perimeter here refers to the length of the outer edge (boundary), not including the inner edges caused by perforations

Shape metric	Description	Range of variation	Raster or vector	Computation	Metric reaction to spatial changes			
					Complexity	Elongation	Perforation	Size
Perimeter-area ratio	Ratio between patch perimeter and patch area	0–unbounded	Both	GIS, Fragstats, APACK, Patch Analyst	+	+	+	+
Shape index	Based on the ratio between patch perimeter and the square root of patch area (e.g. Saura and Carballal 2004)	1–unbounded	Both	GIS, Fragstats, APACK, Patch Analyst	+	+	+	0
Patch fractal dimension	Based on the ratio between the logarithm of patch perimeter and the logarithm of patch area (Kojima et al. 2006)	1–unbounded	Both	GIS, Fragstats, Patch Analyst	+	+	+	+
Circumscribing circle index	One minus the ratio between patch area and the area of the smallest circle circumscribing the patch	0–1	Both	Fragstats	0	+	+	0
Convex hull area index	One minus the ratio between patch area and the area of the convex hull of the patch	0–1	Both	Script/ programming	+	0	+	0
Largest axis index	Based on the ratio between the length of the straight line connecting the two furthest-apart points in the patch and the square root of patch area (Saura and Carballal 2004)	1–unbounded	Both	Script/ programming	0	+	0	0

Table 10.1 (continued)

Shape metric	Description	Range of variation	Raster or vector	Computation	Metric reaction to spatial changes			
					Complexity	Elongation	Perforation	Size
Linearity index	Based on the ratio between patch area and the average cell value of the medial axis transformation of a patch, where each pixel value represents the distance (in pixels) to the nearest edge (Gustafson and Parker 1992)	0–1	Raster	Fragstats	+	+	–	0
Number of shape characteristic points	Minimum number of points (polygon vertices) needed to describe a patch boundary (Moser et al. 2002)	1–unbounded	Vector	Script/ programming	+	0	0	0
Twist number	Total number of twists that divide the patch perimeter in a set of straight segments (Bogaert et al. 1999)	4–unbounded	Raster	Script/ programming	+	0	0	0
Contiguity index	Assesses the spatial connectedness, or contiguity, of cells within a patch to provide an index of patch boundary configuration and patch shape (Lagro 1991)	01	Raster	Fragstats	–	–	–	+

the shape index (Table 10.1) but treating all the area and perimeter of all the patches in the landscape as a single large patch. However, unlike the shape index, if the set of patches comprises multiple circular patches of different sizes, the landscape shape index will not be equal to 1 (Saura and Carballal 2004). The perimeter-area fractal dimension derives from the scaling properties of self-similar fractal objects, and equals 2 divided by the slope of the linear regression of the logarithm of patch areas against the logarithm of patch perimeters (McGarigal et al. 2002). There are some other shape-related metrics as well, but they will not be considered here since they are rarely used or are obtained after slight variations of some of the metrics already described in Table 10.1, measuring the landscape shape in quite a similar way.

In practice the term 'shape' is used quite loosely in many landscape pattern analysis applications, with metrics with quite different characteristics being equally regarded as measuring 'shape'. However, the concept of shape involves several different aspects, and not all the available metrics are characterising landscape shape in the same way. Hereafter, we will use the term 'shape irregularity' to refer to patterns with complex and/or elongated shapes, considering as regular those shapes that are both compact (isodiametric) and with simple boundaries (e.g. circles and squares), as in Saura and Carballal (2004).

10.2.2 Spatial Characteristics and Limitations of Landscape Shape Metrics

At least in some cases it would be interesting to differentiate between shape complexity (lobulated, convoluted, dendritic boundaries) and shape elongation (Figure 10.1). A shape can be elongated but with simple, straight and regular boundaries (e.g. rectangles), and it can also be relatively isodiametric but still present considerably complex boundaries (Figure 10.1). However, most of the metrics are not able to differentiate these two shape characteristics, and are sensitive both to complexity and elongation (Table 10.1). As noted by Moser et al. (2002) 'narrow, elongated landscape structures are systematically classified as highly complex even if they are quite simple in shape'. In fact, only two metrics (the number of shape characteristic points and the twist number) are only affected by the shape complexity and not by any other of the spatial changes considered in Figure 10.1. Only the values of the largest axis index can be exclusively attributed to shape elongation (Table 10.1). However, also the circumscribing circle index and the convex hull area index would be specifically related to shape elongation and complexity, respectively, if the area enclosed by the patch perimeter (including the inner holes caused by perforations) was used instead of the patch area for their computation (Table 10.1).

On the other hand, some metrics may perform poorly in various applications if they are sensitive to other landscape pattern characteristics different from shape itself. Several metrics that are regarded as measuring shape are in fact sensitive and more largely affected by patch size and pattern fragmentation (Table 10.1). This is the case for the perimeter-area ratio, which increases for smaller patches even if the

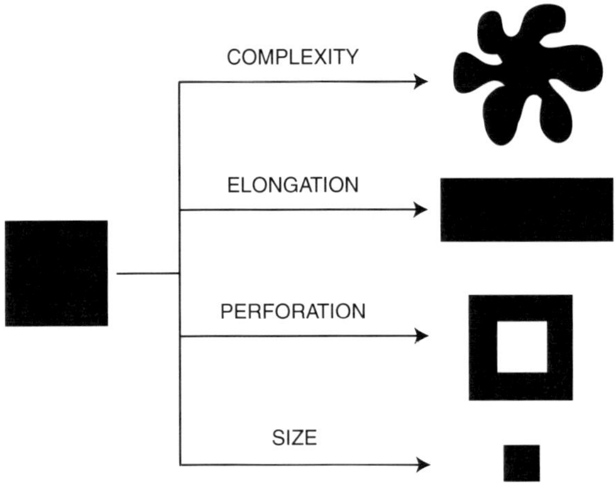

Fig. 10.1 Four spatial changes in a forest patch that may affect the values of different shape-related metrics (forest area shown in *black*). The four changes are: increased complexity (with patch area and elongation remaining constant), increased elongation (with patch area remaining constant), perforation (with patch perimeter and shape remaining constant) and decrease in the size of the patch (with shape remaining constant)

shape is held constant, and for the landscape shape index. This latter metric increases for more fragmented patterns and actually conveys the same information than the AI metric proposed by He et al. (2000) to measure landscape pattern aggregation, as noted by Bogaert et al. (2002). Many of the shape metrics are also very sensitive to patch perforations, even when both the outer perimeter and the inner holes may present the same shape (Figure 10.1, Table 10.1).

In general, landscape metrics are much easier to interpret if they have a predefined and bounded range of variation (independent of the particular analysed landscape), especially if the metric is relative, ranging from 0 to 1 or from –1 to 1 (Li and Wu 2004). However, many of the shape metrics are unbounded in their original definition (Table 10.1), although in many cases help is available in the form of standardisation operations (Li and Wu 2004).

Other considerations (apart from those just described and reported in Table 10.1) should be taken into account for several shape metrics before deciding their operational use. For example, the patch 'fractal' dimension, despite the name given to this metric in the literature, is only loosely connected with geometric fractals, and is not a true fractal measure (Saura and Carballal 2004; Kojima et al. 2006). In fact, although the range of variation of this metric is regarded to be restricted between 1 and 2 (e.g. McGarigal et al. 2002), the values of the patch fractal dimension, unlike true fractal measures, can in fact be higher than 2, and are also dependent on the units adopted to evaluate the area and perimeter of the patches in the landscape (Kojima et al. 2006). The other fractal-related metric (perimeter-area fractal dimension) is usually regarded as a measure of shape complexity of sets of patches with different sizes (e.g. McGarigal et al. 2002). However, this metric is only a true measure of shape complexity when the pattern under analysis is really self-similar (as perfect fractals), which is not always the case in real world landscape patterns (Saura and Carballal 2004). The perimeter-area fractal dimension should be better interpreted as the rate at which the shape index of the patches increases with their size (Saura and Carballal 2004). The twist number, or the normalised metric derived from it (Bogaert et al. 1999), may appear promising as a measure of landscape shape complexity in raster data. However, it has been noted that the twist number may be largely affected by the aliasing (staircase) effect occurring when representing the spatial patterns in raster data, therefore not adequately reflecting the true complexity of patch boundaries (Saura and Carballal 2004).

10.2.3 Computation of Landscape Shape Metrics and Scale Issues

A relevant issue to consider for the practical application of these metrics is how easily they can be calculated by end users. Many of these metrics are computed just from the area and perimeter of the patches, and therefore can be easily obtained through the information and basic functionalities available in any GIS or image processing software (Table 10.1), including the landscape shape index and the perimeter area fractal dimension. Other metrics are more complex but are included (as well as the former) in free available software that has been specifically developed for automatically computing landscape pattern metrics (Table 10.1). Two of them, Fragstats 3.3 (McGarigal et al. 2002) and APACK 2.23 (Mladenoff and DeZonia 2004), are stand-alone programs that compute pattern metrics in raster format. Patch Analyst 3.12 (Rempel 2006) is a free extension to ArcView 3.x that computes some of these metrics in vector format. Finally, some other metrics are not easy to compute and are not included in any available software, and currently require some additional programming or scripts developed specifically for that purpose (Table 10.1). This may limit their widespread use by many analysts, despite its potential interest for particular applications.

Landscape pattern data are either available as raster data (e.g. per-pixel classification of satellite images) or vector data (e.g. interpretation or segmentation of remotely sensed images). Ideally, a shape metric should not be limited by the type of available landscape data, in order to be widely applicable without need of data transformation (e.g. vector to raster), since those transformations may produce considerable distortions in landscape shape and other pattern characteristics (Bettinger et al. 1996; Congalton 1997). However, some metrics are only suited for either vector (number of shape characteristic points) or raster data (linearity index, twist number, contiguity index), as shown in Table 10.1.

On the other hand, the increasing availability of a wide variety of GIS and remotely sensed data allows for the characterisation of landscape shape at multiple spatial scales. However, it is important to note that the values of most shape metrics are largely dependent on the scale of the analysed data, which prevents their direct comparison in many multitemporal studies or cross-site comparisons. Only a few landscape shape metrics may be considered robust to spatial resolution or extent effects, such as the area-weighted average of the shape index (Saura 2002), or the perimeter-area fractal dimension (Saura and Martínez-Millán 2001). For some other metrics help may be available through specific landscape pattern scaling functions (Wu et al. 2002; Wu 2004; Saura and Castro 2007).

Apart from this, it is important to note that some raster-based metrics are largely affected by spurious subpixel resampling, which increases the total number of pixels but does not vary the underlying spatial pattern (Saura 2004; García-Gigorro and Saura 2005). This important limitation occurs for example for the contiguity index, and should be taken into account for a reliable use of this metric.

For space limitations, in the rest of this chapter we will just present the results for three of the metrics described in Table 10.1 (shape index, circumscribing circle index, and number of shape characteristic points), adapted to be computed at the class or landscape level as described below. These metrics have been shown to be the best performing ones in previous subject-related studies (Moser et al. 2002; Saura and Carballal 2004), are free of different problems or limitations reported above, and can be directly computed in the original vector format of the landscape data to be analysed, as described in the next sections.

10.3 Linking Shape Irregularity with Forest Biodiversity

10.3.1 Study Area

In order to evaluate the relationship between shape irregularity metrics and forest biodiversity, we analysed a large study area comprising the Spanish regions of Galicia (provinces of A Coruña, Lugo, Ourense and Pontevedra) and Asturias (comprising a single province with the same name), with a total area of about 40178 km^2 (Figure 10.2). Galicia and Asturias are characterised by an Atlantic climate with mild temperatures: mean annual temperature is about 13°C and mean annual precipitation is above 900 mm, rising to more than 2000 mm in the mountainous areas, especially in Galicia. Nevertheless, the interior areas of Lugo and Ourense present a more continental character than the rest of the study area, with summer drought and more frost days. Galicia and Asturias present predominantly acid soils and a complex topography, with altitudes ranging from sea level up to 2648 m in Asturias, and up to 2124 m in Galicia. The population density of Galicia and Asturias is above the Spanish average, with about 94 and 102 inhabitants per km^2, respectively. In Galicia most of the population is concentrated in the coastal areas, while in Asturias the most important cities are located in the central zone of this region, with a decreasing population trend from the coast to the interior and following the mining valleys. According to the Third Spanish National Forest Inventory (NFI), the percentage of forests (land covered by a forest tree canopy cover above 20%) in Galicia and Asturias is about 43% and 40%, while 28% and 26% are agricultural lands, respectively. The forest landscape of Galicia and Asturias has been

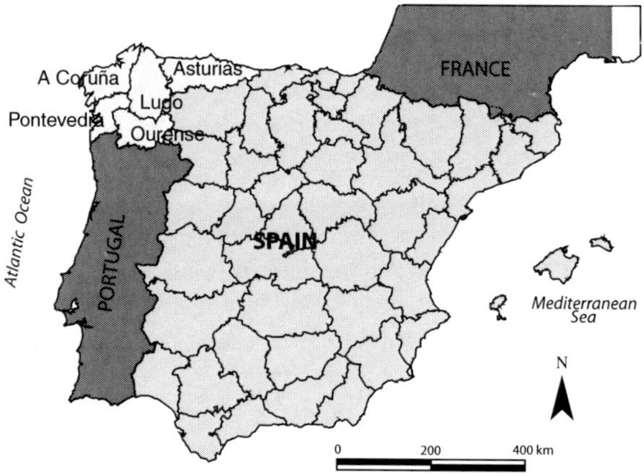

Fig. 10.2 Location of the five Spanish provinces comprising the study area

deeply modified by human action, especially during the last centuries (Manuel and Gil 2002; Manuel et al. 2003). The most abundant forest tree species in the studied provinces are, according to the NFI, *Eucalyptus globulus*, *Pinus pinaster*, *Quercus robur*, *Castanea sativa*, *Quercus pyrenaica*, *Pinus radiata*, *Fagus sylvatica*, *Betula* spp., and *Pinus sylvestris*.

The study was performed at two different levels in order to provide more generality to the analysis and to adequate to the characteristics of the different available data: (1) forest types with homogenous composition and different total areas, and (2) equally-sized but heterogeneous (comprising several forest and land cover types) UTM 10 × 10 km cells.

10.3.2 Forest Type Analysis

We discriminated 50 different forest types in Galicia and Asturias from the information provided by the Spanish Forest Map, which was developed within the recent NFI (Ministerio de Medio Ambiente 2003). The Spanish Forest Map (SFM, scale 1:50000) is a vector format map developed from the interpretation of high-resolution aerial photographs, combined with pre-existing maps and field inventory data. The minimum mapping unit is in general 6.25 ha, lowering to 2.2 ha in the case of forest patches embedded in a non-forest land use matrix. Forest types were discriminated attending to their tree species composition and the province in which they were present (i.e. *Quercus robur* forests in the province of Pontevedra and *Quercus robur* forests in the province of Lugo were considered as two different forest types), and comprised a total of 26240 forest patches and 1857057 ha. The area of the forest types ranged from 4178 ha to 115068 ha. Our definition of forest includes all areas with forest tree canopy cover ranging from 5 to 100%, as defined in the SFM.

The information from 8787 inventory plots of the Third Spanish National Forest Inventory (NFI) in the five provinces was summarised to characterise the following six NFI vegetation diversity variables for each forest type:

- Tree species density (richness divided by sampled area).
- Shannon-Wiener diversity index (obtained from the abundances of forest tree species).
- Shrub species density (richness divided by sampled area).
- Number of old growth trees per ha, defined as those trees with a diameter at breast height (DBH) above 50 cm.
- Amount of standing dead wood (number of stems per ha).
- Percentage of uneven aged stands (as a measure of stand structure complexity).

Tree and shrub species density were used instead of richness to allow comparisons among forest types with very different total areas. The three first variables are themselves characterising components of forest vegetation diversity, while the latter three are commonly regarded as structural or functional indicators of diversity of other taxonomic groups, and are included for this reason in the NFI reporting (Ministerio de Medio Ambiente 2003). The size of the NFI plots varies depending on the tree DBH, ranging from a plot radius of 5 m for trees with DBH lower than 125 mm up to a maximum radius of 25 m for trees with a DBH of at least 425 mm (Ministerio de Medio Ambiente 2003).

The shape of each forest type was characterised through the mean shape index (MSI), the mean circumscribing circle index (MCCI), and the density of shape characteristic points (DSCP), computed in the original vector format of the SFM. MSI and MCCI were obtained respectively as the average of the shape index and circumscribing circle index for all the patches in a certain forest type, while DSCP was calculated as the ratio between the total number of shape characteristic points and the total area of the forest type. In order to compare the potential performance of shape and fragmentation characteristics as biodiversity indicators, we also calculated several fragmentation metrics for each of the forest types, including number of forest patches (NP), mean forest patch size (MPS) and the size of the largest forest patch (LPS).

10.3.3 Equally-Sized UTM Cell Analysis

Different components of forest biodiversity related to flora and vertebrate fauna (trees, birds, mammals, reptiles and amphibians) were analysed at the scale of 10 × 10 km, at which the information from vertebrate atlas data was available. The following variables were considered and estimated in 489 UTM 10 × 10 km cells covering the study area:

- Forest tree species richness and tree species diversity (quantified through the Shannon-Wiener index), as derived from the Spanish Forest Map. A total of 84 different tree species were found in the whole study area.

- Forest bird species richness and specialist bird species richness, obtained from the presence data of the Atlas of Spanish Breeding Birds (Martí and Moral 2003). We considered a total of 67 forest-dwelling bird species present in the study area, and classified as specialists those species that are strongly associated only with forest habitats (26 species), similarly to Gil-Tena et al. (2007).
- Mammal species richness and specialist mammal species richness, obtained from the presence data of the Atlas of Spanish Terrestrial Mammals (Palomo and Gisbert 2002). We considered a total of 37 forest-dwelling mammal species present in the study area, and classified as specialists those that are strongly associated only with forest habitats (9 species).
- Reptile species richness and amphibian species richness, obtained from the presence data of the Atlas and Red book of Spanish Amphibian and Reptiles (Pleguezuelos et al. 2002). Due to the generalist behaviour of many reptiles and amphibians, we only considered those species that mostly select or appear in forest habitats (6 reptiles and 4 amphibians).

As in the previous analysis, the forest landscape shape and fragmentation in each UTM cell was characterised through MSI, MCCI, the total number of shape characteristic points (NSCP), NP, MPS, and LPS, all of them computed in the vector format of the SFM.

10.4 Performance of Shape Irregularity Metrics as Forest Biodiversity Indicators

10.4.1 Forest Landscape Shape and Vegetation Diversity

More irregular shapes were significantly associated with a higher richness (density) and diversity of forest tree species, for the three metrics and both for the forest types and UTM cell analyses (Tables 10.2 and 10.3). These correlations remained high and significant after controlling for the effect of forest area, which indicates that these shape metrics are providing information on the landscape different from that conveyed by forest area itself. Plant species diversity was also related to shape irregularity in Honnay et al. (2003), with MSI being an important factor influencing both total and native species richness in 4 × 4 km cells. Saura and Carballal (2004) also found that mixed forest types presented more irregular shapes than monospecific ones in NW Spain, at least when comparisons were made among forests with all native tree species.

The number of shape characteristic points (NSCP) was the best performing metric for tree species richness for the UTM cell analysis, with a correlation coefficient of 0.72 (Table 10.3). This agrees with the results by Moser et al. (2002), who also analysed equally sized landscape units (although at a finer resolution, 600 × 600 m) in agricultural landscapes in Austria, and found that NSCP was the best predictor (among a set of ten common shape metrics) for the species richness of vascular

Table 10.2 Pearson's correlations (r) and partial correlations controlling for the effect of forest area (r_{area}) between the NFI vegetation diversity variables and shape irregularity (DSCP, MCCI, MSI) and fragmentation metrics (NP, MPS, LPS) for the forest type analysis

		DSCP	MCCI	MSI	NP	MPS	LPS
Tree species density	r	0.57*	0.63*	0.52*	0.27	−0.67*	−0.49*
	r_{area}	0.48*	0.54*	0.54*	−0.15	−0.17	−0.10
Shannon diversity index	r	0.44*	0.47*	0.47*	−0.08	−0.26	−0.29
	r_{area}	0.41*	0.44*	0.45*	0.15	−0.20	−0.24
Shrub species density	r	0.32	0.50*	0.39*	−0.67*	−0.34	−0.47*
	r_{area}	0.12	0.34	0.34	−0.28	0.02	−0.12
Old-growth trees (number/ha)	r	0.13	0.21	0.25	−0.08	0.09	−0.12
	r_{area}	0.13	0.22	0.25	−0.12	0.13	−0.14
Dead wood (standing stems/ha)	r	−0.07	−0.13	−0.15	0.10	0.04	0.06
	r_{area}	−0.03	−0.09	−0.13	−0.01	−0.03	−0.01
Percentage of uneven aged stands	r	0.34	0.53*	0.56*	−0.04	−0.08	−0.15
	r_{area}	0.34	0.56*	0.56*	0.02	−0.05	−0.14

$*p < 0.01$.

plants and bryophytes. However, for the forest type level, DSCP did not perform significantly better than MSI or MCCI for any of the NFI forest vegetation diversity variables (Table 10.2).

Landscapes with irregular shapes (higher MCCI and MSI values) also tend to present more structurally complex forests (uneven aged stands), as shown in Table 10.2. This is consistent with the results by Saura and Carballal (2004), who reported significantly more complex and elongated boundaries for native forests than for exotic forest plantations (typically monospecific, even-aged and single-layered stands) in NW Spain, and found that MCCI was the only metric that perfectly discriminated both types of forests.

However, for the rest of the NFI vegetation diversity variables there were no significant correlations with shape after controlling for forest area, even when some metrics presented significant positive Pearson's correlation with the number of shrub species per unit area (Table 10.2). The lack of significant correlations for the old-growth trees and the standing dead wood may in part respond to the fact that these elements are relatively rare in the analysed landscapes, with an average of only 10.3 dead standing stems per ha and 6.4 old growth trees per ha according to the NFI. This is because most of the forests in Galicia and Asturias are considerably young, as a consequence of long-lasting forest management and harvesting, recurring forest fires and other historical factors (Manuel and Gil 2002; Manuel et al. 2003). On the other hand, it should be noted that the NFI, as implemented in these regions, did not include a specific methodology for dead wood measurement. Only the standing trees within the plots that were characterised as dead but still profitable for wood production were inventoried. Further details on this may be provided in the near future, since the next provinces to be inventoried in the NFI will include a comprehensive

Table 10.3 Pearson's correlations (r) and partial correlations controlling for the effect of forest area (r_{area}) between richness and diversity of several taxonomic groups of forest species and shape irregularity (NSCP, MCCI, MSI) and fragmentation metrics (NP, MPS, LPS) for the UTM 10 × 10 km cell analysis

		NSCP	MCCI	MSI	NP	MPS	LPS
Tree species richness	r	0.72*	0.37*	0.56*	0.40*	0.06	0.22*
	r_{area}	0.65*	0.45*	0.57*	0.36*	–0.25*	–0.25*
Tree Shannon diversity	r	0.54*	0.37*	0.48*	0.36*	0.00	0.14*
	r_{area}	0.47*	0.43*	0.47*	0.32*	–0.24*	–0.24*
Bird species richness	r	0.26*	0.14*	0.26*	0.20*	–0.11	0.16*
	r_{area}	0.13*	0.18*	0.25*	0.15*	–0.09	–0.09
Specialist bird species richness	r	0.33*	0.20*	0.33*	0.13*	0.10	0.16*
	r_{area}	0.25*	0.24*	0.31*	0.09	–0.06	–0.06
Mammal species richness	r	0.36*	0.31*	0.37*	0.07	–0.04	–0.02
	r_{area}	0.44*	0.31*	0.37*	0.07	–0.04	–0.07
Specialist mammal species richness	r	0.34*	0.37*	0.38*	0.05	–0.03	–0.05
	r_{area}	0.43*	0.37*	0.38*	0.06	–0.02	–0.08
Reptile species richness	r	–0.12*	–0.18*	–0.19*	–0.05	0.06	0.09
	r_{area}	–0.21*	–0.17*	–0.20*	–0.07	0.01	0.08
Amphibian species richness	r	–0.01	–0.16*	–0.16*	0.07	0.15*	0.20*
	r_{area}	–0.20*	–0.13	–0.19*	0.02	–0.01	0.04
Total vertebrate species richness	r	0.33*	0.20*	0.31*	0.16*	0.08	0.13*
	r_{area}	0.26*	0.24*	0.30*	0.12*	–0.08	0.08

*$p < 0.01$.

and specific assessment of the dead wood and its decay stage (including different types of snags, logs, branches, stumps, etc.).

10.4.2 Forest Landscape Shape and Vertebrate Species Richness

The richness of birds, mammals and total vertebrate species were all significantly correlated with shape complexity and irregularity (Table 10.3). Results were considerably consistent for the three metrics and also significant after controlling for the effect of forest area in each UTM cell (Table 10.3). Correlations of the shape metrics with the richness of specialist birds and mammals were higher than for the total number of species in each of these two groups (Table 10.3), but the differences were not statistically significant. These results agree with Radford et al. (2005), who obtained that the area-weighted version of MCCI was consistently included in multivariate models explaining bird species richness. On the other hand, Brennan and Schnell (2005) found that the fractal dimension (as a measure of shape complexity of habitat patches) was associated with most of the bird species studied. Irregular forest shapes were also found as a factor related to the increase of forest bird diversity in Barbaro et al. (2005).

On the contrary, for amphibians and reptiles we found much lower correlations (in absolute value) than for the rest of the vertebrate species (Table 10.3). This could be due in part to the fewer forest species for these two taxonomic groups in the analysed datase (only 4 reptiles and 6 amphibians, compared to 37 mammals and 67 bird species in the same area), which may not allow for large differences in the diversity of these taxonomic groups in the UTM cells. On the other hand, most amphibian and reptile species have small home ranges, are sedentary and seem to depend more on the availability of specific habitats (Atauri and de Lucio 2001; Guerry and Hunter 2002; Nogués-Bravo and Martínez-Rica 2004) than on landscape configuration or land cover diversity at broad landscape scales. This may be due to the limited dispersal abilities that these taxa generally present, related to their relatively low body mass, although there are some exceptions (Smith and Green 2005). Amphibians and reptiles may perceive the landscape at a scale much finer than the 10 × 10 km cells, and therefore they may not react to the landscape pattern characteristics at our scale of analysis as much as other more vagile organisms. Finally, note that reptile and amphibian species richness were negatively correlated to the three shape irregularity metrics (Table 10.3), which agrees with a previous study by Gray et al. (2004), who pointed out the negative effect of inter-patch complexity as one of the most important factors affecting amphibian species composition.

10.4.3 Comparative Performance of Fragmentation and Shape Irregularity Metrics

Shape metrics presented higher correlations than fragmentation ones (in absolute value) for most of the forest biodiversity variables (Tables 10.2 and 10.3). This was true both for Pearson's and partial correlations, although the differences tended to be even more pronounced after controlling for the effect of forest area. Pearson's correlations were significantly higher with shape irregularity than with fragmentation metrics ($p<0.01$) for most of the biodiversity variables in the UTM cell analysis (tree Shannon diversity and the richness of trees, specialist birds, total and specialist mammals, and total vertebrate species). For the NFI vegetation indicators in the forest type analysis the differences in Pearson's correlation coefficients with shape and fragmentation metrics were significant ($p<0.01$) only for the Shannon diversity, the percentage of uneven aged stands, and the shrub species richness.

In fact, none of the three fragmentation metrics presented significant partial correlations (r_{area}) with any of the NFI vegetation diversity variables for the forest type analysis (Table 10.2), and the same was true for the species richness of all the vertebrate taxons, with the exception of NP for bird and total vertebrate species richness (Table 10.3). Results were similar for other fragmentation metrics different from NP, MPS or LSP (not shown). This indicates that fragmentation metrics are not really conveying here new independent information (non-redundant with forest area) on the forest landscape pattern that may be valuable to explain forest biodiversity distribution at this scale, which seems not to be the case for shape irregularity metrics. This agrees with Fahrig (2003), who noted that the empirical evidence to date suggests that habitat fragmentation *per se* (independent of habitat amount) has rather weak effects on biodiversity, which in addition are as likely to be positive as negative.

The only significant partial correlations (r_{area}) for the three fragmentation metrics were obtained for tree species in the UTM cell analysis (Table 10.3), indicating that more fragmented landscapes presented a higher richness and diversity of forest trees. This agrees with Honnay et al. (1999) who found that many small forest patches contain more plant species than one large patch of the same total area. Many other previous studies cited therein gave no evidence of habitat subdivision reducing total plant species richness in forests at the landscape scale. Honnay et al. (1999) suggested that this was the result of the probability of higher habitat diversity being present in two geographically separated small forests than in one large forest of the same size (high inter-patch diversity), and concluded that to maximise forest plant species richness at the landscape scale it is important to spread new afforestation geographically to encompass as many different habitat characteristics as possible and not to concentrate them locally in large units. However, this should be interpreted with caution, because for other taxonomic groups several studies have shown that small forest patches are not able to maintain the regional pool of species (Díaz et al. 1998; Santos et al. 2002).

To our knowledge, and despite the studies cited above regarding forest shape, there has been comparatively much more focus on forest fragmentation than on shape irregularity for assessing the relationships of landscape pattern configuration with forest biodiversity (e.g. Drolet et al. 1999; Trzcinski et al. 1999; Villard et al. 1999; Howell et al. 2000; Boulinier et al. 2001; Mitchell et al. 2006; Sallabanks et al. 2006). However, in our study shape irregularity was a considerably more relevant forest landscape pattern feature than fragmentation, and shape metrics were also less correlated and redundant with forest area than fragmentation ones. This suggests that more attention should be paid to forest shape irregularity measures in further research intending to develop indicators of forest biodiversity at the landscape scale. Nonetheless, and agreeing with Trzcinski et al. (1999), we cannot discard that at finer scales forest fragmentation affects forest biodiversity in a more prominent manner, particularly when forest area is low (Andrén 1994; Radford et al. 2005).

10.5 Conclusions and Implications for Forest Landscape Analysis and Biodiversity Monitoring

The rapid development of quantitative methods in landscape ecology has provided a large amount of shape-related metrics and offers new perspectives to explore the relationships between forest landscape pattern and ecological processes. However, it is important to be aware of the characteristics, spatial behaviour and limitations of the existing shape metrics for landscape analysis applications. Otherwise there is a considerable risk of misuse and arbitrary selection of inappropriate metrics, which may lead for example to a poor performance when addressing the relationships between landscape shape and biodiversity distribution.

The shape of landscape patterns may be linked to the imprint of the factors that have configured the boundaries and influenced the diversity of forest patches. As the human influence and the land use intensity increase, the shapes in the landscape may become simpler and more rectilinear, and the richness of different taxonomic groups may also decrease (Moser et al. 2002; Saura and Carballal 2004). Indeed, we have shown that several shape irregularity metrics may serve as spatial indicators of forest biodiversity at the landscape scale, which concurs with some previous studies. Shape irregularity metrics were significantly correlated with different components of forest vegetation diversity and with the richness of several groups of vertebrate species, and these correlations remained significant after accounting for the effect of forest area. Shape metrics correlated more with forest biodiversity variables than fragmentation metrics at our scale of analysis. This suggests that, despite most of the focus on landscape configuration has been devoted to fragmentation processes, more attention should be paid to shape as a key characteristic of landscape patterns. As noted by Forman (1995), 'the literature on the ecological effects of patch shape is sparse; this is a frontier area for research'.

Shape irregularity may be a key variable to be considered for the development of further landscape-level biodiversity indicators. This represents a new approach and perspective for characterising human imprint in the landscape and mapping forest biodiversity patterns, since it relies on landscape-level information that may be obtained at a low cost for large areas. Shape information may be used as ancillary data for the estimation of forest biodiversity and the identification of potential biodiversity hotspots, complementing field information derived from forest inventory plots. Shape metrics may be as well an aid for estimating temporal changes in biodiversity, where long term field-based historical records on species richness are not available, but there are longer temporal series of aerial or satellite images from which shape information can be derived. However, we recognise that further research on the use of shape metrics as biodiversity indicators is needed before being operational, including other study areas and different scales of analysis, since the explanatory factors of landscape biodiversity distribution have been shown to vary at different spatial scales (e.g. Mitchell et al. 2001, 2006).

Acknowledgments This research was financially supported by the Ministerio de Educación y Ciencia (Spain) and the European Union (FEDER funds) through the IBEPFOR project (CGL2006-00312/BOS). The Spanish Forest Map, the Third National Forest Inventory and the Spanish Atlas of Vertebrates were provided by the Dirección General para la Biodiversidad (Ministerio de Medio Ambiente, Spain). A. Gil-Tena benefited from a predoctoral research grant (2007FIC-00874) from the Departament d'Educació i Universitats (Catalan Government) and the European Social Fund. We also thank José Antonio Villanueva and Iciar Alberdi for their helpful assistance with the NFI data, and Begoña de la Fuente, Montse Raurell and David Guixé for their valuable advice in the elaboration of this manuscript.

References

Allen RB, Bellingham PJ, Wiser SK (2003) Developing a forest biodiversity monitoring approach for New Zealand. New Zeal J Ecol 27(2):207–220

Andrén H (1994) Effects of habitat fragmentation on birds and mammals in landscapes with different proportions of suitable habitat–A review. Oikos 71:355–366

Atauri JA, de Lucio JV (2001) The role of landscape structure in species richness distribution of birds, amphibians, reptiles and lepidopterans in Mediterranean landscapes. Landscape Ecol 16:147–159

Barbaro L, Pontcharraud FV, Vetillard F, Guyon D, Jactel H (2005) Comparative responses of bird, carabid, and spider assemblages to stand and landscape diversity in maritime pine plantation forests. Écoscience 12(1):110–121

Bettinger P, Bradshaw GA, Weaver GW (1996) Effects of geographic information system vector-raster-vector data conversion on landscape indices. Can J For Res 26:1416–1425

Bogaert J, Myneni RB, Knyazikhin Y (2002) A mathematical comment on the formulae for the aggregation index and the shape index. Landscape Ecol 17:87–90

Bogaert J, Van Hecke P, Moermans R, Impens I (1999) Twist number statistics as an additional measure of habitat perimeter irregularity. Environ Ecol Stat 6:275–290

Boulinier T, Nichols JD, Hines JE, Sauer JR, Flather CH, Pollock KH (2001) Forest fragmentation and bird community dynamics: inference at regional scales. Ecology 82:1159–1169

Brennan SP, Schnell GD (2005) Relationship between bird abundances and landscape characteristics: the influence of scale. Environ Monit Assess 105(1–3):209–228

Congalton RG (1997) Exploring and evaluating the consequences of vector-to-raster and raster-to-vector conversion. Photogramm Eng Rem S 63:425–434

Díaz M, Carbonell R, Santos T, Tellería JL (1998) Breeding bird communities in pine plantations of the Spanish plateaux: biogeography, landscape and vegetation effects. J Appl Ecol 35: 562–574

Drolet B, Desrochers A, Fortin MJ (1999) Effects of landscape structure on nesting songbird distribution in a harvested boreal forest. Condor 101:699–704

Duelli P, Obrist MK (2003) Biodiversity indicators: the choice of values and measures. Agric Ecosyst Environ 98:87–98

Fahrig L (2003) Effects of habitat fragmentation on biodiversity. Annu Rev Ecol Evol S 34: 487–515

Failing L, Gregory R (2003) Ten common mistakes in designing biodiversity indicators for forest policy. J Environ Manag 68(2):121–132

Forman RTT (1995) Land mosaics: the ecology of landscapes and regions. Cambridge University Press, Cambridge

García-Gigorro S, Saura S (2005) Forest fragmentation estimated from remotely sensed data: is comparison across scales possible? Forest Sci 51(1):51–63

Gil-Tena A, Saura S, Brotons L (2007) Effects of forest composition and structure on bird species richness in a Mediterranean context: implications for forest ecosystem management. Forest Ecol Manag 242(2–3):470–476.

Gray MJ, Smith LM, Leyva RI (2004) Influence of agricultural landscape structure on a Southern High Plains, USA, amphibian assemblage. Landscape Ecol 19(7):719–729

Guerry AD, Hunter ML (2002) Amphibian distributions in a landscape of forests and agriculture: an examination of landscape composition and configuration. Conserv Biol 16:745–754

Gustafson EJ, Parker GR (1992) Relationships between landcover proportion and indexes of landscape spatial pattern. Landscape Ecol 7:101–110

Hagan JM, Whitman AA (2006) Biodiversity indicators for sustainable forestry: simplifying complexity. J Forest 104(4):203–210

Haydon TD, Friar JK, Pianka ER (2000) Fire-driven dynamic mosaics in the Great Victoria Desert, Australia. Landscape Ecol 15:373–381

He HS, DeZonia BE, Mladenoff DJ (2000) An aggregation index (AI) to quantify spatial patterns of landscapes. Landscape Ecol 15:591–601

Honnay O, Hermy M, Coppin P (1999) Effects of area, age and diversity of forest patches in Belgium on plant species richness, and implications for conservation and reforestation. Biol Conserv 87:73–84

Honnay O, Piessens K, Van Landuyt W, Hermy M, Gulinck H (2003) Satellite based land use and landscape complexity indices as predictors for regional plant species diversity. Landscape Urban Plan 63(4):241–250

Howell CA, Latta SC, Donovan TM, Porneluzi PA, Parks GR, Faaborg J (2000) Landscape effects mediate breeding bird abundance in midwestern forests. Landscape Ecol 15:547–562
Hulshoff RM (1995) Landscape indexes describing a Dutch landscape. Landscape Ecol 10: 101–111
Iverson LR (1988) Land-use change in Illinois, USA: the influence of landscape attributes on current and historic land use. Landscape Ecol 2:45–61
Kojima N, Laba M, Liendo XMV, Bradley AV, Millington AC, Baveye P (2006) Causes of the apparent scale independence of fractal indices associated with forest fragmentation in Bolivia. Isprs J Photogramm 61:84–94
Krummel JR, Gardner RH, Sugihara G, Oneill RV, Coleman PR (1987) Landscape patterns in a disturbed environment. Oikos 48:321–324
Kupfer JA (2006) National assessments of forest fragmentation in the US. Global Environ Change 16(1):73–82
Lagro J (1991) Assessing patch shape in landscape mosaics. Photogramm Eng Rem S 57:285–293
Li HB, Wu JG (2004) Use and misuse of landscape indices. Landscape Ecol 19:389–399
Lindemann JD, Baker WL (2001) Attributes of blowdown patches from a severe wind event in the Southern Rocky Mountains, USA. Landscape Ecol 16:313–325
Lindenmayer DB (1999) Future directions for biodiversity in managed forests: indicator species, impact studies and monitoring programs. Forest Ecol Manag 115:277–287
Lindenmayer DB, Cunningham RB, Donnelly CF, Lesslie R (2002) On the use of landscape surrogates as ecological indicators in fragmented forests. Forest Ecol Manag 159:203–216
Lindenmayer DB, Margules CR, Botkin DB (2000) Indicators of biodiversity for ecologically sustainable forest management. Conserv Biol 14:941–950
Loehle C, Wigley TB, Rutzmoser S, Gerwin JA, Keyser PD, Lancia RA, Reynolds CJ, Thill RE, Weih R, White D, Wood PB (2005) Managed forest landscape structure and avian species richness in the southeastern US. Forest Ecol Manag 214:279–293
Manuel C, Díaz-Fernández P, Gil L (2003) Tercer inventario forestal nacional 1997–2006. La transformación histórica del paisaje forestal en Asturias. Dirección General para la Conservación de la Naturaleza, Madrid
Manuel C, Gil L (2002) Tercer inventario forestal nacional 1997–2006. La transformación histórica del paisaje forestal en Galicia. Dirección General para la Conservación de la Naturaleza, Madrid
Martí R, Moral JC (eds) (2003) Atlas de las aves reproductoras de España. Dirección General de Conservación de la Naturaleza – SEO, Madrid
McAfee BJ, Malouin C, Fletcher N (2006) Achieving forest biodiversity outcomes across scales, jurisdictions and sectors with cycles of adaptive management integrated through criteria and indicators. Forest Chron 82:321–334
McGarigal K, Cushman SA, Neel MC, Ene E (2002) FRAGSTATS: spatial pattern analysis program for categorical maps. Computer software program produced by the authors at the University of Massachusetts, Amherst. Available at http://www.umass.edu/ landeco/research/fragstats/fragstats.html
Ministerial Conference on the Protection of Forests in Europe (2002) Improved Pan-European indicators for sustainable forest management. http://www.mcpfe.org/publications/pdf. Cited 15 March 2007
Ministerio de Medio Ambiente (2003) Tercer Inventario Forestal Nacional, Asturias. Dirección General para la Conservación de la Naturaleza, Madrid
Ministerio de Medio Ambiente (2005) Tercer Inventario Forestal Nacional, Lleida. Dirección General para la Conservación de la Naturaleza, Madrid
Mitchell MS, Lancia RA, Gerwin JA (2001) Using landscape-level data to predict the distribution of birds on a managed forest: effects of scale. Ecol Appl 11:1692–1708
Mitchell MS, Rutzmoser SH, Wigley TB, Loehle C, Gerwin JA, Keyser PD, Lancia RA, Perry RW, Reynolds CJ, Thill RE, Weih R, White D, Wood PB (2006) Relationships between avian richness and landscape structure at multiple scales using multiple landscapes. Forest Ecol Manag 221:155–169

Mladenoff DJ, DeZonia B (2004) APACK 2.23 analysis software user's guide. Forest Landscape Ecology Lab. Dept. of Forestry, UW-Madison, Madison. Available at http://forestlandscape.wisc.edu/Projects/apack

Mladenoff DJ, White MA, Pastor J, Crow TR (1993) Comparing spatial pattern in unaltered old-growth and disturbed forest landscapes. Ecol Appl 3:294–306

Moser D, Zechmeister HG, Plutzar C, Sauberer N, Wrbka T, Grabherr G (2002) Landscape patch shape complexity as an effective measure for plant species richness in rural landscapes. Landscape Ecol 17(7):657–669

Newton AC, Kapos V (2002) Biodiversity indicators in national forest inventories. In: Unasylva 53(3) Num. 210. Available at http://www.fao.org/docrep/005/y4001e/Y4001E09.htm#P0_0. Cited 15 March 2007

Nogués-Bravo D, Martínez-Rica JP (2004) Factors controlling the spatial species richness pattern of four groups of terrestrial vertebrates in an area between two different biogeographic regions in northern Spain. J Biogeo 31:629–640

Noss RF (1990) Indicators for monitoring biodiversity, A hierarchical approach. Conserv Biol 4:355–364

Noss RF (1999) Assessing and monitoring forest biodiversity: a suggested framework and indicators. Forest Ecol Manag 115(2–3):135–146

Palomo LJ, Gisbert J (eds) (2002) Atlas de los mamíferos terrestres de España. Dirección General de Conservación de la Naturaleza–SECEM–SECEMU, Madrid

Pleguezuelos JM, Márquez R, Lizana M (eds) (2002) Atlas y libro rojo de los anfibios y reptiles de España. Dirección General de Conservación de la Naturaleza, Asociación de Herpetología Española, Madrid

Radford JQ, Bennett AF, Cheers GJ (2005) Landscape-level thresholds of habitat cover for woodland-dependent birds. Biol Conserv 124(3):317–337

Rempel R (2006) Patch Analyst 3.12: a tool for quantifying landscape structure. Sustainable Forest Management Network and Centre for Northern Forest Ecosystem Research. Lakehead University Campus, Thunder Bay, Ontario. Available at http://flash.lakeheadu.ca/~rrempel/patch/

Rex KD, Malanson GP (1990) The fractal shape of riparian forest patches. Landscape Ecol 4(4):249–258

Sallabanks R, Riggs RA, Cobb LE, Dodson SW (2006) Bird-habitat relationships in grand fir forests of the Blue Mountains, Oregon. Forest Sci 52:489–502

Santos T, Tellería JL, Carbonell R (2002) Bird conservation in fragmented Mediterranean forests of Spain: effects of geographical location, habitat and landscape degradation. Biol Conserv 105:113–125

Saura S (2002) Effects of minimum mapping unit on land cover data spatial configuration and composition. Int J Remote Sens 23(22):4853–4880

Saura S (2004) Effects of remote sensor spatial resolution and data aggregation on selected fragmentation indices. Landscape Ecol 19(2):197–209

Saura S, Carballal P (2004) Discrimination of native and exotic forest patterns through shape irregularity indices: an analysis in the landscapes of Galicia, Spain. Landscape Ecol 19(6): 647–662

Saura S, Castro S (2007) Scaling functions for landscape pattern metrics derived from remotely sensed data: are their subpixel estimates really accurate? ISPRS J Photogramm Rem Sens 62(3): 201–216

Saura S, Martínez-Millán J (2001) Sensitivity of landscape pattern metrics to map spatial extent. Photogramm Eng Rem S 67(9):1027–1036

Smith MA, Green DM (2005) Dispersal and the metapopulation paradigm in amphibian ecology and conservation: are all amphibian populations metapopulations? Ecography 28: 110–128

Tischendorf L (2001) Can landscape indices predict ecological processes consistently? Landscape Ecol 16(3):235–254

Trzcinski MK, Fahrig L, Merriam G (1999) Independent effects of forest cover and fragmentation on the distribution of forest breeding birds. Ecol Appl 9:586–593

Villard MA, Trzcinski MK, Merriam G (1999) Fragmentation effects on forest birds: relative influence of woodland cover and configuration on landscape occupancy. Conserv Biol 13:774–783
Wu J (2004) Effects of changing scale on landscape pattern analysis: scaling relations. Landscape Ecol 19(2):125–138
Wu J, Shen W, Sun W, Tueller PT (2002) Empirical patterns of the effects of changing scale on landscape metrics. Landscape Ecol 17(8):761–782

Chapter 11

Land Suitability for Short Rotation Coppices Assessed through Fuzzy Membership Functions

Piermaria Corona, Riccardo Salvati, Anna Barbati and Gherardo Chirici

Abstract In the land planning process, land suitability can be regarded as a bridging phase linking land resources assessment to the decision-making process. The inherent conflicts and the complex network of socio-economical and ecological constraints affecting land use planning call for a flexible decision-making support tool able to incorporate multiple evaluation criteria, including several stakeholders point of views. Land suitability gives transparent indications to decision-makers concerning land uses which can be sustainable fostered in the land under consideration, allowing areas to be ranked according to their degree of suitability for a specific land use. The resulting maps lend efficient support to negotiation when addressing issues of sustainable development as well as economic competitiveness. Within this framework this paper aims to: (i) outline relevant LSA experiences, with a distinctive focus on MCE GIS-based techniques; (ii) present a case study of MCE applying fuzzy modelling to predict the productive response of land when afforested with target forest plantation species. Drawing from the case study experience, the MCE proved to be a valuable tool for decision-making (i.e. to quantify on a broad scale land physical suitability for an expansion of SRF plantations for energy biomass production in Italy).

11.1 Introduction

Land evaluation is a powerful tool to support decision-making in land use planning: it deals with the assessment of the (most likely) response of land when used for specified purposes; it requires the execution and interpretation of surveys of climate, soil, vegetation and other aspects of land in terms of the requirements of alternative forms of land use. *Land suitability assessment (LSA)* can be regarded as a specific case of land evaluation: it is an appraisal of land characteristics in terms of their suitability for a specific use (FAO 1976). The basic concept behind LSA is that suitability for a specific and sustainable use of the land is the synthetic result of

P. Corona
Department of Forest Environment and Resources, University of Tuscia, Italy
e-mail: piermaria.corona@unitus.it

R. Lafortezza et al. (eds.), *Patterns and Processes in Forest Landscapes*,

complex relationships between different land environmental qualities (e.g., climate, soil characteristics and slope). Suitability for a specific use is therefore evaluated by matching requirements for that use with characteristics and qualities of land components. Land suitability is usually expressed by a hierarchical system organised into *orders* and *classes* (FAO 1976). The *orders* indicate whether the land is suitable or not for a given land use; two main orders are distinguished:

N (not suitable): land whose qualities appear to preclude sustainability for the considered land use; the limitations are so severe that they preclude the successful application of the given land utilisation type;

Classes reflect degrees of decreasing suitability within the order 'suitable'. Most often three classes are applied:

- S (suitable): land on which sustained use is expected to yield benefits which justify the inputs, without unacceptable risk of damage to land resources.
- S1 (highly suitable): land which has no significant or only minor limitations to the sustained application of a given land use;
- S2 (moderately suitable): land which has limitations that are moderately severe for sustained application of the given land use; the limitations will reduce productivity or benefits and will increase the required inputs;
- S3 (marginally suitable): land which has severe limitations for sustained application of a given land use.

LSA applications are common in many fields of land planning. In urban planning, for example, LSA is undertaken to determine the suitability of land for housing within a municipality territory; in rural planning LSA is mainly used to assist the zonation of a rural region into a mosaic of land units, each being suitable for supporting a given farming system. The same holds for forest planning; when forest trees ecological requirements are sufficiently known to be reliably plotted against environmental factors, then land suitability maps can be drawn from environmental geodatabases. LSA is also used to support nature conservation activities: e.g., zoning or delimitation of protected areas. In biological conservation, LSA is applied to habitat evaluation for endangered fauna or flora species; applications of this nature rely upon expert-based knowledge of the habitat preferences of species, as well as on data relating to actual habitat use; such an approach is also referred to as *habitat suitability modeling* (Hirzel et al. 2006).

LSA can be efficiently performed within *Geographic Information Systems* (GIS): in the past few decades, LSA has become one of the most common GIS-based analysis in land planning and management (McHarg 1969; Hopkins 1977; Brail and Klosterman 2001; Collins et al. 2001; Malczewski 2004). In the first applications, GIS-based LSA was basically carried out by overlaying thematic maps in a digital format. Land suitability was derived from the input thematic layers, using simple map algebra or map logic operators. Over the last two decades, LSA issues have been addressed with increasingly complex conceptual models. Specific solutions of data handling and processing have been developed: amongst others, the so-called multi-criteria evaluation (MCE) or multi-criteria analysis (MCA) are widely used procedures (e.g., Banai 1993; Jankowski and Richard 1994; Joerin 1995; Barredo 1996; Beedasy and Whyatt 1999; Malczewski 1999; Barredo et al. 2000; Mohamed et al. 2000; Bojorquez-Tapia et al. 2001; Dai et al. 2001; Joerin 2001, Church 2002).

Within this framework, this chapter focuses on: (i) examining the pros and cons of different GIS-based LSA techniques; (ii) presenting a concrete example of LSA, applied to the evaluation of farmland available for short rotation coppice (SRF) in Italy. The Sections 11.2 and 11.3 review approaches for GIS-based land suitability analysis and mapping and provides a summary of applications. The case study is then outlined in Section 11.4, focusing on the use of fuzzy membership functions in MCE. The benefits of this approach are discussed and recommendations for further work are outlined in Section 11.5.

11.2 GIS-Based Land Suitability Assessment

A GIS can be defined as a computer-assisted system for the acquisition, storage, analysis and display of geographic data (Eastman 2006). As such, the GIS can provide essential management information or be used to develop a better understanding of environmental relationships. In recent years, considerable interest has grown around the use of GIS for land suitability mapping and modelling; two main groups of approaches to GIS-based land suitability can be distinguished: (i) overlay mapping and (ii) multi-criteria evaluation methods (Collins et al. 2001).

11.2.1 Overlay Mapping

The computer-assisted overlay techniques were developed in response to limitations of manual methods of mapping and combining large datasets in paper format (Mac Dougall 1975; Steinitz et al. 1976). Overlay mapping is quite simple. Input thematic layers are acquired and transformed into input factors (or criteria). These are stored in the form of interval data: pixels (raster data) or polygons (vector data) are assigned with relative ranking values, based on well-known relationships between land use requirements and land qualities. For example, to assess land suitability for a specific crop, single environmental factors (e.g., rainfall, soil drainage, soil texture, pH and temperature at germination) are ranked into classes of suitability (cf. Section 11.1); different criteria are then combined using logic or algebraic functions to obtain the final suitability map (Lyle and Stutz 1983). Such a simple approach is basically the reproduction in a computer environment of techniques developed for the overlay of paper maps (Tomlin 1990). The major criticism to the conventional map overlay approach concerns the inappropriate use of methods for standardizing suitability maps and untested or unverified assumptions of independence among suitability criteria (Hopkins 1977; Pereira and Duckstein 1993). This limitation can be overcome by integrating GIS and multi-criteria decision making (MCDM) methods.

11.2.2 Multi-Criteria Decision Making

The MCDM procedures (or decision rules) define a relationship between the input thematic information and the output suitability map that is more complex than logic or algebraic relationships. The suitability model can include decision maker's preferences, which are turned into decision rules. All the input thematic layers are transformed into *constraint* or *factor* criteria. A *constraint* limits the land use options (or alternatives) under consideration (Eastman 2006): e.g., the exclusion of protected areas from housing development or the exclusion from farming of areas with slopes exceeding a 30% gradient. A *factor* is a criterion that enhances or limits land suitability for a specified land use option (alternative): for example, the steeper the slope, the more severe are the limitations for the establishment and sustainable management of forest productive plantations. A number of multi-criteria decision rules have been implemented in the GIS environment for tackling land-use suitability issues. The decision rules used to aggregate the input criteria (constraints and factors) into a final suitability map can be classified into *multi-criteria* (or multi-attribute) and *multi-objective* decision-making methods (Carver 1991; Malczewski 1999). In the *multi-criteria* analysis, several input factors must be aggregated to derive one final suitability map for a single specific objective. *Multi-objective* methods assess land suitability for different goals, whether conflicting or otherwise: they can be regarded as the GIS answer to address the inherent conflicts of land planning.

11.2.2.1 Multi-Criteria Methods

In the last decade, a number of multi-criteria evaluation methods (MCE) have been proposed for GIS based land suitability analysis (Eastman 1999). Amongst others, the most common are (i) Weighted Linear Combination (WLC) or simple additive weighting and (ii) Ordered Weighted Averaging (OWA).

Weighted Linear Combination

In WLC, each input thematic layer (criterion) is assigned with a weight indicating the relative degree of importance each criterion plays in determining the suitability for an objective. Input criteria are standardized to a numeric range, quantifying scores of suitability, and then combined by means of a weighted average. Criteria weights are assigned according to decision-maker preferences and determine how each criterion will trade-off relative to other factors: a criterion with a high weight can *tradeoff* or compensate for poor scores on other factors (Eastman 2006). In contrast, a criterion with a high suitability score but a small weight can only weakly compensate for poor scores on other factors. There are, however, fundamental limitations associated with the use of WLC in a decision-making process, which are comprehensively discussed by Jiang and Eastman (2000); these authors regard the WLC approach as just an extension to, and generalization of, the conventional map overlay methods in GIS.

Ordered Weighted Average

OWA is a more complex class of multi-criteria methods (Yager 1988) that enables the degree of tradeoff between the criteria to be governed; in the OWA, two sets of weights (*criteria* and *orders*) are applied: criteria weights are defined as in the WLC, while order weights determine the overall level of tradeoff allowed. Unlike criteria weights, order weights do not apply to any specific criterion but are defined on a pixel-by-pixel basis according to the ranking order of criteria scores (order weight 1 is assigned to the lowest ranked criteria of a given pixel, order weight 2 to the next higher-ranked criteria for that pixel, and so forth). To understand how order weights influence final results, consider the case where criterion weights are equal for three criteria A, B and C (factor weights = 0.33); the score of these factors for one pixel is respectively 100, 50 and 200. When ranked from minimum to a maximum value, the order of these factors for a given pixel is B, A, C. Any combination of order weights can be defined in OWA that sum to 1. In the combination [1, 0, 0] the criterion with the minimum value in the set (B) will receive all the possible weight and no trade off is possible with other higher ranked criteria (result = 50); this solution is from a decision-making standpoint *risk-averse*: the final aggregated score is more or less close, depending on criteria weights, to the suitability value of the lowest ranked criteria; similarly, the combination [0, 0, 1] makes irrelevant the contribution of lower ranked criteria (result = 200); it can be regarded as a *risk-taking* solution because the final aggregate score is more or less close to the suitability value of the highest ranked criteria. As all the combinations in the continuum between these two extremes are possible, different degrees of tradeoff between the ranked criteria can be set for determining the suitability for an objective. The OWA is therefore a flexible method: it provides a continuum of decision strategies, ranging from a risk-averse through all the intermediate neutral towards risk positions (corresponding to the conventional WLC) to a risk-taking solution. Thus, OWA can be considered as a generalization of WLC.

11.2.2.2 Multi-Objective Methods

Multi-objective methods assess suitability for different land uses (or more in general for different objectives) in a given area. The objectives may be non-conflicting (complementary) or, much more frequently, conflicting (Eastman 2006). They are complementary when different objectives may share the same land unit (e.g., a forest plantation in a protected area); they are conflicting when they are mutually exclusive (for instance, the cultivation of two different agricultural crops in the same parcel). The solution of multi-objective complementary problems is, in general, quite easy: a number of land suitability analyses equal to the number of the objectives is performed and the results are aggregated to find the optimal solution by analysing the degree to which each land unit meets the considered objectives (Voogt 1983). Multi-objective conflicting models are often tackled by converting them to single objective problems, which are then solved using the standard linear programming methods (Feiring 1986; Diamond and Wright 1988; Aerts 2002; Malczewski 2004).

An advantage of the model (and of the linear programming approaches in general) is the ability to map the patterns and opportunity costs in addition to the optimal land suitability pattern. This added information could be used for evaluating the robustness of land suitability patterns and identifying areas where modifications could be made without significant impacts (Cromley 1994; Cromley and Hanink 1999).

11.3 Examples of GIS-Based LSA Applications

A comprehensive review of LSA applications goes beyond the scope of this chapter. Instead, this section provides an overview of the applications developed so far, focusing specifically on agriculture and forestry, habitat suitability, urban and land use planning.

11.3.1 Agriculture and Forestry

In agriculture and forestry, land suitability has a very long tradition; most applications aim at supporting sustainable agricultural development in term of selection of suitable crops and cultivation techniques (Table 11.1).

Many studies (e.g. Liengsakul et al. 1993; Kalogirou 2002) focused attention on land suitability for agricultural crops or forest plantations (e.g. Chirici et al. 2002). LSA is frequently applied to support the improvement of the agricultural sector in relatively poor rural areas (e.g. Thailand, Mexico and NW Spain) by introducing new crops or new cultivation methods. GIS overlay techniques are most common, although MCE with fuzzy classifications is increasingly being used over the past 5 years. Data input layers relate typically to the physical environment (mainly climate, topography and soils). In many cases, expert knowledge or farmers' opinion has been incorporated into LSA (e.g. Cools et al. 2003), by means of fuzzy modelling (cf. Section 11.4). In general, the outputs from such modelling has led to improvements in agricultural land use and also changes in related policies. The definition of the weight to be applied to different factors is always a tricky question. The most common approach is to use questionnaires to experts/farmers to define factors ranking. Answers are then transformed into weights by simple linear relationships (the higher is the average importance associated to a given factor and the higher is its weight in the MCE).

11.3.2 Habitat Suitability

Models predicting the spatial distribution of species (Boyce and McDonald 1999; Guisan and Zimmermann 2000; Manly et al. 2002; Pearce and Boyce 2006) – sometimes referred as resource selection function or habitat suitability models or habitat evaluation procedure – are currently gaining interest in wildlife management

Table 11.1 Examples of land suitability studies in agriculture and forestry.

Author	Purpose	Techniques	Data layers	Results
Cambell et al. (1992)	Planning of agricultural land use strategies and estimation of economic potential of agricultural sector	Linear programming	Labour, market forecasts, technology and cost	Suggestions for resource allocation, farm size, policy application and implementation projects
Liengsakul et al. (1993)	Locating new sites for cropland for highland people in N Thailand after resettlement	Overlay based on vector polygons Thematic Mapping Units (TMU)	Satellite image interpretation, land use/cover, soil maps, accessibility	To locate physically suitable, accessible and not yet used areas for potential new cropland sites
Bydekerke et al. (1998)	Land suitability for growing cherimoya (*Annona cherimola* Mill.) in the Loja province	Map overlay	Expert knowledge, climate, altitude, soil type, ecotype	Land suitability map for cherimoya
Kalogirou (2002)	Implementation of a land suitability evaluation model	GIS-based expert systems and Boolean overlay; mapped values as suitability scores	Physical FAO land classification for crops	Model outputs to identify areas suitable for agriculture and for five selected crops (wheat, barley, maize, seed cotton, sugar beet)
Chirici et al. (2002)	Land suitability for European walnut (*Juglans regia* L.) and Douglas fir (*Pseudotsuga menziesii* Mirb. (Franco)) plantations	Multi-criteria evaluation and fuzzy membership functions	Climate, soil, land use/land cover, relief	Land suitability fuzzy maps

Table 11.1 (continued)

Author	Purpose	Techniques	Data layers	Results
Ceballos-Silva and Lopez-Blanco (2003a)	Land suitability for maize and potato crops in Central Mexico	Multi-criteria evaluation and fuzzy membership functions	Climate, relief, soil	Compared actual and potential areas for maize and potato production
Ceballos-Silva and Lopez-Blanco (2003a)	Land suitability for oat (*Avena sativa* L.) crop production in Central Mexico	Multi-criteria evaluation and fuzzy membership functions	Climate, topography, soil at different spatial and temporal resolutions	Standardized land suitability map
Cools et al. (2003)	To promote adoption of new land use systems in a village in NW Spain	Map overlay	Farmer- and expert-based suitability assessment	Promotion of new land use systems
Carranza et al. (2005)	Agricultural land suitability based on Farmer Knowledge (FK)	Multi-criteria evaluation and fuzzy membership functions	Cropping season, soil colour, texture and depth, and slope	FK-based suitability maps indicating agreement or conflict with land resources development plans

issues (Hirzel et al. 2006). As these models often help understanding species niche requirements and predicting species potential distribution, their use has been especially promoted for conservation issues, such as managing species distribution, assessing ecological impacts of various factors (e.g. pollution, climate change), risk of biological invasions or endangered species management (Scott et al. 2002; Guisan and Thuiller 2005). Habitat suitability models are developed on the basis of a large variety of methods (multi-variate analyses, logistic regression, Gaussian logistic regression, discriminant analysis, nearest neighbours technique, neural networks). For an overview of applied examples we recommend, among others, Manel et al. (1999), Fielding and Bell (1997), McCullagh and Nelder (1989), Smith et al. (2007).

Habitat suitability evaluation is generally based on multi-variate analysis when absence data are lacking; analyses of this nature allow comparison, in the multi-dimensional space of ecological variables, of the distribution of the sites where the species of interest was observed to a reference set describing the whole study area (Hirzel et al. 2002). When the relationships between ecological variables and habitat suitability are well-known, LSA methods can also be applied to habitat suitability mapping. For instance, Store (2003) applied MCE on the basis of habitat preferences of various animal species and Clark et al. (2002) used the same approach to study red squirrel populations. Bayliss et al. (2005) developed a multi–species targeting approach for eight threatened bird species in the UK while Smeins and Wu (2000) performed a similar analysis to develop landscape scale models for assessing the potential and present habitat suitability of eight rare plant species found in southern Texas.

11.3.3 Urban Planning

In urban planning LSA is frequently used to identify future settlements areas (Sui 1992; Kaiser 1995). In such applications, both environmental and socio-economic input factors are used to map possible future development of periurban areas, frequently handling multi-objective problems of conflicting interests amongst stake-holders. The methods used in these applications range from linear programming (Cromley and Hanink 1999), MCE (Bannet et al. 2005) and Analytical Hierarchical Process (Dyer 1990; Thapa and Murayama 2008) to mixed approaches (Ligtenberg et al. 2001). It is noteworthy that LSA is routinely applied in urban planning, though with simple map overlay approaches, to support decision-making.

11.3.4 Land Planning

The applications of LSA in land planning are, in principle, similar to those of urban planning; however, larger areas and more heterogeneous issues are covered (Overmars et al. 2007): e.g. supporting the selection of alternative land uses solutions in decision-making (Bognar et al. 1998; Malczewski et al. 2003; Guo et al. 2006), locating suitable sites for new infrastructures, like greenways (Collins et al. 2001) or wind turbines (Meentemeyer and Rodman 2006).

11.4 Case Study: Assessing Land Suitability for SRF Combining MCE and Fuzzy Membership Functions

11.4.1 Benefits of Fuzzy Set Classification in LSA

As seen before (cf. Section 11.3.1), land suitability evaluation for specific crops/forest plantations is one of the fundamental fields of application of LSA. The conventional methodological steps adopted in suitability assessments of this nature are:

(1) Definition of the target species, be they crop or forest species.
(2) Identification of the ecological requirements of species (i.e., environmental factors affecting productivity or precluding sustainability for the considered cultivation techniques).
(3) Determination of the quantitative relationship between each considered environmental factor and the potential productivity (in terms of yields or economic return) of the considered target species; this relationship is conventionally expressed by suitability classes (quantified as integer scores) corresponding to given ranges of the environmental factor values (cf. Section 11.2.1).
(4) Calculation of a suitability class and score for single environmental factors within the study area on a pixel-by-pixel (raster approach) or land unit-by-land unit basis (vector approach; land units are polygons which can be regarded internally homogeneous as to environmental factors considered).
(5) Combination of the scores from all the considered environmental factors and assignment of the overall suitability class values to pixels or land units.

A central and critical issue of the methodology is how to parameterize and combine land qualities/limitations of a different nature in order to model the productive response of the target species to a given set of environmental factors. The rigid-data model consisting of discrete, sharply bounded, internally uniform entities, is unable to represent important aspects of reality: the continuous nature of environmental factors variation and their small-scale spatial heterogeneity. Considerable loss of information may also occur when data classified according to a rigid-data model are then retrieved or combined using Boolean methods.

The fuzzy set theory offers a useful alternative in this respect; it permits the gradual assessment of the membership of elements in a set with the aid of a continuous scale of membership (Burrough and McDonnell 1998), the so-called *membership function*, valued in the real unit interval [0, 1]. For example, consider the classification of an area according to a continentality index (Ic; i.e., the yearly thermic average interval expressing the range between the average temperature of the warmest and coldest month of the year, where Ic = Tmax-Tmin). The conventional crispy classes of the continentality index are 'Oceanic' ($11^\circ < \text{Ic} < 21^\circ$) and 'Continental' ($21^\circ < \text{Ic} < 65^\circ$); however, it makes no sense to assume the transition between 'Oceanic' and 'Continental' as a sharp boundary; instead, it should be viewed as an intersection between the two classes, within which an area has partial degrees of membership to each class. The fuzzy set classification allows transition from one class to another to be described by means of a membership function. Basic principles of fuzzy sets, operations on fuzzy sets and the derivation of membership functions applied to land evaluation can be found in Groenemans et al. (1997).

In the LSA, the use of a fuzzy set classification is particularly helpful to model the productive response of the target species (in terms of yields or economic return) to single environmental factors. This can be better expressed as a gradual transition (soft classification), rather than as abrupt shifts from one class to another (hard classification). Such a gradual transition can be quantified according to fuzzy membership functions valued in the interval [0, 1], where 1 means a complete suitability (the environmental factor matches the ecological requirements of the target species; the so called *optimum* of the species) and 0 means no suitability. The fuzzy membership function is shaped according to the best available knowledge on crop/species ecological requirements, as drawn from literature.

In the following, we present an application of MCE analysis where fuzzy membership functions are used to assess how much farmland could be technically available for establishing forest plantations for energy biomass production in Italy. This issue has high relevance in the context of European policies to reduce greenhouse gas emissions and secure energy supply, without increasing pressures on the environment (cf. e.g. EEA 2007). In a scenario of increasing market demand of woody biomass, farmers will be pushed to convert arable lands to forest plantations; but the establishment of forest plantations, mainly managed as short rotation coppices (SRF), must be carefully planned on a national and regional scale considering multiple environmental dimensions. The first step is to assess the physical (static) suitability of farmland for SRF.

11.4.2 Material and Methods

The assessment is based upon the basic concept of land suitability/land capability by FAO (1976), as specified by Booth and Saunders (1985). Spatial suitability models were derived from the multi-variate relationships between the ecological requirement of four target tree species and selected environmental factors. The selected target species proved to be, by field experimentation, the most potentially successful for short rotation coppice plantations in Italy (Minotta 2003): white poplar (*Populus alba*), hybrid euro-american poplar (*Populus euroamericana*), black locust (*Robinia pseudoacacia*), white willow (*Salix alba*).

11.4.2.1 Environmental Factors

Environmental factors were selected after an in-depth literature review, screening among those for which mapped information is homogeneously available on a national scale in Italy: climatic data (annual mean temperature, annual mean rainfall, absolute minimum temperature); topographic data (distance from the sea coast); soil data (texture, soil depth). In addition to ecological factors, slope was also considered because of its relevance as constraining factor for the mechanization of SRF management operations.

Georeferenced data were acquired for each factor with a spatial resolution of 250 m and projected under the UTM 32 N WGS84 geographic system. Climatic data were retrieved from Blasi et al. (2007). GIS-calculated distance from the seacoast was derived by applying a linear distance algorithm to official national terrestrial borders. Slope data were derived from a national DEM with an original spatial resolution of 75 m. Soil texture and depth data were derived from the Climagri database (Salvati et al. 2005).

11.4.2.2 Fuzzy Membership Functions and Multi-Criteria Analysis

The suitability assessment was based on fuzzy multi-criterial evaluation (MCE): fuzzy memberships functions were used to score land suitability by the single environmental factors (Van Ranst et al. 1996; Groenemans et al. 1997), while MCE was used to aggregate scores to generate a final suitability value (see Section 11.2.2.1). Fuzzy membership functions were defined based on data from scientific literature and were reviewed by an independent panel of five experts (Table 11.2).

Table 11.2 Fuzzy membership functions. Soil texture is classified according to five texture classes: (1) clay, (2) loam and clay, (3) loam, (4) silt, (5) sand

Factor	Salix alba	Robinia pseudoacacia	Populus euroamericana	Populus alba
Annual mean temperature	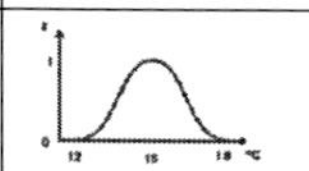	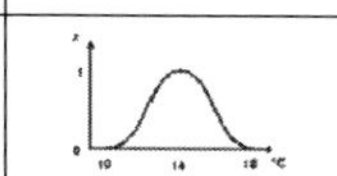	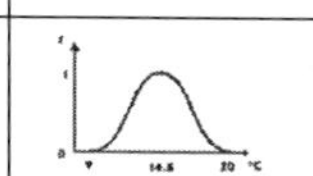	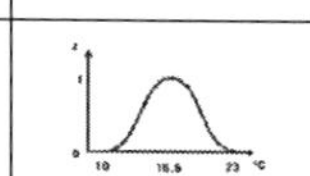
Annual mean rainfall	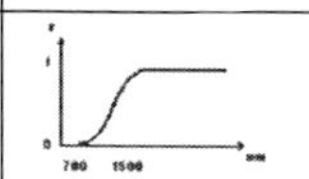	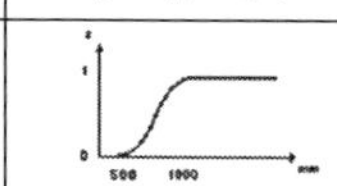	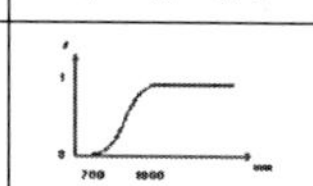	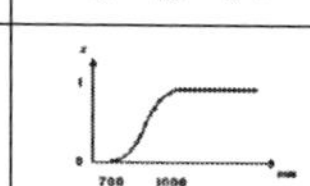
Winter minimum temperature (absolute minimum)	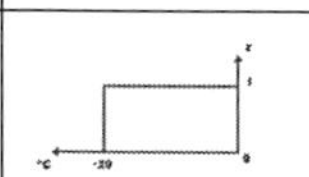	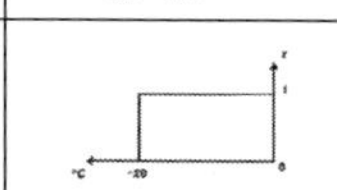	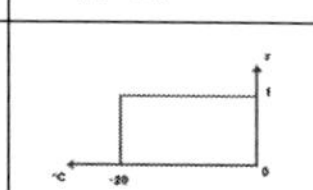	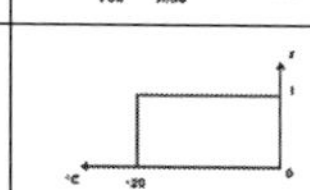
Distance from the sea coast	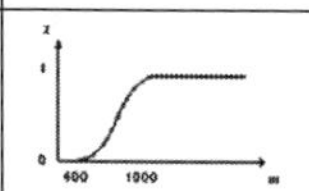	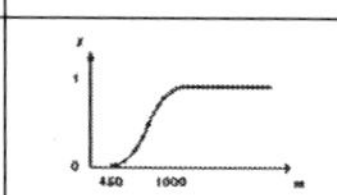	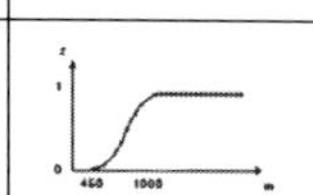	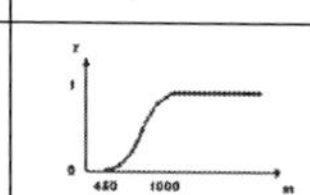
Slope	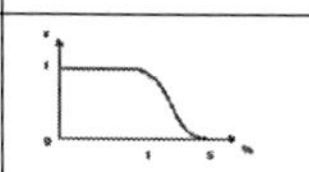	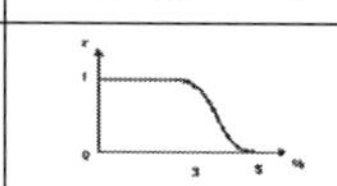	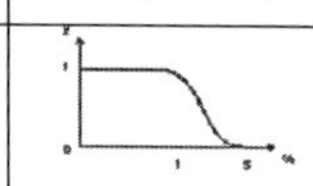	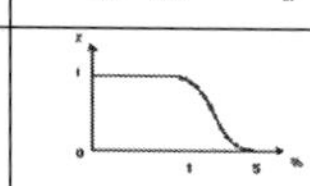
Soil depth	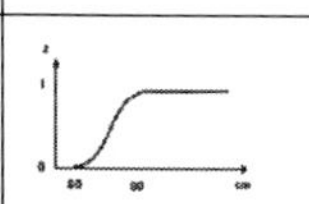	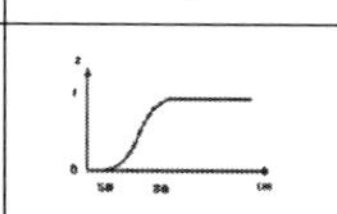	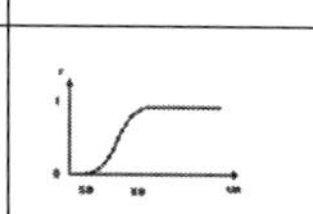	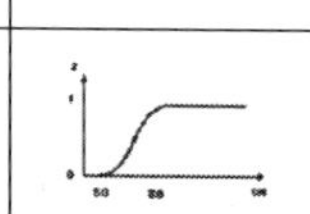
Soil texture	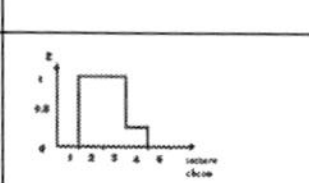	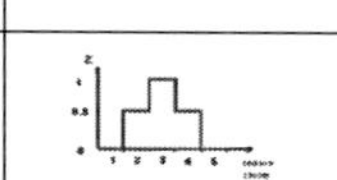	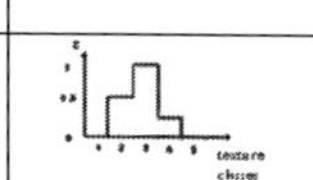	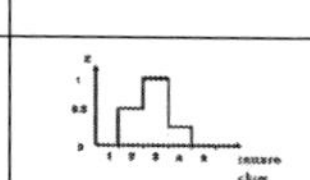

Table 11.3 Weights adopted for the aggregation of the fuzzy suitability values relative to each factor and to each species

Factor	Populus alba	Populus euroamericana	Robinia pseudoacacia	Salix alba
Annual mean temperature	0.3058	0.2418	0.1546	0.0567
Annual mean rainfall	0.0487	0.0496	0.0247	0.1029
Winter minimum temperature (absolute minimum)	0.3004	0.4006	0.0208	0.0223
Distance from the sea coast	0.0563	0.0411	0.1562	0.1127
Slope	0.0862	0.0897	0.2412	0.2535
Soil depth	0.1172	0.1141	0.1210	0.3209
Soil texture	0.0854	0.0631	0.2815	0.1310

WLC was applied to combine the scores of single environmental factors: weight assignment was carried out by factors cross-comparison through a Saaty matrix (Eastman 1999), compiled by the independent panel of five experts (Table 11.3). The Saaty is a square matrix representing pair-wise relationships between criteria valued in the interval [1, 9] and their reciprocals; values are assigned based on experts judgments: e.g., 1 means two criteria contributing equally to land suitability, while a value of 3 signifies that one factor is three times more important than another; and so forth. The method allows consistency between the judgments to be evaluated; if the values in the Saaty matrix are consistent, a global priority index is assigned to each factor quantifying its relative importance within the decision model.

11.4.2.3 Field Validation and Mapping

Field validation was carried out to validate MCE results and to pinpoint thresholds within the continuous range of fuzzy model predictions, to get classes of practical comprehension for land planning. The first class (suitable land, with reference to a potential average annual biomass productivity higher than 8 t ha^{-1}) conceptually corresponds to the classes S1 and S2 of the FAO framework (cf. Section 11.1); the second (marginally suitable land, with reference to a potential average annual biomass productivity between 4 and 8 t ha^{-1}) conceptually corresponds to the class S3; the third (not suitable land, with reference to a potential average annual biomass productivity lower than 4 t ha^{-1}) to the order N.

The thresholds of fuzzy suitability values were defined on the basis of a ground survey covering 34 SRF plantations of northern and central Italy (Fig. 11.1), established between 2002 and 2006. The productive performances of the SRF plantations were assessed and compared to the predicted fuzzy values, giving rise to the

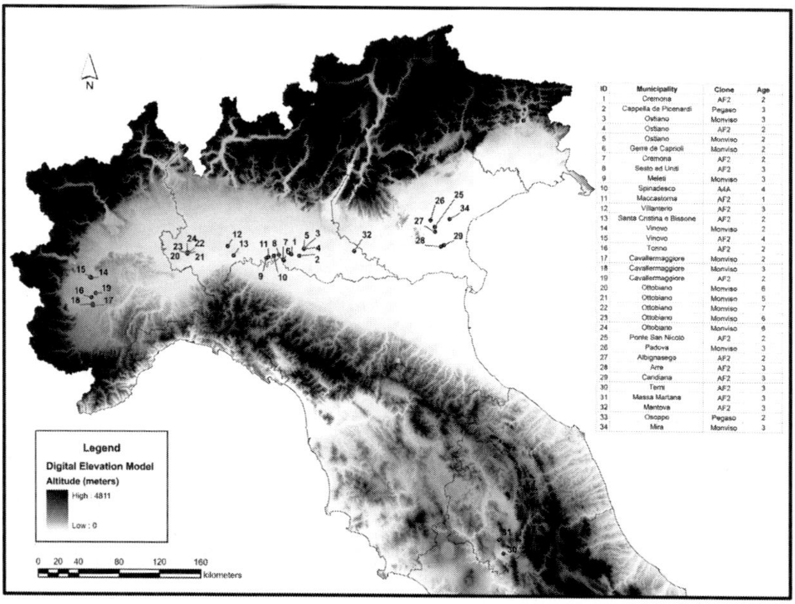

Fig. 11.1 Geographical location of the SRF plantations surveyed

following classification: (i) suitable land [fuzzy value = 0.9–1]; (ii) marginally suitable land [fuzzy value = 0.7–0.89]; (iii) not suitable land [fuzzy value = 0–0.69].

The last elaboration phase concerned the derivation of a vector map of land suitability compatible with the standard land use/cover national Corine Land Cover 2000 map (CLC, minimum mapping unit = 25 ha; see European Environmental Agency 2000). The CLC classes whose conversion to SRF plantations is unlikely or impossible (urban settlements, forests, rocky outcrops, water or agricultural crops more profitable than biomass crops like vineyards) were set as *constraints* of the analysis (cf. Section 11.2.2). Hence, suitability classes were assigned only to pixels falling within polygons of the following farmland classes: non irrigated arable land (CLC code = 211); pastures (CLC code = 231); annual crops associated with permanent crops (CLC code = 241); complex cultivation patterns (CLC code = 242); agro-forestry areas (CLC code = 244). Given the high management intensity of SRF, farmland areas included within designated protected areas (national and regional natural parks, Natura 2000 sites, etc.) were also excluded from the LSA. Suitability class statistics were calculated for each eligible CLC polygon. The polygons without suitable or marginally suitable pixels were eliminated. The polygons with all the pixels belonging to a given suitability class (pure polygons) were assigned to that class. Polygons including suitable and marginally suitable areas of at least 10 ha, were split into pure sub-polygons; otherwise, the polygon was assigned to the suitability class occurring most often on a pixel-by-pixel basis.

11.4.3 Results and Discussion

The fuzzy maps of land suitability for SRF plantations in Italy are reported in Fig. 11.2.

Suitable farmland for SRF plantations in Italy, resulting from the overlap of the suitable areas for all the target species, amounted to 1301051 ha, whereas marginally suitable farmland was 5380431 ha. The geographical distribution of such figures is summarized in Table 11.4.

This case study demonstrates how the fuzzy approach can address the critical issue of modelling the response of target species to environmental qualities and limitations of the land. The LSA fuzzy digital maps are products that can be used by land planners as decision-support tools. For instance, they can be easily used by regional agricultural and forest services and rural planning decision-makers to outline the most suitable land areas for subsidising SRF plantations. The adopted modelling approach is intentionally empirical and planning-oriented: the purpose is to predict, on a large scale, farmland suitability to SRF taking into consideration environmental factors for which geodatabases, of appropriate and comparable spatial resolution, are available. As shown, model outputs can be considered sufficiently sound when compared against field data. The modelling process can be easily replicated at finer scales, provided that appropriate data are available.

Table 11.4 Environmental suitability of farmland (expressed in hectares) for SRF plantations in Italy

Region	Suitable farmland	Marginally suitable farmland
Abruzzo	741	102853
Basilicata	33	66570
Calabria	4545	45716
Campania	1440	316571
Emilia-Romagna	2079	998156
Friuli-Venezia Giulia	540	262263
Lazio	58125	291568
Liguria	1217	14018
Lombardia	453042	712401
Marche	4644	418454
Molise	3913	59156
Piemonte	46601	564785
Puglia	0	20742
Sardegna	0	0
Sicilia	0	0
Toscana	121412	510778
Trentino-Alto Adige	1163	15280
Umbria	7888	287229
Valle d'Aosta	0	2438
Veneto	593668	691453

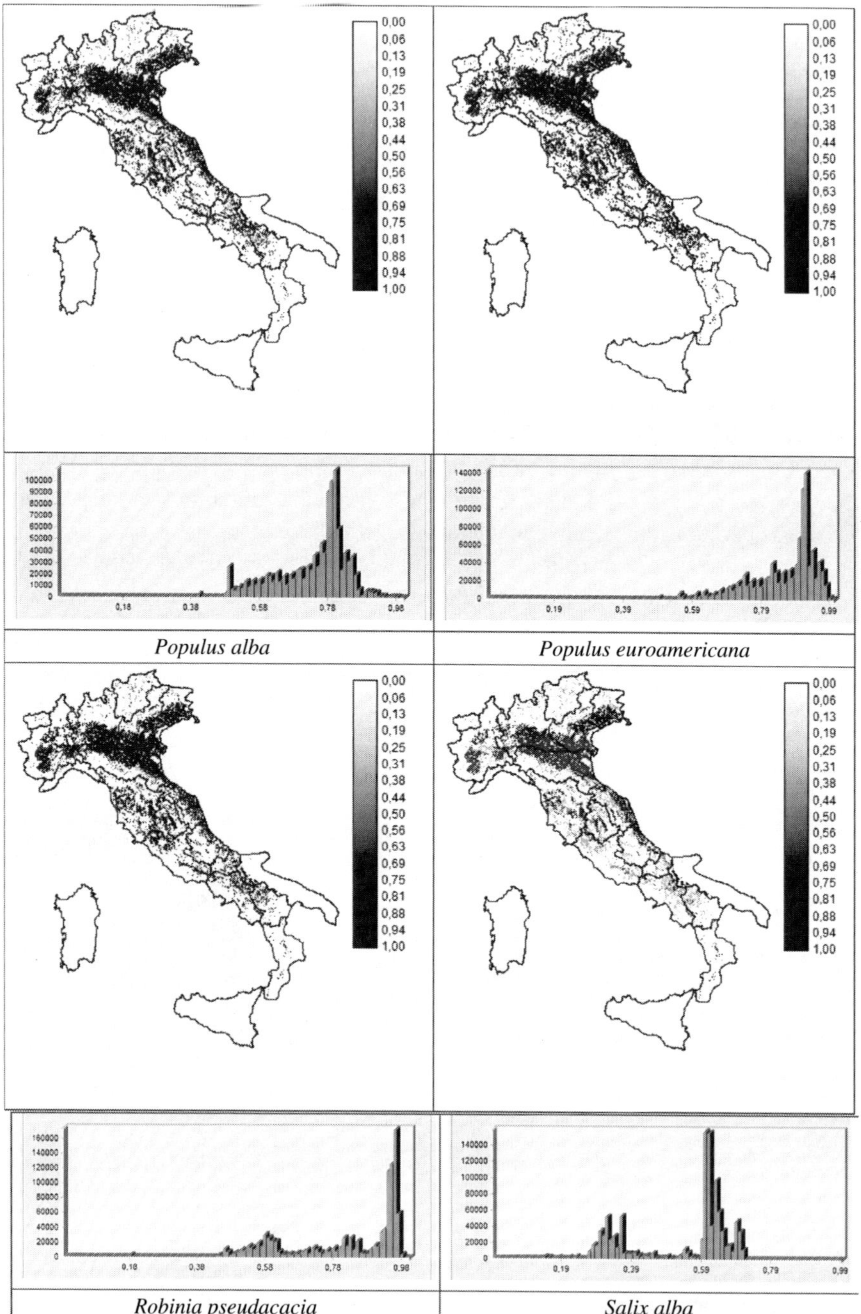

Fig. 11.2 Fuzzy maps of land suitability for the considered tree species; the digital values of fuzzy suitability ranges from a minimum of 0 (no suitability) to a maximum of 1 (optimal suitability). The diagram reports the number of pixels vs. suitability fuzzy values for each target tree species

11.5 Summary and Conclusions

In the land planning process, land suitability can be regarded as a bridging phase linking land resources assessment to the decision-making process. The inherent conflicts and the complex network of socio-economical and ecological constraints affecting land use planning call for a flexible decision-making support tool able to incorporate multiple evaluation criteria, including several stakeholders point of views. Land suitability gives transparent indications to decision-makers concerning land uses which can be sustainably fostered in the land under consideration, allowing areas to be ranked according to their degree of suitability for a specific land use. The resulting maps lend efficient support to negotiation, when addressing issues of sustainable development as well as economic competitiveness.

The spreading of technologies of remote sensing, GIS and global positioning systems (GPS) and of high speed computers enabling the acquisition and effective management of data on spatially distributed land resources, with relatively limited financial resources, has increased enormously in the last years. Consequently, the importance of developing efficient LSA procedures has been emphasized.

In order to address complex decisions, MCE techniques have been developed, incorporating high functionality and an ability to work seamlessly with both raster and vector data structures. A critical point for a successful application of MCE is the quantitative knowledge available concerning the relationships amongst environmental factors affecting land response for specified purposes; no data integration model will work in absence of such knowledge.

Within this framework this paper aimed to: (i) outline relevant LSA experiences, with a distinctive focus on MCE GIS-based techniques; (ii) present a case study of MCE applying fuzzy modelling to predict the productive response of land when afforested with target forest plantation species; drawing from the case study experience, the MCE proved to be a valuable tool for decision-making (i.e., to quantify on a broad scale land physical suitability to an expansion of SRF plantations for energy biomass production in Italy).

More in general, the following operational recommendations are suggested for a sound application of LSA:

- Use of geodatabases of comparable scale and resolution, with the same quality standards and that reflects consistent and complete area coverage.
- Application of MCE methods, usually capable of soundly integrating in a GIS environment information from multiple factors; two MCE techniques (WLC, OWA) are available to model the tradeoff between multiple factors in determining land suitability, enabling the decision maker to select a variety of land use strategies ranging from a risk-averse to a risk-taking strategy.
- Adoption of a fuzzy approach to model the ecological relationship between the productive response of a given crop species and single environmental factors; it offers a sound alternative to hard classification methodologies since such relationships are intrinsically characterized by zones of gradual transition rather than sharp boundaries.

- Robust field validation to verify model predictions and to examine their sensitivity to variations in model parameters (criteria scores, weights, constraints).

From a more general standpoint, it is fundamental to frame LSA experiences within an overall adaptive approach, that continually reassesses information needs, incorporates multi-disciplinary perspectives and responds to society's changing values. To this end, a promising research issue is LSA modelling in a dynamic perspective: i.e., to predict changes in land use aptitude within the context of a changing environment.

Acknowledgments This paper was carried out with equal contributions by the Authors, under the project 'Innovative forest management models for energy biomass production', funded by the Italian Ministry of University and Research (PRIN 2005 funds, Working Unit: Università della Tuscia; local coordinator: P. Corona; national coordinator: O. Ciancio).

References

Aerts J (2002) Spatial decision support for resource allocation: integration of optimisation, uncertainty analysis and visualization techniques. PhD Thesis, Faculty of Science, University of Amsterdam, Amsterdam

Banai R (1993) Fuzziness in geographic information systems: contributions from the analytic hierarchy process. Int J Geogr Inform Syst 7:315–329

Bannet T, Bar P, Svoray T (2005) Urban land-use allocation in a Mediterranean ecotone: Habitat Heterogeneity Model incorporated in a GIS using a multi-criteria mechanism. Landscape Urban Plan 72:337–351

Barredo JI (1996) Sistemas de Informacion Geografica y Evaluation Multi-criterio en la Ordenacion del Territorio. Ra-Ma, Madrid

Barredo JI, Benavidesz A, Hervhl J, Van Westen CJ (2000) Comparing heuristic landslide hazard assessment techniques using GIS in the Tirajana basin, Gran Canaria Island, Spain. International J Appl Earth Observ Geoinform 2:9–23

Bayliss JL, Simonite V, Thompson S (2005) The use of probabilistic habitat suitability models for biodiversity action planning. Agriculture. Ecosyst Environ 108:228–250

Beedasy J, Whyatt D (1999) Diverting the tourists: a spatial decision support system for tourism planning on a developing island. J Appl Earth Observ Geoinform 3/4:163–174

Blasi C, Chirici G, Corona P, Marchetti M, Maselli F, Puletti N (2007) Spazializzazione di dati climatici a livello nazionale tramite modelli regressivi localizzati. Forest@ 4:213–219

Bognar JA, Lathrop Jr, Richard G (1998) Applying GIS and landscape ecological principles to evaluate land conservation alternatives. Landscape Urban Plan 41:27–41

Bojorquez-Tapia LA, Diaz-Mondragon S, Ezcurra E (2001) GIS-based approach for participatory decision-making and land suitability assessment. Int J Geogr Inform Sci 15:129–151

Booth TH, Saunders JC (1985) Applying the FAO guidelines on land evaluation for forestry. Forest Ecol Manag 12:129–142

Boyce MS, McDonald LL (1999) Relating populations to habitats using resource selection functions. Trends Ecol Evol 14:268–272

Brail RK, Klosterman RE (2001) Planning support systems, ESRI Press, Redlands

Burrough PA, McDonnell RA (1998) Principles of geographical information System. Oxford University Press, Oxford

Bydekerke L, Van Ranst E, Vanmechelen L, Groenemans R (1998) Land suitability assessment for cherimoya in southern Ecuador using expert knowledge and GIS. Agriculture, Ecosyst Environ 69:89–98

Cambell JC, Radke J, Gless JT, Whirtshafter RM (1992) An application of linear programming and geographic information systems: cropland allocation in antique. Environ Plann 24:535–549

Carranza EJM, Nidumolu UB, Sicat RS (2005) Fuzzy modelling of farmers' knowledge for land suitability classification. Agr Syst 83:49–75

Carver SJ (1991) Integrating multi-criteria evaluation wuth Geographical Information Systems. Int J Geogr Inform Syst 5(3):321–339

Ceballos-Silva A, Lopez-Blanco J (2003a) Delineation of suitable areas for crops using a Multi-Criteria Evaluation approach and land use/cover mapping: a case study in Central Mexico. Agr Syst 77:117–136

Ceballos-Silva A, Lopez-Blanco J (2003b) Evaluating biophysical variables to identify suitable areas for oat in Central Mexico: a multi–criteria and GIS approach. Agriculture, Ecosyst Environ 95:371–377

Chirici G, Corona P, Marchetti M, Travaglini D, Wolf U (2002) Modello di valutazione dell'attitudine fisica del territorio per la realizzazione di piantagioni di noce comune e di douglasia in Italia meridionale. Monti e Boschi 6:25–31

Church RL (2002) Geographical information systems and location science. Comp Oper Res 29:541–562

Clark MJ, Gurnell J, Lurz P, Mark DF, Rushton S, Steve P (2002) Conserving red squirrels (Sciurus vulgaris): mapping and forecasting habitat suitability using a Geographic Information Systems Approach. Biol Conserv 105:53–64

Collins MG, Steiner FR, Rushman MJ (2001) Land-use suitability analysis in the United States: historical development and promising technological achievements. Environ Manage 28: 611–621

Cools N, De Pauw E, Deckers J (2003) Towards an integration of conventional land evaluation methods and farmers' soil suitability assessment: a case study in north-western Syria. Agr Ecosyst Environ 95:327–342

Cromley R (1994) A comparison of suitability mapping and multi-objective programming techniques for performing spatial searches. The Pennsylvania Geographer 32:37–46

Cromley RG, Hanink DM (1999) Coupling land use allocation models with raster GIS. J Geograph Syst 1:137–153

Dai FC, Lee CF, Zhang XH (2001) GIS-based geo-environmental evaluation for urban land-use planning: a case study. Eng Geol 61:257–271

Diamond JT, Wright JR (1988) Design of an integrated spatial information system for multi–objective land-use planning. Environ Plann 15:205–214

Dyer JS (1990) Remarks on the Analytic Hierarchy Process. Manage Sci, 36(3):249–258

Eastman JR (1999) Idrisi 32 – Guide to GIS and image processing. Clark University, Worcester

Eastman JR (2006) Idrisi Andes – Guide to GIS and image processing. Clark University, Worcester

EEA (2007) How much bioenergy can Europe produce without harming the environment? EEA Report No 7/2006, Copenhagen

European Environmental Agency (2000) CORINE land cover technical guide – Addendum 2000. Technical report 40, Copenhagen

FAO (1976) A framework for land valuation. FAO Soil Bulletin 32, Rome

Feiring BR (1986) Linear programming. An introduction. Quantitative applications in the social sciences, vol. 60, Sage Publications, London

Fielding AH, Bell JF (1997) A review of methods for the assessment of prediction errors in conservation presence/absence models. Environ Conserv 24:38–49

Groenemans R, Van Ranst E, Kerre E (1997) Fuzzy relational calculus in land evaluation. Geoderma 77:283–298

Guisan A, Thuiller W (2005) Predicting species distribution: offering more than simple habitat models. Ecol Lett 8:993–1009

Guisan A, Zimmermann NE (2000) Predictive habitat distribution models in ecology. Ecol Model 135:147–186

Guo L, Liu Y, Wang J (2006) GIS-Based assessment of land suitability for optimal allocation in the Qinling Mountains, China. Pedosphere 16:579–586

Hirzel AH, Hausser J, Chessel D, Perrin N (2002) Ecological-niche factor analysis: how to compute Habitat-suitability maps without absence data? Ecology 83(7):2027–2036

Hirzel AH, Le Lay G, Helfer V, Randin C, Guisan A (2006) Evaluating the ability of habitat suitability models to predict species presences. Ecol Model 199:142–152

Hopkins L (1977) Methods for generating land suitability maps: a comparative evaluation. J Amer Inst Plan 34:19–29

Jankowski P, Richard L (1994) Integration of GIS-based suitability analysis and multi-criteria evaluation in a spatial decision support system for route selection. Environ Plann 21:326–339

Jiang H, Eastman JR (2000) Application of fuzzy measures in multi-criteria evaluation in GIS. Int J Geogr Inform Syst 14:173–184

Joerin F (1995) Methode multi-critere daide a la decision et SIG pour la recherche d'un site. Rev. Int. Geomatique 5:37–51

Joerin F, Theriault M, Musy A (2001) Using GIS and outranking multi-criteria analysis for land-use suitability assessment. Int J Geogr Inform Sci 15:153–174

Kaiser (ed) (1995) Urban land use planning, 4th edition). University of Illinos, USA

Kalogirou S (2002) Expert systems and GIS: an application of land suitability evaluation. Comput Environ Urban 26:89–112

Liengsakul M, Mekpaiboonwatana S, Pramojanee P, Bronsveld K, Huizing H (1993) Use of GIS and remote sensing for soil mapping and for locating new sites for permanent cropland: a case study in the "highlands" of northern Thailand. Geoderma 60:293–307

Ligtenberg A, Bregt AK, Van Lammeren R (2001) Multi-actor-based land use modelling: spatial planning using agents. Landscape Urban Plan 56:21–33

Lyle J, Stutz F (1983) Computerized land use suitability mapping. Cartogr J 20:39–49

Mac Dougall EB (1975) The accuracy of map overlays. Landscape Plann 2:23–30

Malczewski J (ed) (1999) GIS and Multi-criteria Decision Analysis. John Wiley & Sons Inc., New York

Malczewski J (2004) GIS-based land-use suitability analysis: a critical overview. Progr Plann 62:3–65

Malczewski J, Chapman T, Flegel C, Walters D, Shrubsole D, Healy MA (2003) GIS-multi-criteria evaluation with ordered weighted averaging (OWA): case study of developing watershed management strategies. Environ Plan 35:1769–1784

Manel S, Dias JM, Buckton ST, Ormerod SJ (1999) Alternative methods for predicting species distribution: an illustration with Himalayan river birds. J Appl Ecol 36:734–747

Manly BF, McDonald LL, Thomas DL, McDonald TL, Erickson WP (eds) (2002) Resource selection by animals: statistical design and analysis for field studies, 2nd ed. Kluwer Academic Publishers, Dordrecht

McCullagh P, Nelder JA (1989) Generalized linear models, 2nd ed. Monographs on statistics and applied probability, vol 37. Chapman and Hall, London

McHarg IL (1969) Design with nature. Wiley, New York

Meentemeyer RK, Rodman LC (2006) A geographic analysis of wind turbine placement in Northern California. Energ Policy 34:2137–2149

Minotta G (2003) L'arboricoltura da legno: un'attività produttiva al servizio dell'ambiente. Edizioni Avenue Media, Bologna

Mohamed ABAA, Sharifi MA, Van Keulen H (2000) An integrated agro-economic and agro-ecological methodology for land use planning and policy analysis. Int. J Appl Earth Observ Geoinform 2:87–103

Overmars KP, Verburg PH, Veldkamp TA (2007) Comparison of a deductive and an inductive approach to specify land suitability in a spatially explicit land use model. Land Use Policy 24(3):584–599

Pearce JL, Boyce MS (2006) Modelling distribution and abundance with presence-only data. J Appl Ecol 43:405–412

Pereira JMC, Duckstein L (1993) A multi-ple criteria decision-making approach to GIS-based land suitability evaluation. Int J Geogr Inform Syst 7:407–424

Salvati L, Ceccarelli T, Brunetti A (2005) Geo-database sul rischio di desertificazione in Italia. CRA, Ufficio Centrale di Ecologia Agraria, Roma

Scott JM, Heglund PJ, Morrison ML, Haufler JB, Raphael MG, Wall WA, Samson FB (2002) Predicting species occurences: issues of scale and accuracy. Island Press, Washington
Smeins FE, Wu XB (2000) Multi-ple-scale habitat modelling approach for rare plant conservation. Landscape Urban Plan 51:11–28
Smith CS, Howes AL, Price B, McAlpine CA (2007) Using a Bayesian belief network to predict suitable habitat of an endangered mammal – The Julia Creek dunnart (Sminthopsis douglasi). Biol Conserv 139(3–4):333–347
Steinitz C, Parker P, Jordan L (1976) Hand drawn overlays: their history and prospective uses. Landscape Architect 9:444–455
Store R, Jokimäki J (2003) A GIS-based multi-scale approach to habitat suitability modelling. Ecol Model 169:1–15
Sui DZ (1992) A fuzzy GIS modeling approach for urban land evaluation. Comput Environ Urban, 16:101–115
Thapa RB, Murayama Y (2008) Land evaluation for peri-urban agriculture using analytical hierarchical process and geographic information system techniques: a case study of Hanoi. Land Use Policy, 25(2):225–239
Tomlin CD (1990) Geographical information systems and cartographic modelling, Prentice-Hall, Englewood Cliffs
Van Ranst E, Tang H, Groenemans R, Sinthurahat S (1996) Application of fuzzy logic to land suitability for rubber production in peninsular Thailand. Geoderma 70:1–19
Voogt H (1983) Multi-criteria evaluation for urban and regional planning. Pion, LTD, London
Yager RR (1988) On ordered weighted averaging aggregation operators in multi-criteria decision making. IEEE Trans Syst Man Cybernet 18:183–190

Chapter 12

Assessing Human Impacts on Australian Forests through Integration of Remote Sensing Data

Richard Lucas, Arnon Accad, Lucy Randall, Peter Bunting and John Armston

Abstract Prior to and since European settlement, humans have impacted on the vegetation of Queensland, Australia, primarily by changing fire regimes and clearing forests for agriculture but also by introducing flora and fauna. Such changes have been mapped and monitored in the past through the use of airborne (e.g., aerial photography) and spaceborne optical (e.g., Landsat) remote sensing data. However, with the increased provision of data in different modes (radar, lidar) and at various spatial resolutions (<1–> 250 m), opportunities for detecting, characterizing, mapping and monitoring such changes have been increased. In particular, the combination of radar and optical data has allowed better assessment of deforestation patterns (clear felling, stem injection), regeneration and woody thickening, tree death from climatic change, and biomass/biomass change. Such information also provides new insights into the associated changes in carbon dynamics and biodiversity. Using a series of case studies, these advances in technology and the benefits for Statewide and national mapping and monitoring of forest extent and condition are reviewed.

12.1 Introduction

For landscape ecologists, timely information on the spatial and temporal state and dynamics of forests increases knowledge and understanding of the distribution of flora and fauna and also their response to change, whether natural or human-induced. Key attributes of interest include the extent and spatial arrangement of forests of differing structure, growth stage, biomass and tree species composition as this information can be integrated to better understand distributions, behaviour and/or adaptation. Furthermore, scientists, governments and landholders now require data on a regular basis and at local to global scales to support, for example, conservation of biological diversity, quantification of carbon stocks and fluxes and sustainable land use. For this reason, many programmes have focused on the use of remote sensing data acquired by both airborne and spaceborne sensors.

R. Lucas
Institute of Geography and Earth Sciences, Aberystwyth University, Llandinam Tower, Penglais Campus, Aberystwyth, Ceredigion, SY23 3DB, United Kingdom
e-mail: rml@aber.ac.uk

R. Lafortezza et al. (eds.), *Patterns and Processes in Forest Landscapes*,

When man first started viewing the earth remotely, much of the forested landscape (particularly in tropical and boreal regions) was pristine or relatively undisturbed. Large area aerial photographic coverage of these forests became increasingly available from the 1940s and, whilst these data are largely black and white, they provide an important baseline of forest extent against which to assess change. Regular ($\sim$ 16–18 days) satellite sensor observations at relatively moderate ($\sim$ 30 m) spatial resolution commenced in the early 1970s with the launch of the Landsat series of sensors. Since then, the number and diversity of sensors observing the earth's surface has increased substantially. Between the 1970s and the present, these advances in remote sensing technology have presented three main benefits to landscape ecologists. First, the various observation strategies (e.g., associated with the Landsat program) have provided data over a 30–40 year period which allow for the detection of change. Second, airborne and/or spaceborne sensors are observing at spatial resolutions ranging from $<$ 1 m to several km, allowing detail to be resolved across a range of scales (e.g., from individual trees to entire forested regions). Third, sensors are increasingly providing information on forests in three dimensions, particularly with the advent of multi-frequency polarimetric and interferometric Range Detection and Ranging (RaDAR) and Light Detection and Ranging (LiDAR). Collectively, a substantial amount of information on the World's forested landscapes has and continues to be gathered over a period where human impacts have been at their greatest.

This chapter provides an overview of the way in which remote sensing data have been used to characterise, map and monitor wooded savannas, focusing on those occurring in Queensland, Australia. These forested ecosystems have been subject to anthropogenic change prior to but particularly since European settlement. Change has also accelerated since the mid 20th century when remote sensing instruments operating in various modes and at different spatial and temporal resolutions have collected data over this landscape. The chapter also outlines how remote sensing data from both airborne and spaceborne instruments can provide unique information on human-induced changes to forests. Lastly, the actual or potential benefits of using remote sensing data for landscape assessment, particularly in relation to assessments of greenhouse gas (carbon) emissions and biodiversity, are presented.

12.2 Forest Cover Changes in Queensland in Relation to Human Activity

Prior to European settlement, grasslands occupied 20% of the land surface of Queensland but the remaining 80% was comprised of forests and associated vegetation (shrubs and heaths), with most occurring towards the north and east of the State (Wilson et al. 2002, Fig. 12.1a). The forests were structurally and floristically diverse as a consequence of climatic and pedologic variation, and ranged from open mallee scrub to tropical rainforest (Webb 1959; Tracey 1982). The structural classification developed by Specht (1970) reflected the diversity of these forests (Table 12.1), dividing them into categories relating to the life form (trees, shrubs,

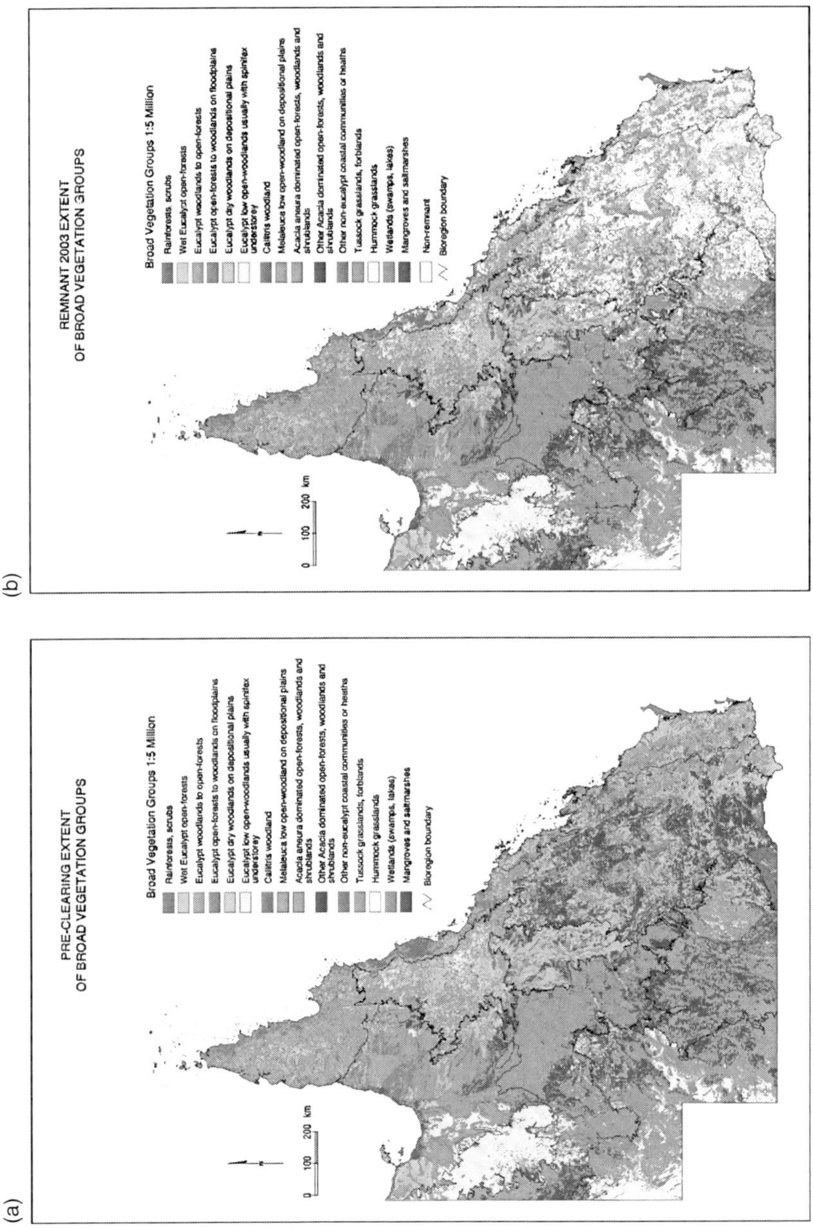

Fig. 12.1 (**a**) Pre-clearing extent of broad vegetation groups and (**b**) remnant broad vegetation groups (Accad et al. 2006)

Table 12.1 Structural classification of forests in Queensland, Australia (Specht 1970)

Life form and height of tallest stratum	Foliage projected cover of tallest stratum			
	Dense (70–100%)	Mid dense (30–70%)	Sparse (10–30%)	Very sparse (<10%)
Trees >30m	Tall closed-forest	Tall open-forest	Tall woodland	Tall open-woodland
Trees 10–30 m	Closed forest	Open-forest	Woodland	Open-woodland
Trees 5–10 m	Low closed-forest	Low open-forest	Low woodland	Low open woodland
Shrubs 2–8 m	Closed-scrub	Open-scrub	Tall shrubland	Tall open shrubland
Shrubs 0–2 m	Closed-health	Open-health	Low shrubland	Low open-shrubland
Herbs (including: moss, ferns, hemicryptophytes, geophytes, therophytes, hydrophytes, helophytes)	Closed-herbland	Herbland Herbland	Open-herbland	
	Closed-tussock grassland	Tussock grassland	Open-herbland	
	Closed-grassland	Grassland Herbfield	Open-tussock grassland	
	Closed-herbfield	Sedgeland Fernland	Open-grassland	
	Closed-sedgeland	Mossland	Open-herbfield	
	Closed-fernland		Open-sedgeland	
	Closed-mossland		Open-fernland	
			Open-mossland	
Hummock grasses 0–2 m			Hummock grassland	Open hummock grassland

grasses, herbs, moss etc.) and height and Foliage Projected Cover (FPC)[1] of the tallest stratum. For example, tropical rainforests were typically classified as tall closed forest with trees $\geq$ 30 m in height and a FPC of between 70 and 100%. Most forests were dominated by species unique to Australia, including *Eucalpytus*, *Corymbia* and *Acacia* (Common and Norton 1992). The extent of natural disturbance within these forests was variable between structural types and was typically associated with the cycles of drought and flooding as well as extreme events (e.g., cyclones). As with today, fires were an integral and frequent component of the disturbance regime (Bowman 1998; Russell-Smith et al. 2003).

Human influence on the forest environment commenced with the arrival of the aboriginal Australians at least 40,000 years ago (Walker et al. 1993; Common and Norton 1992). At the time of European settlement in 1788, the aboriginal population was between 250,000 and 500,000. These populations were largely nomadic hunters and gatherers and lived with the environment. However, many exploited fire to increase opportunities for food gathering. Together with natural fires, those associated with aboriginal burning maintained a savanna structure to much of the vegetation.

European settlement followed the discovery of Australia and clearing of the forests began in earnest, initially adopting the agricultural practices used successfully in Europe. Within Queensland, the reasons for clearing were manifold and related primarily to increased demand for food by an expanding national and, later, a global population. For example, increasing demand for meat led to a corresponding increase in the numbers of stock from a few million in the late 1880s to over 12 million in 2007, assisted by the introduction of tropical breeds, artificial water supplies and feeding supplements (Fensham and Holman 1999). The area under crop production (e.g., maize and grain) also expanded and by 1920, much of the available land had been exploited using technology available at the time (Common and Norton 1992). Other reasons for clearance included increases in incomes for some landholders, the provision of tax liability offsets, agricultural economics, and the development and greater affordability of heavy machinery. Clearing was particularly rapid from the mid 1990s to 2000s (Accad et al. 2006) which prompted the enforcement of clearing restrictions in the mid 2000s (Henry et al. 2005).

In Queensland, most of the clearance has historically occurred within the south central and south east (Fig 12.1b) and initially in areas most suitable for agriculture in terms of soil fertility and topographic position. Clearing was generally limited by physical factors such as slope and drainage and was often confined to particular vegetation types and structures, largely because of associations with suitable soil types or their ease of clearance. In terms of extent, clearance has mainly been within the open forests, woodlands and shrublands dominated by *Acacia* species and woodlands and open forests on floodplains and depositional plains dominated by *Eucalyptus* species (Table 12.2). Less than 1 million ha of other broad vegetation groups has been cleared. The loss of vegetation has been particularly extensive within the Brigalow Bioregion, a fertile area of south central Queensland. Most of

[1] FPC is defined as the percentage cover of leaves projected over a unit area.

this vegetation was cleared in the 1960s and, in the mid 2000s, resulting in less than 15% of forests with brigalow (*Acacia harpophylla*) and gidgee (*Acacia cambagei*) as a major component remaining (Fairfax and Fensham 2000).

Queensland can claim that, in 2003, 81% of the State supported vegetation that was remnant (Accad et al. 2006). Under the 1999 Vegetation Management Act, woody vegetation is considered remnant where the dominant canopy has greater than both 70% of the height and 50% of the cover relative to the undisturbed height and cover of that specific community and that stratum is still dominated by species characteristic of the vegetation's undisturbed canopy (Butler and Fairfax 2003).

Table 12.2 The changing extent of vegetation groups in Queensland (NLWRA 2002)

Major vegetation group	Area Pre-European (km^2)	Area (circa 1997) (km^2)	% of total extent remaining
Eucalypt woodlands	473,272	367,293	77.6
Tussock grasslands	294,662	282,547	95.9
Eucalypt open woodlands	165,065	134,421	81.4
Acacia shrublands	104,368	100,660	96.4
Hummock grasslands	92,009	91,809	99.8
Acacia forests and woodlands	182,089	91,534	50.3
Chenopod shrubs, samphire shrubs and forblands	82,070	81,944	99.8
Melaleuca forests and woodlands	72,173	70,014	97
Other forests and woodlands	49,692	49,266	99.1
Acacia open woodlands	39,861	36,734	92.2
Eucalypt open forests	62,646	35,150	56.1
Tropical eucalypt woodlands/grasslands	20,684	20,653	99.9
Rainforest and vine thickets	30,055	19,558	65.1
Other shrublands	16,780	16,419	97.8
Mangroves, tidal mudflats, samphires and bare areas, claypans, sand, rock, salt lakes, lagoons, lakes	15,442	15,143	98.1
Other grasslands, herblands, sedgelands and rushlands	4,963	4,771	96.1
Callitris forests and woodlands	5,601	4,134	73.8
Casuarina forests and woodlands	11,951	1,545	12.9
Heath	633	470	74.2
Low closed forests and closed shrublands	449	445	99.1
Eucalypt tall open forests	3,976	429	10.8
Eucalypt low open forests	111	111	100
Mallee woodland and shrublands	14	14	100

Much of this area includes the large expanses of forests in the north of the State and most of the grasslands. However, in the last few decades, major clearing of woody vegetation has occurred in the southern, central and eastern parts of the State and particularly within the Brigalow, South-eastern Queensland, Desert Uplands and Mulga Bioregions. Whilst direct clearing is the most visible sign of human impacts, more subtle changes in forest structure have also occurred as a consequence of fragmentation (McIntyre and Hobbs 1999). The introduction of grazing animals and exotic plants has further modified the structure of the forest over time.

Forest loss from direct human activity is anticipated to decrease in Queensland following the introduction of legislation that prohibits further clearance of remnant vegetation from December 2006. Forest is also likely to return to many previously cleared areas because of the difficulty in maintaining cleared areas in productive agriculture. Despite this, forests in Queensland are becoming increasingly threatened from climatic change. Future predictions include a 3.5–4.0 °C maximum temperature increase in the interior or Queensland and more El-Nino like conditions with increased drought frequency and decreased average soil moisture and rainfall. The intensity of rainfall events is also anticipated to increase. For these reasons, the future state of forests is uncertain and will require appropriate and regular observations to establish the nature and extent of change.

12.3 The Use of Remote Sensing for Forest Assessment in Queensland

Queensland covers a vast area (1.73 million km^2) and so approaches to the detailed mapping and monitoring of forests have relied largely upon satellite sensor observations. Even so, aerial photographs were acquired for some areas in the 1940s and for much of the State in the 1950s and 1960s (Fensham et al. 2002). These are used primarily by the Queensland Herbarium of the Environmental Protection Agency (EPA) to establish a historical baseline of pre-clearing and remnant vegetation distributions. Within the stereo overlap area of the photography, which allows for three dimensional viewing of the vegetation and landscape, trained interpreters manually delineate unique map areas (UMAs) (Gunn et al. 1988; Neldner et al. 2005). Following interpretation, selective ground truthing occurs to associate samples of UMAs with vegetation and landscape characteristics. Wider association with Regional Ecosystem types is then undertaken using a combination of interpreted UMAs and existing information on landforms, geology and soils. By 2007, and using 1:80,000 black and white Queensland and Commonwealth photography, the Queensland Herbarium of the EPA had mapped 81% of Queensland vegetation and regional ecosystems for pre-clearing and remnant at 1:100,000 or better (Neldner et al. 2005). Maps at 1:50,000 scale had also been generated for the coastal areas of south east Queensland and the Wet Tropics.

For mapping the changing extent of vegetation at a statewide level with a view to assessing impacts and trends, Landsat sensor data have been the primary data source. These data have been preferred for both regional and national land cover mapping primarily because of the regular data capture and the length of the historical archive. For the periods 1982–1984 and 1990–1991, Graetz (1998) compared maps of woody and non-woody vegetation generated from Landsat Multi-Spectral Scanner (MSS) data to detect and map change and establish levels of disturbance. The Australian Land Cover Change (ALCC) data compared Landsat TM data for the nominal period 1990 (1991 for Queensland and New South Wales) to 1995 to detect changes in woody vegetation (including native forest, plantations and orchards). The previous and replacement land covers were identified together with the reasons for change (e.g., clearing for grazing management). The Queensland Department of Natural Resources and Water (QDNRW 1999, 2004, 2006) Statewide Landcover and Trees Study (SLATS) program also compared Landsat TM and ETM+ data (87 scenes in total) for 1989–1991, 1991–1995, 1995–1997, 1997–1999, 1999–2001 and 2001–2003 to map and monitor land cover change. SLATS products identified similar attributes to the ALCC, although with an expanded list of land uses. The Australian Greenhouse Office's (AGO) National Carbon Accounting System (NCAS) requires woody/non-woody data as input to modelling greenhouse gas emissions for the period 1972–2004 covering 14 epochs and has, for this purpose, utilised Landsat MSS data for 1972, 1977, 1980, 1982, 1985, 1988 and Landsat TM data acquired on a near annual basis thereafter (e.g., 1989–1992, 1995, 1998, 2000, 2002, 2004 etc.).

These woody vegetation change datasets have served several purposes. For example, data on the changing land uses in Queensland have provided input to Australia's National Greenhouse Gas Inventory (NGGI) which is required under the terms of the United Nations Framework Convention on Climate Change (UNFCCC). Areas of change identified through the SLATS change analysis are also cross-checked against the pre-clearing vegetation datasets by the Queensland Herbarium and EPA botanists to establish whether change is substantive and sufficient in terms of height, cover or composition of the canopy species to transfer the vegetation affected to a non-remnant category. Maps of remnant vegetation are updated on at least a biannual basis such that areas moving into a non-remnant category are erased from the pre-clearing vegetation maps and associated instead with a clearing category. Such information has been used to support legislation relating to clearing of native vegetation and regrowth (Henry et al. 2005). With operational programs, continued provision of satellite sensor data is essential and so with recent problems in the operation of the Landsat-7 Enhanced Thematic Mapper (ETM+) and doubts about the continued operation of its predecessor, the Landsat-5 TM, data acquired by other satellite sensors (e.g., France's SPOT High Resolution Geometric (HRG) are increasingly being considered. Spaceborne Synthetic Aperture Radar (SAR), which is an active sensor observing in the microwave region of the electromagnetic spectrum, also represents an alternative or complement to optical sensors but, to date and despite the availability of several global observing sensors, these data have not been used operationally at a regional level (Lucas et al. 2000). A particular advantage of the current (and also proposed) configuration of SAR sensors, however, is their capacity for all weather viewing regardless of the time of day. Furthermore, whereas high frequency C-band (~6 cm wavelength) SAR provides information on the upper canopy of vegetation (as do optical remote sensing data), the lower frequency L-band (~25 cm wavelength) and P-band SAR (~68 cm wavelength) can penetrate through the leaves and retrieve information on the woody components and hence forest structure and biomass (Lucas et al. 2004). Spaceborne SAR have been operating at a global level since the early 1990s, but the launch of the Japanese Space Exploration Agency's (JAXA) Advanced Land Observing Satellite (ALOS) Phased Arrayed L-band SAR (PALSAR) in January 2006 has provided new opportunities for forest mapping and monitoring in Queensland. In particular, this sensor (which has a recurrent cycle of 46 days) is providing L-band SAR data in single (horizontally transmitted and received; HH), dual (HH and HV; where V indicates vertically received) and/or fully polarimetric modes (HH, VV and HV; Rosenqvist et al. 2007). As L-band HH and HV are sensitive primarily to the moisture content, size and geometry of the trunks and branches respectively (Lucas et al. 2004), ALOS PALSAR data are anticipated to provide greater information on the amount and distribution of woody components of vegetation. The proposed launch of spaceborne P-band SAR sensor is expected to provide complementary data for forest characterisation. SAR instruments currently operating in interferometric mode, such as the Shuttle Radar Topographic Mission (SRTM) and spaceborne LiDAR, including the ICESAT Geoscience Laser Altimeter System (GLAS), are also increasingly demonstrating potential for retrieving forest stand height at local to regional scales.

By integrating data from optical sensors, SAR and LiDAR, advances in the characterisation, mapping and monitoring of forests are more likely. For example, the measures of canopy cover and height which can potentially be retrieved from these data effectively define the forest structural types listed in Table 12.1. Potential exists also for providing separate information on the leaf and woody components and retrieving forest biomass up to certain limits. Better information on the floristic composition of forests can be obtained, although characterisation is complex largely because of the large diversity of species across the region, the occurrence of many in the subcanopy and the complex mix of species within an area relative to the spatial resolution of the observing sensor. These datasets can also collectively provide new information that can be utilised to better understand the impacts of humans on the forest environment, whether direct or indirect (e.g., through climate change).

12.4 Remote Sensing for Assessing Human Impacts

The detection of human impacts from remote sensing data is dependent upon the extent to which the land cover is transformed and the relative contrast of the disturbed landscape relative to that which is undisturbed or surrounding. The most direct human impact is the clearance of the forest, which is either complete or partial depending upon the mechanisms used. Following clearance, land can be used for variable time periods and experience different management regimes in terms of, for example, stocking density, weeding and fertilising. When abandoned or unmanaged, regenerating forests are often quick to return but the structural development of the regrowth, biomass and species composition are often influenced by the prior land use. The structure of the forest is also influenced by changes in fire regimes, (which are often human-induced) and the different tolerances of tree species to disturbance (which also influences the floristic composition). Many of these impacts can be detected using remote sensing data but separation from the indirect (e.g., woody thickening) or natural causes of change (e.g., tree mortality following drought or waterlogging) is often problematic. The following sections provide an overview of how human activities and changes in forest structure, biomass and species composition are manifested within remote sensing data.

12.4.1 Clearing Patterns and Processes in Queensland

In the past, the clearance of forests has largely been confined to land which was more accessible, such as flats and lower slopes (Common and Norton 1992). However, with time, forests occurring on less suitable land were increasingly cleared (Randall et al. 2006). Clearing was also common on freehold land as restrictions were imposed on leasehold land and often directed towards those that are of lower commercial value and biomass. A number of different mechanisms for clearing the forest have been implemented, but these have changed over time (Fairfax and Fensham 2000). During periods of cheap labour supply, ring-barking and stem injection were commonplace whilst more mechanised techniques were introduced followed the Second World War. Ring barking involves the stripping of a ring of bark from a tree to kill the tree and prevent further growth. Stem injection involves making horizontal cuts with a narrow bladed axe (5–7 cm wide) through the bark into the sapstream. Herbicide (e.g., "Tordon" picolenic acid) is then applied via the axe blade to ensure that a full dose of herbicide enters the sapstream. In both cases, trees lose leaves and small branches but remain standing. The existing pasture is retained, although surface seeding with the pasture legume Seca stylo (*Styloanthes scabra*) is a recent innovation. These types of clearing have been generally restricted to areas of sparse vegetation. For extensive clearing of less sparse vegetation, *chaining* has commonly been used. This typically involves stringing a linked chain between two bulldozers to uproot all trees in the path. Alternatively, trees are pushed over directly with the bulldozers. The fallen material is usually burned at the end of winter and improved pastures are commonly associated with exotic grasses such as buffel grass (*Cenchrus ciliaris*). Loss of above-ground material is complete and sometimes root stumps are cleared, with these accounting for a large proportion of the below-ground biomass (Burrows et al. 2002). Using this technique, large areas could be cleared and the brigalow lands with their fertile soils were especially targeted for development in the 1960s (Nix 1994). *Burning* only tends to be used in vegetation clearance where there is sufficient grass biomass to retain a burn. However, natural fires caused by lightning strikes are a common feature of the landscape.

Many areas were cleared of vegetation before the advent of Landsat and other moderate spatial resolution sensors (ca. 1972). Whilst aerial photography has provided some insight into clearance patterns and mechanisms, successful analysis has been compromised by image availability at appropriate scales and quality. Using Landsat sensor data, areas that are wholly cleared of forest are generally detected by the rapid decrease in the derived FPC product (Danaher et al. 2004; Lucas et al. 2006a) and change in spectral indices between successive dates (QDNRW 2006). However, identifying the clearing mechanism using optical remote sensing observation can be problematic for several reasons. Many trees left standing after stem injection or ring-barking cannot be resolved. Spectral differences between cleared vegetation (which is typically lying dead) and the background are also typically small. For example, where vegetation is chained, the non-photosynthetic woody debris is commonly stick raked into piles or parallel strips which exhibit a similar reflectance to the background soil. Finally, regeneration and the associated reduction in spectral differences between the cleared and non-cleared areas is often rapid following clearance. High temporal resolution observations from the Moderate Resolution Imaging Spectrometer (MODIS) can overcome this latter problem (Gill et al. 2006), although the spatial resolution of the sensor is generally too coarse to allow detection of some forms of clearing.

Using SAR data, wholesale clearance of forests can generally be detected by a reduction in σ^0 in all wavelength regions, regardless of polarization, which occurs because of a decrease in the number and diversity of surface scatterers. For example, C-band HV data are particularly sensitive to the amount of leaves and small branches which depolarize microwaves through a process known as volume scattering (Lucas et al. 2004). At L-band and P-band HH, double bounce scattering between the ground surface and the tree trunks is dominant although at HV polarizations, volume scattering from the larger branches occurs. The loss of leaves, larger branches and trunks therefore leads to the reduction in σ^0 relative to that of the intact forest. Greater confusion does occur, however, where cut stumps are not uprooted or debris is left on site as scattering from these elements can still occur.

Compared to optical data, SAR data provide better opportunities for detecting the clearing mechanisms employed. For example, within areas identified as non-forest using Landsat-derived FPC data (Fig 12.2a), linear patterns of enhanced backscatter, particularly at lower (e.g., L-band) frequencies and HH polarizations, indicates the use of chaining to clear the forest. In these cases, there is a strong double-bounce interaction between the piles of woody debris, particularly where the radar pulse is perpendicular to the dominant orientation of the lines. Trees that have been stem-injected can also be identified as these typically exhibit a Landsat-derived FPC typical to non-forest areas but a pronounced backscatter at lower frequencies, with this again being most pronounced at HH polarizations (Fig 12.2b–d). This occurs because of greater double bounce scattering from dead standing trunks and low volume scattering (particularly at C-band HV polarizations) resulting from the lack of leaves and smaller branches within the canopy (Lucas et al. 2004).

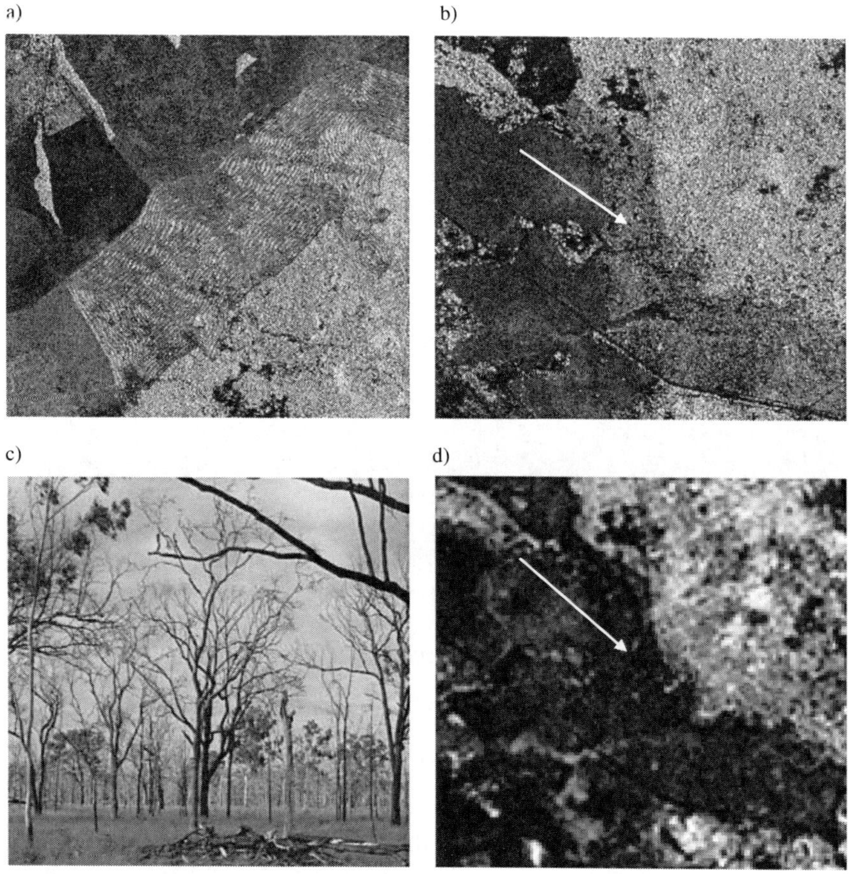

Fig. 12.2 (**a**) Land cleared through chaining/blade ploughing, as observed from airborne SAR, (**b**) stem injected forests (*dark blue*) exhibiting a high L- and P-band HH and low C-band HV (This reflects the presence of dead, standing trees with no foliage cover (**c**)), and (**d**) an FPC image where low values over the stem injected areas reflect the loss of foliage

12.4.2 Using the Land

Following clearance, a number of mechanisms are used to maintain agricultural productivity. For example, *blade ploughing* is carried out to disrupt regrowth in the early stages of establishment and is commonly used in the Brigalow belt where vigorous regeneration of this species occurs. The blade plough cuts to a depth of approximately 25 cm, lifts the trees and then drops them to prevent suckering of the roots following exposure to light. The uprooted trees die and decompose over variable time periods. Pastures or crops are generally seeded at the time of ploughing and burning is not required. *Offset ploughing* is also used to clear regrowth as well as to till the soil and areas treated are typically used for planted pastures and crops. This approach uses two sets of discs set at opposing angles; these cut the roots and require seasonal applications to successfully clear the fields. The detection of such treatments is difficult but can be manifested as a reduction in the variation in the spectral reflectance or σ^o from a cleared area.

Time-series comparisons of Landsat sensor or derived datasets (e.g., FPC) have primarily been used to detect the increase in the deforested area over time or age the subsequent regrowth. However, the use of these data for tracking the periods of active land use and land management practices (e.g., the use of fire or the frequency of reclearance of regrowth) has rarely been undertaken, even though such management practices can impact on the capacity of forest to regenerate subsequently. For example, continued burning of areas can promote fire resistant species in the regenerating forest community. Enhanced growth of some species (e.g., *E. populnea*) can also occur where tree density is artificially reduced through, for example, stem injection (Fensham et al. 2005). Extending the use of remote sensing data to better understand the impacts of land management on the capacity of forests to regenerate is therefore advocated. This approach also provides an opportunity to identify areas that are most suitable for regeneration and to predict the likely changes in forest structure and biomass as the regeneration proceeds.

12.4.3 Regenerating the Forest

Where large areas of forest are cleared, many farmers are often unable to control the regeneration that follows, although reclearing of forests is nevertheless commonplace. The Queensland Vegetation Management and Other Legislation Amendment (2004) also provides incentives for land owners to protect areas of regrowth in environmentally sensitive landscapes (e.g., riparian zones). The amount of regrowth occurring across the state is therefore likely to be extensive over the next few decades, particularly given the high rate of clearance in the 1990s and 2000s and subsequent restrictions on clearing.

Detecting regenerating forests in Queensland using optical remote sensing data has proved problematic because of spectral confusion with remnant forests. Time-series comparisons of Landsat-derived FPC are also being investigated for their potential use in mapping regrowth, although the trajectories are complicated by, for example, seasonal flushes of non-woody vegetation. However, by integrating SAR data of different frequency and polarization with optical data, many of these issues can be overcome. To illustrate, within the Brigalow Bioregion, extensive areas of regeneration occur and are typically dominated by *Acacia harpophylla*. The shoots of this species emanate from a root system, and these stands typically consist of clusters of stems that are generally of small diameter but collectively form a relatively closed canopy because of the large amount of foliage associated with each. This can lead to an FPC and also C-band HV σ^o that is equivalent to or greater than that of remnant forest and therefore separation from regrowth using these data layers alone is difficult. At L- and P-band, confusion occurs instead between Brigalow-dominated regrowth and non-forest areas as these lower frequency microwaves do not interact with foliage and the dimensions of stems are also often too small to evoke a scattering response. However, by integrating these datasets, regrowth can be mapped as stands support an FPC/C-band HV σ^o and a L- and P-band σ^o equivalent to forested and non-forested areas respectively (Lucas et al. 2006a). Where regrowth is more established, and hence stems are larger, differentiation using the lower frequency P-band rather than L-band data is necessary as the former provides better penetration of the forest volume. L-band and P-band HV data can be useful also for describing the forest structure, as scattering is more a function of branch dimensions (Lucas et al. 2004) which vary in size and number over time and in proportion to the trunks. The potential therefore exists for tracking the structural development of regrowth forest if coincident P and L-band data are available from satellite sensors. Whilst the ALOS PALSAR is providing the required L-band data, P-band missions are not, as yet, operational.

12.4.4 Changing the Fire Regime

Fire is an inherent component of the Australian landscape and is caused by both humans (e.g., indigenous burning, fuel load reduction) or by natural events (e.g., lightening strikes). Fires are also regular and extensive across the northern part of Australia, and up to half of the region can be burnt annually. However, in central Queensland, average fire frequency has reduced from 1.2 fires decade^{-1} in the 1950s to 0.9–1.0 decade^{-1} in the 1960s through to the 1990s (Fensham and Fairfax 2003), although emissions from fires increased after 2000 (Henry et al. 2005). The reduction in fire intensity and frequency is considered to be a major cause of woody thickening that has been reported in many parts of Queensland since pastoral occupation. Woody thickening is associated with an increase in the density and biomass of woody shrubs and trees and also a decline in grass and other herbage (Henry et al. 2005; Witt et al. 2006) and has been reported from other regions in Australia and also worldwide (e.g., South America). Other factors leading to thickening include grazing and browsing pressures, changes in the availability of water and nutrients, climate and land use (Silva et al. 2001). For example, grazing can promote woody plant proliferation by directly reducing competition from perennial grasses and spreading seeds and indirectly by reducing fuel loads and thereby fire frequency and intensity. However, browsing can suppress the growth and cover of plants and encourage grass production and fires (Fensham et al. 2005). Fensham et al. (2005) also suggested that low amounts of vegetation cover observed in early aerial photographs coupled with higher rainfall in subsequent periods suggested that higher rates of vegetation thickening were associated with a recovery of the forest from previous adversities and that land management was less influential. Indeed, many of the areas thickening in the latter half of the 20th century might have been recovering from the protracted and intense droughts in the first half (Witt et al. 2006).

As forests are extensive across northern Australia, minor changes in carbon associated with processes such as thickening can be significant at a regional level

(Fensham and Holman 1999). For example, Burrows et al. (2002) estimated a 0.53 t carbon ha^{-1} annual increase in the live and dead biomass of grazed woodlands dominated by *Eucalyptus* species and, on this basis, calculated a sink of 35 Mg C yr^{-1} for Queensland (although see Fensham et al. 2005). This sink was considered sufficient in magnitude to largely offset the emissions associated with vegetation clearing. Even so, such extrapolations are difficult because the extent of thickening varies regionally (Witt et al. 2006; Witt and Beeton 1995; Fensham and Holman 1999).

To establish the extent of woody thickening, most approaches have analysed time-series of aerial photography. For example, Fensham et al. (2005) compared 1:25,000 and 1:40,000 scale aerial photography covering the period 1945–1999 (with an average elapse time of 17 years; range 8–41 years). However, despite achieving some success in detecting woody thickening, the aerial photographs were found to be limited without field observations as woody plant dynamics could not be related directly to plant population processes (e.g., recruitment, mortality) or species interactions.

As thickening is associated with an increase in woody shrubs and trees and declines in herbaceous plants, potential exists to link long term increases in woody FPC, and hence basal area, to this process in some areas. However, the time-series of cloud-free Landsat sensor imagery is often quite sparse and the trajectories of FPC data are often difficult to interpret because of seasonal variability (Danaher et al. 2004). Time-series decomposition of high temporal resolution MODIS data may prove useful for separating seasonal and long term changes FPC in (e.g. Gill et al. 2006). However, the limited spatial resolution and temporal extent of the MODIS time-series makes integration with the Landsat time-series difficult. Vegetation thickening is also linked to a change in the size class distribution and density of stems over time, and detection might therefore be achievable using time-series of lower frequency SAR data. In particular, the JERS-1 SAR acquired L-band HH data over Queensland from 1992 to 1998 whilst the ALOS PALSAR has acquired similar data (e.g., in terms of incidence angle; Lucas et al. 2006b) since 2006. Any changes in these data should relate to an increase in the amount of woody vegetation over the intervening time period. Since SAR and Landsat are sensitive to different but complementary structural attributes that change with woodland thickening, integration of these data might also lead to better detection.

12.4.5 Tree Death – Natural or Human-Induced?

Natural drivers of tree death include drought and waterlogging. In particular, lack of precipitation leads to widespread mortality, particularly of adult plants. For example, extensive droughts in north Queensland in the 1990s resulted in mortality of 85% of the overstorey and a 29% reduction in the basal area of woodlands within an area of 55,000 km^2 (Fensham and Holman 1999). Even so, the distribution of affected areas varied with, for example, geology and the also the competitive influence of trees of certain species or taxonomic groups. Droughts in the early 20th century would also have resulted in substantial declines in woody cover and the creation of a relatively open woodland structure. However, rainfall conditions were more favorable thereafter and woody plants proliferated in the more open woodlands, particularly during the period 1951–1965. From 1985 onwards, this reduced because of the onset of density-dependent interactions and fluctuations in rainfall around the long-term average.

In many situations, areas associated with tree death are visually similar to those that have been stem-injected. Differentiating these using remote sensing data is problematic although natural tree death is often more extensive and does not conform to property boundaries. From spaceborne optical data, tree death over large areas can be detected along with clearing by a change in spectral indices and FPC between successive dates (QDNRW 2006), although field observations are required to verify whether the changes are associated with natural or human-induced event. Many trees remain standing and, as with those that are stem injected, can be identified using a combination of optical and SAR data. However, both the response of different tree species to adverse conditions and the extent of tree death depend on their different tolerances and both are often more evident within aerial photography than in moderate resolution satellite imagery.

An alternative approach is to monitor the impacts of drought at an individual tree level using crowns delineated and differentiated to species from digital fine spatial resolution airborne or spaceborne datasets. As an example, Bunting and Lucas (2006) used eCognition processes (Definiens 2005) to delineate tree crowns within open forests and woodlands in Queensland. Once delineated, each crown or crown cluster

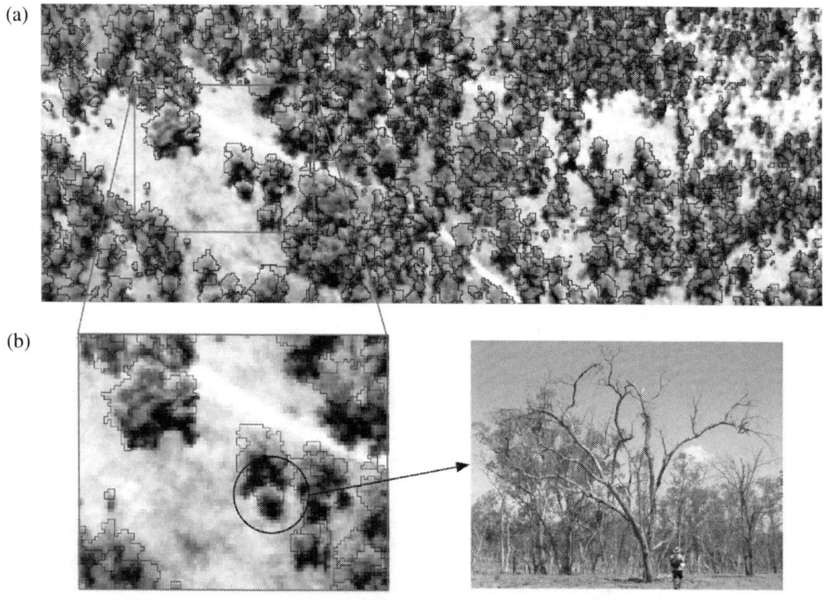

Fig. 12.3 (**a**) Individual tree crowns delineated within hyperspectral CASI data and (**b**) observed in 2006

was assigned subsequently to a species type by extracting the meanlit spectra (i.e., the average spectra from the upper half of crown) and applying a supervised multiple stepwise discriminant classification. The algorithm was developed on and applied to Compact Airborne Spectrographic Imager (CASI) data acquired in 2000 when many of the trees identified, regardless of species, were healthy (Fig. 12.3a). However, in 2006 and because of prevailing drought conditions, individuals of the species *Angophora floribunda* (rough barked apple) were experiencing death or dieback whilst others (e.g., *A. leiocarpa* and other *Eucalyptus species*) were less affected (Fig. 12.3b). Fensham and Holman (1999) also observed that *E. crebra* and *E. xanthoclada* were prone to dieback whilst other species (e.g., *Corymbia clarksoniana* and *Melaleuca nervosa*) were more tolerant. These observations therefore suggested the value of comparing crown and associated species maps at the individual tree to better indicate the impact of natural events (e.g., drought) on the forest stand as a whole but also on different species. However, such an approach necessitates a sampling approach within a strategic mapping and monitoring framework because of the impracticalities of acquiring fine spatial resolution datasets across the State.

12.5 The Implications of Change

Whereas aboriginal burning has led to some changes in the structure and species composition of forests (Common and Norton 1992), the main changes in extent have occurred since European settlement. Satellite sensor data have been particularly useful in tracking such change but alternations in the structure, biomass and species composition, particularly in the past few decades, have proved more difficult to quantify at the statewide and continental scales. This has introduced uncertainties in the estimates of greenhouse gas emissions and losses of biodiversity associated with land use and cover changes (Henry et al. 2005). Nevertheless, advances in remote sensing technology and algorithms for retrieving forest attributes have increased opportunities for quantifying and understanding some of the implications of these changes.

12.5.1 Carbon Stocks

The environmental impacts of clearing are well known but have been particularly adverse in Queensland where the majority of Australia's deforestation occurred in the period 1988–2002 (Henry et al. 2005). This clearing contributed significantly to Australia's total greenhouse emissions, with the annual National Greenhouse Gas Inventory reporting emissions ranging from $\sim$ 45–60 Mt CO_2yr^{-1} after 1990 with these increasing noticeably after 1995 (Henry et al. 2002).

Within forests, carbon is stored in the soil but also within the above ground (leaves, branches and stems) and below ground (roots) components. From remote sensing data, only the above ground biomass can be retrieved directly whilst the below ground can only be inferred. The retrieval of biomass also requires the use of sensors that provide information on the three-dimensional structure of vegetation, namely polarimetric/interferometric SAR and/or LiDAR. Nevertheless, optical data can be used to infer the three dimensional structure (and hence biomass) of forests by considering, for example, shade fractions derived from spectral unmixing or semi-empirical bi-directional reflectance distribution function (BRDF) model parameters derived from multiple view angle data such as MISR (Armston et al. 2007). Information on the species composition of forests at various scales (from the tree to the stand), which is best determined from optical data, can also be useful for inferring the above and below ground allocation of biomass within communities (e.g., those dominated by *Callitris* or *Eucalyptus* species; Lucas et al. 2004).

At the level of individual trees, the biomass can be retrieved from knowledge of species type and size (e.g., height, diameter) as appropriate allometric equations can be applied. Individual tree crowns can be classified to species using optical data and their height retrieved from co-registered LiDAR data. Such information can be scaled across local areas. However, scaling is currently impractical for statewide mapping and the use of moderate resolution satellite sensor data is therefore required. Nevertheless, the combination of fine spatial resolution optical and LiDAR data for biomass retrieval is still useful for supporting the calibration and validation of satellite-based models of biomass retrieval. As a close relationship exists between FPC and basal area and between basal area and biomass, biomass can be approximated from the Landsat-derived FPC but the accuracy of estimates is lower where this relationship breaks down. For example, FPC may be high for grasslands, crops, heathlands and regrowth forests even through the basal area is zero or very low. By contrast, the FPC may be lower for forests with more open canopies supporting leaves that are more vertically orientated (e.g., Eucalyptus) or of lower density (e.g., because of seasonal or drought conditions). An alternative approach is to utilize SAR data because of the sensitivity of $\sigma^o$to biomass (including that of the wood) at different frequencies and polarizations. A limitation of SAR data, however, is that saturation of the relationship occurs depending on frequency, polarization and the incidence angle of observation. Previous research in most closed forest environments (e.g., plantations or tropical forests) has demonstrated saturation of SAR backscatter at certain levels of AGB (typically $\sim$ 20 Mg ha^{-1}, 60–70 Mg ha^{-1} and 100–150 Mg ha^{-1} for C, L and P band respectively). However, within open forests and woodlands, the levels have been observed to differ in that saturation at C-band (and particularly HV polarisations) is higher at $\sim$ 50 Mg ha^{-1} and P-band is lower at $\sim$ 65 Mg ha^{-1} (Lucas et al. 2004). The higher saturation level at C-band is explained by the higher return associated with increases in tree density and hence the number of leaves and small branches with which these microwaves interact (Lucas et al. 2006b). The reduced saturation at P-band may be attributable to the size of many woody components being insufficient to evoke a response at this lower frequency. For example, for a forest supporting an AGB of 100 Mg ha^{-1}, the biomass of the woody components providing a return at P-band HH might only be 60 Mg ha^{-1}. Evidence of this has been both (a) the progressive disappearance of smaller trees observed within fine spatial resolution optical data acquired over the same forests and (b) the lack of or reduced $\sigma^o$from regrowth forests within AIRSAR data of decreasing frequency (Lucas et al. 2006a). These observations have also been supported by modelling studies (Lucas et al. 2004; Woodhouse 2006). A number of alternatives for more reliable retrieval have been considered including (a) inversion of SAR backscatter models (Moghaddam and Lucas 2003; Lucas et al. 2006a), (b) integration of SAR data with optical data (which relates to leaf biomass and crown cover) and height from spaceborne LiDAR and/or interferometric SAR data (which relates to stem volume in some forests) and (c) quantifying relative changes in data (e.g., σ^0 and derived products such as FPC) over time that can be better related to changes in biomass. This latter approach negates the need to generate absolute estimates of biomass over time-separated periods.

These approaches have been limited, until recently, by the availability (in time and space) of appropriate datasets. However, this has been addressed in part by the provision of new sensor configurations in the form of the ICESat GLAS, SRTM and ALOS PALSAR at a statewide level and increased provision of time-series datasets (e.g., Landsat-derived FPC from 1987 through to 2007 and the 1995/96 JERS-1 SAR mosaic of northern Australia which is comparable to those generated from ALOS PALSAR data from 2007 onwards).

Knowledge of clearing practices is also important for establishing the dynamics of greenhouse gas emissions associated with land use change. In most clearing operations, fallen debris is generally not burnt and many standing trees (e.g., those that are stem injected) are left in place and may remain there for several decades. Debris is often raked into piles which are often left to decay. Where forests are ring-barked or stem injected, the turnover time of coarse woody debris is extended compared to when these are burned and depends on factors such as species (wood density and composition), temperature, rainfall history, moisture and also the presence of termites. The level of disturbance to the soil (e.g., through blade and offset ploughing) also influences the release of carbon (Hill et al. 2006).

12.5.2 Changes in Biodiversity

Australia, both past and present, has supported a large diversity of flora and fauna, many of which are distinct and unique. This diversity has resulted largely because of the contributions from different regions (Godwana and Laurasian) and the relative isolation of the continent. Almost 50% and 15% of plant species are endemic in the temperate and tropical zones respectively (Common and Norton 1992). Whilst extinctions of species (e.g., megafauna) occurred during the period of aboriginal settlement, most followed the arrival of Europeans and, for some groups (e.g., mammals), are the highest worldwide.

When forests are cleared, losses of floral and faunal diversity occur because of a reduction in numbers which can result in extinctions to differing degrees (local to total). The loss of abundance rather than species type was highlighted by Bennett (1993) who estimated that for every 100 ha of woodland cleared, between

1,000 and 2,000 birds permanently lost their habitat. Other studies (e.g., Loyn 1987; Barrett et al. 1994) have also reported a reduction in bird species with habitat loss and degradation, which might mirror that of mammals, and an increase in aggressive species such as the introduced noisy miner (*Manorina melanocephala*).

Biodiversity is also lost from remnant forest over differing time periods through a number of mechanisms including degradation, which can be associated with a change in structure (e.g., reduction in the size class distribution of trees by species because of a lower number of older trees) and microhabitats (e.g., hollows for shelter and breeding).

Fragmentation also isolates species, impacts on basic ecological functions (e.g., water and nutrient cycling) and renders forests more vulnerable to adverse conditions (e.g., fire, drought) and invasion by exotic flora and fauna. Many of these effects influence biodiversity long after the forest has been affected.

The past distributions of faunal and floral diversity can only realistically be assessed from historical records and, to some extent, aerial photography. However, recent advances in remote sensing technology have provided opportunities for directly quantifying the diversity of tree and shrub species, although the associated faunal diversity can only be inferred.

Key indicators of diversity include richness and evenness (Warren and Collins 2007) although quantifying these remotely depends upon the scale of observation. For some decades, information on the number of tree species within a given area (richness) and their relative abundance and proportions (evenness) could be determined from aerial photography and largely through manual interpretation. As well as being able to visually identify and separate different species, judgments were also based on context (e.g., plant associations and biogeographical distributions, soils, landform). However, quantitative assessments of diversity at the tree level are often too demanding of resources and therefore only undertaken for research purposes. Within the advent of fine spatial resolution digital datasets, however, automated approaches to mapping individual trees and discriminating to species have been developed for both open (e.g., Bunting and Lucas 2006) and closed (Culvenor 2002) forests in Australia. Whilst many have been developed on airborne hyperspectral data, these algorithms have been applied successfully to true colour/colour infrared aerial photography, LiDAR and/or spaceborne IKONOS and Quickbird imagery although the spectral information available for species discrimination is lower.

At the stand level, aerial photography has been utilized (e.g., by the Queensland Herbarium) to define UMAs but the spatial variability (e.g., in structure and plant species composition) is difficult to capture. An alternative approach is to link (based on distance which can be weighted) individual tree crowns/crown clusters delineated within digital airborne datasets using a graphs-based minimum spanning tree and subsequently delineate stands of trees which can then be attributed with information on their species composition and structure. The latter approach is unlikely to replace that of aerial photography which has traditionally and successfully been used for vegetation assessment across Queensland. However, the scaling-up of tree-level assessments of species distributions might be achievable in the longer term with the advent of finer spatial resolution spaceborne hyperspectral sensors.

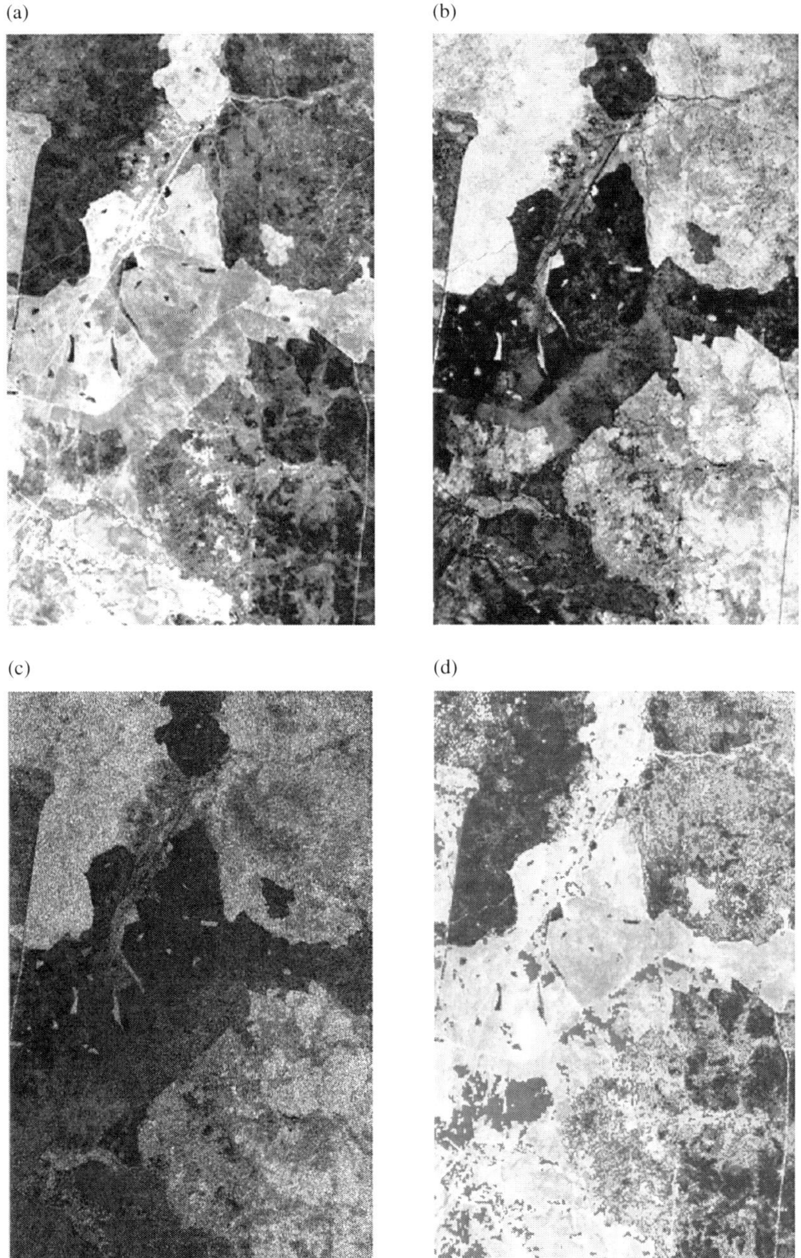

Fig. 12.4 The appearance of different forest communities in Landsat reflectance (**a**) and FPC (**b**) and L-band SAR backscatter (**c**) and (**d**) a rule-based classification of forest communities dominated primarily by *E. populnea* (orange), *E. melanaphloia* (blue), *C. glaucophylla* (dark green) with *E. melanaphloia* (light green) or *E. populnea* (yellow), and *A. harpophylla* (in the early (pink) and later (magenta) stages of regrowth)

The use of satellite sensor data for describing and mapping vegetation communities has been less successful, largely because of the level of detail that can be resolved at moderate spatial resolution and the relatively low dynamic range (typically 0–255 levels) and spectral resolution of the data. Whilst these issues have been resolved partly through the provision of fine spatial resolution sensors such as Quickbird and IKONOS and airborne hyperspectral sensors (e.g., CASI, HyMap), the amount of imagery needed for statewide coverage is too great and interpretation is more complex compared to when aerial photographs are used. Nevertheless, new opportunities for discriminating and mapping the floristic composition of forests has arisen with the increased capacity to retrieve remotely quantitative information on structural attributes as well as reflectance which, in various combinations, are unique to particular species groups. Examples of this are given in Fig. 12.4 where forests dominated by *C. glaucophylla* typically exhibit an FPC greater than those dominated by other species (e.g., *E. populnea* and *E. melanaphloia*) because of

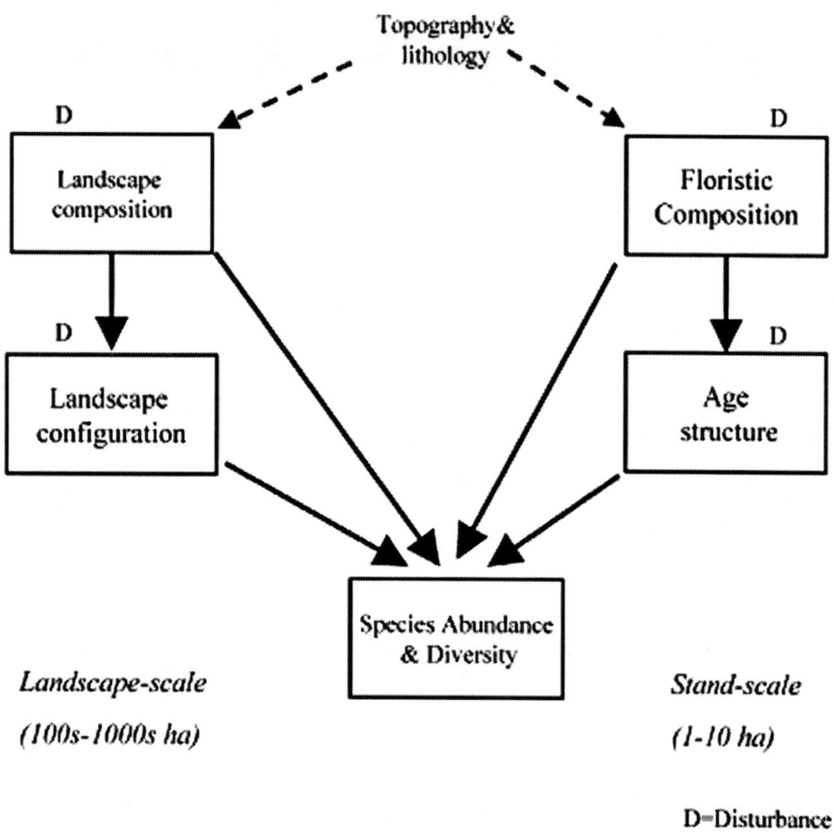

Fig. 12.5 Conceptual model of postulated patterns of influence at the stand scale (1–10 ha) and the surrounding landscape scale (100s–1000s ha) which influence species abundance and diversity (McAlpine and Eyre 2002)

differences in the structure, amount and spatial arrangement of foliage within the crown and the density of trees. Stands dominated by *C. glaucophylla* are further distinguished as they exhibit a higher L-band HH SAR backscatter because the high density of vertically orientated stems results in a strong double bounce return. Forests dominated by *E. populnea* and *E. melanaphloia* are separated by differences in the near infrared and short wave infrared reflectance. Non-regrowth forests dominated by *Acacia* species also exhibit a higher FPC and a lower L-band and P-band backscatter because of high foliage cover and low basal area compared to other woody vegetation which allowed their discrimination. Through integration of this information, maps of broad species types (at least to the genus level) can be generated using, in the case of Fig. 12.4, a rule-based classification within eCognition (Definiens 2005). Interpretation of these data and validation of the resulting species maps has been assisted by referring to tree species maps generated using co-registered finer spatial resolution datasets. From this information, the diversity and abundance of fauna within forests can be inferred at the stand level ($\sim$ 1–10 ha) by considering the floristic composition and age class structure and at the landscape (50–1000s ha) by considering the composition and configuration (spatial arrangement) of the surrounding environment (e.g., the proportion and diversity of different habitat types) (McAlpine and Eyre 2002, Fig. 12.5).

12.6 Conclusions

Throughout Australia, significant changes in the extent but also the species composition, structure and biomass of forests have occurred as a direct result of human activity, with these impacting upon biodiversity and greenhouse gas emissions but also on water supply, agricultural profitability and conservation value. In comparison to other states, extensive clearance of woody vegetation in Queensland has been relatively recent and many of the changes have been observed, using remote sensing data. For this reason, these data have played a key role in quantifying, mapping and monitoring change but assessments have been based primarily on optical sensors. In particular, aerial photographs and data from the Landsat series of sensors have been used primarily to establish the extent and floristic composition of pre-clearing vegetation and track the progression of deforestation through changes in FPC and spectral indices (Danaher et al. 2004; Accad et al. 2006). Whilst these data have been used successfully for this purpose, it has also been recognized that optical sensors are primarily responding to variations in foliage amount and that information on the distribution of woody material within the forest volume can be retrieved with greater confidence using lower frequency SAR data, either singularly or in combination with height information retrieved from LiDAR. Furthermore, to provide a more comprehensive description of the structure and biomass of forests, a combination of optical, SAR and height (whether derived from interferometric SAR or LiDAR) data is required. Already, and as highlighted in this chapter, combinations of these data have proved effective for describing the extent and structural development of forests regenerating on previously cleared areas, identifying clearing mechanisms (e.g., stem injection), and differentiating forest structural types and communities. However, many of these approaches illustrated have been undertaken using airborne sensors and are at the early stages of development. Even so, regional extrapolation is now a realistic option, particularly given the availability of global observations from sensors including the ALOS PALSAR, the SRTM digital surface models and ICESat GLAS data.

The time-series of remotely sensed datasets acquired for Queensland have also provided opportunities for establishing historical descriptions of vegetation type, changes in the extent of forest (including regrowth) and non-forest extent, and structural changes within the forest area (e.g., dieback as a consequence of drought, woody thickening). The integration of such data into models of forest recovery following clearance and as a function of influencing factors such as the methods of forest clearance (e.g., stem injection, chaining) and the types and intensities of land use prior to the regeneration of forests contained within these data is only just being explored. Such information is contained however within much of the existing data and derived products (e.g., FPC) that have been collated across the State and could provide options for predicting the future state of forests when combined with knowledge of ecosystem response to change.

Remotely sensed data acquired at a regional level are critical for statewide characterization, mapping and monitoring. However, sampling of the forested landscape using field and finer spatial resolution data (e.g., aerial photography, hyperspectral sensors and/or LiDAR) is also essential for providing spatial datasets of forest structure, biomass and species composition that can be used subsequently to assist calibration of regional-scale retrieval algorithms and models and validation of outputs. Such data can also be used to assess, for example, the impacts of human-induced and natural change (e.g., drought) on forests and better establish the response of different species to such changes.

Whilst human-induced change has impacted significantly on forests in Queensland, protection is now afforded for many through legislation, including the 1999 Vegetation Management Act and the 2004 Vegetation Management and Other Legislation Amendment. Clearance of non-remnant vegetation is likely to continue but regrowth will continue to establish in many areas. However, climate change is posing a new and previously understated threat to the forests of Queensland. In the midst of the most severe drought for 100 years, large areas of forest are being threatened with drought-related dieback and bushfires are widespread. These changes will lead to losses of carbon but also biodiversity. For example, under minimum and maximum climate change scenarios respectively, Thomas et al. (2004) predicted extinction of 7–13% and 43–58% extinction for species in the montane forests of Queensland. Increases in pest and weed species are also anticipated to occur. Therefore, there is an increasing need to better establish and understand stocks and fluxes of carbon associated with forests and biodiversity distributions, particularly in relation to changes in the structure, biomass and species composition associated with vegetation thickening (Burrows et al. 2002), woody encroachment (e.g., rainforests into savannas; Henry et al. 2005), regrowth (Lucas et al. 2006a) and dieback (as a function of species tolerance to droughts of varying intensity). Better knowledge of the impacts of land use and management on the capacity of forests to restore carbon and biodiversity is also needed. For this purpose, remote sensing data are anticipated to play a key role although will need to be used in combination with field observations and ecosystem and flux (e.g., carbon) models. This chapter has provided examples of how the use of these data may be extended to assist this process.

Acknowledgments The authors would like to thank the Queensland Department of Natural Resources and Water (QDNRW), the Queensland Herbarium, the Australian Research Council and Definiens AG (and particularly Ursula Benz and colleagues). Tony Milne (University of New South Wales) and Alex Lee (Australian National University) are also thanked.

References

Accad, A, Neldner, VJ, Wilson, BA, and Niehus, RE (2006). Remnant Vegetation in Queensland. Analysis of remnant vegetation 1997-1999-2000-2001-2003, including regional ecosystem information. Brisbane: Queensland Herbarium, Environmental Protection Agency

Armston JD, Scarth PF, Phinn SR, Danaher TJ (2007) Analysis of multi-date MISR measurements for forest and woodland communities, Queensland, Australia. Remote Sens Environ 107: 287–298

Barrett G, Ford HA, Recher HF (1994) Conservation of woodland birds in a fragmented rural landscape. Pacific Conserv Biol 1:245–256

Bennett AF (1993) Fauna conservation in box and ironbark forests: A landscape approach. Victorian Naturalist 110:15–23

Bowman DMJS (1998) Tansley Review No. 101. The impact of Aboriginal landscape burning on the Australian biota. New Phytologist 140:89–102

Bunting P, Lucas RM (2006) The delineation of tree crowns in Australian mixed species forests using hyperspectral Compact Airborne Spectrographic Imager (CASI) data. Remote Sens Environ 101(2):230–248

Burrows, W.H., Henry, B.K., Back, P.V., Hoffman, M.B., Tait, L.J., Anderson, E.R. et al. (2002) Growth and carbon stock change in eucalypt woodland stands in north-east Australia: Ecological and greenhouse sink implications. Global Change Biol 8:769–784

Butler DW, Fairfax RJ (2003) Buffel grass and fire in a Gidgee and Brigalow woodland: A case study from central Queensland. Ecol manag restor 4(2):120–125

Common MS, Norton TW (1992) Biodiversity: Its conservation in Australia. Ambio 21(3):61–78

Culvenor DS (2002) TIDA: An algorithm for the delineation of tree crowns in high spatial resolution remotely sensed imagery. Comput Geosci 28:33–44

Danaher TJ, Armston JD, Collett LJ (2004) A regression approach to mapping woody foliage projective cover in Queensland with Landsat data. Int Geosci Rem Sens Sym

Definiens (2005) eCognition Version 5 object orientated image analysis user guide. Definiens AG, Munich, Germany

Fairfax RJ, Fensham RJ (2000) The effects of exotic pasture development on floristic diversity in central Queensland. Biol Conserv 94(1):11–24

Fensham RJ, Fairfax RJ (2003) Assessing woody vegetation cover change in north-west Australian savanna using aerial photography. Int J Wildland Fire 12(4):359–367

Fensham RJ, Fairfax RJ, Holman JE, Whitehead PJ (2002) Quantitative assessment of vegetation structural attributes from aerial photography. Int J Remote Sens 23(11):2293–2317

Fensham RJ, Fairfax RJ, Archer SR (2005) Rainfall, land use and woody vegetation cover change in semi-arid Australian savanna. J Ecol 93:596–606

Fensham RJ, Holman JE (1999) Temporal and spatial patterns in drought-related tree dieback in Australian savanna. J Appl Ecol 36:1035–1050

Gill TK, Armston JD, Phinn SR, Pailthorpe BA (2006) A comparison of MODIS time series decomposition methods for estimating evergreen foliage cover. Proceedings of the 13th Australasian Remote Sensing and Photogrammetry Conference, Canberra, Australia

Graetz RD (1998) The terrestrial carbon pools of the Australian continent: An assessment of their size, dynamics and tractability. Report to the National Greenhouse Gas Inventory. CSIRO Earth Observation Centre, Canberra

Gunn RH, Beattie RE, Reid RE, Van de Graaff RHM (1988) Australian soil and land survey handbook: Guidelines for conducting surveys. Inkata Press, Melbourne, Victoria

Henry BK, Danaher T, McKeon GM, Burrows WH (2002) A review of the potential role of greenhouse gas abatement in native vegetation management in Queensland rangelands. Rangeland J 24:112–132

Henry BK, Mitchell C, Cowie A, Woldring O, Carter J (2005) A regional interpretation of rules and good practice for greenhouse accounting: Northern Australian savanna systems. Australian J Botany 53:589–605

Hill MJ, Roxburgh SH, McKeon GM, Carter, JO, Barrett DJ (2006) Analysis of soil carbon outcomes from interaction between climate and grazing pressure in Australian rangelands using Range-ASSESS. Environ Model Soft 21:779–801

Loyn RH (1987) Effects of patch area and habitat on bird abundances, species numbers and tree health in fragmented Victorian forests. In: Saunders DA, Arnold GW, Burbidge AA, Hopkins AJM (eds) Nature conservation: The role of Remnants of native vegetation. Surrey Beatty, Chipping Norton, New South Wales, pp 65–77

Lucas RM, Cronin N, Moghaddam M, Lee A, Armston J, Bunting P, Witte C (2006a) Integration of Radar and Landsat-derived foliage projected cover for Woody Regrowth Mapping, Queensland, Australia. Remote Sens Environ, 100:407–425

Lucas RM, Cronin N, Lee A, Witte C, Moghaddam M (2006b) Empirical relationships between AIRSAR backscatter and forest biomass, Queensland, Australia. Remote Sens Environ 100:388–406

Lucas RM, Milne AK, Cronin N, Witte C, Denham (2000) The potential of Synthetic Aperture Radar (SAR) data for quantifying the above ground biomass of Australia's woodlands. Rangeland J 22:124–140

Lucas RM, Moghaddam M, Cronin N (2004) Microwave scattering from mixed species woodlands, central Queensland, Australia. IEEE T Geosci Remote 10:2142–2159

McAlpine C, Eyre TJ (2002) Testing landscape metrics as indicators of habitat loss and fragmentation in continuous eucalypt forests (Queensland, Australia). Landscape Ecol 17:711–718

McIntyre S, Hobbs R (1999) A framework for conceptualizing human effects on landscapes and its relevance to management and research models. Conserv Biol 13(6)1282–1292

Moghaddam M, Lucas RM (2003) Quantifying the biomass of Australian subtropical woodlands using SAR inversion models. Proc Geoscie Rem Sens Sym 3:1619–1621

NLWRA, National Land and Water Resources Audit (2002). Australian terrestrial biodiversity assessment 2002. ISBN 0 642 3713

Neldner VJ, Wilson BA, Thompson EJ, Dillewaard HA (2005) Methodology for survey and mapping of regional ecosystems and vegetation communities in Queensland, Version 3.1. Queensland Herbarium Environmental Protection Agency, Brisbane

Nix H (1994) The Brigalow. In: Dovers S (ed) Australian environment history: Essays and cases. Oxford University Press, Melbourne, pp 198–233

Queensland Department of Natural Resources and Water (1999) Land cover changes in Queensland 1999–2001 – a Statewide Landcover and Tree Study (SLATS) Report, Aug 2002. Department of Natural Resources and Water, Brisbane, Australia

Queensland Department of Natural Resources and Mines (2004) Land cover changes in Queensland 1988–1991 – a Statewide Landcover and Tree Study (SLATS) Report, Dec 2004. Department of Natural Resources and Water, Brisbane, Australia

Queensland Department of Natural Resources and Mines (2006) Land cover changes in Queensland 2003–2004 – a Statewide Landcover and Tree Study (SLATS) Report, Feb 2006. Department of Natural Resources and Water, Brisbane, Australia
Randall L, Lucas RM, Austin J, Lee A (2006) Understanding the dynamics of land cover change: Coupling remote sensing data with farmer interviews. Australasian Remote Sensing and Photogrammetry Conference, Canberra, Australia
Rosenqvist A, Shimada M, Ito N, Watanabe M (2007) ALOS PALSAR: A pathfinder mission for global monitoring of the environment. IEEE T Geosci Remote 45(11):3007–3316.
Russell-Smith J, Yates C, Edwards A, Allan GE, et al (2003) Contemporary fire regimes of northern Australia, 1997–2001: Change since Aboriginal occupancy, challenges for sustainable management. Int J Wildland Fire 12:283–297
Silva JF, Zambrano A, Fariñas MR (2001) Increases in the woody component of seasonal savannas under different fire regimes in Calabozo, Venezuela. J Biogeogr 28:977–983
Specht RL (1970) The Australian environment. CSIRO Australia and Melbourne University Press
Thomas CD, Cameron A, Green RE et al (2004) Extinction risk from climate change. Nature 427:145–148
Tracey JG (1982) The vegetation of the humid tropical region of North Queensland, CSIRO, Melbourne, p 124
Vegetation Management Act (1999) Queensland Government Act No. 90 of (1999). http://legislation.qld.gov.au/LEGISLTN/CURRENT/V/VegetManA99.pdf, [accessed:18 May 2008]
Walker J, Bullen F, Williams BG (1993) Ecohydrological changes in the Murray-Darling Basin. The number of trees cleared over two centuries. J Appl Ecol 30:265–273
Warren A, Collins M (2007) A pixel-based semi-empirical system for predicting vegetation diversity in boreal forest. Int J Remote Sens 28:83–105
Webb LJ (1959) A physiognomic classification of Australian rain forests. J Ecol 47:551–570
Wilson BA, Neldner VJ, Accad A (2002) The extent and status of remnant vegetation in Queensland and its implications for statewide vegetation management and legislation. Rangeland J 24:6–35
Witt GB, Beeton RJS (1995) The regional implication of 'naturalness' in protected area management: A case study from the Queensland mulgalands. Ecological research and management in the Mulgalands, conference proceedings, The University of Queensland Gatton College, Australia, pp 177–182
Witt GB, Luly J, Fairfax, RJ (2006) How the west was once: Vegetation change in south-west Queensland from 1930 to 1995. J Biogeogr 33:1585–1596
Woodhouse IH (2006) Predicting backscatter-biomass and height-biomass trends using a macroecology model. IEEE Trans Geosci Remote 44(4):871–877

Chapter 13

Habitat Quality Assessment and Modelling for Forest Biodiversity and Sustainability

Sandra Luque and Nina Vainikainen

Abstract Safeguarding biodiversity has been one of the most important issues in the environmental and forest policies since 1990s. The problem remains in terms of decisions and knowledge on where to set appropriate conservation targets. Hence, we need detailed and reliable information about forest structure and composition and methods for estimating this information over the whole spatial domain. The approach presented aims to develop a practical tool for conservation planners and foresters to evaluate alternative conservation plans to expand and connect protected areas while identifying key forest habitats and its associated biodiversity value. In order to reach this goal and learn more about habitat quality for woodland species in boreal forests and spatial characteristics of forest landscape, we used a combination of remote sensing and field data derived from the Multi-source Finnish National Forest Inventory (MS-NFI) Habitat quality assessment and suitability maps constitute a useful approach for designing management plans to improve biodiversity conservation. In this chapter, we present an approach and tools for assessing biodiversity values in both managed and protected forest areas. The approach is intended to assist decision-making concerning protection of valuable habitats and management of natural resources. The different habitat quality models presented are used as a surrogate for biodiversity value. The indicators and the models developed reflect a sound scientific basis that can be implemented in other European countries that invest in national forest inventories. Within this framework, focusing on forests in Finland and on end-user needs, this effort constitutes the first attempt undertaken at the landscape level to use National Forest Inventory data for forest biodiversity monitoring and management.

S. Luque
Cemagref, – Institute for Agricultural and Environmental Engineering Research. UR Ecosystèmes Montagnards, Grenoble, 2 rue de la Papeterie, BP 76, F-38402 Saint-Martin-d'Hères cedex, France
e-mail: sandra.luque@cemagref.fr

R. Lafortezza et al. (eds.), *Patterns and Processes in Forest Landscapes*,

13.1 Introduction

Historically, habitat destruction and fragmentation have been viewed as the major factors driving biodiversity loss (Moore 1962; Webb and Haskins 1980; Burgess and Sharpe 1981; Wilcove et al. 1986; Lord and Norton 1990; Gigord et al. 1999; Luque et al. 1994; Luque 2000; Noss 2001). However, much of the change in the contemporary European landscape is now attributed to changes in the management of semi-natural habitats. The challenge then for conservation is to ensure that this management of often complex landscapes retains and enhances biodiversity values. One key issue is whether to prioritize extensive landscape management or prioritise management of the wider landscape and focus on intensive management on designated sites.

One of the fundamental problems that remain is the identification of areas that have a certain biodiversity "value" for conservation. In particular, conceptualizing and assessing biodiversity and setting conservation priorities are not narrowly-defined biological problems – they are broad-based human enterprises with a large social and political component. Furthermore, human resources are insufficient to protect biodiversity in all its various guises, and difficult choices on how to prioritise conservation lay ahead (Perlman and Adelson 1997; Vitousek et al. 1997).

In many European countries, forest and forestry are important parts of nature, society and economy but perhaps more so in Scandinavia and central Europe. Here, forestry activities such as harvesting, drainage, scarification and even re-forestation/afforestation impacts on soil, water and biota systems (Schmidt 2005; Strandberg et al. 2005; Uotila and Kouki 2005; Vellak and Ingerpuu 2005; Winter et al. 2005). – are components of the intensive management that have altered the structure of forest stands and landscapes in Europe since centuries.

Biodiversity issues have gained importance in forestry as a result of the increased awareness of forest landscape changes, but still there is much to do before forest management meets reasonable goals for forest protection and renewal of biodiversity (Spence 2001). In order to achieve efficient monitoring systems that focus on the understanding of changes and its linkage to ecological processes, a thorough detailed-spatial knowledge of the landscape is needed. Despite the increasing importance of forest management in the last ten years, most of the studies on the functioning of forest ecosystem have targeted natural forest systems and have typically excluded managed forest areas. However, there is an increasing need to develop tools for assessing the forest system as a whole. This holistic approach will consider biodiversity value at the same time that the needs of forestry activities are also addressed. Sustainable forest management goals include the conservation of biological diversity and its constituent elements (Junninen et al. 2006; Kuusinen and Siitonen 1998). As an example, dying and dead trees, have been recognised as being of prime importance as they represent a resource and habitat for a diverse range of faunal and floral species (e.g. Esseen et al. 1997; Martikainen et al. 2000; Magura et al. 2004; Moretti et al. 2004; Odor and Standovar 2001).

In order to illustrate this chapter we chose the boreal forests of Finland as they represent a typical forest management case. Finland is the most heavily forested country in Europe, as 78% of Finland's total area is forestry land.

With 24.0 million forest hectares in Finland the main biodiversity conflicts are related to forests. The forest has also an important impact within the economic sector for the country (Finnish Forest Research Institute 2006). There are almost one million forest owners in Finland, whilst the state owns approximately one third of the forests. The forest industry is one of the cornerstones of the national economy. Its annual turnover is approximately 20 billion euros, which corresponds to 27 percent of net exports. Of the total amount of coniferous timber resources, 50 percent is Scots pine and 30 percent is Norway spruce. Of the broadleaved trees, birch is predominant, representing 16 percent of the timber resources.

Forest management in Finland has undergone fundamental changes during the last 15 years. By the end of the 1980s the criticism against the intensive forest management practices was aroused among the public debate, and more and more emphasis was addressed to the environmental aspects and biodiversity related to forests and forestry. During the 1990s ecosystem management even replaced forest management in the vocabulary of foresters (Mielikäinen and Hynynen 2003). The new biodiversity oriented principles set to the forestry were also stated in international agreements, e.g. in the documents of Helsinki process 1993–1995 (Ministerial Conference on the Protection of Forests in Europe 1994). Today, biodiversity aspects are taken into account on a regular basis in Finnish forest management as is also the case for Sweden, Austria and Danemark, among others countries where forest and forestry are key elements on the overall landscape matrix. The key issue in forest management is the modern concept of sustainability which includes the ecological, economical and socio-economic aspects. This new approach is stated in Finnish forestry legislation, reformed totally in 1997. In practical management of the commercial, multifunctional forests of Finland, maintaining of ecological biodiversity is an equally emphasized goal together with the maintaining of the sustainable yield. The intensive silvicultural methods aiming at maximization of wood production were practically abandoned by the end of 1990s, when, for example, new drainage of peatlands for forestry purposes and nitrogen fertilization of forests ceased almost completely (Finnish Statistical Yearbook of Forestry 2001). However, the intensive management has altered the structure and tree species composition of forests stands and the amount of coarse woody debris (Essen et al. 1997; Kouki 1994; Löfman and Kouki 2001). Also, regional characteristics, such as the spatial structure of forest landscapes, have been changed (Luque et al. 2004; Löfman 2006). A younger more dense even forest stands are dominating in detriment of old growth-forests (Luque et al. 2004; Andrén 1994; Andrén 1997; Rassi et al. 2001). Consequently, fragmentation and loss of old-growth forests are primary threats to forest dwelling animal and plant species, up to the point that many have become locally extinct (Hanski and Hammond 1995; Hildén et al. 2005). The threatened species include 50 vertebrates, 759 invertebrates, 180 vascular plants, 142 cryptogams, and 374 fungi or lichens. Some 37% of the threatened species are primarily associated with forest habitats, particularly herb-rich woodland and old growth heathlands forest habitats (Raisi et al. 2001; Hildén et al. 2005). It is likely that with the present forest management, the existing conservation-area network will not provide adequate protection for the biodiversity of boreal forests in the long term up to the point that many species have become already locally extinct (Hanson and Larsson 1997; Hanski 2000). To maintain habitats and viable populations of species typical of old-growth forests, the network of reserves should be completed with new conservation areas and corridors to connect small existing areas (Siitonen et al. 2002).

In order to preserve forest biodiversity in the long run and improve forest protection and management we propose first to locate and assess habitat quality. It is crucial to develop a method to manage the forest in accordance to ecological sustainability aims and principles at the level of habitat quality for certain key species in Southern Finland. To reach a sustainable balance we need to find a good balance of protection within a network of favourable conservation sites to protect forest species and habitats. Nevertheless, protection needs to be implemented within the context of a strategy that looks at a combination of conservation areas and appropriate management of commercial forests The approach aims to develop a practical tool for conservation planners and foresters to evaluate alternative conservation plans to expand and connect protected areas while identifying key forest habitats and its associated biodiversity value.

We look at the quality of the forest habitats that have an importance for certain species according to the research produced up to date for the region. Two of the forest habitat models presented depict habitats in heath forests with a large amount of coarse woody debris (CWD) vitally important for several animal and plant species (Vallauri et al. 2005; Vellak and Ingerpuu 2005; Kappes 2005; Ponge 2003). The third one describes biodiversity values in herb-rich forests where the species richness is the highest, but the amount of representative conservation areas is low in Finland. The data includes forest stand properties, area cover and geographical allocation of the sites. The originality of the work resides in the spatial approach that allows a geographical location of the habitats with different potential for protection which is essential for planning and policy making.

13.2 Habitat Quality Models

13.2.1 Habitat Index as a Surrogate for Biodiversity Value

With the rise of new powerful statistical techniques and GIS tools, the development of predictive habitat distribution models has rapidly increased in ecology. Such models are static and probabilistic in nature, since they statistically relate the geographical distribution of species or communities to their present environment. The analysis of species–environment relationship has always been a central issue in ecology. The quantification of such species–environment relationships represents the core of predictive geographical modelling in ecology. These models are generally based on various hypotheses as to how environmental factors control the distribution of species and communities (Jongman et al. 1995; Schuster 1994; Guisan and Zimmermann 2000). We depart from the concept of the species' habitat, considering the "optimal" habitat, the area where the presence of a species is due to suitable conditions for its survival. This "quality" concept conceived for this modelling approach represents the optimal environmental conditions for a particular type of forests known to sustain an important numbers of species considered as a surrogate for biodiversity value. The biodiversity value we use to support the forest habitat quality models is based upon the concept of "naturalness" developed by several authors see: Bartha et al. 2006; Nitare and Norén 1992; Sverdrup-Thygeson 2002.

The problem we still have is that in the case of biodiversity conservation, empirical evaluation models based on real field data for all species of interest cannot be expected to become available. Up to date habitat models based on species census data are too restricted in area and time consuming. Then, the challenge is to develop methods and practices of locating and evaluating suitable sites for threatened species. The main interest within the framework of this research is to improve the existing forest management planning, in this sense it is essential that the decision alternatives be assessed with respect to a combination of expert knowledge and habitat models. The one way of dealing with this problem, as proposed in this work, is to use habitat quality indices that reflect the quality of the forest by identifying possible causal relationships between forest structure, environmental data, and ecological conditions. We depart from the hypothesis that all species have specific habitat requirements, which can be described by habitat factors. These factors are connected to the critical characteristics of the habitat, e.g. to those of vegetation and soil, but also areas surrounding the habitat can influence the habitat quality (e.g. spatial structure of landscape elements). Habitat factors can also be classified according to the particularity of the factor: a deterministic habitat factor has to be present in a high-quality habitat, but a non-deterministic factor has a trade-off with some other factor (Store and Kangas 2001; Romero-Calcerrada and Luque 2006). Thus, a deterministic factor can be taken as a non-compensatory habitat characteristic whereas a decrease in a non-deterministic factor can be compensated by an increase in the value of another non-deterministic factor, as expressed in the habitat-evaluation model (Store and Kangas 2001; Store and Jokimaki 2003).

The habitat index reflects the value and importance that an area potentially possesses in terms of biodiversity. The calculated habitat index for the different forest habitats were used as the sole ecological variable in optimizing the site selection for conservation. The first step in assessing the quality is to determine the forest habitat factors on the basis of an analysis of existing studies and expert knowledge and forest inventory data. Here, judgements made by experts, in particular in terms of key species requirements of forest habitat suitability, were applied to develop the tresholds for the different habitat models (Kouki 1994; Väisänen and Järvinen 1996; Martikainen et al. 2000; Siitonen 2001; Hildén et al. 2005; Romero-Calcerrada and Luque 2006).

13.2.2 Habitat Quality Assessment for Conservation

Habitat quality assessment and habitat suitability maps constitute an useful approach in the design of management plans that seek to expand or create new protected areas (Rautjärvi et al. 2004; ReVelle et al. 2002; Rodrigues and Gaston 2002; Romero-Calcerrada and Luque 2006). Angelstam and Anderson 2001 This approach helps to satisfy a number of conservation goals based on habitat characterisitics for certain species of particular importance (Anglestam and Pettersson 1997; Angelstam and Anderson 2001: Reid 2006). In this sense, habitat modelling generated using spatial statistics and GIS can help in the characterizations of habitat requirements and the localization of suitable habitats (Guisan et al. 1998).

A particular challenge is to develop methods and practices for locating and evaluating suitable sites for threatened species at a regional level in order to improve conservation planning. It is extremely important for conservation purposes to develop methods that can use existing data because collecting data from large areas is time and resource consuming, and using existing data likely saves limited funds for the conservation actions. This is particularly the case when the data-source fulfils high-quality standards and avoids error propagation that is common in many multi-source large scale data (Burrough and McDonnell 1998).

Our proposal is to develop habitat quality indices as a surrogate for biodiversity values. The method helps to decide where to protect forest biodiversity based on the habitat value of the forest. The work relies and builds upon data from the Finnish National Forest Inventory (NFI) and other related databases from permanent inventories. These indices should reflect the quality of the forest by identifying possible causal relationships between forest structure, environmental data, and ecological conditions. This departs from the hypothesis that all species have specific habitat requirements, which can be described by habitat factors. These factors are connected to the critical characteristics of the habitat (e.g., vegetation or soil, but also areas surrounding that can influence the habitat quality (e.g., spatial structure of landscape elements).

13.3 Study Area and Data Sources

13.3.1 Southern and Central Finland

Finnish boreal forests are dominated by coniferous trees. There are about twenty indigenous tree species growing in Finland, the most common being Scots pine (Pinus sylvestris), Norway spruce (Picea abies) and Birch (Betula pendula and B. pubescens). Naturally pure pine stands are found in rocky terrain, on top of arid eskers and on pine swamps. Natural spruce stands are found on richer soil. Birch is commonly found as an admixture, but it can occasionally form pure birch stands. About half of the forest land area consists of mixed stands. Rarer species are found mostly as solitary trees. The south-western corner and the south coast of Finland are touched by a narrow zone growing oak, maple, ash and elm.

The focus area for this study was chosen in accordance with recent discussion and new conservation efforts on nature conservation in Finland (Ministry of Environment 2004). Finland seeks new innovative ways to conserve forests on a voluntary basis (Mininstry of Environment 2004). Within this framework, concerns over the accelerating loss of biodiversity have led governments worldwide to focus their attention to better address sustainability in their natural-resource policies (CBD 2002). In Finland, a special emphasis in this area is currently set on protection of forest biodiversity in the southern parts of the country. In the south of the country, the conservation share from the forestry land area is relatively small, 2.2% (Finnish Statistical Yearbook of Forestry 2006, p. 91) when compared to the respective share in Northern Finland (15.8%). In particular, more herb-rich and low herb heath forests should be conserved (Virkkala et al. 2000, p. 26).

To provide answers at a regional level, the study was performed for a total of 16.7 million hectares of forestry land within the whole of southern Finland (Fig. 13.1). Following the needs of the Metso Program (Ministry of Environment 2004;MOSSE 2007) we developed a regional approach that has an application as an operational method for decision making.

The needs from the METSO programme for Finland were to define ecological criteria to cover the most significant forest habitats in terms of biodiversity (MOSEE 2007) in order to monitor the biodiversity status and develop a sustainable

Fig. 13.1 Shaded region shows the study area in southern Finland. The study was performed for a total of 16.7 million hectares of forestry land

conservation planning. Therefore, the habitats depicted and the variables used in the habitat quality models were chosen in accordance with that aims of the program and policy makers needs. In particular, we focused on two important issues: the scarcity of protected herb-rich forests in Southern Finland and the importance of dead wood for many threatened forest species.

13.3.2 Thematic Maps from Finnish Multi-source National Forest Inventory (MS-NFI)

All decision-making requires information. In forestry, this information is acquired by means of forest inventories, systems for measuring the extent, quantity and condition of forests. More specifically, the purpose of forest inventories is to estimate means and totals for measures of forests characteristic over a defined area (Kangas and Maltamo 2006). Such characteristics include the volume of the growing stock, the area of forest type and nowadays certain inventories include biodiversity related measures such as dead wood, site characteristics, understory vegetation type, among others variables.

Since the beginning of the 1920s, Finnish forests have been largely inventoried and monitored, which allowed the development of the Finnish National Forest Inventory (NFI) and the multi-source Finnish National Forest Inventory (MS-NFI) (Tomppo 2006). The Finnish NFI has been producing large-scale information on Finnish forests since the beginning but forests statistics for small areas have been computed since 1990 using satellite images and digital map data in addition to field measurements by means of MS-NFI (Multi-Source National Forest Inventory) (Tomppo 2006).

The multi-source thematic maps used for the development of habitat models represent estimates of the volume of the growing stock and different tree species characteristics; stand age, potential productivity of the site and site quality at a spatial resolution of 50 m. The use of National Forest Inventories in building up forest quality habitats is key to our approach. In this sense, the volume and age of the growing stock are basic attributes describing the forest structure. When combining these attributes with the volume estimates for individual tree species we gain in considerable knowledge of stands structure. Volume of growing stock is the stem volume of all living trees above stump height (with a minimum height of 1.3 m) derived from field plot level measurements, and predicted for pixels (m^3/ha). Stand age is the weighted mean age of the trees of the main tree storey. Site quality is an ordinal variable depicting the fertility of the stand based on the vegetation composition and structure on the stand (Cajander 1926). The productivity of a site is the average increment of the growing stock of the corresponding site.

The pixel level predictions were produced in the MS-NFI based on k-nearest-neighbour (k-nn) estimation and its improved version (Tomppo and Halme 2004). MS-NFI procedure assigns field data of forest inventory to all satellite image pixels using a multi-source approach (Tomppo and Halme 2004; Tomppo 2006b); digital maps are used to delineate forestry land from other land use classes. The input data for the Finnish multi-source inventory are thus NFI field data, satellite images and digital map data of different types, e.g. basic map data, soil data for stratifying between mineral soil, spruce mires, pine mires and open bogs, and a digital elevation model (Tomppo 2006). We used the pixel level predictions of selected forest variables as input data for the models in this study in addition to interpolation layers calculated for some of the NFI field plots data from the 9th rotation of the NFI (in years 1996–2003). Note that the pixel level predictions used in map format were produced by the multi-source Finnish National Inventory (MS-NFI) as explained above (Tomppo 2006; Tomppo and Halme 2004).

13.3.3 Variables from Field Plot Data

13.3.3.1 Coarse Woody Debris

Coarse woody debris (CWD) is considered one of the key attributes indicators of biodiversity in boreal forests (e.g. Esseen et al. 1997) and large differences have been observed in the volume of CWD between managed forests and natural or semi-natural forests (Esseen et al. 1997; Magura et al. 2004; Magura et al. 2003; Siitonen 2001; Junninen et al. 2006; Kuusinen and Siitonen 1998; Moretti et al. 2004; Odor and Standovar 2001). It has been estimated that 20%–25% of all forest species are to some degree dependent on dead and decaying wood, in particular in Finland, i.e., 4000–5000 forest species, rely on dead wood habitats (Siitonen 2001). The average volume of dead wood in forests outside the protected areas is as low as 2.5 m^3/ha in Southern Finland.

In the Finnish NFI, the volume, quality, and roughness of CWD have been measured on all field plots on forest or other wooded land. We generated a 50 m spatial resolution wall-to-wall layer of the total volume of CWD per hectare and per pixel. The dead wood layer for this study was produced using volume measurements on the 53,464 circular plots from NFI with a radius of 7 m landing on our study area. Ordinary kriging interpolation was applied using the Geostatistical Analyst in ArcGIS (Johnston et al. 2001). For each NFI plot, the average volume of CWD per hectare was calculated and those values were used with a spherical model for the empirical semivariogram. Figure 13.2 ilustrates the map produced following the explained approach. It is important to point out that 3 m^3/ha is considered the average CWD for Southern Finland. Therefore for the habitat models the threshold for dead wood is set for areas where amounts of equal or more than 3 m^3/ha of CWD are present.

13.3.3.2 Local Density of Natural-Like Stands

High biodiversity value forests are often in a natural or semi-natural state in terms of human influence (e.g., silvicultural operations). In this study, observations of silvicultural history were used as a surrogate for the degree of naturalness (Bergstedt 1997; Reif 1999/2000; Bartah et al. 2006; Nitare and Norén 1992;

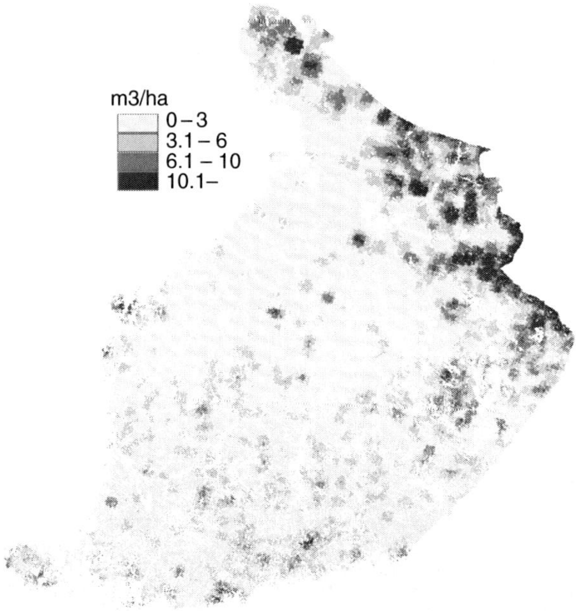

Fig. 13.2 Volume of dead wood, map produced by kriging interpolation of NFI plots data. Values for the whole country varied from 0 m^3/ha to over 200 m^3/ha. Light yellow represents areas of CWD bellow the average for Southern Finland (3 m^3/ha)

Sverdrup-Thygeson 2002; Uotila et al. 2002). Silvicultural history has also been evaluated on all plots on forest and scrub land (Tomppo 1992). Measured variables are previous fellings and their date, previous soil preparation and its date and previous silvicultural measure and its date. These variables were used to identify on one hand all the areas where fellings had not been done at all and on the other areas where fellings had not been done during the past 30 years. The map then is derived from field observations where no fellings or other operations took place for at least 30 years or more. The observations were used to calculate a kernel density for the occurrence of natural-like stands using ArcGIS Spatial Analyst (McCoy and Johnston 2001). The search radius was adjusted to the plot design (Tomppo 2006). The resulting density layer (sites/area) was reclassified so that class 1 included pixels where the density was below the overall mean for Southern and Central Finland, class 2 included pixels where the density was between the overall mean and mean +1 standard deviation (SD) and class 3 the pixels where the density was higher than the overall mean +1 SD (Fig. 13.3)

13.3.3.3 Area Contribution of Key Biotopes

Key biotopes (Fig. 13.4) have been observed on all NFI plots on forestry land. Recordings have been made of all biotopes that are protected under the Forest Act (1096/1996) or the Nature Conservation Act (1096/1996). At most, three different

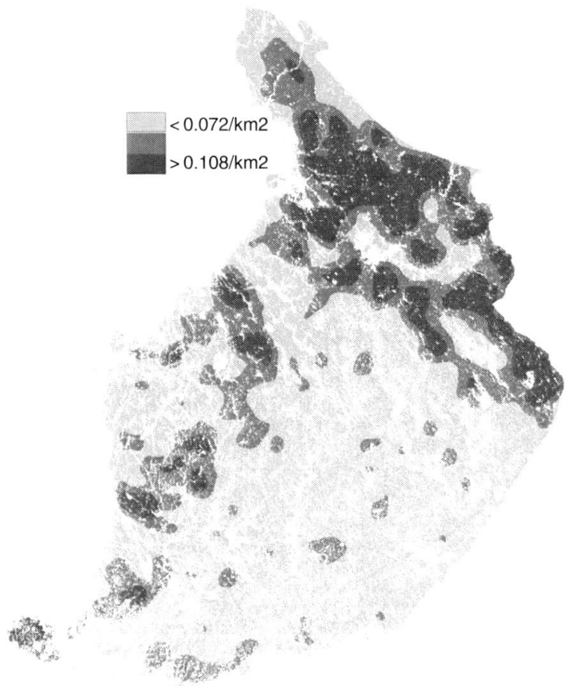

Fig. 13.3 *Naturalness* based on NFI field observations that were used to calculate a kernel density for the occurrence of natural-like stands. The values represent the density of this natural like stands having no fellings or other procedures for the last 30 years or more

biotopes could be recorded on the circular plot with 30 m radius (Tomppo 2006). In addition to the biotope type and its area, the state of naturalness, management history and ecological value of the biotope have been observed within the NFI sampling framework. For this study, we use recordings made on herb-rich forests, which are divided into six different habitat classes (naturally regenerated stands of rare hardwood species, and gorge, ravine and precipice habitats). These observations were used to generate a map layer showing the density of herb-rich forests that is used in the habitat model. The gorge, ravine and precipice habitats were handled separately from the other biotopes due to their different nature and spatial extent (i.e. very constrained in its spatial extent). For both habitat groups, the contribution of forest area (%) was used as the value to be interpolated using kriging and a spherical model for the empirical semivariogram.

13.4 Habitat Quality Model

The models include two types of input layers derived from the NFI: (i) multi-source thematic maps representing estimates for the volume of growing stock and

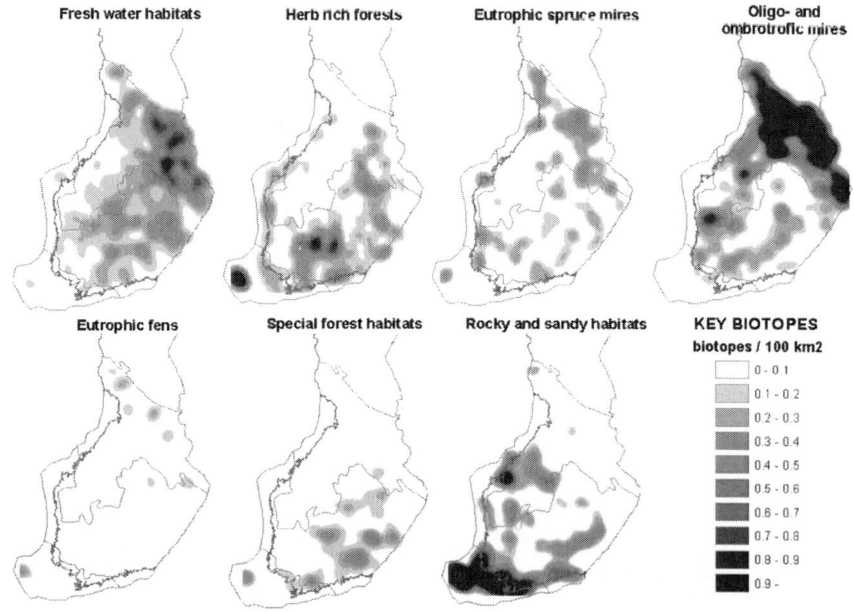

Fig. 13.4 Key Biotopes types and its area spatially represented by density (Biotopes type/100 km^2)

of individual tree species, stand age, productivity of the site and site quality and (ii) interpolation layers derived directly, without any auxiliary information, from the field plot data representing estimates for the volume of dead wood (CWD), area contribution of key biotopes and density of sites with no forest management actions for the last 30 years or more.

All data were integrated into a GIS platform using Geostatistical Analyst (Johnston et al. 2001) and Spatial Analyst in ArcGIS (Johnston et al. 2001). All input layers were reclassified according to specific thresholds values based on literature review, expert knowledge, forest characteristics based on NFI data and MS-NFI and on landscape patterns analysis derived from the input maps (Fig. 13.5).

The conditions for each of the models are presented in the flowcharts (Fig. 13.5a, b, c). The flowcharts for the three conditions represent the logical analysis performed once the input layers were produced as explained before. All input maps have the same extent, the same resolution (50 m) and values for forestry land. However, the thresholds and conditions differ according to the type of habitat modelled (Fig. 13.5 and Fig. 13.6).

It must be considered that since no coherent species data are available for this extensive area, we are not aiming to predict habitat suitability to any particular species, but rather to produce a spatially-based habitat quality model at a regional level as aforementioned.

Some sites were deemed high quality sites by all or two of the models, and the overlapping was removed so that the high quality sites according to the constrained model were kept as they are and those sites were excluded from the sites pin-pointed by the other models. This way, the sites selected according to the herb-rich model rules were excluded from the sites established by the general model. Pixels in an individual input layer meeting or exceeding the threshold value were assigned the value 1; pixels not meeting the threshold value were assigned value 0 (i.e., a Boolean approach). All input layers were then added resulting in a map with values ranging from 0 to 7. The resulting layer was reclassified so that pixels whose value was below the average were assigned a value of "Low" habitat quality, pixels whose value was between the average and one standard deviation from the average were assigned a value of "Average" quality, and pixels whose value exceeded the average by at least 1 standard deviation were assigned a "High" value (Fig. 13.6).

The models produced aim at representing forest biodiversity and hence the modelling was confined to forested biotopes. The habitats to depict and the variables used were chosen in accordance with the habitats and criteria mentioned in accordance to the Forest biodiversity programme for Southern Finland (Ministry of Environment 2004) as explained before.

We produced three habitat quality situations: the habitat quality models produced (Fig. 13.6) will be called: ***Herb-rich Habitat Quality Model*** ("Herb-rich"), ***General Model for High Biodiversity Value Forests*** ("General") and ***Constrained Habitat Quality Model*** (also "Constrained") (Fig. 13.6). With the *constrained model* we aim at depicting forests in natural or semi-natural state with large amounts of coarse woody debris, dominant volumes of broad leaf and a patch size of equal or more of 0.75 ha. Considering the dissected forest habitat the patch size considered constitutes an important threshold. Those habitats are regarded as the most important heath forest habitats for biodiversity in the Finnish forests (Ministry of Environment

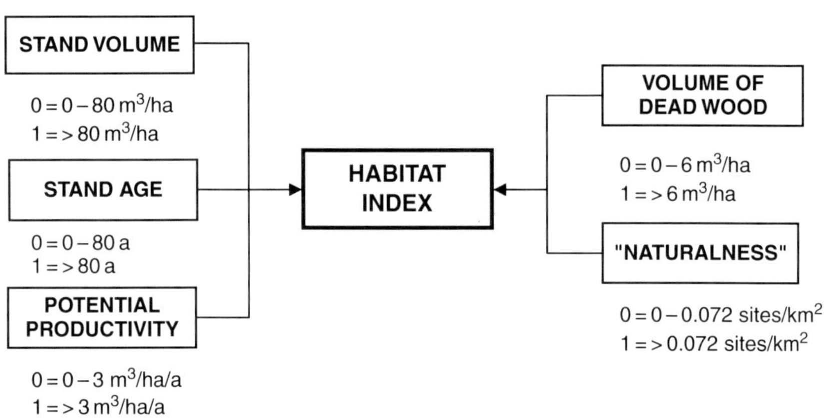

Fig. 13.5 Flowchart of the general model for high biodiversity forests presenting the model inputs, their thresholds and the operators and conditions applied. All input maps have the same extent, the same resolution (50 m) and values only on forestry land for Southern Finland. Figure 13.5a Flowchart of the general model; 5b Flowchart of the herb-rich habitat quality model; 5c Flowchart of the constrained habitat quality model

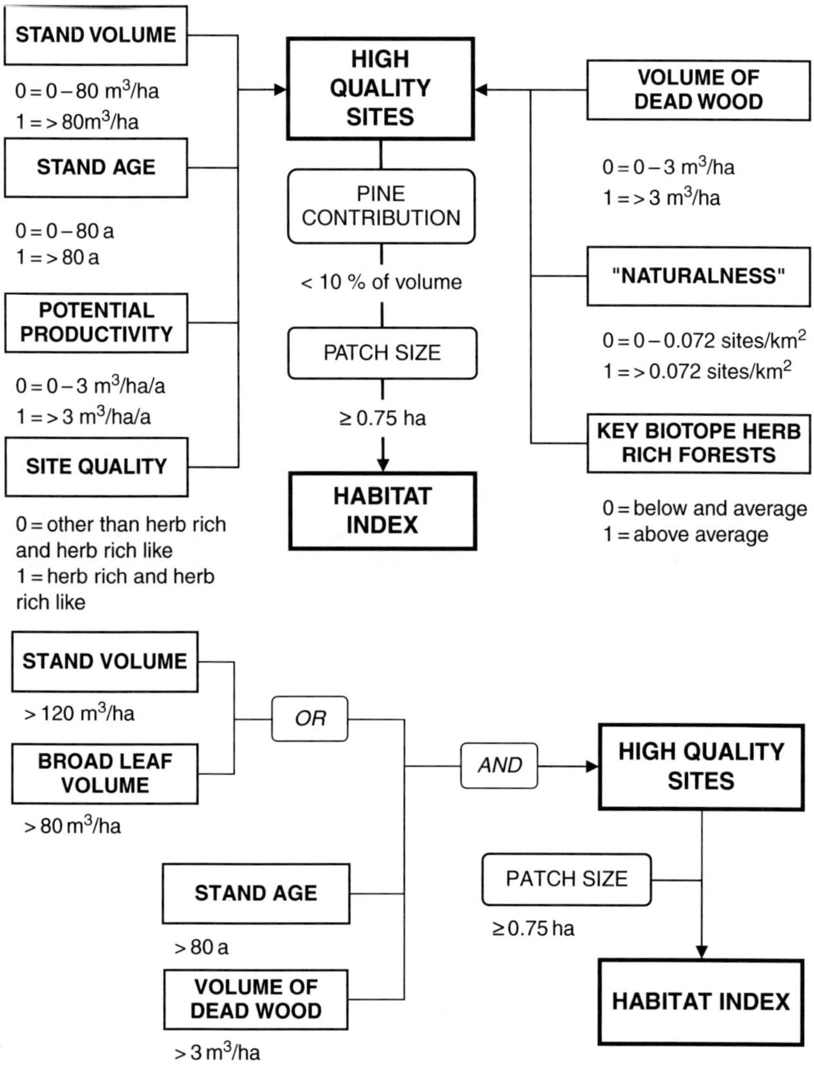

Fig. 13.5 (continued)

2004). The *herb-rich model* depicts herb-rich forests which are the most fertile forest habitats in Finland with the highest species richness. For this habitat type, class 1, showing low forest habitat quality covered 52.0% of forestry land, class 2 for an intermediate quality accounted for 32.1 % of forestry land and class 3 showing the highest quality of forest habitat based on the constraints considered for the model accounted for only 15.9% in the result layer. After this, a 4th class was separated from class 3 by first setting a further constraint for the pine contribution to be a maximum of 10 % of total volume. This provided a four class with a higher constraint setting

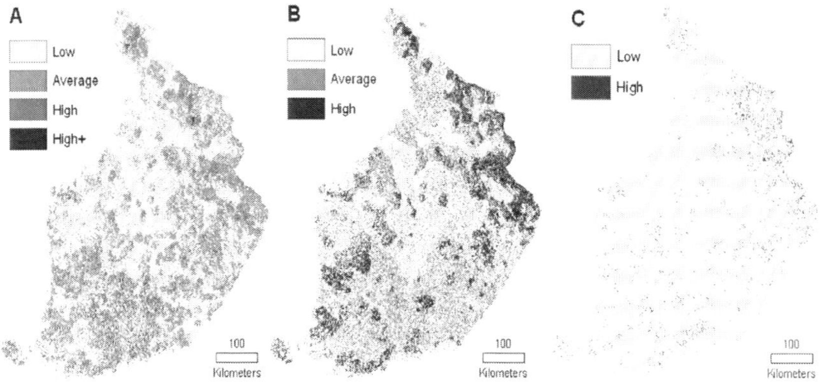

Fig. 13.6 Maps of habitat indices resulting from the habitat models: **A**. herb-rich habitat model, **B**. general habitat model, **C**. constrained habitat index. Light colors show areas of low habitat quality while dark areas are related to a high quality

a minimum patch size of 0.75 ha for areas being classified into that 4th class of high habitat quality (Fig. 13.6 A-High+). The *general model* aims at depicting the same habitat type as the constrained model but is less restrictive thus focusing on slightly different habitats that present a minimum of spatial continuity, we considered then a minimum forest patch area.

13.5 Results

The models can now be used for evaluating existing conservation areas and finding potential regions for expanding the conservation area network. The flexibility of the GIS system created to develop the habitat quality models, has a potential to be adjusted and refined to other specific needs. Table 13.1a shows general results for each of the three habitat quality model situations in relation to all forestry land and all protected areas in southern Finland. When looking at how the models behave on (i) all forestry land, (ii) near protected areas (within 2 km), and (iii) inside protected

Table 13.1a Forestry land divided into habitat index classes derived from the General model for high biodiversity value forests

Quality class	Area contribution, %		
	Inside protected areas	Inside 2 km buffer around protected areas	In all forestry land
Low	45.4	45.8	52.0
Average	30.3	35.1	32.1
High	22.9	17.9	15.0
High %	1.4	1.2	0.9

Table 13.1b Forestry land divided into habitat index classes derived from the Herb-rich habitat quality model

Quality class	Area contribution, % Inside protected areas	Inside 2 km buffer around protected areas	In all forestry land
Low	36.6	44.7	50.4
Average	28.9	35.8	33.7
High	34.5	19.4	15.9

Table 13.1c Forestry land divided into habitat index classes derived from the Constrained habitat quality model

Quality class	Area contribution, % Inside protected areas	Inside 2 km buffer around protected areas	In all forestry land
Low	86.1	96.9	97.5
High	13.9	3.1	2.5

areas, great differences in habitat quality can be seen. In general, low percentages of high quality habitat can be found inside protected areas. (Table 13.1a, b, c).

When looking at all protected areas by its status in Southern Finland it was found that in particular Special Protected areas followed by Natural Parks, Old forest Conservation Programme and National Parks are the ones with the highest proportion of areas with high biodiversity value according to the habitat quality model developed.

If we look in detail at the distribution of the Herb Rich forest model classes according to categories of protected land (Fig. 13.7) we find that 50% of the forestry land protected by the "Finnish Forest and Park Service" has high quality of herb rich forest habitat. Other protected categories with over 40% of the forestry land classified as herb rich forests are "Herb Rich Forest Protected Areas", "Special Protected Areas", "Natural Parks", and "Old Forest Conservation Programs"(Fig. 13.7).

These findings provide a good tool to look into the implementation measures of protection for the aforementioned areas so to follow the example for implementation of future expansion on protected areas.

13.5.1 4.1 Herb-Rich Habitat Quality Model

The result layer for the herb-rich habitat quality model showed the probability of an area to be herb-rich forest (Fig. 13.8). This figure is derived from the results obtained of the 4 classes calculated as explained in the methods section. In this way, we obtained the forest area that represented larger patches of dominant herb rich forest from low to high habitat quality. The figure aims at illustrated in a spatial format the result showed in Table 13.1b. Figure 13.8 provides a good communication tool for policy makers to show the potential for further conservation actions. The details showed in the windows allow the comparison of the red polygons boundaries of conservation areas with the dark blue shaded areas representing high values of "Herb Rich Habitat Quality". Considering the low percentage of protected areas of herb

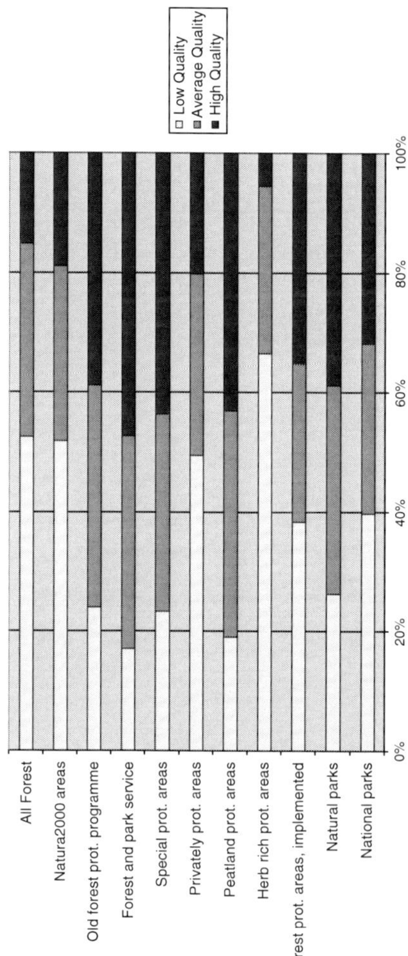

Fig. 13.7 Distribution of the habitat quality Herb Rich forest model classes according to categories of protected land in Southern Finland

rich habitat, it may be important to expand existing conservation areas and/or create corridors to enlarge and connect existing protected areas. The Herb Rich model allows a rapid evaluation of good candidates' areas to target enlargement and protection, take a look at all the blue areas (High quality habitat) that are not receiving any protection (Fig. 13.8). Furthermore, when examining 2 km wide buffer areas around conservation areas it can be noted that some types of conservation areas have more valuable herb rich forests in their vicinity than others (Fig. 13.8). We assessed the connectivity between protected areas and areas with high probability of containing herb rich forests patches and high quality mineral soil forest patches through a multiple distance ring buffer analysis (Fig. 13.9).

A more connected area can be highlighted to the northeast of the study area (Fig. 13.9). While a good candidate area to improve conservation of new areas and expand the conservation network can be identified towards the south (Fig. 13.9)

A recent report advocates that herb rich forests should be conserved especially in know high-density herb rich forest areas (Suomen Ympärsitö 2000; Virolainen et al. 2001) independent of size. In this sense, the spatial model produced shows regions of special interest that should be considered in future conservation planning efforts.

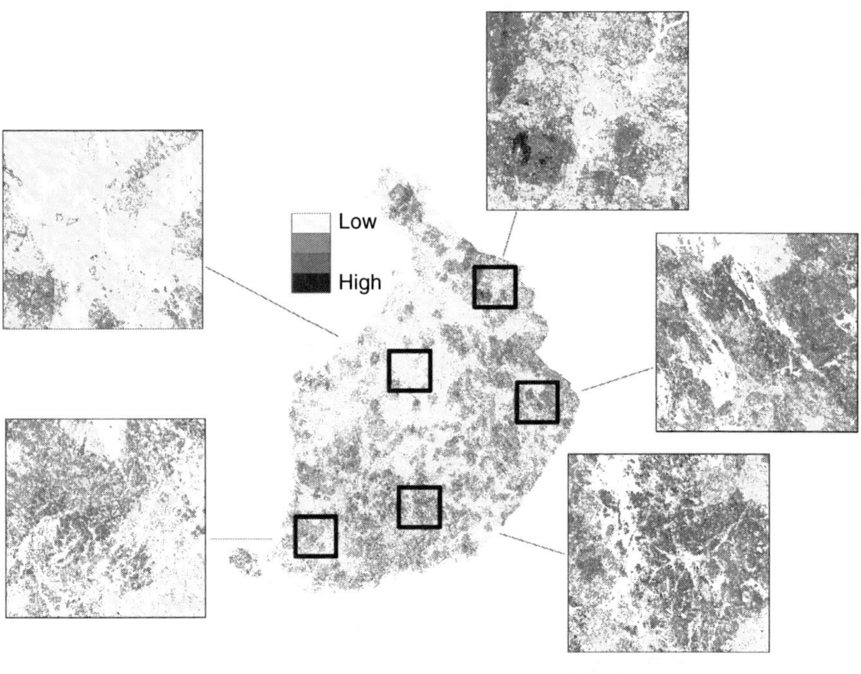

Fig. 13.8 The result layer of the herb-rich habitat quality model. Blue areas denote high habitat quality according to the variables and criteria used. In red the boundaries of protected areas to be compared against the habitat quality delineated areas

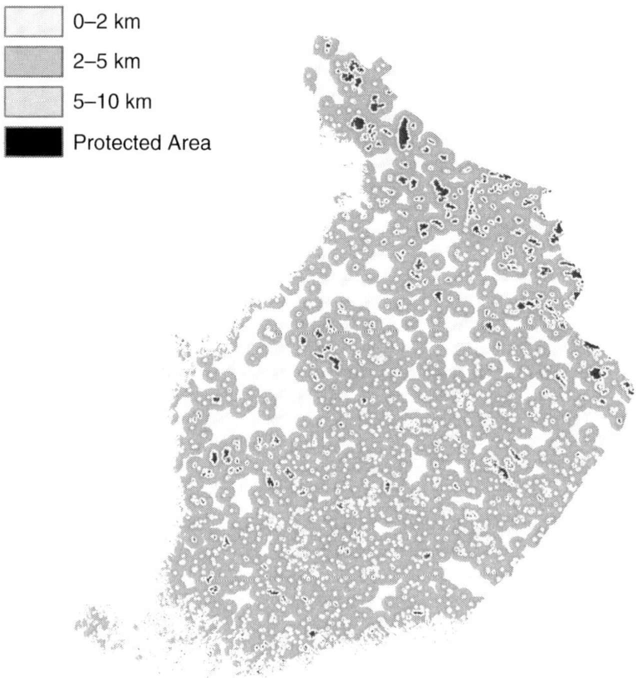

Fig. 13.9 Map of protected areas and multiple distance buffers

13.6 Discussion

The model approach presented, based on the development of a GIS framework to handle the many layers, provided a flexible and practical tool for policy makers, conservation planners and foresters to perform rapid assessment to elaborate alternative conservations plans and select individual candidate reserves in a cost effective way. Trends from biodiversity indicators show regional differences as well as different patterns within southern Finland that reveal different management history and different driving environmental factors. During the twentieth century, substantial changes in landscape level characteristics due to changing forestry practices have led to the present situation, where a significant number of forest dwelling species are considered as threatened (Rassi et al. 2001) in Finland. Herb-rich forests are among the most threatened habitats; according to the 9th Finnish National Forest Inventory data, they cover as little as 2.4% of forest area in Southern Finland (Ministry of Environment, 2004). Some 50% of herb-rich forests have been converted into fields during the past centuries (Alanen et al. 1995). Furthermore, ditching and the favouring of Norway spruce in forestry have caused further changes in vegetation. At the present, only 1.3% of the herb-rich forest in Southern Finland is protected (Ministry of Environment, 2004).

The approach applied so far was proven useful for the development of "operational methods" to monitor biodiversity and to help build up a conservation network for key forest habitats that needs to be protected. The approach is of particular interest for other countries having NFI data. Nevertheless, protection needs to be implemented within a framework of a strategy that looks at a combination of conservation areas and appropriate management of commercial forests

The model showed protected areas with a highest biodiversity quality forest habitat as compared to protected areas that seem not to have such an important habitat quality value. Work is still underway in tandem with Appendix of the habitat directorate to produce a list of habitats of importance that needs to be protected and at the present do not receive the protection needed.

This approach can be enhanced by considering the location of selected areas that may be target for protection in the analysis. For this purpose, it would be important to identify which landscape elements are the most critical for the maintenance of overall forest landscape continuity and connectivity (Pascual-Hortal and Saura 2006; Saura and Pascual-Hortal 2007). Spatial configuration of protected stands may be an important issue in fragmented landscapes where individual dispersal among habitat patches is limited, and a rule-of thumb recommendation is to spatially aggregate selected areas whenever possible (Wilson and Willis 1975). However, in boreal forest landscapes, where forest succession continuously alters stand and landscape characteristics, there is not much evidence that fragmentation affects species persistence (e.g., Schmiegelow and Mönkkönen 2002). Therefore, habitat availability, not the spatial configuration, is the primary concern (Andrén 1994; Fahrig 1998; Fahrig 2003). It is possible to extend our approach to cover also the spatial configuration of protected areas, but one needs more sophisticated methods to solve explicitly spatial site selection problems .

In all, the method and the tools presented can be applied in assessing biodiversity value of both managed and protected forest areas to help decision-making concerning valuable habitats protection and consequently manage natural resources. This effort constitutes the first attempt done at the landscape level, focusing on end users' needs, to use NFI data for biodiversity monitoring and management. The main purpose is to be able to learn about habitat quality at a regional level for planning without the high costs and time consuming that requires extra good quality census species data. Many countries have National Forest Inventories that are in many cases sub-utilized. Therefore, the message is to use NFIs to plan for forest biodiversity protection in a sustainable way. Furthermore, many countries have multisource inventories as presented here for the case of Finland, using a combination of field measurements and satellite sensor data to produce spatial information about forest characteristics. The accuracy of such data may be in some cases too coarse for small ecological scale analysis, but nevertheless the inventories do provide repeated coverage of nationwide information on an array of forest measurements, and the data can be analysed as presented here using spatial statistics to produce information on landscape characteristics and to monitor forest quality.

Data collections for forest inventories are constantly improving because of the increased availability of Light Detection and Ranging (LiDAR) and other laser scanning techniques. The other advantage we showed, regarding the use of NFI data, is that instead of presenting a category of pine-dominated forest, for example, the proportion or volume of each tree species for each unit element, in our case a pixel, can be expressed precisely. Therefore, this information produced from MS-NFI and field data allows regional planning with a certain precision depending on research objectives and/or users needs. The habitat models can be improved with individuals' species data that will help to refine the thresholds. But also the development of this type of habitat quality models helps to detect where to focus sampling efforts for particular species of local to regional importance. In this way, gradient analysis and a multiscale approach will be the next steps that can follow this first modelling phase presented here to improve sustainable forest management.

References

Alanen A, Leivo A, Lindgren L, Piri E (1995) Lehtojen hoito-opas. Metsähallituksen luonnonsuojelujulkaisuja, sarja B, no 26, pp120

Andrén H (1994) Effects of habitat fragmentation on birds and mammals in landscapes with different propositions of suitable habitat: a review. Oikos 71:355–366

Andrén H (1997) Habitat fragmentation and changes in biodiversity. Ecol Bull 46:171–181

Angelstam P, Anderson L (2001) Estimates of the needs for forest reserves in Sweden. Scand J For Res Supplement 3: 38–51

Angelstam P, Pettersson B (1997) Principles of present Swedish forest biodiversity management. Ecol Bull 46;191–203

Bartha D, Ódor P, Horváth T, Tímár G, et al (1997) Relationship of tree stand heterogeneity and forest naturalness. Acta Silv Lign Hung, 2:7–22

Burgess RL, Sharpe DM (eds) (1981) Forest island dynamics in man-dominated landscapes. Springer Verlag, New York

Burrough P, McDonnell A (1998) Principles of geographical information systems. Oxford University Press, New York

Cajander AK (1926) The theory of forest types. Acta Forest Fenn 29:1–108

CBD (2002) Convention on Biological Diversity. Conference of the Parties (COP) http://www.biodiv.org/convention/cops.sht

Esseen PA, Ehnström B, Ericson L, Sjöberg K (1997) Boreal forests. Ecol Bull 46:16–47

Fahrig L (1998) When does fragmentation of breeding habitat affect population survival? Ecol Model 105:273–292

Fahrig L (2003) Effects of habitat fragmentation on Biodiversity. Annu Rev Ecol Evol Syst 34: 487–515

Finnish Forest Research Institute (2006) Finnish Statistical Yearbook of Forestry. Vammalan Kirjapaino Oy, Vammala, Finland.

Finnish Statistical Yearbook of Forestry (2001) Finnish Statistical Yearbook of Forestry, 2001. Finnish Forest Research Institute, pp374

Gigord L, Pico, F, Shykoff J (1999) Effects of habitat fragmentation on Dombeya acutangula (Sterculiaceae), a native tree on La Réunion (Indian Ocean). Biol Conserv 88 43–51

Guisan A, Zimmermann ZE (2000) Predictive habitat distribution models in ecology. Ecol Model 135:147–186

Guisan A, Theurillat JP, Kienast F (1998) Predicting the potential distribution of plant species in an alpine environment. J Veg Sci 9: 65–74

Hanski, I. 2000. Extinction debt and species credit in boreal forests: modelling the consequences of different approaches to biodiversity conservation. Annales Zoologici Fennici 37: 271–281.

Hanski I, Hammond P (1995) Biodiversity in boreal forests. Trends Ecol Evol 10: 5–6

Hanson L, Larsson TB (1997) Conservation of boreal environments a completed research program and a new paradigm. Ecol Bull 46: 9–15
Hildén M, Auvinen AP, Primmer E (2005) Suomen biodiversiteettiohjeman arviointi. (Evaluation of the Finnish national action plan for biodiversity.) Suomen ympäristö 770. Edita, Helsinki, pp251
Johnston K, Ver hoef JM, Krivoruchko K, Lucas N (2001) Using ArcGIS Geostatistical Analyst. ESRI Press, Redlands, pp300
Jongman RHG, ter Braak CJF, van Tongeren OFR (1995) Data analysis in community and landscape ecology, Cambridge University Press, Cambridge. Salisbury, E.J., 1926. The geographical distribution of plants in relation to climatic factors. Geogr J 57:312–335
Junninen K, Simila M, Kouki J, Kotiranta H (2006) Assemblages of wood-inhabiting fungi along the gradients of succession and naturalness in boreal pine-dominated forests in Fennoscandia. Ecography 29:75–83
Kangas A, Maltamo M (eds) (2006) Forest inventory. Methodology and applications. Managing Forest Ecosystems. Vol 10. Springer, Dordrecht, pp363
Kappes H (2005) Influence of coarse woody debris on the gastropod community of a managed calcareous beech forest in western Europe. J Mollusc Stud 71: 85–91
Kouki J (1994) Biodiversity in the Fennoscandian boreal forests: natural variation and its management. Ann Zool Fenn 31:1–217
Kuusinen M, Siitonen J (1998) Epiphytic lichen diversity in old-growth and managed Picea abies stands in southern Finland. J Veg Sci 9:283–292
Löfman S (2006) Changes in forest landscape structure in southern Finland in the late 1900's. Dissertationes Forestales 32, University of Joensuu, Finland, pp30
Löfman S, Kouki J (2001) Fifty years of landscape transformation in managed forests of southern Finland. Scand J For Res 16:44–53
Lord JM, Norton DA (1990) Scale and the spatial concept of fragmentation. Conserv Biol 4: 197–202
Luque S (2000) Evaluating temporal changes using multispectral scanner and thematic mapper data on the landscape of a natural reserve: The New Jersey pine barrens, a case Study. Int J Remote Sens. Special Issue Remote Sensing and Landscape Ecology: Landscape Patterns and Landscape Change 21(13&14):2589–2611
Luque S, Lathrop RG Jr, Bognar JA (1994) Temporal and spatial changes in the New Jersey pine barrens landscape. Landscape Ecol 9(4):287–300
Luque S, Riutta T, Joensuu J, Rautjärvi N, Tomppo E (2004) Multi-source forest inventory data for biodiversity monitoring and planning at the forest landscape level. In Marchetti M (ed) Monitoring and indicators of forest biodiversity in Europe – from ideas to operationality. EFI – IUFRO Proceedings, pp. 430–444
Magura T, Tothmeresz B, Bordan Z (2004) Effects of nature management practice on carabid assemblages (Coleoptera : Carabidae) in a non-native plantation. Biol Conserv 93:95–102
Martikainen P, Siitonen J, Punttila P, Kaila L, Rauh J (2000) Species richness of Coleoptera in mature managed and old-growth boreal forests in southern Finland. Biol Conserv 94:199–209
McCoy, J., and K. Johnston. 2001. Using ArcGIS Spatial Analyst. ESRI Press, Redlands. 240 pp.
Mielikäinen K, Hynynen J (2003) Silvicultural management in maintaining biodiversity and resistance of forests in Europe–boreal zone: case Finland J Environ Manage 67:47–54
Ministerial Conference on the Protection of Forests in Europe (1994) Criteria and Indicators for Sustainable Forestry, Ministry of Agriculture and Forestry, Helsinki, Finland, pp20
Ministry of the Environment (2004) METSO – The forest biodiversity programme for southern Finland 2002–2007. http://wwwb.mmm.fi/metso/TIEDOTUSMATERIAALI/esitteet_ja_kalvosarjat/metso_booklet.pdf
Moore NW (1962) The heaths of dorset and their conservation. J Ecology 50:369–391
Moretti M, Obrist MK, Duelli P (2004) Arthropod biodiversity after forest fires: winners and losers in the winter fire regime of the southern Alps. Ecography 27:173–186

MOSSE - Biodiversity and Monitoring Programme: METSO 2003–2007 (2007) Forest Biodiversity Programme for Southern Finland http://wwwb.mmm.fi/metso/ international/research/LIST_and_PRESENTATIONS_ of_RES_PROJECTS.html

Nitare J, Norén M (1992) Woodland key-habitats of rare and endangered species will be mapped in a new project of the Swedish National Board of Forestry. Sven. Bot. Tidskr. 86, pp. 219–226 (In Swedish, with English summary)

Noss RF (2001) Forest Fragmentation in the Southern Rocky Mountains. Landscape Ecol 16: 371–372

Nuutinen T, Anola-Pukkila A, Haara A, Kärkkäinen L, et al (2001) Team report from Finnish Forest Research Institute, MELA Team. In Nordic Trends in Forest Inventory, Management Planning and Modelling Proceedings of SNS Meeting, Slovalla, Finland April 17–19, 2001, pp21–28

Odor P, Standovar T (2001) Richness of bryophyte vegetation in near natural and managed beech stands: the effects of management-induced differences in dead wood. Ecol Bull 49: 219–229

Pascual-Hortal, L, Saura S (2006) Comparison and development of new graph-based landscape connectivity indices: towards the priorization of habitat patches and corridors for conservation Landscape Ecol 21:959–967

Perlman DL, Adelson G (1997) Biodiversity: Exploring Values and Priorities in Conservation. Blackell Science, MA., USA.pp192

Ponge JF (2003) Humus forms in terrestrial ecosystems: a framework to biodiversity. Soil Biol Biochem 35:935–945

Pressey RL, Nicholls AO (1989) Efficiency in Conservation Evaluation: Scoring versus Iterative Approaches. Biol Conserv 50:199–218

Rassi P, Alanen A, Kanerva T, Mannerkoski I (eds) (2001) The Red List of Finnish Species. Ministry of the Environment & Finnish Environment Institute, Helsinki, pp432 (English Summary)

Rautjärvi N, Luque A, Tomppo E (2001) Mapping spatial patterns from national forest inventory data: a regional conservation planning tool. Schriften aus Fortslichen Falcultät der Universität Göttingen und der Niedersächsischen Forstlichen Versuchsanstalt. band 138: 293–302. J.D. Sauerländer's Verlag Frankfurt am Main, pp407

Reid WV (2006) Nature: the many benefits of ecosystem services, Nature,.443:749

ReVelle CS, Williams JC, Boland JJ (2002) Counterpart models in facility location science and reserve selection science. Environ Model Asses 7:71–80

Rodrigues ASL, Gaston KJ (2002) Optimization in reserve selection procedures-why not? Biol Conserv 107:125–19

Romero-Calcerrada R, Luque S (2006) Habitat quality assessment using weights-of-evidence based GIS modelling: the case of Picoides tridactylus as keystone species indicator of the biodiversity value of the Finnish forest. Ecol Model 196:62–76

Saura S, Pascual-Hortal L (2007) A new habitat availability index to integrate connectivity in landscape conservation planning: Comparison with existing indices and application to a case study, Landscape Urban Planning 83 (2–3): 91–103

Schmidt W (2005) Herb layer species as indicators of biodiversity of managed and unmanaged beech forests. For Snow Landsc Res 79 (1/2): 111–125

Schmiegelow FKA, Mönkkönen M (2002) Fragmentation issues in dynamic landscapes: avian perspectives from the boreal forest. Ecol Appl 12:375–389

Schuster A (1994) Regional distribution of breeding birds elaborated by a geographic information system – Possibilities and limitations. In Hagemeijer EJM, Vertrael TJ (eds) Bird Numbers 1992 – Distribution, monitoring and ecological aspects, Statistics Netherlands, Voorburg/Heerlen and SOVON, Beek-Ubbergen, pp493–501

Siitonen J (2001) Forest management, coarse woody debris and saproxylix organisms: Fennoscandian boreal forests as an example. Ecol Bull 49:1–31

Siitonen P, Tanskanen A, Lehtinen A (2002) Method for selection of old-forest reserves. Conserv Biol 16:1398–1408

Spence JR (2001) The new boreal forestry: adjusting timber management to accommodate biodiversity. Trends Ecol Evol 16(11):591–593
Store R, Jokimaki J (2003) A GIS-based multi-scale approach to habitat suitability modelling. Ecol Model 169:1–15
Store R, Kangas J (2001) Integrating spatial multi-criteria evaluation and expert knowledge for GIS-based habitat suitability modelling. Landscape Urban Plan 55:79–93
Strandberg B, Kristiansen SM, Tybirk K (2005) Dynamic oak-scrub to forest succession: effects of management on understorey vegetation, humus forms and soils. For Ecol Manage 211:318–328
Suomen ympärsitö 437 (2000) Metsien suojelun tarve Etelä-Suomessa ja Pohjanmaalla. Etelä-Suomen ja Pohjanmaan metsien suojelun tarve -työryhmän mietintö.(Forest protection in southern Finland and Ostrobothnia.) Edita, Helsinki, pp284
Sverdrup-Thygeson A (2002) Key habitats in the Norwegian production forests: A case study. Scand J For Res 17:166–178
Tomppo E (1992) Satellite image aided forest site fertility estimation for forest income taxation purposes. Acta Forestalia Fennica 229, pp70
Tomppo E (2006) The Finnish National Forest Inventory. In Kangas A, Maltamo M (eds) Forest inventory. Methodology and applications. Managing Forest Ecosystems. Vol 10. Springer, Dordrecht, pp179–194
Tomppo E, Halme M (2004) Using coarse scale forest variables as ancillary information and weighting of variables in k-NN estimation: a genetic algorithm approach. Remote Sens Environ 92:1–20
Uotila A, Kouki J (2005) Understorey vegetation in spruce-dominated forests in eastern Finland and Russian Karelia: Successional patterns after anthropogenic and natural disturbances. Forest Ecol Manage 215 (1–3): 113–137
Uotila A, Kouki J, Kontkanen H, Pulkkinen P (2002) Assessing the naturalness of boreal forests in eastern Fennoscandia. Forest Ecol Manage 161: 257–277
Väisänen R, Järvinen O (1996) How are extensive, human-caused habitat alterations expressed on the scale of local bird populations in boreal forests? Ornis Scandinavica 17:282–292
Vallauri D, André J, Dodelin B, Eynard-Machet R, Rambaud D (2005) Bois mort et à cavités : une clé pour des forêts vivantes. Lavoisier, Paris
Vellak K, Ingerpuu N (2005) Management effects on bryophytes in estonian forests. Biodiver Conserv 14:3255–3263
Virkkala R, Korhonen KT, Haapanen R, Aapala K (2000) Protected forests and mires in forest and mire vegetation zones in Finland based on the 8th National Forest Inventory. Finnish Environment Institute, Forest Research Institute. The Finnish Environment n.395
Virolainen K, Nättinen K, Siitonen J, Kuitunen M (2001) Selecting herb-rich forest networks to protect different measures of biodiversity. Ecol Appl 11:411–420
Vitousek PM, Mooney HA, Lubchenco J, Melillo JM (1997) Human domination of Earth's ecosystems. Science 277: 494–499
Webb NR, Haskins LE (1980) An ecological survey of the Heathlands in the Poole Basin, Dorset, England. Biol Conserv 17:281–296
Wilcove DS, McLellan CH, Dobson AP (1986) Habitat Fragmentation in the Temperate Zone. In Soulé ME (ed.). Conservation Biology: The Science of Scarcity and Diversity. Sinauer Associates, Sunderland, Massachusetts, pp237–256
Wilson EO, Willis EO (1975) Applied biogeography. In Cody ML, Diamond JM (eds) Ecology and Evolution of Communities. Belknap Press, Cambridge, MA, pp522–534
Winter S, Flade M, Schumacher H, Kerstan E, Möller G (2005) The importance of near-natural stand structures for the biocoenosis of lowland beech forests. For Snow Landsc Res 79 1/2: 127–144

Synthesis

Ecological Modelling and Perspectives of Forest Landscapes

Jiquan Chen

1 Modelling – A Necessity in Landscape Study

Computer models have always played a critical role in landscape analysis and projections, primarily because of the complexity of land mosaics and the associated processes in time and space (Chen and Saunders 2006; Green et al. 2006). From spatial analysis of landscape structure, generating spatial mosaics under different regulative processes (e.g., disturbances and management) and projecting landscape dynamics to the exploration of patch interactions and linking patterns and processes, one cannot comprehend the overwhelming amount of information without the assistance of computer models to generate simple pictures for achieving the study objectives (Chen and Saunders 2006). From a management perspective, models are also absolutely needed to predict the future conditions and the consequences of alternative management scenarios for policy making. A suite of models, hence, have been developed, including natural disturbances with human-induced changes and to seek a trade-off between different beneficiaries and local communities (Sayer et al. 2005). Consideration of cultural patterns and ecological processes are therefore required in modern management practices in order to devise a more realistic and relevant foundation for guiding sustainable management and multiple use of forest landscapes.

- FRAGSTATS (McGarigal et al. 2002) and APACK (Mladenoff and DeZonia 2004) for spatial pattern analysis.
- Neutral model for simulating landscape patterns under predefined processes (Gardner et al. 1987).
- HARVEST (Gustafson and Crow 1994), LSPA (Li et al. 1993) and ECOLECON (Liu 1993) for mimicking forest management in time and space.

J. Chen
Landscape Ecology and Ecosystem Science (LEES) Lab, University of Toledo, Department of Environmental Sciences 2801 W. Bancroft St. Toledo OH 43606-3390 USA
e-mail: Jiquan.Chen@utoledo.edu

R. Lafortezza et al. (eds.), *Patterns and Processes in Forest Landscapes*,

- LANDIS (Mladenoff 2004), LEEMATH (Li et al. 2000), SEPM (Dunning et al. 1995) and PATCH (Schumaker 1998) for exploring stand dynamics, disturbances, land mosaics, ecological consequences and conservation.

Depending on whether spatial relationships of data (or pixels) are used or not, these models are labelled as "spatially-explicit" or "spatially-inexplicit", respectively. Authors of this section have applied some of these models to address the applications in their studies. Convincingly, models can be used efficiently to provide landscape-level tools and information for understanding the unseen world of landscape changes and processes. For example, Luque and Vainikainen (Chapter 13) provides a good example of model projections in selecting habitats for conservation and management of biodiversity in a Finnish landscape. Unlike conventional ecological models (e.g., Shugart 1984), landscape models contain spatial information. This requirement brings uniqueness, as well as several challenges, for modellers. The nature of spatial information and the processes involved in all landscape models has hindered the development in early landscape modelling prior to the 1990s due to a lack of mandatory computing technology and geographic information systems (GIS). For example, many remote sensing products were available since the early 1970s, but the effective use of them in landscape modelling did not surface until 20 years later when storage space and computing speed began to meet the needs. However, advancement in landscape modelling and projections has not been frictionless primarily due to: (1) availability and proper uses of spatial data (see next section) and (2) theoretical challenges on scaling issues (e.g., projections at multiple or proper spatial scales).

Scaling has always been one of the key research foci in landscape analysis and modelling (e.g., Saunders et al. 2002). Regardless of the significant empirical and conceptual development on this topic over the past two decades, there lacks a clear set of instructions in model development and projections. One would be immediately challenged with questions such as: What are the proper scales (i.e., pixel resolution or time interval) to construct and run a model? Is "scale-free" a preferred feature for a model? Do we have the proper datasets at similar resolutions for model parameterization and validation? At what scale(s) should model projections to be made? Projections may be better received for some parts of the landscape, but not for other sections (i.e., increasing uncertainties in projection); how can one assess the model for its applications? Is the model easy to use for managers? These questions were directly or indirectly explored in all five chapters, yet no obvious answers were reached (i.e., they remain as challenges).

A clear message from the authors of this section is that models should not be used blindly, but with extra caution on model assumptions, capability and quality of input data. Echoing previous suggestions on the misuse of quantitative metrics in landscape ecology (Li and Wu 2004), Saura et al. (Chapter 10) argued that shape irregularity should be used as an indicator of biodiversity, rather than using the other 60+ measurements produced from FRAGSTATS. Even with this precaution, they conclude "there is a considerable risk of misuse and arbitrary selection of inappropriate metrics, which may lead, for example, to a poor performance when addressing the relationship between landscape shape and biodiversity distribution".

2 Data – The Foundation

Most landscape models require multiple types of spatial data to make projections. Often, multiple variables are needed as input and intermediate variables for model parametrization and validation. Yet, available data may not always match the model requirements (e.g., incomparability among types). First, we have to admit that data at broader spatial (e.g., region and continent) and temporal (long-term) scales are very scarce because of the required resources and long-term commitments of landscape researchers. Significant efforts are needed for the founding agencies and the scientific community to build these kinds of databases. Secondly, some available data was collected to answer different scientific and management questions (i.e., not for the specific projections of a model). Finally, data was recorded at different temporal and spatial resolutions, with different qualities, and often did not have the same quality because of the knowledge and technology limitations when data were collected. For example, it is widely known that surveyors of GLO crews did not differentiate between conifers but recorded them as "pine".

Proper processing and use of the data in model projections cannot be successful without careful quality control and analysis (QA/QC). An equally challenging task, meanwhile, comes from scaling during model parameterization, processing and projections. A foremost challenge is that one does not have the "right" input data layers to parameterize the model. This is because the datasets do not exist (scarce), are in poor quality or are with unmatched spatial-temporal resolutions. Moser and colleagues (Chapter 9) faced the challenges of bringing historical GLO notes and FIA databases to understand the landscape change and concluded that it "requires other sources of information that are rarely systematic and conclusive". Whereas, Lucas et al. (Chapter 12) had a wide variety of remotely-sensed datasets to use, but a major effort was still needed to select the "right" layers in their study.

Bringing large datasets, sometimes incompatible and from different sources, together is not an easy task (Chapter 11). In some cases, there can be too much data to choose from (e.g., application of remote sensing products). It is not only that data may be collected at different scales, but also the fact that each dataset can provide different information. Some are good for detecting biodiversity, while others are suitable for soil moisture (e.g., SAR) or canopy roughness (Lidar). For example, Landsat images have been widely used in landscape studies, yet Landsat MSS were with 80 m resolution while recent Landsat TM since 1982 were with 30 m resolution. These datasets have been used to examine landscape dynamics (e.g., Bresee et al. 2004), but the issue of infusing datasets of different resolutions remains unsolved (Turner et al. 2000). Development of sound methodology to infuse "right" datasets needs to be continuously explored for any model projection. Fortunately, the authors convinced us that multiple variables can be successfully brought together for meaningful interpretations. The fuzzy modelling exercise used by Corona et al. (Chapter 12) in multi-criteria evaluation (MCE) is a good example in this regard.

In conclusion, landscape modellers can only work with available datasets and can only be based on the best science. Applied historical GLO and FIA databases in the USA (Chapter 9) or MS-NFI in Finland (Chapter 13) are successful case studies showing that important projections can be made. It is also clear that installing long-term research projects to collect the right data-matching models should be a priority in future research. Efforts of manipulative long-term ecosystem experiments such as MOFEP (Shifley and Kabrick 2002) and the Teakettle Experimental Project (North et al. 2004) or observatory networks such as LTER (http://lternet.edu/) and NEON (http://www.neoninc.org/) should be strongly supported to ensure that the "right" data is collected for model projections. Clearly, we should all keep in mind that today's data collection should be concerned with future use, not just for current studies. Along a similar line, efforts are also needed to properly archive the data such as the Carbon Dioxide Information Analysis Center (CDIAC) at the Oak Ridge National Lab (http://cdiac.ornl.gov/) and to encourage open-sharing of data among the broader scientific community.

3 Projections – A High Priority

A landscape model is developed to answer specific scientific questions, to provide predictions for changes (with different inputs) and/or to be used as a tool for land managers (Perera et al. 2006). In recent years, landscape managers appear to be more acceptant of computer models, with particular interests in projections of landscape under alternative management scenarios. Many of them also understand that models are developed with the best available information (i.e., not perfect science). This principle of developing adaptive management protocols based on the best science and available tools of resource management is the same for both managers and modelers. They also understand that some questions related to landscape-level management can only be answered through modeling because one cannot go back through history or to the future. For example, modeling may be the only way to select "optimal" habitats (see Chapter 13) when different management options are tried. These attitudinal changes and good willingness are encouraging for modelers. Yet, model projections should be focused on current plans and new protocols by evaluating options and by using instantaneous models to evaluate the consequences of different scenarios.

However, model users need "cookbooks" or easy-to-use manuals to manage their landscapes, while policy makers are more interested in predicted results (Chapter 1). Unfortunately, many landscape models are "research-based", with an overwhelming number of parameters and complicated processes and are hard to run. While these models were developed to ensure the quality of sciences, they do often prevent landscape managers from using the models. An effective application of our knowledge and computer models requires the use of simplified and less-parameterized models, along with graphical representations and biological explanations to describe model output. Computer visualization is becoming more common in all areas of science to visually interpret data and processes and to bring greater understanding to complex problems (Wang et al. 2006). All too often, technology transfer is limited to a passive delivery of an overview of research results and techniques. Inclusion of managers and policy makers throughout the model projections (i.e., active technology transfer) would greatly enhance the influences of a model.

References

Bresee M, Le Moine J, Mather S, Crow T, Brosofske KD, Chen J, Crow TR, Radmacher J (2004) Disturbance and landscape dynamics in the Chequamegon National Forest Wisconsin, USA, from 1972 to 2001. Landscape Ecol 19(3):291–309

Chen J, Saunders SC (2006) Ecology of multiple ecosystems in time and space. In: Chen J, Saunders SC, Brosofske KD, Crow TR (eds) Ecology to hierarchical landscapes: from theory to application. Nova Publishing, Carbondale, IL, USA, pp 1–34

Dunning JB, Stewart DJ, Danielson BJ, Noon BR, Root TL, Lamberson RH, Stevens EE (1995) Spatially explicit population models: current forms and future uses. Ecol Appl 5:3–11

Gardner RH, Milne BT, Turner MG, O'Neil RV (1987) Neutral models for the analysis of brad-scale landscape pattern. Landscape Ecol 1:19–28

Green DG, Klomp N, Rimmington G, Sadedin S (2006) Complexity in landscape ecology. Springer, New York, USA, p 208

Gustafson EJ, Crow TR (1994) Modeling the effects of forest harvesting on landscape structure and the spatial distribution of cowbird brood parasitism. Landscape Ecol 9:237–248

Li H, Franklin JF, Swanson FJ, Spies TA (1993) Developing alternative forest cutting patterns: a simulation approach. Landscape Ecol 8:63–75

Li H, Gartner DI, Mou P, Trettin CC (2000) A landscape model (LEEMATH) to evaluate effects of management impacts on timber and wildlife habitat. Comp Electron Agric 27:263–292

Li H, Wu J (2004) Use and misuse of landscape indices. Landscape Ecol 19:389–399

Liu J (1993) ECOLECON: a spatially-explicit model for ECOLogical-ECONomics of species conservation in complex forest landscapes. Ecol Model 70:63–87

McGarigal K, Cushman SA, Neel MC, Ene E (2002) FRAGSTATS: spatial pattern analysis program for categorical maps. Computer soft-ware program produced by the authors at the University of Massachusetts, Amherst

Mladenoff DJ (2004) LANDIS and forest landscape models. Ecol Model 180:7–19

Mladenoff DJ, DeZonia, B (2004) APACK 2.23 analysis software user's guide. Forest landscape ecology lab. Department of Forestry, UW-Madison, Madison. http://forestlandscape.wisc.edu/Projects/apack

North M, Chen J, Oakley B, Song B, Rudnicki M, Gray A (2004) Forest stand structure and pattern of old-growth western hemlock/Douglas-fir and mixed-conifer forests. Forest Sci 50(3): 299–311

Perera AJ, Buse LJ, Crow TR (eds) (2006) Forest landscape ecology: transferring knowledge to practice. Springer, New York, p 223

Saunders SC, Chen J, Drummer TD, Crow TR, Brosofske KD, Gustafson EJ (2002) The patch mosaic and ecological decomposition across spatial scales in a managed landscape of northern Wisconsin, USA. Basic Appl Ecol 3:49–64

Sayer J, Maginnis S, Laurie M (eds) (2005) Forests in landscapes: ecosystem approaches to sustainability, Earthscan Forestry Library, Earthscan Publications, London, p 257

Schumaker NH (1998) A user's guide to the PATCH model. EPA/600/R-98/135. US Environmental Protection Agency, Environmental Research Laboratory, Corvallis, Oregon, USA

Shifley SR, Kabrick JM (2002) Proceedings of the second Missouri Ozark forest ecosystem project symposium: post treatment results of the landscape experiment. USDA Forest Service GTR NC-227, St Paul, MN, USA

Shugart HH (1984) A theory of forest dynamics: ecological implications of forest succession models. Springer-Verlag, New York

Turner DP, Cohen WB, Kennedy RE (2000) Alternative spatial resolutions and estimation of carbon flux over a managed forest landscape in Western Oregon. Landscape Ecol 15:441–452

Wang W, Song B, Chen J, Zheng D, Crow TR (2006) Visualizing forest landscapes using public data sources. Landscape Urban Plan 75:111–124

Part IV
Long-Term Sustainable Plans and Management Actions

Chapter 14

The Role of the Sustainable Forestry Initiative in Forest Landscape Changes in Texas, USA

João C. Azevedo, X. Ben Wu, Michael G. Messina, Jimmy R. Williams and Richard F. Fisher

Abstract We studied the changes in landscape pattern and function resulting from the application of the Sustainable Forestry Initiative (SFI) in East Texas, USA. Changes in landscape structure were studied by comparing landscapes with different management histories. A methodology to integrate landscape and stand pattern dynamics with processes was developed based upon modeling and simulation. The effects of pattern on processes were analyzed with this methodology considering the quality, quantity and configuration of vertebrate habitat and hydrological processes.

Comparisons among landscapes revealed that forest management has a strong influence on landscape structure. The SFI program has increased overall fragmentation with an increase in number of patches, length of edges and shape complexity and a decrease in patch size, and number and size of core areas.

Management according to the SFI program resulted generally in higher habitat suitability for many of the species analyzed and higher habitat diversity in the landscape. The SFI program induced fragmentation of the habitat of pine warbler and the establishment of narrow and elongated habitats in a network structure for most of the remaining species. Landscapes managed under the SFI program showed lower sediment yield at the watershed level than those under the non-SFI program due to lower channel erosion. The effects of the SFI program at the landscape level are related to the network of buffer strips.

In general we conclude that relevant measures at the landscape level improve the sustainability of forested landscapes in East Texas.

14.1 Introduction

The landscapes we see today are the outcome of the combination of natural, economical, and political elements acting through time. Before human expansion

J.C. Azevedo
Departamento de Ambiente e Recursos Naturais/CIMO-Centro de Investigação de Montanha, Escola Superior Agrária, Instituto Politécnico de Bragança, Portugal
e-mail: jazevedo@ipb.pt

R. Lafortezza et al. (eds.), *Patterns and Processes in Forest Landscapes*,

in North America during the Holocene, landscape change was driven by natural disturbances and climatic change. Growing populations modified considerably the structure and function of the landscape until the arrival of European settlers to the continent (Denevan 1992). Landscape change then became dominated by the expansion of agriculture (Meyer 1995) and later by growth of urban centers and infrastructure (Olson and Olson 1999). Forests decreased in area until the early twentieth century and have increased since then with the abandonment of agriculture and regrowth of cut areas (Meyer 1995).

In ancient forested landscapes recent change has been marked by intensive cutting and conversion of old growth into second growth forests (Ripple et al. 1991) as well as by fire suppression (Baker 1992). In East Texas, USA, the existing forested landscapes result mostly from reforestation campaigns that took place during the twentieth century and from natural establishment of forest in abandoned agriculture areas following the intensive exploitation of the nineteenth century. They are also the product of the forest management philosophy and practices followed during the past century.

Today, forest management is dominated by sustainable forestry. This concept, and the correspondent practice, was developed worldwide in the 1990s to integrate economical, environmental, and social objectives in forest management. It was strongly influenced by landmark events of the 1980s such as the World Conservation Strategy of 1980, The World Commission on Environment and Development Report ("Our Common Future") of 1987, and by the establishment of organizations such as the International Tropical Timber Agreement (ITTA), in 1983, and the Tropical Forestry Action Programme (TFAP), in 1985 (Upton and Bass 1996). After the UN Conference on Environment and Development (UNCED) held in Rio de Janeiro in 1992, sustainability of forests became a global goal. Two documents approved in the Rio summit, the "Statement of Forest Principles" and the "Convention on Biodiversity", defined broadly the concepts of actual sustainable forest management. International initiatives such as the Montréal Process in North and South America, Russia, Asia, and Oceania, and the Helsinki Process in Europe developed the criteria and standards for implementation of sustainable forestry at the national level.

In North America, forest sustainability has become the goal and the practice in public, nonindustrial private and industrial forests. The United States Department of Agriculture Forest Service, the Canadian Forest Service and the State and Provincial Forest Services adopted sustainable forestry concepts and practices in national and state forests in the US and Canada. Several programs are available to nonindustrial private forest owners such as the American Tree Farm System, the Forest Stewardship Program, and Green Tag Forestry, among others. The forest products industry follows mainly the standards of the Sustainable Forestry Initiative (SFI). This program was launched in 1994 by the American Forest and Paper Association (AF&PA) based upon the initial SFI Principles and Implementation Guidelines. In 1998 SFI became an industry standard and in 2001 a certification scheme. It has been a fully independent forest certification program since the beginning of 2007.

SFI is the most important certification program in North America and is currently followed on more than 61 million hectares of forestland (AF&PA 2005a).

Table 14.1 Principles of the sustainable forestry initiative (AF&PA 2005b)

Principle	Description
1. Sustainable Forestry	To practice sustainable forestry to meet the needs of the present without compromising the ability of future generations to meet their own needs by practicing a land stewardship ethic that integrates reforestation and the managing, growing, nurturing, and harvesting of trees for useful products with the conservation of soil, air and water quality, biological diversity, wildlife and aquatic habitat, recreation, and aesthetics.
2. Responsible Practices	To use and to promote among other forest landowners sustainable forestry practices that are both scientifically credible and economically, environmentally, and socially responsible.
3. Reforestation and Productive Capacity	To provide for regeneration after harvest and maintain the productive capacity of the forestland base.
4. Forest Health and Productivity	To protect forests from uncharacteristic and economically or environmentally undesirable wildfire, pests, diseases, and other damaging agents and thus maintain and improve long-term forest health and productivity.
5. Long-Term Forest and Soil Productivity	To protect and maintain long-term forest and soil productivity.
6. Protection of Water Resources	To protect water bodies and riparian zones.
7. Protection of Special Sites and Biological Diversity	To manage forests and lands of special significance (biologically, geologically, historically or culturally important) in a manner that takes into account their unique qualities and to promote a diversity of wildlife habitats, forest types, and ecological or natural community types.
8. Legal Compliance	To comply with applicable federal, provincial, state, and local forestry and related environmental laws, statutes, and regulations.
9. Continual Improvement	To continually improve the practice of forest management and also to monitor, measure and report performance in achieving the commitment to sustainable forestry.

In the US more than 90% of the industry-owned forest is managed under this program (AF&PA 2005a). The current SFI standard is based upon nine principles (Table 14.1) and 13 objectives (Table 14.2) for which a set of performance measures and indicators were established (AF&PA 2005b).

SFI relates directly and indirectly to the landscape. Firstly, the landscape scale is conceptually implicit in the program since sustainability and sustainable management of forests is addressable only when considered at this scale. Processes that are essential in terms of productivity and diversity in ecological systems, namely hydrological and biological processes, operate at landscape scales and their conservation necessitates landscape scale considerations. Also, the economical component of sustainability requires a broad scale approach to be properly addressed.

The implementation of SFI is landscape dependent and the landscape scale is directly or indirectly considered throughout the program standard. This is particularly noticeable in principles and objectives dealing with conservation of biological diversity including the promotion of diversity of wildlife habitats, forest types, and

Table 14.2 Objectives for the sustainable forestry standard (AF&PA 2005b)

Objective	Description
Objectives for Land Management	
Objective 1	To broaden the implementation of sustainable forestry by ensuring long-term harvest levels based on the use of the best scientific information available.
Objective 2	To ensure long-term forest productivity and conservation of forest resources through prompt reforestation, soil conservation, afforestation, and other measures.
Objective 3	To protect water quality in streams, lakes, and other water bodies.
Objective 4	To manage the quality and distribution of wildlife habitats and contribute to the conservation of biological diversity by developing and implementing stand- and landscape-level measures that promote habitat diversity and the conservation of forest plants and animals, including aquatic fauna.
Objective 5	To manage the visual impact of harvesting and other forest operations.
Objective 6	To manage Program Participant lands that are ecologically, geologically, historically, or culturally important in a manner that recognizes their special qualities.
Objective 7	To promote the efficient use of forest resources.
Objectives for Procurement	
Objective 8	To broaden the practice of sustainable forestry through procurement programs.
Objective for Forestry Research, Science, and Technology	
Objective 9	To improve forestry research, science, and technology, upon which sound forest management decisions are based.
Objective for Training and Education	
Objective 10	To improve the practice of sustainable forest management by resource professionals, logging professionals, and contractors through appropriate training and education programs.
Objective for Legal and Regulatory Compliance	
Objective 11	Commitment to comply with applicable federal, provincial, state, or local laws and regulations.
Objective for Public and Landowner Involvement in the Practice of Sustainable Forestry	
Objective 12	To broaden the practice of sustainable forestry by encouraging the public and forestry community to participate in the commitment to sustainable forestry and publicly report progress.
Objective for Management Review and Continual Improvement	
Objective 13	To promote continual improvement in the practice of sustainable forestry and monitor, measure, and report performance in achieving the commitment to sustainable forestry.

ecological or natural community types, such as objective for management no. 4. Wildlife conservation, which includes landscape level considerations, is also part of other objectives such as objective for procurement no. 8 and objective for forestry, research, science, and technology no. 9. It is also noticeable in principles and objectives dealing with visual impacts of forest operations (objective no. 5).

Additionally, there are particular measures implemented within SFI that are likely to have a strong effect on landscapes both structurally and functionally. Examples of these measures are the establishment of streamside buffer strips, the definition of green-up intervals and the limitation of the size of clearcuts. The establishment of streamside buffer strips is an important component of SFI. Although not directly stated in the standard, these buffers are mainly implemented according to management objective no. 3 in compliance with federal, state or province regulations and best management practices (BMPs). Both performance measures of this objective support the establishment of streamside buffer strips. These buffers are also an indicator of the performance measure 2.2 ("*minimize chemical use required to achieve management objectives while protecting employees, neighbors, the public, and the forest environment*"), part of objective for management no. 2).

The definition of green-up intervals is a performance measure of the SFI objective no. 5, defined as 3 years old or 5 feet high between adjacent clearcut areas. Also size of clearcuts is addressed as a performance measure in objective 5 along with clearcut shape and location. Only size, however, is directly considered as an indicator of the performance measure. Clearcut average size should not exceed 49 ha (AF&PA 2005b). Some companies further restrict the size of clearcuts according to their own sustainable forestry policy or according to the state or province regulations where they operate.

All the measures described above based on the SFI program are relevant at the landscape scale and can profoundly affect current landscapes. Previous studies where the implementation of sustainable forestry measures was simulated indicate that the structure of the landscape is affected by the types of management changes introduced (Hagan and Boone 1997; Cissel et al. 1998). Changes in function are also to expect from the application of sustainable forestry. Both changes in structure and function caused by sustainable forestry programs need to be fully understood.

The goal of this work is to evaluate the implications of changes in forest management on landscape structure and function associated with the SFI program. The specific objective is to detect the types and nature of change in landscape structure and function caused by the application of Sustainable Forestry Initiative measures relevant at the landscape level in intensively managed forested landscapes in East Texas. In this study we addressed the following questions: (i) Is the SFI program changing the pattern of intensively managed forested landscapes in East Texas? (ii) Can changes in structure, if any, affect ecological processes at the landscape level in this region?

14.2 Is SFI Changing the Pattern of Intensively Managed Forested Landscapes in East Texas?

14.2.1 Methods

We analyzed the effect of SFI on landscape pattern comparing landscapes with different management histories. Three areas were chosen. One area (SFI) has been intensively managed according to sustainable forestry principles since 1991. Practices in this area included a reduction in harvest unit size and the establishment of streamside buffer strips and a green up interval. Another area (IM) has been managed according to traditional intensive forest management followed by the timber industry in the region. Although changed in confined parts by more recent application of SFI practices, this landscape still reflects the pattern resulting from past management. The third area (EM) has been managed for wildlife and timber based on extensive forest management. Forest management is essentially based on the selection system applied in small areas. The EM area represents the natural landscape pattern of the region. All the areas are owned and managed by Temple-Inland Forest Products Corporation, Diboll, TX.

14.2.1.1 Areas of Study

The areas of study are located in southeastern Texas, USA (Fig. 14.1) in similar ecological conditions. The SFI area is located in Sabine County and is approximately 5000 ha in size. The IM area (5200 ha) is located in Angelina County and the EM area is 4400 ha in size and located in Trinity County. We consider that differences among areas in terms of geomorphology, pedology, hydrology, and others, do not have a strong influence on differences in landscape pattern. Management at the stand level is intensive in SFI and IM including mechanical site preparation, vegetation control, use of genetically improved vegetative material, fertilization, thinning, and harvesting. Rotation is around 30 years.

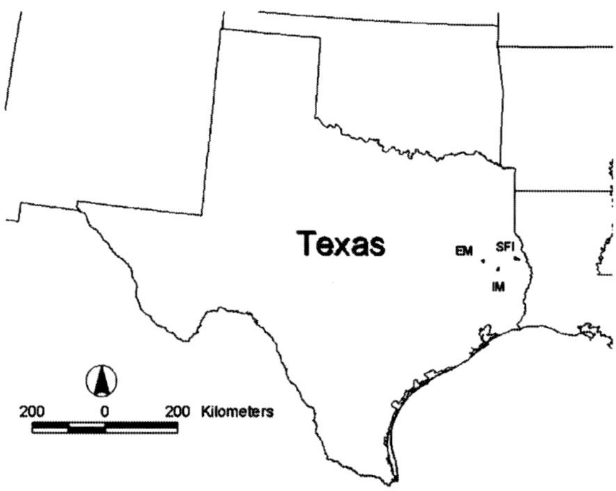

Fig. 14.1 Location of the study areas. SFI: area managed according to the SFI program; IM: area managed according to traditional forest management; EM: area managed by extensive management

14.2.1.2 Descriptive Comparison

We classified GIS coverages from 1999 of the three areas using a system comprised of seven classes, developed in order to differentiate among stands in terms of vertical (height, number of strata) and horizontal (density, basal area) structure (Table 14.3; Fig. 14.2). For that purpose we used graphical and statistical analyses (multivariate discriminant analysis and clustering methods) based on distributions of density,

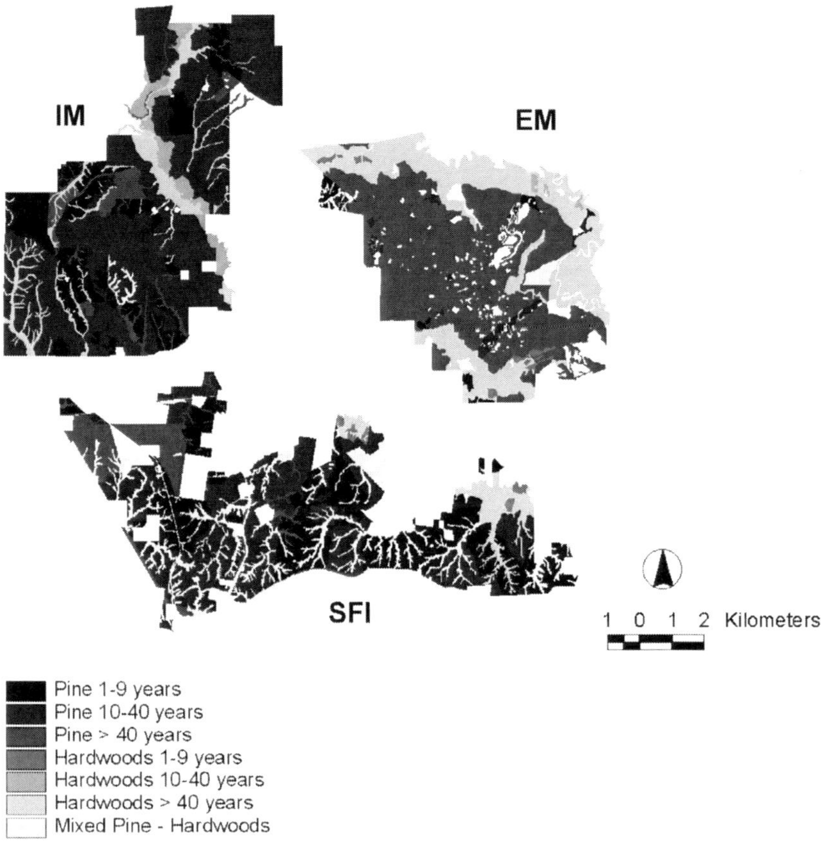

Fig. 14.2 Study areas classified according to stand structure. SFI: area managed according to the SFI program; IM: area managed according to traditional forest management; EM: area managed by extensive management

basal area, height, age and diameter at 1.3 m above ground (DBH) for both loblolly pine and hardwood stands. Raster files (10-m resolution) of the classified study areas were described in terms of landscape metrics with FRAGSTATS (McGarigal and

Table 14.3 Classes in the detailed classification system

Class number	Forest type	Age (years)
1	Pine	0–9
2	Pine	10–40
3	Pine	>40
4	Hardwood	0–9
5	Hardwood	10–40
6	Hardwood	>40
7	Pine-Hardwood	All ages

Marks 1995) at the stand, class, and landscape levels. A distance of 100 m was considered for core area and a distance of 1000 m was considered for proximity index determination.

14.2.1.3 Statistical Comparison

Multivariate analysis of variance (MANOVA) was performed to test for statistical differences in structure among landscapes. Metrics values calculated in watersheds classified according to the system described above were used in the analyses (Table 14.4). The size of the watersheds is small to allow the occurrence of a reasonable number of observations to apply statistical methods. The "Hydrologic Modeling Sample Extension" in ArcView was used in the watersheds delineation using 30 m resolution Digital Elevation Model (DEM) data (United States Geologic Survey).

Table 14.4 Small watersheds considered in the statistical comparison of the landscapes

		Area				
Landscape Name	N	Mean (ha)	St. Dev (ha)	SE (ha)	Min. (ha)	Max. (ha)
SFI	11	163.5	39.8	12.0	100.6	229.7
IM	14	162.7	52.9	14.1	91.8	248.1
EM	10	149.1	35.9	11.3	104.3	234.6

We performed MANOVA sequentially with all the metrics computed by FRAGSTATS, with the variables that graphically showed to be the best discriminants among areas of study in a previously performed hierarchical analysis, and with the variables that presented significant differences among areas of study in univariate analysis of variance (ANOVA) for the 95% and 99% levels. We established simultaneous confidence intervals (Bonferroni approach) for the 0.05 level to identify the variables and components of structure (effects) that contributed most to the observed differences in the multivariate populations.

14.2.2 Results

Both the descriptive and the statistical analysis indicated that there were differences among the landscapes compared. According to the landscape metrics calculated at the overall landscape scale (Table 14.5), SFI was the landscape presenting the highest evenness. Although young and middle age pine stands dominated both SFI and IM landscapes, in IM one single class occupied 60% of the landscape. The maximum area a single class occupied in SFI was 35% (middle age). EM was dominated by stands of the oldest classes of both pine and hardwood species (92% of the area).

SFI presented a much higher number of patches and much smaller patch size than the remaining landscapes (Table 14.5). Differences were in part due to the large average size of class 2 stands in the IM landscape. Core areas in the EM and IM landscapes represented higher proportions of the landscape and were larger in

Table 14.5 Summary of landscape metrics calculated at the landscape level

Variable	Landscape		
	SFI	IM	EM
Total Area (ha)	4943.7	5109.3	4368.6
Largest Patch Index (%)	6.5	23.8	48.3
Number of patches	207	118	77
Patch Density (#/100 ha)	4.19	2.31	1.76
Mean Patch Size (ha)	23.9	43.3	56.7
Total Edge (m)	444050	319540	108140
Edge Density (m/ha)	89.8	62.5	24.8
Landscape Shape Index	20.7	13.5	11.2
Mean Shape Index	2.45	2.67	2.06
Area-Weighted Mean Shape Index	4.4	4.1	5.4
Double Log Fractal Dimension	1.49	1.42	1.35
Mean Patch Fractal Dimension	1.13	1.15	1.12
Area-Weighted Mean Fractal Dimension	1.2	1.17	1.2
Total Core Area (ha)	1014.2	2090.3	2252.1
Number of Core Areas (#)	188	121	66
Core Area Density (#/100 ha)	3.8	2.37	1.51
Mean Core Area 1 (ha)	4.9	17.71	29.25
Mean Core Area 2 (ha)	5.39	17.28	34.12
Total Core Area Index (%)	20.51	40.91	51.55
Mean Core Area Index (%)	5.5	10.6	5.8
Mean Nearest Neighbor (m)	79.5	148	195.5
Mean Proximity Index	1594.5	4205.8	9485.1
Shannon's Diversity Index	1.48	1.21	0.99
Simpson's Diversity Index	0.74	0.59	0.54
Modified Simpson's Diversity Index	1.35	0.9	0.77
Shannon's Evenness Index	0.83	0.67	0.51
Simpson's Evenness Index	0.89	0.71	0.63
Modified Simpson's Evenness Index	0.75	0.5	0.4
Interspersion/Juxtaposition Index (%)	64.4	73.9	67.9
Contagion (%)	52.5	61.6	72.6

size than in SFI (Table 14.5). In SFI the number of core areas was much higher than in the other landscapes and the percentage of patch area in core areas was the smallest of all. In terms of edges, SFI was the landscape presenting highest absolute and relative edges at the landscape level. This was also reflected in shape metrics that indicated SFI as the landscape with more complex shapes.

On average, patches of the same class in SFI were closer to each other than in the other landscapes (Table 14.5). Contagion was much higher in the EM landscape thus reflecting the higher aggregation observed in this landscape. SFI presented the lowest contagion value.

The statistical analyses indicated that SFI had more edges, more complex shapes, and less core area than the remaining landscapes. MANOVA was initially performed with all the computed variables with the exception of Contagion, Simpson's Evenness Index, Modified Simpson's Evenness Index, and Relative Patch Richness due to the impossibility of conducting the analysis in the presence of very highly correlated variables. The null hypothesis (no difference among the groups) was rejected and the alternative hypothesis was accepted at the 0.05 level according to two of the criteria used (Wilk's and Pillai's). Significant differences were also observed having as responses diverse combinations of metrics including the variables that seemed to better discriminate among landscapes in a multiple scales pattern analysis conducted previously (NP, TE, ED, LSI, TCA, NCA, CAD, MCA1, MCA2, and MPI) and the variables that individually showed significant differences among the landscapes with univariate ANOVA at the 0.05 and 0.001 level (Table 14.6).

Throughout the analyses we observed high correlation among variables. Therefore, a smaller number of variables could be used in distinguishing effectively the

Table 14.6 Results of ANOVA for the landscape metrics considering the three areas of study simultaneously

Variable	F	p
Largest Patch Index (%)	7.40	0.002 **
Number of patches	11.12	0.000 ***
Patch Density (#/100 ha)	12.64	0.000 ***
Mean Patch Size (ha)	32.04	0.000 ***
Total Edge (m)	26.32	0.000 ***
Edge Density (m/ha)	70.44	0.000 ***
Landscape Shape Index	13.70	0.000 ***
Mean Shape Index	5.88	0.007 **
Area-Weighted Mean Shape Index	8.17	0.001 **
Double Log Fractal Dimension	11.20	0.000 ***
Mean Patch Fractal Dimension	3.36	0.047 *
Area-Weighted Mean Fractal Dimension	9.84	0.000 ***
Total Core Area (ha)	11.04	0.000 ***
Number Core Areas	3.61	0.039 *
Core Area Density (#/100 ha)	5.27	0.010 *
Mean Core Area 1 (ha)	26.15	0.000 ***
Mean Core Area 2 (ha)	5.88	0.007 **
Total Core Area Index (%)	22.00	0.000 ***
Mean Core Area Index (%)	30.90	0.000 ***
Mean Nearest Neighbor (m)	2.70	0.082 ns
Mean Proximity Index	1.99	0.153 ns
Shannon's Diversity Index	5.03	0.013 *
Simpson's Diversity Index	3.98	0.029 *
Modified Simpson's Diversity Index	3.60	0.039 *
Patch Richness	3.92	0.030 *
Patch Richness Density (#/100 ha)	1.27	0.295 ns
Relative Patch Richness (%)	3.92	0.03 *
Shannon's Evenness Index	4.85	0.014 *
Simpson's Evenness Index	4.02	0.028 *
Modified Simpson's Evenness Index	3.68	0.036 *
Interspersion/Juxtaposition Index (%)	0.17	0.845 ns
Contagion (%)	9.26	0.001 **

* - difference at the 0.05 level;
** - difference at the 0.01 level;
*** - difference at the 0.001 level.

structure of the landscapes. These could be those representing different components of heterogeneity and simultaneously proven useful in discriminating univariately among landscapes: number of patches (or density), mean patch size or contagion for arrangement, landscape shape index for shape, total edge or edge density for edges, total core area index or mean core area index (1 or 2) for core areas, and Shannon's diversity index for composition. Combinations of these variables indicated significant differences among areas of study at the 0.001 level.

Bonferroni intervals were established to compare the three landscapes pairwise for the 26 variables for which univariate ANOVA presented significant differences among areas of study for the 0.05 level (Table 14.7). SFI was different from IM in terms of edges (TE, ED), shape (LSI, AWMPFD) and core area (TCAI). Other core area metrics were very close to a significant difference between the two landscapes. It can be speculated that edges, shapes, and core areas were the major factors differentiating SFI and IM. These factors seemed also to have a great deal of interaction. SFI was different from EM in many other metrics: LPI, NP, PD, MPS, TE, ED, LSI, DFLD, AWMPFD, MCA1, TCAI, MCAI, and CONTAG.

Table 14.7 Lower and upper limits of Bonferroni simultaneous confidence intervals for comparisons among the three landscapes based upon small watersheds. Underlined values indicate significant differences for the 95% confidence level

Variable	SFI- IM		SFI-EM		IM -EM	
	lower	upper	lower	upper	lower	upper
Largest Patch Index (%)	−46.99	14.80	−67.58	−0.57	−49.72	13.77
Number of patches	−4.45	12.28	2.10	20.25	−1.34	15.85
Patch Density (#/100 ha)	−2.79	6.14	1.38	11.07	−0.04	9.14
Mean Patch Size (ha)	−9.22	4.46	−22.01	−7.17	−19.24	−5.18
Total Edge (m)	422.5	13949.9	6718.7	21388.3	−83.2	13817.8
Edge Density (m/ha)	21.0	71.0	57.7	111.9	13.1	64.5
Landscape Shape Index	0.06	2.77	0.46	3.39	−0.88	1.90
Mean Shape Index	−0.12	0.54	−0.04	0.67	−0.23	0.45
Area-Weighted Mean Shape Index	−0.04	2.06	−0.12	2.16	−1.06	1.10
Double Log Fractal Dimension	−0.01	0.29	0.03	0.36	−0.10	0.20
Mean Patch Fractal Dimension	−0.02	0.04	−0.01	0.05	−0.02	0.04
Area-Weighted M. Fractal Dimension	0.00	0.11	0.00	0.12	−0.05	0.06
Total Core Area (ha)	−61.45	1.01	−73.38	−5.65	−41.39	22.80
Number Core Areas	−2.66	6.03	−1.37	8.06	−2.81	6.12
Core Area Density (#/100 ha)	−0.82	3.31	−0.37	4.11	−1.50	2.75
Mean Core Area 1 (ha)	−6.99	1.31	−12.97	−3.97	−9.90	−1.37
Mean Core Area 2 (ha)	−28.74	8.72	−38.66	1.97	−27.59	10.91
Total Core Area Index (%)	−34.83	−4.46	−44.15	−11.22	−23.64	7.56
Mean Core Area Index (%)	−8.22	1.11	−15.43	−5.31	−11.61	−2.02
Shannon's Diversity Index	−0.31	0.47	−0.08	0.76	−0.14	0.66
Simpson's Diversity Index	−0.16	0.32	−0.07	0.45	−0.13	0.36
Modified Simpson's Diversity Index	−0.33	0.56	−0.15	0.82	−0.23	0.68
Patch Richness	−1.79	0.49	−1.10	1.37	−0.39	1.96
Shannon's Evenness Index	−0.13	0.45	−0.06	0.57	−0.20	0.40
Modified Simpson's Evenness Index	−0.17	0.51	−0.11	0.63	−0.26	0.44
Contagion (%)	−27.68	4.19	−36.69	−2.12	−24.04	8.72

The differences analyzed concern landscape fragments of reduced size and the analysis of the results should be cautious for this reason. No. Patches and Mean Patch Size have a strong tendency to differentiate the landscapes when the area of the sample units is large. However, here, sample areas were small thus artificially biasing patch area and number metrics. Average patch density was 10.4, 8.7, and 4.2 patches/100 ha for sample areas in SFI, IM, and EM, respectively, whereas for the total areas it was 4.2, 2.3, and 1.8 patches/100 ha.

14.2.3 Discussion

The results of this work suggest that the application of the SFI program is changing forested landscapes in East Texas. The most important changes can be described as fragmentation. Although fragmentation is often seen as a function of an organism or function taken under consideration (Loyn and McAlpine 2001) it can also be understood in a more general sense as the division of habitats into smaller pieces (Forman 1995; Turner et al. 2001). In such an approach, seral stages, communities, or ecosystems are taken as surrogates of population or physical processes. In this particular case, given the proportion of pine stands in the landscape, fragmentation is centered in this component.

Typical effects of forest fragmentation include increase in number of patches and edge length and decrease in patch size and core area (Franklin and Forman 1987; Ripple et al. 1991). Isolation among patches of interest increases also with fragmentation (Saunders et al. 1991; Andrén 1994). The sustainable landscape (SFI) presented many more and smaller patches than the non-sustainable (IM) or the non-intensively managed (EM) landscapes. It presented also the highest edge length. Isolation was not considered a major differentiating factor among the landscapes of study. Actually, average distances at the landscape level were usually smaller in SFI than in IM.

This fragmentation can be explained mainly by the inclusion in the landscape of Streamside Management Zones (SMZs), stream buffer zones wider than 30 m, and established according to the SFI program. These long, narrow elements break the large blocks of pine forest into smaller units increasing the number of patches, decreasing their size, and simultaneously increasing their edge length. Core areas consequently decrease in size and increase in number. This process corresponds to dissection (Forman 1995). The increase in proximity is also an effect of the introduction of the thin SMZs that make the average separation distance among stands of the same type smaller. Isolation is usually more evident in extreme fragmentation scenarios where area of habitats of interest is smaller (Gustafson and Parker 1992).

Fragmentation in primeval forests as a result of management or land use change is well known. The results of this work indicate that fragmentation results also from the application of sustainable forestry practices in intensively managed landscapes. This kind of process has been described previously. Li et al. (1993) through simulation in theoretical maps have detected increasing fragmentation with decreasing harvesting size expressed by edge density, patchiness, shape, and interior habitat parameters. When less than 40–45% of the landscape was harvested, edge density was higher if stream networks were considered as constraints. Hagan and Boone (1997), simulating the application of the Maine Forest Practices Act program noticed increasing fragmentation measured in terms of edges, core areas, and mature forest remaining. This fragmentation resulted from the reduction in clearcut size and from the establishment of separation distances and separation zones between clearcuts. Cissel et al. (1998) observed that the implementation of a management plan based on the standards, guidelines and assumptions of the Northwest Forest Plan in Oregon resulted in increasing fragmentation compared to the existing pattern. The plan included the creation of riparian reserves along streams among other measures. Patches increased very significantly in number and decreased in size and edges increased abruptly. The separation zones in the case of Hagan and Boone (1997) and the riparian reserves in the case of Cissel et al. (1998) associated with a reduction in harvest units produce the same type of pattern observed when the SFI program is implemented in East Texas. The effect of the reduction of harvest unit size seems in all cases to be less important than the establishment of buffer strips.

14.3 Can Changes in Structure Affect Ecological Processes at the Landscape Level?

14.3.1 Methods

A landscape model and several forest stand-level models were used to simultaneously simulate the dynamics of landscapes and forest stands as a function of management rules and initial conditions. Wildlife habitat quality and spatial pattern and hydrological processes (erosion and water yield) were selected as processes to evaluate based on habitat suitability models and a hydrological model. The selection of these processes resulted from the water, soil, and biodiversity conservation criteria and the indicators soil loss, water yield, and the amount, quality, and spatial pattern of habitat for vertebrate species, part of sustainable forestry programs.

Landscape dynamics were simulated using the model HARVEST 6.0 (Gustafson and Rasmussen 2002). This model allowed incorporating parameters such as harvest unit size, total area harvested, rotation length, and green-up interval, among others (Gustafson and Crow 1999). Stand-level dynamics was simulated with growth and yield models for the five forest management types applied in the area of study: (1) pine-clearcutting, (2) hardwood-clearcutting, (3) pine-selection, (4) hardwood-selection, and (5) pine-hardwood-selection. We used Compute P-Lob (Baldwin and Feduccia 1987) for planted even-aged loblolly pine stands, SouthPro (Schulte et al. 1998) for uneven-aged pine, hardwood, and mixed pine-hardwood stands, and the southern variant of the Forest Vegetation Simulator (FVS) (Donnelly et al. 2001) for even-aged hardwood stands.

Habitat suitability was modelled at the stand and landscape levels with habitat suitability index (HSI) models (Schamberger et al. 1982). HSI models provide a standardized way of quantification of habitat suitability assuming a direct linear relationship with carrying capacity (US Fish and Wildlife Service 1981). Hydrological processes were simulated with the Agricultural Policy/Environmental eXtender (APEX) model, version 1310 (Williams et al. 2000). This is a mechanistic model that combines the EPIC model (Environmental Policy Integrated Climate) with routing capabilities allowing the analysis of processes occurring simultaneously at the field and watershed levels. The model has been recently modified to describe hydrology in forested areas (Saleh et al. 2002).

The models were run stand-alone and information exchange among them occurred external to individual models. HARVEST produced landscape maps every 2 years of the simulation period using as inputs landscape structure maps prepared in a GIS according to management criteria. Stand ID, age, management type, and site index were used to link individual stands in the GIS coverage with stand structure data simulated in the growth and yield models for the respective management type and site index and with HSI scores calculated according to the HSI models. HSI variables and final scores were calculated at the stand level using data from the growth and yield models. Habitat structure was described in FRAGSTATS from HSI maps created in the GIS. APEX files used information obtained from maps provided by HARVEST and particular characteristics of the stands provided by the growth and yield models.

The changes in processes caused by management were based on the comparison to two landscape management scenarios. An SFI scenario followed on the application of SFI landscape measures, namely SMZs $\geq$30 m wide along streams, limits in harvest unit size (pine 49 ha; hardwoods 12 ha) and a three-year green up interval. A Non-SFI scenario was established in the absence of these rules.

We ran HARVEST for 400 years. For each scenario, five replicate runs were conducted using independently generated random number seeds. Partial studies on the effects of SFI on the landscape processes in intensively managed forested landscapes in East Texas are available in Azevedo et al. (2005a), Azevedo et al. (2005b), and Azevedo et al. (2006).

14.3.1.1 Study Area

The wildlife study was conducted in a 5,773-ha area, corresponding roughly to the IM area of the previous section. It lays in the Yegua Formation of coastal plain sediments of late Eocene origin. Soils were Ultisols (Rosenwall series) and Alfisols (Diboll and Alazan series). Elevation ranged from 41 to 113 m above sea level. Mean annual rainfall was 1,054 mm and mean annual temperature was 19.4 C. Most of the area was owned by Temple-Inland Forest Products Corporation, Diboll, TX, and managed for industrial forestry. For the hydrology study we considered a smaller watershed of this area, 1190 ha in size.

14.3.1.2 Wildlife Habitats

We selected eight species among vertebrates potentially occurring in the region where the study area was located (83 herps, 132 birds, 51 mammals) to represent guilds of breeding and foraging requirements. The species were classified based on vertical stratification of the pine, hardwood and pine-hardwood forest breeding and foraging habitats. We conducted a cluster analysis using the Ward's minimum variance clustering method with distances based upon Jaccard's coefficient of similarity (Lapointe and Legendre 1994). From the twelve guilds initially considered (Fig. 14.3), four were excluded for being comprised of species associated with non-existing local conditions, relying upon parameters difficult to estimate at the resolution of the data used or lacking published habitat models. One species was selected to represent the corresponding habitat requirements: American beaver (*Castor canadensis* Kuhl 1820), American woodcock (*Scolopax minor* J. F. Gmelin 1789), pine warbler (*Dendroica pinus* (Wilson, 1811)), downy woodpecker (*Picoides pubescens* (Linnaeus 1766)), barred owl (*Strix varia* Barton 1799), wild turkey (*Meleagris gallopavo silvestris* Vieillot 1817), fox squirrel (*Sciurus niger* Linnaeus 1758) and gray squirrel (*Sciurus carolinensis* Gmelin 1788). The habitats were modeled with HSI models using data provided by the growth and yield models and in few cases from assumptions based upon published data. Application of the HSI models is described in detail in Azevedo et al. (2006). At the landscape level, HSI was calculated from the GIS coverages resulting from the landscape simulations. Five habitat suitability classes were defined: "unsuitable" (HSI = 0), "low" ($0 < \text{HSI} \leq 0.25$), "medium" ($0.25 < \text{HSI} \leq 0.5$), "high" ($0.5 < \text{HSI} \leq 0.75$), and "very high" ($0.75 < \text{HSI} \leq 1$). Maps of high and very high suitability habitats were analyzed in terms of landscape metrics calculated with FRAGSTATS (McGarigal and Marks 1995).

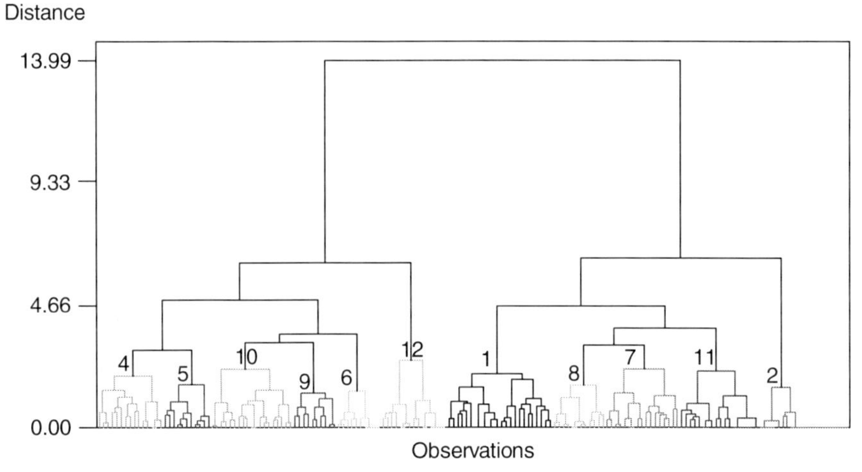

Fig. 14.3 Dendrogram for the clusters analysis with Ward's minimum variance and distances based upon Jaccard's coefficient of similarity. Numbers in the chart indicate cluster number. Cluster 3, comprised of non-forest species is not represented

14.3.1.3 Hydrology

The use of APEX relied on watershed discretization and parameterization of the model components, mainly subareas and operation schedules files. The delineation of subareas in the study area was performed with the watershed delineation module of SWAT2000, ArcView interface (Di Luzio et al. 2002) based on 30 m resolution digital elevation model (DEM) data (United States Geological Survey). Larger sub-basins were manually subdivided to reduce soil and stand variability and to minimize errors in channel length mensuration. Further discretization was made to distinguish among forest stands and buffer zones. For each scenario, routing was schematized in a diagram based on SWAT sub-basin coverages and stand maps derived from HARVEST outputs.

Subareas files were built with soil and operation schedule file codes, area, channel length and slope, upland slope, reach length and slope, when applicable, as inputs. Receiving subarea, operation schedule file, and soil file were also associated to each entering subarea. Soil series distribution in the study area was obtained from a SSURGO digital map for Angelina County (Soil Survey Geographic Data Base, USDA- Natural Resources Conservation Service). The stands were managed by operation schedules according to their composition and age. These files described stand development and management operations in the stands and synchronized APEX with the stand and landscape dynamics simulated in HARVEST.

Evaluation of the model for the study area was performed in controlled subareas for different magnitudes and combinations of parameter values for soil, crop type, density, thinning, age to maturity, partition flow through filter strips, and slope, among others. Different subarea delineations were also used to evaluate the role of discretization on the processes simulated including the effect of buffer strips on runoff and sediment loss.

Weather data were generated based on parameters for Lufkin, Texas. The model was run 30 years prior to the period of interest to allow stabilization of the system and stand growth. Three simulations for each scenario (SFI and non-SFI) were performed. The methods are described in detail in Azevedo et al. (2005b).

14.3.2 Results

All the results refer to a period of 30 years given the fact that the simulated landscapes presented a return interval of this duration.

14.3.2.1 Wildlife Habitats

There were differences between scenarios in terms of habitat suitability for the species analyzed (Table 14.8). Habitat suitability for pine warbler was slightly lower in SFI than in Non-SFI. HSI values for American woodcock and American beaver were slightly higher in the SFI scenario. Given the uniformity of simulation runs all the differences were statistically significant ($p<0.001$; repeated measures ANOVA with management as a fixed effect and runs as random subjects). There were major differences between scenarios in habitat suitability for wild turkey, fox squirrel, and gray squirrel: very low suitability in the Non-SFI scenario and relatively high suitability in the SFI scenario. HSI values for barred owl and downy woodpecker

Table 14.8 Summary statistics of habitat suitability index (HSI) values for selected species under Sustainable Forestry Initiative (SFI) and Non-SFI management scenarios. Values refer to a 30-year simulation cycle

Species	SFI scenario			Non-SFI scenario		
	Mean	Min	Max	Mean	Min	Max
Pine warbler	0.19	0.15	0.23	0.23	0.17	0.28
American woodcock	0.45	0.43	0.46	0.41	0.39	0.44
Eastern wild turkey	0.54	0.52	0.55	0.06	0.03	0.09
Fox squirrel	0.24	0.23	0.24	0.02	0.02	0.03
Gray squirrel	0.21	0.21	0.22	0.03	0.03	0.04
Downy woodpecker	0.03	0.02	0.04	0.03	0.02	0.03
Barred owl	0.04	0.02	0.06	0.002	0.000	0.005
American beaver*	0.63	0.61	0.64	0.55	0.53	0.57

*Calculated for the area within buffers only

were negligible in both scenarios. Habitat suitability was relatively stable during the simulations for all the species in both management scenarios.

Highly suitable habitat for American woodcock was abundant only in the SFI landscape. This habitat was in few patches spread over the landscape with an extremely large edge length, and few and small core areas (Table 14.9). Near 100% of the area of this class was in a single patch. This habitat class corresponded mostly to the SMZs network established in the SFI scenario (Fig. 14.4). Metrics for the high suitability pine warbler habitat, the highest observed for the species, indicated considerable fragmentation in the SFI scenario (more and smaller patches, less aggregated, more edges, less core area, and lower isolation) as compared to the Non-SFI scenario (Table 14.9; Fig. 14.4).

For fox and gray squirrel and wild turkey there was almost no quality habitat in the Non-SFI scenario. Very high suitability habitat for fox squirrel and gray squirrel comprised the majority of suitable habitat in the SFI scenario. High suitability habitat metrics express the structure of the SMZ network: few patches, one patch containing more than 90% of the class area, considerable total area occupied, low aggregation, small core area percentage, and small distances (Table 14.9). For barred owl and downy woodpecker none of the scenarios presented practically suitable habitat patches. Very few, small, and isolated patches provided the only quality habitat for barred owl. In SFI, the SMZ network provided relatively abundant but low suitability class habitat for both species.

14.3.2.2 Hydrology

The results obtained at the subarea level were generally within the expected values for forested watersheds in East Texas under similar conditions. Water and sediment yields were generally small and most of the runoff and erosion observed occurred during intense storm events.

SFI and non-SFI management scenarios originated the same amount of surface runoff and water yield at both subarea and watershed levels (Table 14.10).

Table 14.9 Selected landscape metrics for American woodcock, pine warbler and gray squirrel "high" ($0.5<HSI\leq0.75$), and "very high" ($0.75<HSI\leq1$) suitability habitat classes. Values are averages (three simulations; 15 observation dates)

Variable	American woodcock				Pine warbler		Gray squirrel			
	"high" ($0.5<HSI\leq0.75$)		"very high" ($0.75<HSI\leq1$)		"high" ($0.5<HSI\leq0.75$)		"high" ($0.5<HSI\leq0.75$)		"very high" ($0.75<HSI\leq1$)	
	SFI	Non-SFI	SFI	Non-SFI	SFI	Non-SFI	SFI	Non-SFI	SFI	Non-SFI
Percentage of Landscape (%)	26.8	4.1	4.0	8.8	25.8	32.9	0.7	1.7	24.4	1.7
Patch Density (#/100 ha)	0.2	0.2	0.4	0.1	1.3	0.4	0.1	0.1	0.1	0.1
Edge Density (m/ha)	69.7	4.8	6.8	6.6	37.4	19.7	1.8	2.4	69.7	2.9
Largest Patch Index (%)	26.8	1.6	0.7	5.0	2.9	13.7	0.2	0.8	24.3	0.7
Landscape Shape Index	25.7	4.9	7.0	4.5	15.0	7.5	4.3	3.6	26.9	4.5
Mean Patch Area (ha)	165.4	22.9	10.4	185.7	20.8	89.3	5.4	30.2	169.6	17.8
Core Area Per. of Land. (%)	5.9	1.4	0.5	4.4	4.8	17.3	0.0	0.4	4.7	0.3
Mean Core Area (ha)	36.3	7.8	1.2	93.0	3.9	47.0	0.1	6.7	32.8	3.1
Mean Core Area Index (%)	2.3	15.0	4.8	38.9	8.2	19.9	1.1	16.5	2.3	8.4
Mean Euclidean Nearest Neighbor Distance (m)	153.0	722.8	160.5	192.4	80.6	212.2	641.9	1214.1	126.3	715.4

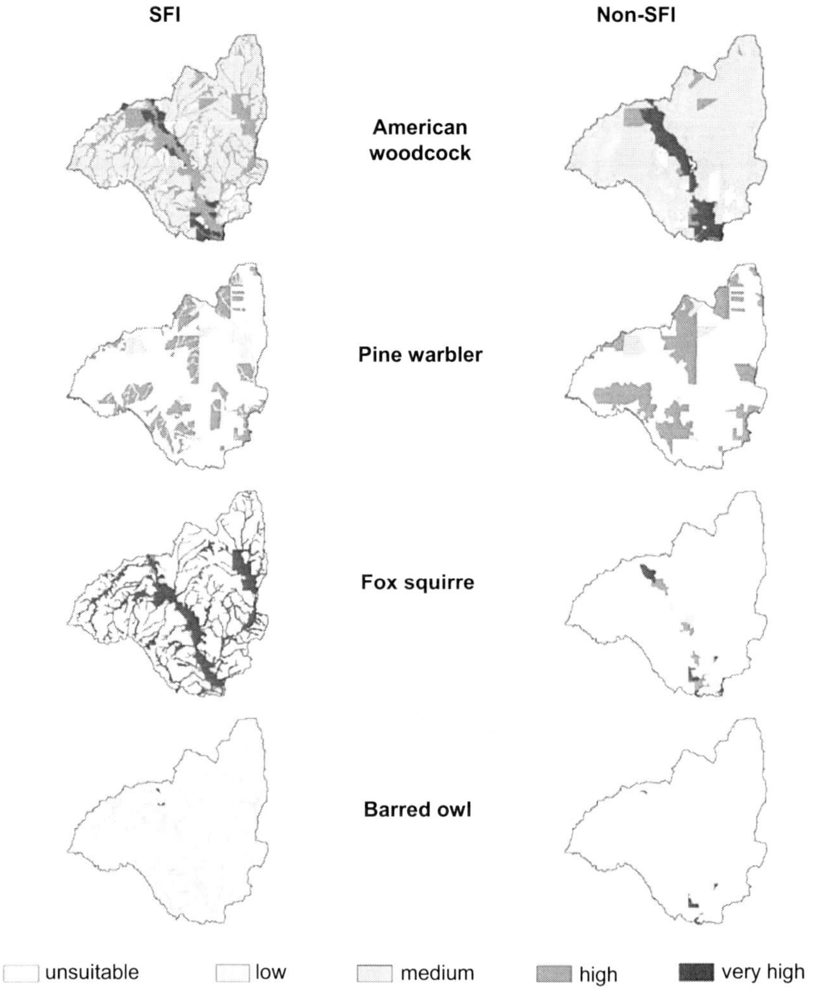

Fig. 14.4 Examples of spatial pattern of habitat suitability classes for the study area in alternative management scenarios. Images refer to a single simulation year

Differences in forest cover between scenarios were attenuated by the nearly level slopes in the study area, the lower annual mean precipitation and by the fact that results are averages for 30 years and for 3 runs.

Sediment yield at the subarea level was approximately the same in both scenarios. At the watershed level, however, the non-SFI scenario presented considerably more sediment yield than the SFI scenario. The difference in watershed sediment yield resulted from the routing processes, mainly channel degradation. Sediment deposition also occurred but in low quantity due to the fact that sediment loss is usually very low in the nearly level slopes of the area. Deposition was appreciable only during intense storm events, mainly in the SFI scenario, when sediment yield

Table 14.10 Annual precipitation, runoff and water and sediment yield in the study watershed. Results are averages from 30 years and three simulations

Scenario	Precipitation (mm)	QSS (mm)	QSW (mm)	QTS (mm)	QTW (mm)	YS (t/ha)	YW (t/ha)
SFI	1074.7	20.64	20.27	26.98	26.52	0.09	0.16
Non-SFI	1074.7	20.58	20.40	26.84	26.59	0.08	0.38

QSS-average subarea surface runoff;
QSW-average watershed surface runoff;
QTS-average subarea water yield;
QTW-average watershed water yield;
YS-average subarea sediment yield;
YW-average watershed sediment yield

was high. Channel degradation was common in both scenarios but higher in the non-SFI scenario (annual average values of approximately 0.3 t/ha against 0.08 t/ha in the SFI scenario). Channel degradation was responsible for the differences in watershed sediment yield between the two landscapes. The Non-SFI scenario presented fewer buffer zones and was also less fragmented than the SFI landscape. Degradation occurred mostly in periods of intense precipitation.

14.3.3 Discussion

The results above indicated that changes in forest management of the type included in the SFI program affect processes at the landscape level. Wildlife habitats of the species selected to indicate particular habitat conditions changed in quality, abundance and spatial structure when SFI landscape measures were applied. In general the SFI scenario provided higher habitat suitability. The habitat heterogeneity, expressed by higher diversity and evenness of habitats, also increased which creates the possibility of a more diverse wildlife in the SFI landscape. Spatially, changes caused by SFI can be of the kind indicated by pine warbler that presented an increase in the fragmentation of the most suitable habitat. Changes can also be of the type observed for American woodcock, wild turkey, fox and gray squirrels, where suitable habitat follows the configuration of the SMZs network established in the area according to the SFI program. The landscape structure of the habitat is not limiting for most of these species. The conditions created seem to indicate also the possibility of maintaining large populations of many species. In spite of improvements induced by the program, the SFI landscapes, however, are still insufficient in a larger perspective of maintenance of biodiversity. There are important habitats that are missing in this landscape such as mature pine and hardwood stands. These types of stands are known for the richness and abundance of species they retain and provide particular habitat for species that are exclusively associated with these environments.

Sediment yield also showed that SFI affects hydrological processes. Lower sediment yield at the landscape level was observed in the SFI scenario which was related to the establishment of SMZs along streams.

From the SFI landscape measures simulated, the SMZs seem to have the strongest effects of all. As seen before, they are key elements in landscape structure change caused by the SFI program. SMZs are also essential in the wildlife habitat quality, abundance and configuration, playing a key role in the reduction of channel erosion.

Based on the results obtained in this modeling and simulation exercise we conclude that the changes of the type occurring currently in forested landscapes in East Texas as driven by the SFI program are also changing landscape processes in this region.

14.4 Overall Conclusion

Forest management can be considered as an anthropogenic process that modifies landscape structure, which in turn influences the processes and functions of landscape such as hydrology, soil erosion, availability and quality of wildlife habitat, and species diversity. A key issue related to these complex interactions on managed landscapes is their sustainability.

In the absence of a comprehensive and operational definition of landscape sustainability (Wu and Hobbs 2002) we consider as sustainable a landscape that is able to maintain its essential structures and processes over time in a management context. Sustainability is mainly a management concept and it is particularly useful in testing the capacity of a natural system to support human induced change through resources management. According to the framework established for this work, we compared structure among landscapes managed by different management perspectives and we analyzed, based upon modeling and simulation, soil loss, water yield, and the amount, quality, and spatial pattern of habitat for vertebrate species as indicators of soil, water and biodiversity conservation, usual criteria of sustainable forestry. Based upon the results of this work we consider that SFI improves landscape sustainability. SFI creates landscapes that are better structured and contribute better to the conservation of wildlife and soil.

The SFI landscape had a more complex pattern than the other landscapes, including the non-SFI landscape, presenting more patches and more complex shapes. Evenness and diversity were also higher in this landscape.

SFI scenarios in the simulations presented higher diversity of habitats, higher suitability for most of the species considered, and a configuration that is not generally limiting for the species. The SFI scenarios in the hydrology study indicated that there is a reduction in soil loss from the system when SFI is followed. Therefore, we conclude that SFI contributes to the sustainability of forest landscapes in East Texas by changing these landscapes towards a better structure and function. Whether these changes create sustainable landscapes or not we are not able to verify.

Acknowledgments We thank Temple-Inland Forest Products Corporation, Texas, USA, Instituto Politécnico de Bragança, Portugal, Luso-American Development Foundation (FLAD), and the PRODEP III program (Portugal) for supporting this research. We also thank several anonymous reviewers for suggestions on this and on partial manuscripts published from this project.

References

AF&PA (2005a) Sustainable forestry initiative. 10th Annual Progress Report. AF&PA, Washington, DC

AF&PA (2005b) Sustainable forestry initiative (SFI) standard 2005–2009 edition. AF&PA, Washington, DC

Andrén H (1994) Effects of habitat fragmentation on birds and mammals in landscapes with different proportions of suitable habitat: a review. Oikos 71:355–366

Azevedo JC, Wu XB, Messina MG, Fisher RF (2005a) Assessment of sustainability in intensively managed forested landscapes: A case study in east Texas. For Sci 51:321–333

Azevedo JC, Williams JR, Messina MG, Fisher RF (2005b) Impacts of the sustainable forestry initiative landscape level measures on hydrological processes. Water Resour Manag 19:95–110

Azevedo JC, Wu XB, Messina MG, Fisher RF (2006) Effects of the sustainable forestry initiative on the quality, abundance, and configuration of wildlife habitats. J Sustain For 23:37–65

Baker WL (1992) Effects of settlement and fire suppression on landscape structure. Ecology 73:1879–1887

Baldwin VC, Feduccia DP (1987) Loblolly pine growth and yield prediction for managed west gulf plantations. USDA For. Serv. Res. Pap. SO-236, p 27

Cissel JH, Swanson FJ, Grant GE, Olson DH, Gregory SV, Garman SL, Ashkenas LR, Hunter MG, Kertis JA, Mayo JH, McSwain MD, Swetland SG, Swindle KA, Wallin DO (1998) A landscape plan based on historical fire regimes for a managed forest ecosystem: the Augusta Creek study. USDA For. Serv. Gen. Tech. Rep. PNW-GTR-422, p 82

Denevan WM (1992) The Pristine Myth: the landscape of the Americas in 1492. Ann Assoc Am Geogr 82:369–385

Di Luzio M, Srinivasan R, Arnold JG, Neitsch SL (2002) ArcView Interface for Swat2000-User's Guide. Texas Water Resources Institute Rep. TR-93, p 345

Donnelly D, Lilly B, Smith E (2001) The Southern Variant of the Forest Vegetation Simulator. Forest Management Service Center, Fort Collins

Forman RTT (1995) Land mosaics: The ecology of landscapes and regions. Cambridge University Press, Cambridge

Franklin JF, Forman RTT (1987) Creating landscape patterns by forest cutting: ecological consequences and principles. Landscape Ecol 1:5–18

Gustafson EJ, Crow TR (1999) HARVEST: Linking timber harvesting strategies to landscape patterns. In Mladenoff DJ, Baker WL (eds) Spatial modeling of forest landscape change: Approaches and applications, Cambridge University Press, New York, pp 309–332

Gustafson EJ, Parker GR (1992) Relationships between landcover proportion and indices of landscape spatial pattern. Landscape Ecol 7:101–110

Gustafson EJ, Rasmussen LV (2002) Assessing the spatial implications of interactions among strategic forest management options using a Windows-based harvest simulator. Comput Electron Agric 33:179–196

Hagan JM, Boone RB (1997) Harvest rate, harvest configuration, and forest fragmentation: a simulation of the 1989 Maine Forest Practices Act. Report MODCF-97001, Manomet Center for Conservation Sciences, Manomet, p 17

Lapointe FJ, Legendre P (1994) A classification of pure malt scotch whiskies. Appl Statist 43: 237–257

Li H, Franklin JF, Swanson FJ, Spies TA (1993) Developing alternative forest cutting patterns: a simulation approach. Landscape Ecol 8:63–75

Loyn RH, McAlpine C (2001) Spatial patterns and fragmentation: Indicators for conserving biodiversity in forest landscapes. In: Raison RJ, Brown AG, Flinn DW (eds) Criteria and indicators for sustainable forest management. IUFRO Research Series, No 7. CABI Publishing, Wallingford, pp 391–422

McGarigal K, Marks BJ (1995) FRAGSTATS: Spatial pattern analysis program for quantifying landscape structure. USDA For. Serv. Gen. Tech. Rep. PNW-GTR-351, p 122

Meyer WB (1995) Past and present land use and land cover in the USA. Consequences 1:25–33
Olson RK, Olson AH (1999) Farmland loss in America. In: Olson RK, Lyson, TA (eds) Under the blade, Westview Press, Boulder
Ripple WJ, Bradshaw GA, Spies TA (1991) Measuring forest landscape patterns in the Cascade Range of Oregon, USA. Biol Conserv 57:73–88
Saleh A, Williams JR, Wood JC, Hauck L, Blackburn WH (2002) Application of APEX for forestry. In: Proceedings of the 2002 conference on Total Maximum Daily Load (TDML): Environmental regulations, American society of agricultural engineers, St. Joseph, pp 595–605
Saunders DA, Hobbs RJ, Margules CR (1991) Biological consequences of ecosystem fragmentation: a review. Conserv Biol 5:18–32
Schamberger M, Farmer AH, Terrell JW (1982) Habitat suitability index models: introducion. USDI Fish and Wildl. Serv. FWS/OBS-82/10, p 2
Schulte BJ, Buongiorno J, Lin CR, Skog K (1998) SouthPro: a computer program for managing uneven-aged loblolly pine stands. USDA For. Serv. Gen. Tech. Rep. FPL-112, p 47
Turner MG, Gardner RH, O'Neill RV (2001) Landscape ecology in theory and Practice – pattern and process. Springer Verlag, New York
Upton C, Bass S (1996) The forest certification handbook. Earthscan, London, p 219
US Fish and Wildlife Service (1981) Standards for the development of habitat suitability index models. USDI Fish Wildl. Serv. 103 ESM np
Williams JR, Arnold JG, Srinivasan R (2000) The APEX model. BRC Report No. 00-06. Texas Agricultural Experiment Station, Temple, p 121
Wu J, Hobbs R (2002) Key issues and research priorities in landscape ecology: An idiosyncratic synthesis. Landscape Ecol 17:355–365

Chapter 15

Biodiversity Conservation and Sustainable Livelihoods in Tropical Forest Landscapes

Jean-Laurent Pfund, Piia Koponen, Trudy O'Connor, Jean-Marc Boffa, Meine van Noordwijk and Jean-Pierre Sorg

Abstract In developing countries, much remains to be done to truly integrate the livelihoods of rural people and biodiversity conservation into land use decision-making and management processes. Yet, research institutions can support informed landscape management decisions by communities, conservation agencies and policy-makers. This can be accomplished by developing methods and instruments that facilitate coherent linkages between stakeholders across various spatial and decisional scales. Researchers need to facilitate equitable participation in the planning processes and provide information on the options that best integrate biodiversity conservation and livelihoods. This chapter aims to analyse how research has contributed to this objective and how it could be designed for future integrative activities at the landscape level. It identifies lessons from case studies that combine biodiversity conservation and livelihood aims in tropical regions and reviews methodological issues relevant to transdisciplinary research. In addition to the critical elements emerging from case studies, the article highlights the crucial role of institutions in helping to bridge the gaps between science, planning, decision-making and effective management. Finally, it describes an approach that two international research organizations are developing to promote the sustainable use of forests and trees and biodiversity conservation in fragmented tropical forest landscapes.

15.1 Introduction

Biodiversity faces severe threats in many tropical developing countries and hotspots (Chapin et al. 2000). Tropical forests are still being converted (Chomitz 2006) and socio-economic disparities keep increasing to the detriment of rural areas (Kanbur and Venables 2005). Over the last few decades, forest landscapes have become

J.-L. Pfund
CIFOR, Center for International Forestry Research, Bogor, Indonesia
e-mail: j.pfund@cgiar.org

R. Lafortezza et al. (eds.), *Patterns and Processes in Forest Landscapes*,

increasingly fragmented (Koehler et al. 2003). The resulting mosaic landscapes that have fragments of natural or semi-natural habitats have long been poorly valued in terms of biodiversity conservation (Hanski 2005). At the same time, many national parks and reserves are deteriorating (Jepson et al. 2001) and in some cases, probably do not have the optimal governance structure for biodiversity conservation (Hayes 2006). Even when conservation bodies recognize that protected areas need to be managed as a part of their surrounding bio-cultural matrix (IUCN 2003), conflicting opinions often persist between local people and land use planners. Thus, tradeoffs must be negotiated between local and external interests. In most developing tropical countries, much remains to be done to truly integrate both the livelihoods of rural people and biodiversity conservation into land use decision-making and natural resource management planning (Naughton-Treves et al. 2006). Demographic trends and responses to market demands remain major drivers of long-term land use choices. Depending on the various combinations of agricultural intensification, extensification and migration (Zeller et al. 2000), patterns of land use changes and the potential for biodiversity conservation differ from place to place. Locally, trends are generally guided by concerns over livelihood security and social organization and are externally influenced by policy decisions (Lambin et al. 2003). In deforested environments, communities may restore landscapes to enhance their own livelihoods (for example, in Tanzania; Lamb and Gilmour 2003). However, conservation around large forest areas remains a complex challenge because in the short term, natural resources may seem sufficient for all. Unfortunately, our understanding of conservation in complex mosaics outside protected areas remains limited and local conservation efforts have rarely been truly supported in fragmented landscapes that are likely to support sizeable human populations. This paper explores the challenges research is facing in combining conservation and livelihood aims and identifies opportunities for researchers to improve this situation. We give an overview of the state of knowledge and experience in integrated research and development in order to promote this approach in linking biodiversity and livelihood issues. This paper draws on several types of information in addressing these questions. A systematic literature review supports several sections of this paper. Firstly, it gives an overview of the state of knowledge on applied conservation strategies. Secondly, it provides information on the relationships between integrated and disciplinary science. Finally, it provides the basis for discussion of integrative and transdisciplinary research and how transdisciplinary research results can be used in governance and management. The gaps that appeared in the literature review were addressed in semi-structured interviews. These discussed the engagement of various disciplines and project objectives in real-life situations. The final types of information presented are the results of a focused discussion by a large group of experienced professionals from the field, gathered in a workshop situation. This provides the basis for a proposed new approach that is aimed at addressing the key issues of integrating conservation and development and emphasizing development partnerships and transdisciplinary approaches.

15.2 Methods

The approach for the project called 'landscape mosaics' described in this paper was initiated at a workshop held in Bogor, Indonesia by 30 scientists from CIFOR (Center for International Forestry Research) and ICRAF (World Agroforestry Centre) (Pfund et al. 2006). The workshop underlined the need for multidisciplinary as well as applied research to catalyze the development of new thinking, approaches to the practice of biodiversity conservation and the sustainable use of multifunctional landscapes. This workshop was followed by a systematic literature review to gain an understanding of how researchers report in scientific journals of our domain focusing on integration of landscape ecological science into conservation and development activities. We concentrated on tropical landscapes and used very general ecological terms as an entry point for the search. In the Web of Science and CAB abstract databases, we searched with the following combination of words: (Landscape and tropic* and (biodiversity or conservation) and (patch or forest fragment* or matrix* or corridor or connectivity)) AND ((English) in LANGUAGE). In Web of Science, we searched within the items TS (topic) and TI (title) and in CAB, searched the whole article. In Google Scholar, we had to use a slightly different (but comparable) search, only in titles, given the large number of articles. The search string was 'all in title: landscape (AND) tropical (AND) biodiversity OR conservation OR patch OR forest OR fragment* OR matrix OR corridor OR agroforest'. To evaluate the level of integration of social and ecological disciplines as well as the expressed links between research and development initiatives in the resultant set of papers, we analyzed the articles on the basis of their abstracts. We decided that signs of multidisciplinarity were (i) a report from researchers that they had used assessment methods from different disciplines (or the article itself was a multidisciplinary literature review), (ii) researchers conducted the analyses in an integrated manner, i.e., they combined data from diverse disciplines or (iii) there was a substantial participatory aspect in an otherwise traditional biophysical survey. Only those articles that reported that the integration had started *from the beginning* or, that this, was planned to be an essential part of the study, were considered to be integrated. Thus, we did not accept as integrated articles those that later extrapolated strict ecological findings to broader contexts. In response to the lack of development-oriented journal articles emerging from the literature review, we conducted interviews with experienced practitioners of applied research. Appendix provides the list of questions used for these semi-structured interviews.

15.3 Context of Biodiversity Conservation at Landscape Level: From National Parks to Collaborative Management of Landscape Mosaics

The 20th Century was the era of the National Park. The realization that habitat loss is a major cause of extinction followed after the industrialization and widespread deforestation of western countries. After the creation of Yellowstone National Park in 1872, the park model was applied across the globe, including in tropical colonies, where it has sometimes been interpreted as a strategy of land appropriation or expropriation of hunting resources (Adams 1995). In many cases, areas selected for reservation were inhabited and therefore, local people were displaced and lost the means to meet their livelihood needs (Peluso 1993). By the end of the 20th Century, this strategy had been broadly criticized due to its insensitivity to human needs, while some also claimed that it was inherently ineffective and politically infeasible (Brandon and Wells 1994; Wells et al. 1999; Naughton-Treves et al. 2006). Since the 1970s, the industrial reforestation movement in tropical countries has faced similar criticism (Gerber and Steppacher 2007) so that neither conservation nor intensification were convincing as 'people friendly' forest management approaches. Following the more recent increased understanding of global biodiversity patterns and the consideration of multifunctionality at broader landscape levels, the reserves that had carried the conservation banner for the past 80 years were still considered necessary but no longer sufficient. New paradigms of protecting biodiversity beyond National Parks emerged. These generally tried to combine ecosystem protection, active management of natural resources and even restoration through an integrated and participatory approach to the planning and implementation of conservation within priority landscapes (Dudley and Aldrich 2007). For the past 30 years, habitat loss and fragmentation have largely been studied within the framework of two key theories: the theory of island biogeography (MacArthur and Wilson 1967) and the metapopulation concept (Levins 1969; Hanski and Ovaskainen 2000). However, applications or recommendations for conservation were not straightforward. Difficulties in conserving reserves as well as the need to better consider their spatial arrangement led to a reconsideration of approach. While the metapopulation concept is particularly suitable in highly fragmented landscapes and with habitat specialists, the corridor-patch-matrix model (Forman 1995) acknowledges the need to manage 'areas between' for effective conservation. Both have raised interest in investigating how characteristics and structure of the entire landscape affect the viability of populations.

Landscape ecology provides conservation biologists with tools to address issues such as how habitat loss and fragmentation affect population viability. Within landscape ecology, landscape metrics is a quantitative approach for spatial pattern analysis that has been used extensively (Turner 2005). In addition to this type of structural landscape assessment, new conservation approaches try to better reflect the dynamic nature of populations with local extinctions and colonization of new habitats (Hanski 1999; Siitonen 2003). At the same time, they aim to address the dynamic landscape itself with habitat patches changing due to factors such as human influence and natural succession changing the rates of isolation, attrition and edge–interior relations (Hanski 1999). Besides structural assessment, functional landscape ecological assessment has proven to be a valuable approach to better understand the processes at a landscape level (e.g., Clergue et al. 2005). Unfortunately, spatial landscape pattern analysis is often difficult to link with biophysical or socio-economic factors or processes (Imbernon and Branthomme 2001; Li and Wu 2004). Plans for landscape management are still often based on assumptions while more function and process-oriented understanding of landscapes is needed for theory to be truly integrated into planning (Chen and Saunders 2006). There have been ongoing debates over theories on spatial priorities of conservation design such as 'SLOSS' (single large or several small) and 'integrate-separate' conservation and production areas. However, the need to halt biodiversity loss, coupled with problematic social situations around many protected areas, has required management decisions to be made before the resolution of these debates. To overcome the practical limitations of conceptual models like the landscape continuum (McIntyre and Hobbs 1999) and the corridor-patch-matrix (Forman 1995), Lindenmayer and Franklin (2002; Lindenmayer et al. 2006) proposed, for instance, the use of five general principles to address biodiversity conservation in forest management: connectivity, landscape heterogeneity, stand structural complexity, integrity of aquatic systems and risk-spreading ('don't do the same thing everywhere'). Through such simple general guidelines, biodiversity values are actively integrated into operational management of timber production systems (Brown et al. 2006; Marjokorpi 2006). Practical experience has also sharpened the focus on tradeoffs: between species, land uses and between conservation and people. Conservation landscape approaches with more prominent elements of stakeholder engagement and negotiation have also been developed by international organizations for tropical countries where poverty alleviation is a major goal (see Loucks et al. 2004 for WWF; Brown et al. 2005 for IUCN). According to Ahern (2004), under the sustainability paradigm, sectoral planning is being replaced with multipurpose planning that explicitly acknowledges the integrated continuum of abiotic, biotic and cultural resource goals. Conservation organizations are coming closer to forestry institutions, bringing a wider (landscape) focus to forest issues (Mansourian et al. 2005). Thus, the general trend over the past three decades has been to widen the conservation focus from static reserves, such as national parks, to more dynamic reserves, such as large conservation landscapes (Bengtsson et al. 2003). Further, there has been a change to more active, adaptive and collaborative management for multiple values (Colfer 2005; Carey 2006).

15.4 Scientific and Applied Approaches to Conservation and Development

15.4.1 Integration of Conservation and Development Themes in Scientific Journal Articles

Since 1992, there has been a rapid increase in articles on the ecology of tropical landscapes. However, there is no observable trend toward integration between disciplines of research and development goals (Fig. 15.1). The number of integrated articles (44 of 375 reviewed) occurred in similar proportions of those articles reviewed from each database: Google Scholar 15% (4 of 37 articles), CAB 13% (16 of 122) and Web of Science 11% (24 of 216).

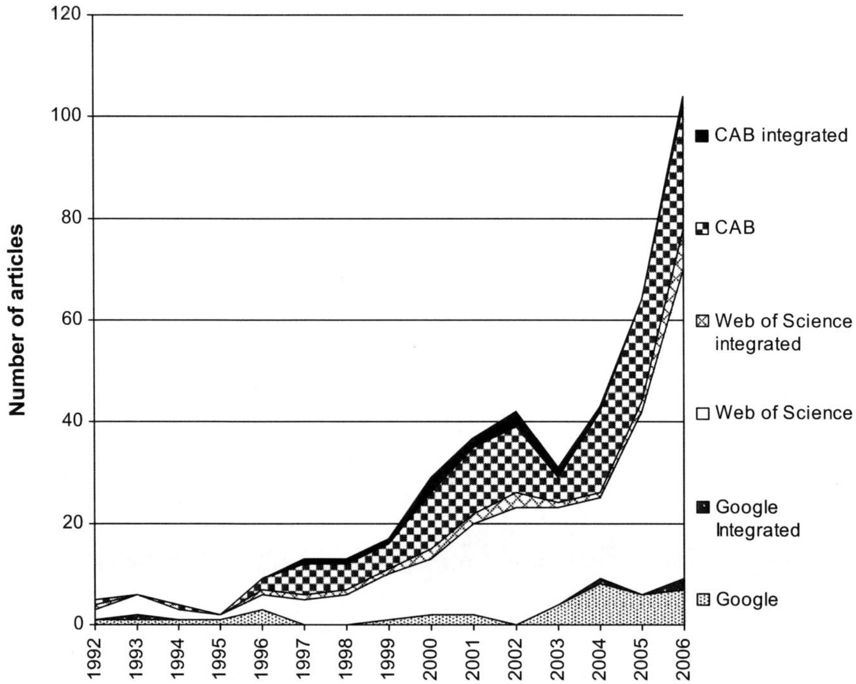

Fig. 15.1 Trend in the number of published articles that can be retrieved from three databases (CAB abstracts, Web of Science and Google Scholar) for journal articles published between 1992 and 2006 with a combination of landscape ecology and biodiversity conservation related keywords; 'integrated' articles include both conservation and development aspects

15.4.2 Issues Encountered in Practice – Lessons Learnt from Case Studies

The interviews conducted after the literature review yielded elements of practical lessons learned from nine sites. In the following section, we will focus on elements that contributed to the strategy of integrating conservation with development in these case studies. The source of statements presented in the text is indicated by the case study and interviewed experts' numbers: 1: Nepal, East Midhills (Laxman Joshi), 2: Indonesia, Sumba Island (Pete Wood and Syarif Indra), 3: Indonesia, Jambi Province (Tom Tomich), 4: Indonesia, Tanimbar Island (Yves Laumonier), 5: Indonesia, Central Sulawesi (Charles Palmer), 6: India, Western Ghats (Gladwin Joseph), 7: Madagascar, Menabe (Clémence Dirac and Lanto Andrimabelo), 8: Brazil, Zona da Mata (Irene Cardoso), 9: Ecuador, Loja (Els Bognetteau).

15.4.2.1 Knowledge on Landscape Patterns and Processes

Landscape mosaics reflect the past drivers of change, history of land uses and accessibility of landscapes. Access to markets, commonly via roads and rivers, influences fragmentation and potential overexploitation of marketable species (3, 4). Interviewed experts agreed that in undertaking conservation projects, past trends must be understood and action tailored to actual threats and opportunities. They acknowledged that biodiversity resource assessments and threat analysis at landscape levels are necessary. Nevertheless, most experts also recognized the insufficient financial resources to fully engage with ecological theory and research while promoting conservation (1, 2, 3, 4, 8). Where social, capital and local readiness to negotiate land use agreements exist, there may be prospects for natural resource-based enterprises. Where protection is urgent and land use planners are interested in collaboration, 'action research' intended to support multi-stakeholder negotiations on land access may be more relevant initiatives for conservation than isolated scientific studies (3, 4).

Topography and relief play a central role in influencing spatial patterns and it is common that in mountainous areas such as the Western Ghats, Nepal, Jambi and Central Sulawesi, well-connected forests occur at higher elevations, whereas lower elevation forests are in isolated fragments (1, 3, 5, 6). Interestingly, Jambi's formerly most intensively used areas, the riverine agroforests, are now important in providing connectivity between patches of forest and it seems that at least long-distance dispersing plants benefit from these 'stepping stones' (3). Studies of bird life and soil macrofauna in Jambi and invertebrates in Nepal show that as intermediate land types, agroforests are not perfect systems but provide important habitats for many organisms (1, 3). Unfortunately, it is still difficult to know if they will be sufficient for the survival of forest interior species in the long term. In India, ATREE has been helping local Non-Government Organizations (NGOs) to identify where to buy land to maximize critical forest connections (6). The installation of corridors for biodiversity conservation is supported by scientific results in some circumstances (Damschen et al. 2006). However, they are not often implemented in the tropics, perhaps because of the uncertainties in designing them and handling the competition with other land uses. Another reason for the limited support of restoration or conservation of fragments and corridors might be the complexity of comparing different landscape situations and various efficiencies of corridor types in assisting the movement of seeds, wildlife or genes. To enable comparisons between landscapes in terms of performance and to facilitate spatial planning, there is a need for a clearer tropical landscape typology (3).

15.4.2.2 Informed and Capable Actors

Better understanding of the various values and motivations to use or conserve biodiversity and environmental services is required. Also, better communication: Sheil et al. (2006) argue that the preferences and perceptions of local stakeholders often remain hidden when conventional biodiversity surveys are conducted and that misunderstandings may lead to irrelevant or short-term decisions. Surveys from Lore Lindu suggest that biodiversity is not very relevant *per se* to local people, especially for those who are poor (5), but that they use environmental services that can rely on biodiversity conservation. In Madagascar, biodiversity resource assessments, even at the landscape or ecoregional level, only partially represent the relevant conservation issues. Further, they should always be linked with solid assessments of the villagers' needs (7). A utilitarian view of biodiversity seems to be common among all case studies. Local people may have detailed knowledge of fodder systems including animal preferences and seasonality as well as tree-crop interactions, as in the Nepal case, or a thorough understanding of the species they harvest, as in Sumba and Tanimbar (1, 2, 4). Yet despite this traditional knowledge, overharvesting may still be a problem. Species such as dugong are locally extinct even in Tanimbar, which is generally recognized as having a 'conservationist' population due to their communal ban of forest access by logging companies (4). In some cases, an unsustainable use of resources can be explained by a lack of internal cohesion, caused, for example, by migration processes (3, 7). In Lore Lindu, migrants coming from recent resettlements lack a strong attachment to the land and local rules (5). In the Central Menabe case, internal movement has led to mixed populations in the villages: natives and migrants often do not share the same understanding of the value of the biodiversity (7, Cabalzar 1996).

In Sumba, BirdLife focused on endemic birds and attempted to raise local pride and responsibility regarding these species. The traditionally strong respect for law in that society has assisted messages about the illegality of hunting the endemic Sulphur-crested Cockatoo subspecies (*Cacatua sulphurea citrinocristata*). BirdLife has also linked the conservation of cockatoos to broader issues of forest ecology and preservation of water sources to give it more immediate appeal to the community (2). In Tanimbar, the unusual success of the awareness program was explained by the many months spent by the socialization team, going between villages to make sure that local people understood what the project was about and its relevance to them (4). This introductory process enhanced the ongoing communication between all of the relevant stakeholders, giving, for the first time an opportunity for local people to have their voices heard. Awareness-raising activities may also be directed at managers and Government decision-makers; Alternatives to Slash and Burn (ASB) program's research in Jambi showing that intermediate land uses may be rich as habitats has had an influence on official perceptions, adding recognition of the value of traditionally managed systems and giving more options for improving biodiversity values at the landscape level (3).

Local empowerment: Poor rural communities may be insufficiently empowered to negotiate with incoming stakeholders such as resource extraction companies (3, 6). There is a need for public advocacy to provide communities with basic information about their options and how arrangements with outside players will work. Similarly, in the Western Ghats, communities need assistance in negotiation. Indeed, the resource companies there are also using NGO consultants for negotiations, as they are not confident of breaching the social divide (6). For many reasons, local people are not in a position to negotiate even with State Forest Service representatives, as in Madagascar. NGOs or private negotiators can play a very positive role but they must be skilled not only in technical matters but also in communication (7). In contrast, in the northern Nepal case, local people were probably empowered to negotiate, as local land tenure is clear (1). In Sumba, BirdLife had an intermediary role in negotiating National Park boundaries and resource use by the community with government agencies, while in Tanimbar, locals were incidentally helped to overcome their reticence to communicate their needs to government agencies (2, 3). Thus, while the effectiveness of the negotiated arrangements varies, it seems that many communities have insufficient capacities to negotiate, even if they have some part in the decision-making processes. In many of the cases explored, NGOs do seem to have a role as negotiators or intermediaries.

Sufficient capacity of NGOs: While aiming to raise capacity of local communities, in some cases, the NGOs themselves have insufficient resources for the tasks at hand. There is skepticism about the effectiveness of NGOs operating around Lore Lindu, where over 30 organizations are working with poor inter-coordination (5). While there has been widespread facilitation of agreements between the community and National Park, in some villages, many local people remain unaware of these agreements. NGOs have tended to be paternalistic and assume that local people needed help (5). It is not clear if their involvement is leading to empowerment when there is neither sufficient financial backing for projects nor encouragement of local initiatives. Some organizations present are not equipped with the skill sets required to integrate societal objectives into conservation schemes (5). Perhaps this shows symptoms of the growing requirement for NGOs to be 'all things to all people', when this is in fact beyond their scope. In a transdisciplinary research context, addressing this problem will need investments in capacity-building and adequate partnerships. NGOs and private negotiators should support the rural communities in implementing the agreements for years after initial negotiation (7).

15.4.2.3 Rules and Incentives

Adequate institutions: The institutional and legal context of resource use in landscape mosaics frequently involves both customary and state rules; sometimes they are complementary, but they are frequently in conflict. In many places, such as in Indonesia, national land legislation is derived and still strongly influenced by colonial laws that sought to appropriate land and exclude local users (3, 4). These rules are often related to land rights and extraction of resources from 'public lands'. Customary rules are common throughout Thailand, the Philippines, Indonesia and the Pacific. They work effectively while everyone is using the same set of rules (3). Many indigenous groups have arrangements to manage forest resources such as through regulations related to religious values (Wadley et al. 2004). As an example, sacred forests were used to conserve forest around vulnerable springs and rivers in dry areas in Tanimbar and Sumba (2, 4). The commitment of people to maintaining natural values may depend on sanctions based either on spiritual beliefs or strong community penalties to be paid in cash or labor (such as in Tanimbar, 4). In rural India, these 'assets' of traditional regulatory systems should be valued and used as local communities and are not ready for national state policies

(6). Multiple interviewees suggested that traditional regulations should be integrated more frequently in modern resource regulation (1, 4, 7). Yet, the integration of traditional and state regulations may not be simple. One problem is the complexity and possible incompatibility of the rule systems that have been developed by disparate kingdoms and the heterogeneity of their contexts (6). Participatory mechanisms are thus needed. The Tanimbar project has been re-negotiating land use designation to reduce further conflict between conservation and development interests (4). The careful involvement of all stakeholders in the process and the lack of current conflict have enabled successful negotiations. By focusing on participatory planning, the Podocarpus program in Ecuador has helped to open dialogue between local and national governments as well as indigenous farmer groups, commercial forest users and NGOs (9). There, co-management committees were successful in negotiating national reserve status for a lowland forest area threatened by gold mining and logging.

Rewards for biodiversity conservation: If conservation in developing countries is not properly resourced with sufficient incentives for locals, it will fail. The national park system may not be applicable in Indonesia and different approaches to conservation are needed (5). One form of incentive has emerged through new markets for cultural ecotourism, although these have often not met expectations. Nevertheless, other types of incentives may be created. Co-management efforts can lead to incentives such as in the Podocarpus program under which colonists are granted legal rights to land if forest cover is maintained and they allow hunting by indigenous groups (9). The translation of local park management plans into more clear and secure land access rights, access to safe drinking water, sustained yields of previously threatened tree products, more sustainable land use practices and ecotourism revenues make local people more aware of the benefits of conservation. In the selected case studies, rewarding local people directly for biodiversity conservation is either in an exploratory phase or seems difficult (3, 5, 7, 8). In Lore Lindu, there is some discussion about direct payments among local NGOs, inspired by reportedly successful schemes already established in Central America (5). Incentives for growing cocoa under shade, thus increasing its habitat value, are also being explored. However, in Zona da Mata, 'green coffee' marketing has been problematic due to a lack of market linkages, while in Jambi, marketing 'green rubber' from jungle rubber agroforests is difficult without official recognition of this land use (8). ATREE is exploring research results in the Western Ghats, with the hope of convincing policy-makers to use State taxes to pay for goods and services that compensate land managers for their environmental services (6). In the context of such rewards, measurable indicators for biodiversity service provision need further development. Currently, transaction costs, especially for monitoring, are frequently very high (3, 7). In Madagascar, the question of how incentives could work in the field of biodiversity conservation is quite new and thus related experiences are very few (7). The prospect of communities receiving payments for carbon sequestration by forests seems promising. However, it remains to be seen to what extent the rural poor will be able to take advantage of these schemes.

15.4.3 Realities of the Integration of Science in Conservation and Development Activities

Science has to adapt to reality: Project managers and project types are changing regularly – every five to ten years. This turnover can create a mismatch between scientific paradigm(s) and practice. As an example, the Integrated Conservation and Development Project (ICDP) was a dominant paradigm or approach before being replaced by the concept of payments for environmental services (Ferraro and Kiss 2002; van Noordwijk 2005). Nowadays, many specialists have become more critical about the potential of such markets (on this debate, see Karsenty 2004; McCauley 2006; Reid's response 2006; Wunder's response 2006). Despite such temporal variations, some common ideas, and sometimes myths, may last. For instance, the 'land use intensification hypothesis' that agricultural intensification will save land for conservation still remains an implicit paradigm in many cases without always being tested. Some theoretical ideas such as the concept of connectivity and corridors as part of mosaic landscapes can excite the imagination of people more easily than others. Ecological corridors are relatively easy to understand and they are already implemented in many parts of the world (for example, by the European ecological network and European Agrienvironmental program). There is no one overriding reason to discourage integrated and dynamic strategies encompassing the role of corridors, yet they are probably not always the most cost-effective way for conservation. Among other potential pitfalls, corridors can also promote edge effects, the filtering of communities, invasions and negative genetic impacts by reconnecting isolated fragments (Hilty et al. 2006). In addition, the concept is not as simple as is often advocated: corridor and matrix each mean a different thing for each organism and interpretations are scale-dependent. Thus, it is difficult to have a meaningful discussion on the overall value of corridors for biodiversity. Difficult decisions and trade-offs are involved in selecting and securing the best conservation investments. This arises from a general lack of knowledge on plant and animal movement within landscapes, concurrent uses of land, limited conservation funding and range of landowners and political entities with whom to negotiate (Morrison and Reynolds 2006). The discrepancy between science and practice, especially concerning biodiversity, was clear in Tanimbar (4). There, the concepts of rewards for environmental services, the principle of adaptive co-management and the influences of multiple scales were not addressed in existing land use plans. Even precautionary principles were difficult to advocate without local examples. This is a challenge for conservation in the developing world where it is critical to demonstrate the importance of biodiversity before attempting to incorporate science into projects involving local people (4).

The task of science is, on the one hand, to develop new concepts in order to tackle increasingly complex challenges and on the other hand, to meet societal needs. This second role is particularly important in tropical developing countries. Addressing social needs may not always need the development of new concepts and theories – it seems that a lack of ability to implement is one of the barriers we face. Instead, it requires sufficient accuracy in the interpretation of the causes and catalysts of social issues. In some cases, research can be seen to have evolved not only technically but to also incorporate better knowledge of 'real-life' social issues. In Madagascar, for example, during the last 50 years, the knowledge about ecology, silviculture, management and exploitation of the dry forest system has increased substantially (Sorg 2006). This can be considered a result of the strong link between technology and applied research. More recently, a better understanding of the needs of the local population has grown, together with increasing international awareness of forest problems and political pressure to resolve these problems. Together with increasing pressure from the international conservation movements, this has led to the reconsideration of the research priorities for the dry forest landscape. Today, local research has three objectives:

- Review and promote the dissemination of the accumulated knowledge with respect to its general application in the forest but also outside the natural forest area (agroforestry, single trees outside the forest, secondary formations).
- Understand the people-forest interface at the level of the villages surrounding the forests, including use of non-wood forest products and potential for compensation for ecological services.
- Enhance multifunctional management of large forested landscapes and their surroundings to meet the different needs of the people.

Thus, the drier technical and theoretical questions that once drove forest science are in this case giving way to an approach more in tune with the local environment, including human populations (8). Donors are another major influence on the way that science is practiced. These have their own motives and driving forces and may strongly affect research approaches, especially in development and conservation projects. Most of the interviewed experts commented on the general disinterest or even discouragement by donors to integrate scientific theories and project activities. Real choices for the design and management of reserves are limited and often, many of the critical decisions have already been taken. Thus, 'ideal' approaches generated from new theoretical models are unlikely to be very relevant (2). Donors do not always demand a sound scientific basis and integration between theories and planned activities, but want to be assured that outputs will be concrete and create immediate recommendations for development (2, 4).

Institutionalizing science: Conservation circles have drifted far from research and sometimes fall to surprising levels of simplification while trying to influence political will (4). On Sumba Island, government and communities did not know about and were not influenced by the cutting-edge theories (2). Since these communities are the ultimate decision-makers in management there, BirdLife acting as a facilitator concentrates on the most influential factors for these managers. At an even more basic level, valuable research data and results are often not used because of a lack of time to interpret or scan the latest literature. Data is even lost because of poor governance and short rotation periods of the staff. There is a need for some NGOs to act as intermediaries between scientists and practitioners to aid in the transfer of science to policy and action. These so-called boundary organizations (Cash et al. 2003) could try to facilitate in a way that might be uncomfortable for traditionally trained scientists. The need for change concerning research and development relationships is also related to scales. Central government officers closely collaborate with scientists in Indonesia, while in the districts, officers lack trust in science (4). This might have been caused partly by bad past experiences with incompetent, short-term advisors. In the case of Tanimbar, a remote area, consultants have allegedly been opportunistic in earning quick money without delivering reports with sound data (4).

Improved monitoring: The Kerinci Seblat National Park in Sumatra was meant to have a solid scientific basis and considered as a showcase ICDP before being widely described as a failure (Sanjayan et al. 1997; Wells et al. 1999). It seems that one of the biggest sources of error was insufficient consideration of external driving forces (for example. international coffee markets and their influence on land use decisions). Unfortunately, only a few analyses have investigated this (4). Of course, time and funding set the limits. Issues such as the success of corridors need to be explored by long-term monitoring. This has rarely taken place. In the Podocarpus project, research still has to establish the importance of particular forest zones for the Páramo bear population (9). An adaptive management approach might be an acceptable option; however, this still requires some baseline information and project monitoring. In reality, this often starts too late or finishes too early to correctly evaluate the efficiency of the initiatives. Even the European Agrienvironmental program has had corridor schemes operating for several years without having corresponding baseline studies and monitoring systems (Kleijn and Sutherland 2003). While optimal solutions may indeed involve combinations of approaches, the contribution of each component may be difficult to isolate and monitor (for example, where there is simultaneous use of large protected areas and contiguous community forestry agreements at the local level). Considering the ongoing experience of Madagascar in forest management, monitoring is a crucial issue that is still not adequately resolved, especially at the landscape level (7, see also Muttenzer 2006).

15.5 Research for Informed Governance and Management of Tropical Forest Landscapes

15.5.1 Integrative or Transdisciplinary Research on Landscape Management in Developing Countries

The concept of transdisciplinarity was developed in the 1970s (Jantsch 1970; Piaget 1972) before the principle of sustainable development (Brundtland 1987) further encouraged integrative approaches. However, it is still being studied at a theoretical level (Naveh 2001; Klein 2004; Nicolescu 2005). The approach combines academic disciplines, takes into account ethical values, implies the participation of various stakeholders, academic or not, and is recognized as useful for landscape-level approaches (Tress et al. 2001). This 'action research' concept has been suggested to address complex societal problems (Horlick-Jones and Sime 2004) as they occur in developing countries. While transdisciplinarity and its related systemic approaches are convincing in addressing sustainable development issues, they place very high demands on both research and development organizations (Tress and Tress 2001). Many authors argue that a first lesson is not to get lost in (or too attracted by) the diversity, complexity and variability of socio-ecological systems (Horlick-Jones and Sime 2004; Hadorn et al. 2006). Other known problems relate to the difficulty in overcoming disciplinary boundaries and related prejudices of different participants (Daily and Ehrlich 1999; Opdam et al. 2002). In practice, a clear definition of the role of each actor is a key element to avoid confusion in a team (Sillitoe 2004) as well as in a network of institutions. Another scientific difficulty lies in the numerous ways participation can be defined if one wants to generalize experiences, which are, by essence, very context-dependent (see the interesting debates on case study generalization, for example, in Flyvbjerg 2006).

The Consultative Group on International Agricultural Research's (CGIAR) research mission is to achieve sustainable food security and reduce poverty in developing countries through scientific research and research-related activities. The approach taken by the centers has evolved in parallel with the transdisciplinary movement as it focused on agricultural productivity in the 1960s and moved toward 'action research' or the Integrated Natural Resource Management framework for sustainability (Campbell and Sayer 2003; Frost et al. 2006). This framework highlights four sets of interrelated linkages between: (1) production and conservation, (2) spatial scales, (3) time scales and (4) research and adoption of results (Harwood and Kassam 2003). In developing countries, the integration of conservation into other priorities, especially economic development, should be a central preoccupation. According to Globescan (2004), the majority of people from developing countries feel that individuals can do little against species loss. These populations must often give more importance to economics and less to conservation. The implication for land use planning is that local people are to be included in natural resource management to avoid the failure or sabotage of measures due to their inability to meet local needs. Further, while poverty alleviation usually depends on local access to resources, state services generally do not have the means to monitor resources in remote areas. Thus, research for biodiversity conservation in tropical landscapes has to be designed to take into account local poverty concerns, local needs and rights to self-determination.

15.5.2 From Research Findings to Informed Governance and Collaborative Management

Within the transdisciplinary framework, scientific disciplines are not the only bodies facing difficulties in reconciling their differences. Conservation and development professional circles have faced similar difficulties. While conservationists are still struggling to understand the dynamics of the landscapes patterns and their driving forces (Rouget et al. 2006), people engaged in human development support might argue that complex systems can be self-organizing and that the need to build adaptive capacity must be prioritized. This ability to adapt may be particularly emphasized given the present concerns regarding the impacts of climate change. At the same time, scientists might question the effectiveness of past efforts when the outcome monitoring of development projects has often been neglected. In terms of priority-setting and monitoring, it seems clear that closer collaborations would benefit both sides.

The link between a plan and its expected impact is crucial for researchers as well as for development practitioners. What is sometimes more complicated is how to transfer information from one 'world' to the other. In reality, policy-makers are unlikely to have a natural interest in research findings. If policy changes are a goal of a scientific team, suitable dissemination of research results must be planned from the beginning. A similar communication and efficiency gap appears between policy and site management. It is acknowledged by many governance specialists that some policies did not have the expected influences on the ground. In the context of landscape management, the key questions are:

- What type of (integrated or aggregated) information can influence policy-makers in making choices that integrate biodiversity conservation into land use planning, rather than opposing it with development?
- What type of policy changes or incentives will influence natural resource managers so that they would change their strategies in the short term, in the interests of biodiversity conservation in the long term?
- What kind of mechanisms will ensure that the new system will be able to correct false assumptions or react to new situations? (In other words, will it be adaptive?)

In this real-life context, domains such as ethics and psychology (Saunders et al. 2006) might have to be considered when planning research. As suggested by our case studies, regular links between 'knowledge' and 'action' are, to date, often poor. To fill this gap, boundary organizations that can form bridges for selected and relevant information are currently receiving attention (Cash et al. 2003). These 'boundary organizations' are those that are able to understand the values of both scientific and non-scientific knowledge to allow a beneficial exchange between the two. These may include quasi-government organizations, Non-Government Organizations or institutions that combine research and development aims and activities. In particular, institutions such as agricultural extension services, which mediate between the needs and interests of local farmers on one hand and the work of researchers on the other. Research that may inspire change could be directed to policy-makers or more directly to managers or to both.

Figure 15.2 illustrates different combinations of 'science' and 'action'. The lower categories of the y axis would include articles done essentially for the sake of pure science or academic ambition as they would not aim to reach decision-makers, managers or implementers. Box A represents specific technical research used for development. For example, research typical of findings provided by scientists about a Non-Timber Forest Product processing technology to the company managers. From a multidisciplinary perspective, the reports that Millennium Ecosystem Assessment

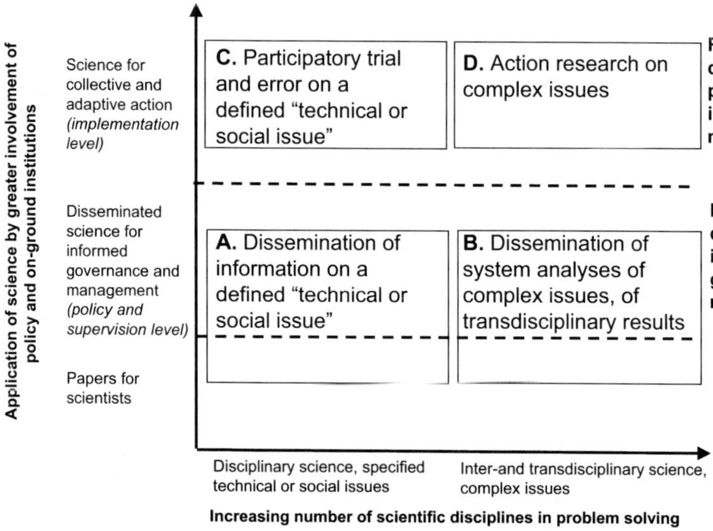

Fig. 15.2 (modified from van Noordwijk et al. 2006): Interface between knowledge and action along a qualitative scale. Increasing integration of scientific disciplines in problem solving is along the horizontal axis, while the involvement of institutions such as NGOs increases on the vertical axis

(WRI 2005) provided to high-level decision-makers would illustrate the type of research of Box B. If one goes further in the willingness to collaborate directly with operational partners, participatory research on mahogany planting might be an example of Box C, while an action research project focusing on negotiation support systems, aiming at rewarding upland managers for environmental services (RUPES, van Noordwijk 2005), could fall in Box D.

If we consider that biodiversity conservation at the landscape level is urgent, institutional changes in the context of research for conservation and development are needed in the short term. New ways of communicating and partnering are required – between disciplines in research as well as between science, policy and active, adaptive management. In reality, integration of multiple forms of knowledge and multiple partially independent decision-makers acting on the same landscape and on its drivers of changes will be challenging. Thus, the bridges between science, policy and implementation need to be built early, well-designed and empowered in adaptive and collaborative management systems (Cash et al. 2003; Tomich et al. 2006).

15.5.3 An Example of a New Research Approach

This section describes a recent example of research design that is being developed for enhanced tropical landscape management. In 2006, CIFOR and ICRAF launched a 'Biodiversity Platform' during a workshop of over 30 scientists from various disciplines (ecological and social sciences, focus on environmental services, livelihoods and governance) interested in development and conservation. The Platform aims to identify principles and practices that promote conservation, sustainable use and equitable sharing of biodiversity goods and services in landscape mosaics. Participants debated the scientific (research gaps) and development (impact pathways) aspects of conservation and production in mosaic landscapes. Participatory action research in multiple sites was accepted as the general approach but the risk of reduced scientific quality was also highlighted. Ideally, the initiative will allow the collection of a large number of site experiences and collaboration between research institutions over time to advance knowledge on fragmented landscape mosaics. Empirical evidence demonstrates that forest fragments and intermediate-intensity land uses such as agroforestry systems provide important biodiversity conservation services that complement those of dedicated reserves (Forman 1995; Lindenmayer and Franklin 2002). The platform will thus focus on intermediate intensity land uses: remnant, managed and secondary forests, agroforests and plantations in selected landscapes. A combined approach of hypothesis-driven and participatory action research is proposed to both provide international public goods and support negotiations for improved and adaptive landscape management.

15.5.3.1 Hypotheses, Common Analytical Framework and Flexibility

To allow cross-landscape comparisons and to deliver internationally applicable results, four main assumptions were defined: (1) external conservation values and local values of biodiversity goods and services of natural and semi-natural fragments vary non-linearly in time depending on the landscape patterns and overall intensity of use (possible trajectories are suggested for various landscape types), (2) timely empowerment of local populations through integration of scientific and local knowledge will mitigate biodiversity loss and maintain or increase livelihood security, (3) reward mechanisms will only work where external value exceeds local values of land use systems and local regulations based on local environmental services can constrain individual decisions, if external commitment is serious and follows up on promises made, and (4) overall landscape sustainability is enhanced if public policies are informed by and allow for customary or local rules and practices. These assumptions provide a preliminary common thematic framework for participating scientists from different disciplines and sites. For each, a 'thematic' group of researchers will gather information and experience. Some of the assumptions (2, 4) do not allow direct testing but the thematic groups will gather information that tends to confirm or cast doubt on the assumption and will form common, narrower research questions.

15.5.3.2 Assessment of Landscape- and Local Level Facts and Values

Based on the hypotheses, spatial analyses and a set of common aggregated data will be standardized across the Platform's landscapes (see matrices and methods developed by other site networks in Ostrom 1995; Tomich et al. 1998; Colfer 2005). A scientific analysis of landscape modifications over time and of the driving forces of changes will drive discussion about general threats to biodiversity and their causes according to different stakeholders. Such a reconnaissance phase can give a good overview of the understanding and perception of general trends, build confidence with partners and provide hints to stratify landscapes and select plots for further local surveys. Local livelihood perspectives are intentionally emphasized in the approach but perspectives of external stakeholders in biodiversity are also taken into account. Livelihood needs will be surveyed and locally appropriate mechanisms that may lead to adaptive and collaborative landscape management identified. Three foci (local people, external stakeholders and scientists) will guide field biodiversity surveys. Biodiversity products are found important by the *local population* (e.g. timber, non timber forest products, and game), species or habitats have special existence values for *conservationists* and finally, data such as tree diversity and their linkages to dispersal mode and life history will interest *scientists* for cross-site standardized comparisons.

15.5.3.3 Facilitation of Collective Planning

Based on this multidisciplinary landscape analysis, tools will be developed for collective planning through an open discussion of future management scenarios. To be able to project various landscape developments, CIFOR and ICRAF have developed and currently use participatory scenario modeling (Wollenberg et al. 2000; Vanclay et al. 2003; Purnomo et al. 2004). This allows planning and discussion of management options with the community and other stakeholders. Dynamic spatial models and qualitative soft and hard systems approaches along with multi-agent modeling provide a framework. This will allow for participatory analysis of stakeholders' (or agents') interactions and facilitates problem solving and decision making (Purnomo and Guizol 2006). Once the various stakeholders' perceptions and options are known, negotiation support tools, sometimes with games, may facilitate the search for compromises between groups, and when needed, the discussion of incentives (van Noordwijk 2001; Hartanto 2003).

15.5.3.4 Early Participation and Monitoring

The potential for successfully implementing the project will obviously depend on the uptake of ideas by local, regional and national institutions involved in land use planning and management. To ensure this, the potential users should be identified early and involved in both study design and implementation. As part of this joint initiative, the institutional design and the link to ongoing development or conservation initiatives will be carefully analyzed. Partnerships are sought with stakeholders who may be posing biodiversity threats as well as those who currently offer benefits to conservation.

In order to analyze the efficiency of the approach as well as its effects, a systematic monitoring of the landscape mosaics project, implementation and outcomes will be set up from the beginning. In the short term, this should allow discussion of transaction costs linked to complex partnerships and transdisciplinary settings of the project. In the long term, it should allow discussion about real-life outcomes and facilitate regular monitoring of landscape management activities performed by local populations.

Summarized, the proposed steps to be taken by the Platform for its transdisciplinary research are:

- Creation of institutional partnerships, identification and involvement of output users
- Landscape definition and spatial analysis of land use changes and their drivers
- Collection of data of local and external biodiversity relevance and biodiversity indicators linked to understanding the degradation processes of habitats
- Scenario development and possible use of models simulating stakeholders' decisions
- Support to negotiations through partnerships and promotion of long-term collaborations
- Regular monitoring and evaluation of progress and outcomes for adaptive management

15.6 Conclusion

Natural and social processes change rapidly in tropical contexts and the sometimes implicit assumption that ecosystem responses to human use are linear has been revised (Folke et al. 2002). Social and political dynamics, tenure uncertainties and financial constraints on land management make the field application of conservation and landscape ecology theories, at best, uncertain (Wu and Hobbs 2002; Sayer and Campbell 2003). In real life, planning and implementation, practicality, flexibility and potential for adaptation may be more important factors to sustainably integrated conservation and development than achieving optimal landscape ecology (e.g., see Margules and Pressey 2000; Brown et al. 2006; Rouget et al. 2006). Yet, in terms of biodiversity conservation, research in tropical forest landscapes has tremendous gaps to address in order to better understand the potential of managed semi-natural landscape patches and of sustainable use of wild species in landscape mosaics. The real issues are how to prioritize research and how to conduct it. A possible standpoint is to analyze what issues currently influence the potential of research and development outcomes. In developing contexts, outcomes will be greatly dependent on:

- The communication channels and boundary organizations that will enable and facilitate fair exchanges between local and administrative, market or conservation actors,
- The understanding of the different development and conservation trajectories in order to project realistic scenarios,
- The availability of understandable criteria and indicators related to prioritized conservation objectives,
- The agreements, commitments and incentives that can be decided among key stakeholders and that can be realistically enforced and monitored and

- The local rights and capacities to manage natural resources as well as the possibilities of involving other stakeholders fairly through community-based or co-management schemes to implement, monitor and adapt the agreements.

Biodiversity conservation must be promoted according to the development contexts and research must be able to provide 'bundled' and understandable recommendations to reach key actors such as decision-makers, extension services and local managers. Success factors will rely on the capacity to interest people, induce action in the short term and launch long-term adaptive collaborative mechanisms. Be it for poverty alleviation or for biodiversity conservation, rural people must generally be better supported to improve or change management practices. The way to achieve more support is still greatly debated, especially on the topic of payments for environmental services. Generally, if landscape ecology research is designed according to the local contexts (including livelihood needs, stakeholders' perceptions at various levels and institutional systems), its findings have good chances to be considered relevant and the potential to encourage new commitments and supporters.

In tropical landscapes, linkages between disciplines and a research-development continuum must be ensured to effectively combat poverty and environmental degradation. Moreover, linkages between science and policy must be realized and new knowledge must presented in a way that will influence decision-makers. Cash et al. (2003) distinguish credibility, salience or legitimacy as principles to ensure impacts of research. Scientists can thus combine the search for local impacts (legitimacy and salience) with cross-site analyses (broadness of application and credibility) to extrapolate results. However, in the field, scientists face challenges in integrating disciplines and involving multiple actors with differing values. Typically, they cannot act alone; success requires clear and strong partnerships which may need facilitation by third parties. For long-term success of such complex research approaches, scientists must go beyond academic norms. Currently, incentives for transdisciplinarity are rare in a system which emphasizes scientific paper production. In the field, the goodwill and openness of many actors are needed for tangible improvements and acceptable compromises between conservation, private sector, Government and local interests. Nonetheless, the provision of biodiversity-relevant information and efficient planning tools to key players may help to facilitate communication and achieve better landscape-level outcomes in the tropics.

Appendix : List of Questions Used for the Semi-structured Interviews

Landscape Mosaics

1. What were the landscape elements where you worked?
2. What was the dominant land use – spatially, economically?
3. How much 'natural' vegetation remained? Where and why?
4. How connected was it?

5. How important do you think this connection was for organism movement and reproduction? (Or how limiting?)
6. How difficult was it to maintain connectivity within the production systems present?
7. In your opinion, in the landscapes where you worked would conservation be better facilitated by an integrated or separated landscape mosaic?
8. What was the history of land use in the area?
9. What was the apparent role of intermediate land uses (e.g. agroforests, managed forests) in conservation?
10. For what organisms? What were the limitations? Were the organisms entire needs fulfilled within the intermediate land uses? How dependent was the patch species composition (inc. Fauna) on nearby forest? How sustainable is the ecology of the systems? What options were present to increase the biodiversity value of the systems? Were these acted upon? Was it successful? What were the limitations on biodiversity improvement?

Methods

1. What are your experiences of using mixed (multidisciplinary) datasets?
2. How were different data types combined?
3. How did you come to terms with issues of scaling and mixed units?
4. Did your project involve action research (as opposed to pure research)? Research and development?

Rewards

1. What are your experiences of the use/attempted use of reward mechanisms for environmental services?
2. Did these include biodiversity as a consideration?
3. How was it measured?
4. What reward type was used?
5. How was adherence to the system monitored?
6. How successful do you consider the program?
7. What were the difficulties or limitations?
8. What advice would you offer?
9. Was biodiversity the only service involved or was it bundled with others? e.g. Water or carbon?
10. What potential and limitations do you see for binding of service rewards?

Livelihoods and ecological knowledge

1. What were the main sources of income and products?
2. Did the biodiversity products (non timber forest products, agroforestry products etc) produced act as a safety net?
3. Were people able to meet their own needs from local sources?

4. Was market access sufficient for people to benefit from selling their biodiversity products?
5. Was there potential to improve production processing and commercialisation of biodiversity products?
6. Have you recorded and analysed traditional ecological knowledge related to biodiversity products?
7. How? How did you use this knowledge? Was it compared with scientific knowledge? If so, were they consistent? –> Lessons learned (win-win or lose-less), recommendations?

Participation

1. How were local people involved in the research – as groups, individuals and community?
2. Had they (communities) been involved in resource negotiations with outside stakeholders? Were you involved in negotiations? Who else participated? What was the result of this?
3. Were the local community sufficiently empowered to negotiate?
4. How did you involve difficult players? At what stage?
5. What are the perceptions of local people towards conservation? Had they had previous experiences of dealing with conservation agencies?
6. Lessons learned about 'efficient' or 'difficult' partners? Recommendations?

Governance

1. Did the community have local rules for use of biological resources and sharing of benefits? (Access, management rights)
2. Were those effective?
3. In what parts of the landscape did they operate?
4. How were they enforced?
5. Were they consistent with rules of outside agencies?
6. Were they recognized by outside agencies?
7. If not, was there conflict over this?
8. What state agencies were involved in biodiversity use and conservation?
9. What other players?
10. What were the private and common resources?
11. What lessons regarding effective and ineffective governance can be learned from this landscape?
12. What recommendations?

Combining practice and scientific theories on conservation ecology

1. Did the project have strong basis in scientific theory?
2. What theories gave the basis for the hypotheses?
3. Did donors or funding agencies of the project demand both sound scientific basis on conservation part (basis in ecology) and clear development outcomes?

References

Adams WM (1995) Green development theory? Environmentalism and sustainable development. In: Crush J (ed) Power of development, London, Routledge

Ahern J (2004) Theories, methods and strategies for sustainable land use planning. In: Tress B, Tress G, Fry G et al (eds) Proceedings of the Frontis workshop from landscape research to landscape planning: Aspects of integration, education and application, Wageningen

Bengtsson J, Angelstam P, Elmqvist T et al (2003) Reserves, resilience and dynamic landscapes. Ambio 32(6):390–396

Brandon KE, Wells M (1994) Planning for people and parks: design dilemmas. World Dev 20: 557–570

Brown J, Mitchell N, Beresford M (eds) (2005) The protected landscape approach – linking nature, culture and community. IUCN, The World Conservation Union. Gland, Cambridge

Brown S, Palola E, Lorenzo M (2006) The possibility of plantations: Integrating ecological forestry into plantation systems. National Wildlife Federation. Available via Plantations 2020. http://www.plantations2020.com.au/assets/acrobat/PossibilityofPlantations.pdf.

Brundtland GH (ed) (1987) Our common future. United Nations World Commission on Environment and Development, Oxford Univ. Press, Oxford

Cabalzar G (1996) Le milieu humain. In: Ganzhorn JU, Sorg J-P (eds) Ecology and economy of a tropical dry forest in Madagascar, primate report, special issue 46–1, pp 13–19

Campbell B, Sayer JA (eds) (2003) Integrated natural resources management: Linking productivity, the environment and development. CAB International with Center for international forestry research. Wallingford, Cambridge

Carey AB (2006) Active and passive forest management for multiple values. Northwest Nat 87:18–30

Cash DW, Clark WC, Alcock F et al (2003) Knowledge systems for sustainable development. P Natl Acad Sci 100:8086–8091

Chapin III FS, Zavaleta ES, Eviner VT et al (2000) Consequences of changing biodiversity. Nature 405:234–242

Chen J, Saunders SC (2006) Ecology of multiple ecosystems in time and space. In: Chen J, Saunders SC, Brosofske KD et al (eds) Ecology to hierarchical landscapes: from theory to application, Nova Publishing, Carbondale

Chomitz K (2006) At Loggerheads? Agricultural expansion, poverty reduction, and environment in the tropical forests. World Bank, Washington, DC

Clergue B, Amiaud B, Pervanchon F et al (2005) Biodiversity: function and assessment in agricultural areas, a review. Agron sustain dev 25:1–15

Colfer CJP (2005) The complex forest: Communities, uncertainty and adaptive collaborative management. Resources for the Future and Center for international forestry research, Washington DC

Daily GC, Ehrlich PR (1999) Managing earth's ecosystems: an interdisciplinary challenge. Ecosystems 2:277–280

Damschen EI, Haddad NM, Orrock JL, Tewksbury JJ, Levey DJ (2006) Corridors increase plant species richness at large scales. Science 313:1284–1286

Dudley N, Aldrich M (2007) Five years of implementing forest landscape restoration – lessons to date. WWF International. http://assets.panda.org/downloads/flrlessonslearntbooklet.pdf

Ferraro PJ, Kiss A (2002) Direct payments to conserve biodiversity. Science 298:1718–1719

Flyvbjerg B (2006) Five misunderstandings about case-study research. Qual Inq 12(2):219–245

Folke C, Carpenter S, Elmqvist T (2002) Resilience and sustainable development: building adaptive capacity in a world of transformations. Ambio (31):5

Forman RTT (1995) Land mosaics: The ecology of landscapes and regions. Cambridge University press, Cambridge

Frost P, Campbell B, Medina G, Usongo L (2006) Landscape-scale approaches for integrated natural resource management in tropical forest landscapes. Ecol Soc 11(2):30. http://www.ecologyandsociety.org/vol11/iss2/art30/

Gerber JF, Steppacher R (2007) Propriété contre possession: les conflits liés aux plantations industrielles d'arbres. Swiss Forestry J 158(3–4):64–69
Globescan (2004) Results of first-ever global poll on humanity's relationship with nature. http://www.globescan.com/news_archives/IUCN_PR.html
Hadorn GH, Bradley D, Pohl C et al (2006) Implications of transdisciplinarity for sustainability research. Ecol Econ 60:119–128
Hanski I (1999) Metapopulation ecology. Oxford University Press, Oxford
Hanski I (2005) Landscape fragmentation, biodiversity loss and the societal response. EMBO Rep 6(5):388–392
Hanski I, Ovaskainen O (2000) The metapopulation capacity of a fragmented landscape. Nature 404:755–758
Hartanto H (2003) Facilitating collaboration and partnerships: lessons from adaptive collaborative management in the Philippines. ETFRN News 39–40:118–120
Harwood RR, Kassam AH (eds) (2003) Research towards integrated natural resources management: examples of research problems, approaches and partnerships in action in the CGIAR. FAO, Rome
Hayes TM (2006) Parks, people and forest protection: an institutional assessment of the effectiveness of protected areas. World Dev 34(12):2064–2075
Hilty JA, Lidicker WZ, Merenlender AM (2006) Corridor Ecology: The science and practice of linking landscapes for biodiversity conservation. Island Press, Washington DC
Horlick-Jones T, Sime J (2004) Living on the border: knowledge, risk and transdisciplinarity. Futures 36:441–456
Imbernon J, Branthomme A (2001) Characterization of landscape patterns of deforestation in tropical rain forests. Int J Remote Sensing 22(9):1753–1765
IUCN (2003) The Durban Link. Strengthening protected areas. Ten target areas for action in the next decade. http://www.iucn.org/themes/wcpa/wpc2003/pdfs/outputs/durbanlink/action_plan_eng.pdf
Jantsch E (1970) Inter- and transdisciplinary university: a systems approach to education and innovation. Policy Sci 1:403–428
Jepson P, Jarvie JK, MacKinnon K, Monk KA (2001) The end for Indonesia's lowland forests? Science 292(5518):859–861
Kanbur R, Venables AJ (2005) Rising spatial disparities and development. UNU Policy Brief. http://www.unu.edu
Karsenty A (2004) Des rentes contre le développement? Les nouveaux instruments d'acquisition mondiale de la biodiversité et l'utilisation des terres dans les pays tropicaux. Mondes dev 127(3):1–9
Kleijn W, Sutherland DJ (2003) How effective are European agri-environment schemes in conserving and promoting biodiversity? J Appl Ecol 40:947–969
Klein JT (2004) Prospects for transdisciplinarity. Futures 36:515–526
Koehler P, Chave J, Riéra B et al (2003) Simulating the long-term response of tropical wet forests to fragmentation. Ecosystems 6(2):114–128
Lamb D, Gilmour D (2003) Rehabilitation and restoration of degraded forests. IUCN, The World Conservation Union, Gland
Lambin EF, Geist HJ, Lepers E (2003) Dynamics of land-use and land-cover change in tropical regions. Annu Rev Env Resour 28:205–41
Levins R (1969) Some demographic and genetic consequences of environmental heterogeneity for biological control. B Entomol Soc Am 15:237–240
Li H, Wu J (2004) Use and misuse of landscape indices. Landscape Ecol 19:389–399
Lindenmayer DB, Franklin JF (2002) Conserving forest biodiversity: a comprehensive multiscaled approach. Island Press, Washington DC
Lindenmayer DB, Franklin JF, Fischer J (2006) General management principles and a checklist of strategies to guide forest biodiversity conservation. Biol Conserv 131:433–445
Loucks C, Springer J, Palminteri S et al (2004) From the vision to the ground: A guide to implementing ecoregion conservation in priority areas. WWF US Conservation Science Program, Washington DC

MacArthur RH, Wilson EO (1967) The theory of island biogeography. Princeton University press, Princeton

Mansourian S, Vallauri D, Dudley N (eds) in cooperation with WWF International (2005) Forest Restoration in landscapes. Beyond planting trees. Springer, New York

Margules CR, Pressey RL (2000) Systematic conservation planning. Nature 405:243–253

Marjokorpi A (2006) Biodiversity management in fast-growing tree plantations: a case study from West Kalimantan, Indonesia. Dissertation, Annales Universitatis Turkuensis, Series All, part 200, Turku

McCauley DJ (2006) Selling out nature. Nature 443:7

McIntyre S, Hobbs RJ (1999) A framework for conceptualizing human effects on landscapes and its relevance to management and research models. Conserv Biol 13:1282–1292

Morrison SA, Reynolds MD (2006) Where to draw the line: integrating feasibility into connectivity planning. In: Crooks KR, Sanjayan M (eds) Connectivity conservation. Conserv Biol 14, Cambridge University Press, pp 536–554

Muttenzer F (2006) Déforestation et droit coutumier à Madagascar. L'historicité d'une politique foncière. Dissertation, Institut universitaire d'études du développement, Université de Genève

Naughton-Treves L, Alvarez N, Brandon K et al (2006) Expanding protected areas and incorporating human resource use: a study of 15 forest parks in Ecuador and Peru. Sustainability: Science, practice and policy 2(2):32–44

Naveh Z (2001) Ten major premises for a holistic conception of multifunctional landscapes. Landsc Urban Plan 57:269–284

Nicolescu B (2005) Towards transdisciplinary education. The Journal of Transdisciplinary Research in Southern Africa 1:5–16

Opdam P, Foppen R, Vos C (2002) Bridging the gap between ecology and spatial planning in landscape ecology. Landsc Ecol 16:767–779

Ostrom E (1995) A framework relating human 'driving forces' and their impact on biodiversity. Paper presented at the Smithsonian/Man and the Biosphere Biodiversity Program International Symposium "Measuring and monitoring forest biological diversity: The international network of biodiversity plots, Washington, DC, May 23–25, 1995

Peluso NL (1993) Coercing conservation? The politics of state resource control. Global Environ Chang 3:199–217

Pfund J-L, Boffa J-M, Koponen P, O'Connor T (2006) Report on workshop to launch the CIFOR and ICRAF biodiversity platform. 2nd – 5th March 2006. http://www. worldagroforestry.org/SEA/Publications/searchpub.asp?publishid=1453

Piaget J (1972) The epistemology of interdisciplinary relationships. In: Interdisciplinarity: problems of teaching and research in universities. Organization for Economic Cooperation and Development, Paris

Purnomo H, Guizol P (2006) Simulating forest plantation co-management with a multi-agent system. Math Comput Model 44:535–552

Purnomo H, Mendoza RA, Prabhu R (2004) Model for collaborative planning of community managed resources based on soft systems approach. J Trop Sci 16:106–131

Reid WV (2006) Nature: the many benefits of ecosystem services. Nature 443

Rouget M, Cowling RM, Lombard AT et al (2006) Designing large-scale conservation corridors for pattern and process. Conserv Biol 20(2):549–561

Sanjayan MA, Shen S, Janssen M (1997) Experiences with integrated-conservation development projects in Asia. World Bank technical paper 368

Saunders CD, Brook AT, Myers OE (2006) Using psychology to save biodiversity and human well-being. Conserv Biol 20(3):702–705

Sayer JA, Campbell BM (2003) Research to integrate productivity enhancement, environmental protection, and human development. In: Sayer JA, Cambpell BM (eds) Integrated natural resource management: linking productivity, the environment and development. CABI Published in association with the CIFOR, Trowbridge, Wallingford, Oxon, Cambridge

Sheil D, Puri R, Wan M et al (2006) Recognizing local people's priorities for tropical forest biodiversity. Ambio 35:17–24

Siitonen P (2003) Reserve network design in fragmented forest landscapes. Dissertation, University of Helsinki

Sillitoe P (2004) Interdisciplinary experiences: working with indigenous knowledge in development. Interdiscipl Sci Rev 29:1–18

Sorg J-P (2006) Orientations nouvelles de la recherche dans la zone des forêts denses sèches à Madagascar. In Schwitzer C, Brandt S, Ramilijaona O et al (eds) Proceedings of the German-Malagasy research cooperation in life and earth sciences. Concept Verlag, Berlin, pp 67–71

Tomich T, Alegre J, Areskoug V et al (2006) The challenges of integration: report of an online consultation among researchers of the alternatives to slash-and-burn (ASB) programme. http://www.maweb.org/documents/bridging/papers/Tomich.tom.pdf

Tomich TP, van Noordwijk M, Vosti S et al (1998) Agricultural development with rainforest conservation: methods for seeking best bet alternatives to slash-and-burn, with applications to Brazil and Indonesia. Agr Econ 19:159–174

Tress B, Tress G (2001) Capitalising on multiplicity: a transdisciplinary systems approach to landscape research. Landsc Urban Plan 57:143–157

Tress B, Tress G, Décamps H, d'Hauteserre A-M (2001) Bridging human and natural sciences in landscape research. Landsc Urban Plan 57:137–141

Turner M (2005) Landscape ecology: what is the state of science? Annu Rev Ecol Evol Syst 36:319–344

Vanclay J, Sinclair F, Prabhu R (2003) Modelling interactions amongst people and forest resources at the landscape scale small-scale forest economics. Manage Pol 2:117–121

van Noordwijk M (2001) Understanding local action and its consequences for global concerns in a forest margin landscape: the FALLOW model as a conceptual model of transitions from shifting cultivation. In: van Noordwijk M, Williams S, Verbist B (eds) Towards integrated natural resource management in forest margins of the humid tropics: local action and global concerns. Lecture Note 11. ASB Teaching Material

van Noordwijk M (2005) RUPES typology of environmental services worthy of reward. The World Agroforestry Centre, Bogor

van Noordwijk M, Villamor G, Leimona B et al (2006) Criteria and indicators for ecosystem reward and compensation mechanisms: realistic, voluntary, conditional and pro-poor. SSLWM paper B6, Luang Prabang

Wadley RL, Colfer CJP (2004) Conserving nature in culture: case studies from Southeast Asia. Hum Ecol 32(3):313–338

Wells M, Guggenheim S, Khan A et al (1999) Investing in biodiversity: A review of Indonesia's integrated conservation and development projects. The World Bank report, Washington DC

Wollenberg E, Edmunds D, Buck L (2000) Anticipating change: Scenarios as a tool for adaptive forest management, a guide. Center for international forestry research, Bogor

WRI (2005) Millenium ecosystem assessment. Ecosystem and human well-being: synthesis. Island Press, Washingtom DC

Wu J, Hobbs R (2002) Key issues and research priorities in landscape ecology: An idiosyncratic synthesis. Landsc Ecol 17:355–365

Wunder S (2006) Are direct payments for environmental services spelling doom for sustainable forest management in the tropics? Ecol Soc 11(2):23. http://www.ecologyandsociety.org/vol11/iss2/art23/

Zeller M, Lapenu C, Minten B et al (2000) The critical triangle between environmental sustainability, economic growth, and poverty alleviation. In: Minten B, Zeller M (eds) Beyond market liberalization. Ashgate, Aldershot

Chapter 16

Forest Management and Carbon Sink Dynamics: a Study in Boreal and Sub-Alpine Forest Regions

Chao Li, Shirong Liu, Yuandong Zhang, Jianwei Liu and Chuanwen Luo

Abstract Increased human activities have changed global ecosystems that include the dynamics of carbon (C) stocks in forest lands, which are determined by the sizes of living biomass, dead woody materials and soil C pools. This study focuses on C dynamics in living biomass that has been used in international reporting. Based on the forest conditions of Fort A La Corne (FALC) in central Saskatchewan, Canada and Miyaluo in Sichuan Province, P.R. China, this study employed a strategic model to simulate C stock dynamics under various combinations of forest fire and harvest alternatives. Our simulation results suggest that the forest C sink size is less likely to be sustained with a simple strategy of complete protection against all disturbances. Changes in the C sink size are largely attributed to the dynamics of forest age distribution. Forest management options that keep forests within a certain range of mean forest age could result in both higher mean annual increment (MAI) and C sequestration rates than the default and average values used by the Intergovernmental Panel on Climate Change (IPCC). The specific range of mean forest age will be region-specific, depending on the tree species composition and physical conditions. Our simulation results suggest that, in most cases, the FALC forests will function as C sinks except when the fire cycle becomes very short; hence, the area has the potential to positively contribute to the C sink. For Miyaluo forests, a strategy of regulated harvest activities can enhance the C sink.

16.1 Introduction

The world's forests store about 60 gigatons (Gt) of carbon (C) from the atmosphere every year through photosynthesis during their growth process (Brown 1996). Increased human activities have changed global ecosystems including the dynamics of C stocks in forest lands. Research results have shown that climates across the world

C. Li
Canadian Wood Fibre Centre, Canadian Forest Service, Natural Resources Canada, 5320-122 Street, Edmonton, Alberta, Canada T6H 3S5
e-mail: cli@nrcan.gc.ca

R. Lafortezza et al. (eds.), *Patterns and Processes in Forest Landscapes*,

have changed in an acceleratory manner and the CO_2 concentration in the atmosphere has already increased by about 30% since the industrial revolution started; this change is largely attributed to the increased use of fossil fuels (coal, oil and natural gas) for energy (IPCC 2001; Metz et al. 2005; Eggleston et al. 2006). These observations have been supported and explained by a number of general circulation models and global climate models (GCMs) (Euskirchen et al. 2002; Pregitzer and Euskirchen 2004; Noormets et al. 2007). Despite the uncertainties involved, the international communities have been working intensively on a consensus of how the environmental conditions could be managed for a healthy and sustainable development. In all of these efforts, the central issue is to understand how different human activities might influence an ecosystem's dynamics such as changes in C stocks in forest lands.

The dynamics of C stocks in forest lands are determined by the sizes of living biomass, dead woody materials and soil C pools (Penman et al. 2003). Among these C pools, results from intensive research on the dynamics of the living biomass C pool have been reported in the literature and it has become the requirement in some of the international reporting such as in Kyoto Protocol (KP) (Penman et al. 2003). This chapter shall focus on the C dynamics in living biomass in forest lands and explore how it could be influenced by different management operations.

Normal forest growth and decline processes will be influenced by various natural and anthropogenic disturbance regimes. In the boreal forests of Canada, major natural disturbances include fire, insect, disease, wind and climate-induced mortality. In western Canada, fire remains the most destructive natural disturbance for forests. Timber harvest is the major anthropogenic disturbance regime that has a direct influence on forest dynamics and the management of natural disturbance regimes has an indirect influence on forest dynamics. The major insect pest species in this region are the mountain pine beetle (MPB, *Dendroctonus ponderosae* Hopkins), the spruce budworm (*Choristoneura fumiferana* (Clem.)) and the forest tent caterpillar (*Malacosoma disstria* Hubner). Among these species, MPB has the most devastating impact on forest dynamics and has been in an outbreak phase for the last several years in British Columbia (British Columbia Ministry of Forests 2003). With the climate warming in western Canada, this species has rapidly expanded its infestation area into northern British Columbia and western Alberta where lodgepole pine is distributed (Alberta Sustainable Resource Development 2004; Li and Barclay 2004; Li et al. 2005a). In this chapter, however, we focus on fire and harvest regimes.

In the sub-alpine forests of China, forest growth is generally described to follow the classical Clementsian succession concept (Clements 1916; Odum 1969) (i.e., forest growth with age according to an orderly, predictable and deterministic direction, which eventually reaches climax or steady equilibrium status at about 170 years, in this case). The climax, which is uneven-aged stand structure and sustained via gap regeneration, will then last forever without the impact of any catastrophic disturbance. Furthermore, no observation was reported on major natural disturbances in the region (Liu et al. 2001, 2002, 2003; Zhang et al. 2006). However, large-scale harvest activities from 1953 to 1978 have significantly changed forest conditions. The available volume for harvest has sharply declined and ecosystem degradation, in terms of loss of biodiversity, soil erosion and hydrological change, has created significant challenges in ecosystem services and environmental conservation (Li 1990; Liu et al. 2001, 2006). Therefore, development of methods and criteria for a sustainable forest resource management strategy is urgently needed.

The objective of this chapter is to present a comparative case study to demonstrate how the long-term C stock dynamics at the landscape scale, which is scaled up from stand scale understanding, could be influenced by various levels of effort in managing disturbance regimes, thus providing supporting evidence for defining suitable management strategies. The study is based on the current forest conditions of a boreal forest area in Canada (Fort A La Corne (FALC), central Saskatchewan) and a sub-alpine forest area in China (Miyaluo, Sichuan). Based on the model investigation using the Woodstock (Walters and Cogswell 2002), our results suggested that a simple strategy of complete forest protection from any disturbance might not be the best approach for enhancing C stock in forest lands and the existence of disturbances might not necessarily be entirely negative for C dynamics. We show that forest age distribution and its dynamics are important in determining the forest growth rate and thus C sequestration, in which a certain range of mean forest age will result in values higher than the default and average values recommended by the IPCC.

16.2 Materials and Methods

16.2.1 Study Areas

The FALC area (53°6′–53°25′ N, 104°11′–105°11′ W), with a total size of 132,502 ha, is a forested landscape surrounded by agricultural lands located in central Saskatchewan, Canada (Fig. 16.1A). This area is within the Boreal Transition Ecoregion that represents the gradation from the grasslands of the south to the boreal forest of the north (Saskatchewan Environment 1999). Fine sands dominate the soils of the FALC area. The area has about 60% of timber-producing land and the major tree species are jack pine (*Pinus banksiana* Lamb), trembling aspen (*Populus tremuloides* Michx.), black spruce (*Picea mariana* (Mill.) B. S. P.) and white spruce (*Picea glauca* (Moench) Voss). This area is about 500 m above sea level with a generally flat terrain. Fire has been the key natural disturbance that controls forest species composition, age structure and vegetation patterns. The forests in the FALC are particularly susceptible to fire because of light rainfall, the lack of moisture-retaining soil, exposure to adjacent farmland and burning permit areas, long-term exploitation of forest products, abundant dry fuels and infestation of disease such as the dwarf mistletoe. The fire cycle (defined as the number of years required to burn over an area equal to the entire area of interest) has been changed from about 105 years before 1945 to about 213 years after 1945 (Li et al. 2005b).

The Miyaluo forest area is at the upper reach of the Zagunao watershed of Minjiang Valley, located in the Li County of the Sichuan Province, China

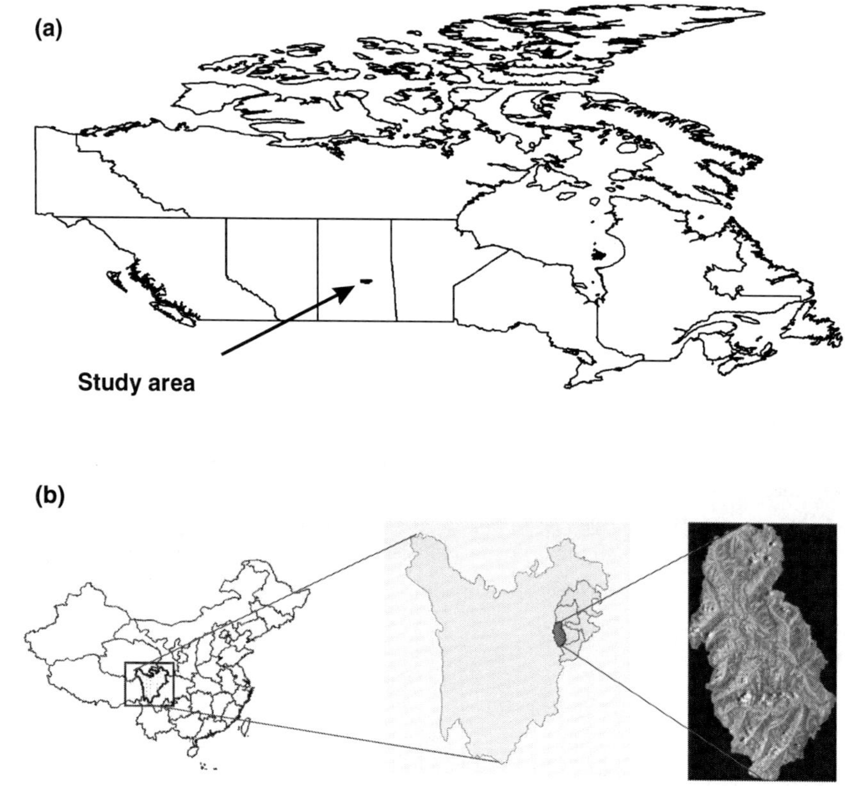

Fig. 16.1 Locations of the study areas: (**A**) FALC in central Saskatchewan, Canada. (**B**) Miyaluo in Sichuan, PR. China

(31°24′–31°55′ N, 102°35′–103°4′ E) (Fig. 16.1B). With 62,331 ha covered by forests, this sub-alpine area has elevations ranging from 2,200 m to 5,500 m above sea level. The forested areas are dominated by dark coniferous species such as fir (*Abies faxoniana* Rehd. et Wils) and spruce (*Picea purpurea* Mast and *Picea asperat*a Mast). Other tree species include larch (*Larix potaninii* Batalin), birch (*Betula albo-sinensis* Burk), oak (*Quercus aquifoliodes* Rehd. et Wils), pine (*Pinus densata* Mast and *Pinus tabulaeformis* Carr) and hemlock (*Tsuga chinensis* Pritz) (Liu et al. 2003). Before the large-scale harvest activities started in the early 1950s, the forest regenerations were mainly self-replacement naturally via gaps. The major harvest period was from 1953 to 1978, with an average of 329, 000 m^3 cut annually and the peak years were from 1958 to 1960, with an annual harvest higher than 550, 000 m^3. This over-utilization resulted in the forest ecosystem degradation resulting in loss of biodiversity, soil erosion and water capacity reduction that have created challenges in environmental conservation. The harvest activities were stopped in 1998 when the national key programs of the “Natural Forest Protection Program (NFPP)” and the “Sloping Land Conversion Program (SLCP)” (from agricultural land) were started (China Council for International Cooperation on Environment and Development 2002). A significant plantation period was started from the mid-1950s, with a cumulative area of 16,700 ha by the year 2000. Spruce was used as the major species for plantation and was planted on the clear-cut site and the natural generation of birch (*Betula albo-sinensis* Burk) also occurred in other harvest residue lands. Consequently, the forests have two main age groups: one in 20–40 years, resulting from plantation and the other in 160–210 years, with natural origins distributed within the patches on high elevations with less accessibility. No other significant natural disturbances were observed in this study area.

In this study, the operational forest inventory data in both study areas was used with main variables of forest cover type, stand age, tree density and site index (Fig. 16.2).

Forest inventory records were not available for about 40% of the total FALC area because several large fires had occurred in 1995 including the English fire event (Fig. 16.2A). This created difficulty with the fire simulations because the burned area has almost separated the FALC into two parts and thus a fire starting in one part could barely spread to the other. To overcome this shortfall, a map of provincial fuel types was placed over the forest cover type layer and the corresponding forest cover types were converted for those pixels with missing data. This provided a reconstructed forest cover map for the requirement of the model simulations (Fig. 16.2B). For these pixels, we set the stand age for the burned area to zero (Fig. 16.2C) and the tree densities to an intermediate level of C that is 51–75% of that of the crown closure (Fig. 16.2D). These reconstructed layers might contain bias in estimated forest productivity but the qualitative conclusions drawn from this simulation investigation should not be affected.

16.2.2 Woodstock Model Description

Woodstock software of Remsoft (Walters and Cogswell 2002) (a commercial software package for timber supply analysis and harvest planning) appears appropriate for use in our current modeling investigation due to its wide usage in various resource management agencies in west-central and Atlantic Provinces of Canada. The Chinese Academy of Forestry (CAF) has also introduced this software package for developing sustainable forest management standards and regulations. Woodstock inputs forest inventory data into a geographical information systems (GIS) format in order to calculate timber supply over multiple planning periods based on user-defined yield equations for different tree species or species association. The calculation is then optimized through linear programming to allow for a relatively constant annual allowable cut (AAC) determined for different periods of harvest planning. This relatively constant AAC is ideal for the stable timber production and processing industries.

The Woodstock model considers how the timber resource is utilized in an optimized manner based on average forest growth; however, other disturbances, such as fire, are not incorporated. In modeling the investigation of how C dynamics could

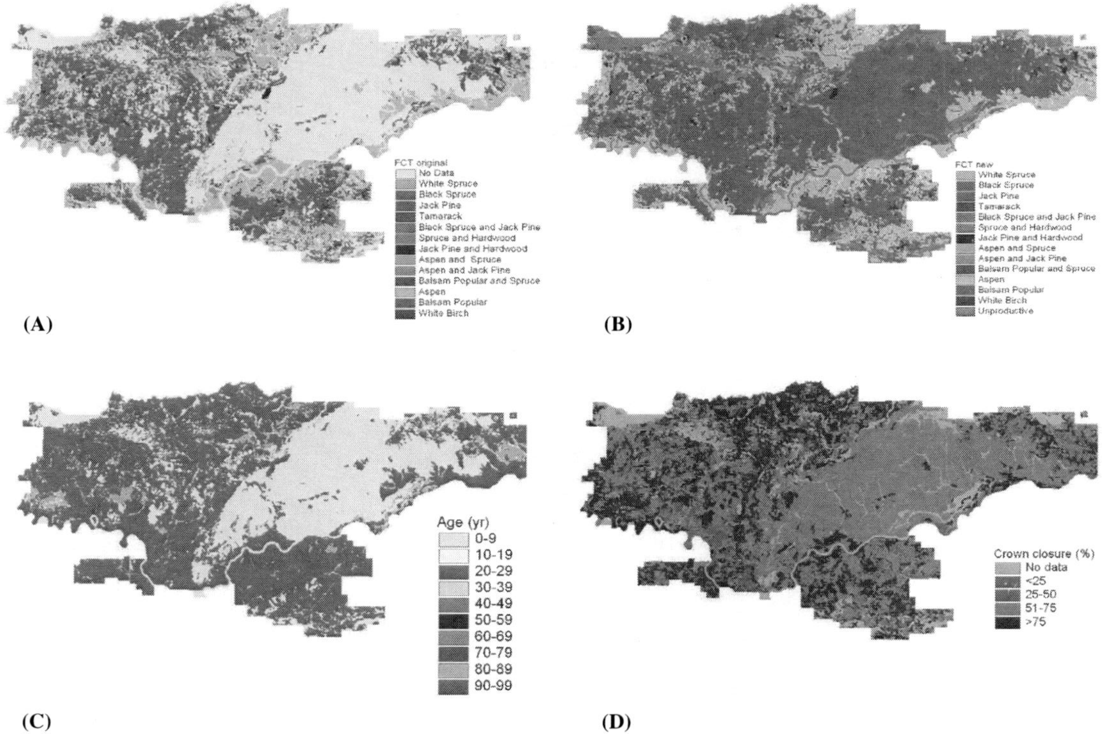

Fig. 16.2 Maps of the Fort A La Corne study area, Saskatchewan. (**A**) Original land cover type. (**B**) Reconstructed land cover type. (**C**) Forest age. (**D**) Stand density

be influenced by fire and harvest regimes, we assumed that the burned area in a given forest type is proportional to its total area within the landscape. Therefore, the estimated annual area burned from a given fire regime can be allocated to different types of forests. This assumption may or may not hold in different regions. Nevertheless, this algorithm provided an approximation of the fire effect on the AAC determination.

16.2.3 Forest Growth Patterns

The forest growth patterns in boreal Canada and sub-alpine China are different, primarily due to the differences in tree species, climate and physical site conditions. For the FALC area of Canada, characterization of growth patterns are based on observations from Permanent Sampling Plot (PSP) and Temporary Sampling Plot (TSP) data. For the Miyaluo area of China, the growth patterns are adapted from Yang (1985) by assuming the maximum stand volumes are reached at about 170 years of forest age and the stand volume would be in climax conditions afterward, unless disturbed by human activities such as harvesting (Fig. 16.3).

16.2.4 Model Experiments and Data Analysis

We explored the forest C dynamics under all possible fire and harvest regimes from a long-term perspective. This has resulted in the model experimental design in which we included complete ranges of possible harvest AAC levels (50,000, 75,000, 100,000, 125,000, 150,000, 175,000 and 190, 000 m^3) and fire cycles (50, 75, 100, 125, 150, 200, 300 and 400 years), mimicking different efforts in fire management. This model experiment is used to provide an estimate on whether C stock could continue to increase when a complete forest protection is achieved and where forests can grow normally and regenerate when reaching their longevity (200 years) in the FALC area. To reflect the regional condition without a significant impact of fire disturbances in the Miyaluo sub-alpine forest area in China, only a full range of possible harvest AAC levels (50,000, 100,000, 150,000, 200,000, 250,000, 280,000 and 300, 000 m^3) was included in the model experiment.

The Woodstock model was run in order to investigate forest C stock dynamics under two different scenarios: (1) without the effect of fire and harvest disturbance regimes for both study areas and (2) different combinations of fire cycles and AAC levels for the FALC area and different AAC levels for the Miyaluo area. Each simulation lasted 200 years with a planning period of every 10 years. Ten replications were performed for each of the model simulation scenarios where forest age and volume in each forest type at different planning periods were recorded.

The simulated forest volume dynamics are converted to C contents using the method summarized in Von Mirbach (2000), in which the conversion factors are based on the IPCC's Greenhouse Gas Inventory Guidelines Reference Manual

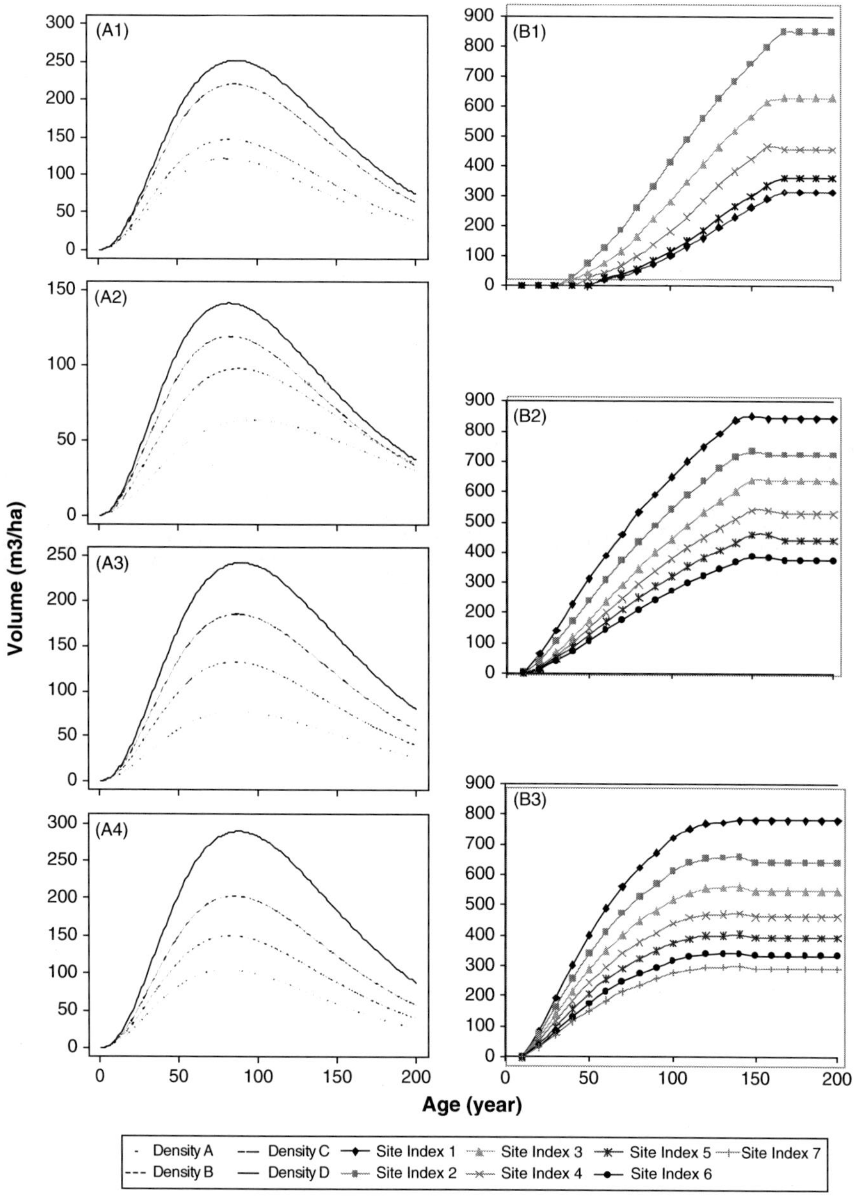

Fig. 16.3 Forest growth patterns in the two study areas: column A denotes examples in FALC area, and column B denotes examples in Miyaoluo area. (**A**1) Species association 11 dominated by white spruce. (**A**2) Species association 12 dominated by black spruce. (**A**3) Species association 13 dominated by jack pine. (**A**4) Species association 71 dominated by aspen. (**B**1) Natural stands of fir forests. (**B**2) Natural stands of spruce forests. (**B**3) Plantation stands of fir forests

and modified by Environment Canada using Canadian data. They are considered acceptable for national and international reporting requirements. According to Von Mirbach (2000), the above-ground forest volume (m^3) is estimated by multiplying the merchantable volume by a factor of 1.454 and the below-ground volume is estimated as 0.396 of the merchantable volume. The total wood volume is then converted to dry matter biomass (tonnes) by a factor of 0.43. The C stock is considered to be one half of the dry matter biomass. Biomass consumption by crown fire for boreal forests is 25.1 ton/ha. In the IPCC-approved method, crown fires in boreal forests would have a 0.43 combustion factor defined as the proportion of pre-fire biomass consumed and a 0.15 factor for surface fires. Based on 12,345 historical fire records in the adjacent province of Alberta during the period from 1961 to 1995, crown fires burned 56.3% of the total burned area on 38.3 million ha of Alberta's forest land, while surface fires burned 43.4% (Li 2004). Thus, we used an average value for the combustion factor of 0.29 in the calculation.

16.3 Results

16.3.1 C Stock Dynamics Without Disturbance

We found that the mean forest age can increase gradually until 180 years old at the simulated 120th year and then decline (Fig. 16.4A) and that the mean forest C stock per hectare can increase in the first 50 years and then gradually decrease for the following 100 years (Fig. 16.4B). After that, the C stock will increase again to

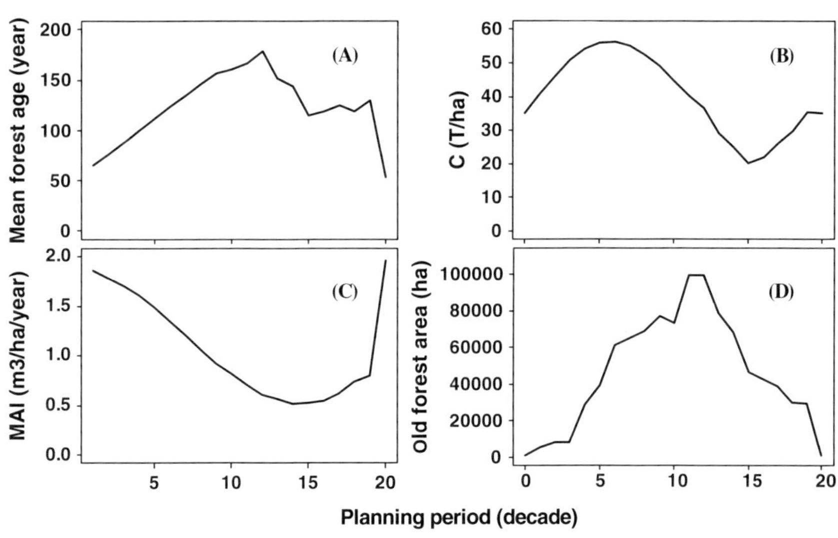

Fig. 16.4 Forest and C stock dynamics without disturbances in FALC area. (**A**) Mean forest age. (**B**) C stock per hectare. (**C**) Mean annual increment (MAI). (**D**) Old forest area

recover the initial C stock level at the last planning period. The different dynamic patterns in mean forest age and mean C stock are probably caused by the non-constant mean annual increment (MAI) of volume over time (Fig. 16.4C). We also examined the dynamics of areas older than 120 years (Fig. 16.4D) and found a similar pattern with mean forest age (i.e., the old forests can increase in age until 100–110 years old and then decrease).

For the FALC area, the forest C stock cannot increase without limit, even when complete protection can be achieved. This is probably due to the site's carrying capacity limit and the longevity of the forest will be reached at the final simulation year. C sequestration could decrease when certain mean forest age is reached and thus the total C stock could decline. This was a surprise result. This result suggested that a complete protection might not be the best strategy to increase C sequestration and MAI (Fig. 16.4C) and old forest area (Fig. 16.4D) from a long-term strategic perspective.

For the Miyaluo sub-alpine forest area of China, no stand age can be identified as longevity; therefore, the forest growth can continue without decline after reaching an age of 170 years or so and the oldest forest in the inventory data was about 210 years old (Fig. 16.5A). Therefore, no natural forest stand mortality was applied in

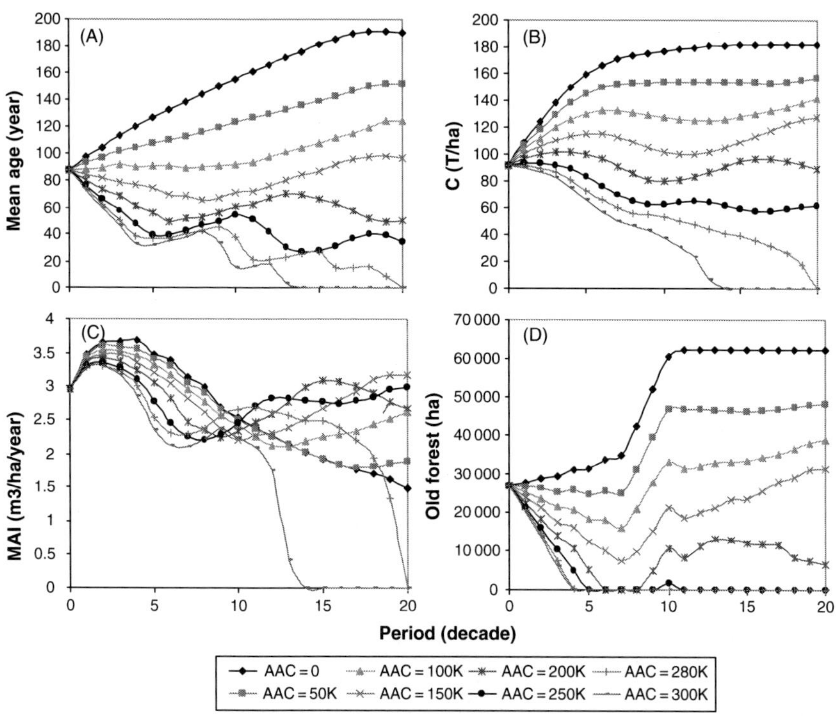

Fig. 16.5 Forest and C stock dynamics in Miyaluo area under different AAC levels. (**A**) Mean forest age. (**B**) C stock per hectare. (**C**) Mean annual increment (MAI). (**D**) Old forest area

the model simulations. We found that the total biomass C stock per ha gradually increased in the first 140 years to reach a maximum level and this maximum C stock continued until the end of the simulation (Fig. 16.5B). The old forest area experienced a slow increase period for the first 70 years, followed by a rapid increase period lasting until about 100 years of age to reach the maximum level (Fig. 16.5D). However, the MAI curve (Fig. 16.5C) indicates that the C sequestration was different. It increased for the first 40 years and then steadily decreased without bouncing back, which showed in the simulations of the FALC area.

Our simulation results for the Miyaluo area also suggested that the forest C stock did not increase without limit under complete protection, simply because the carrying capacity of the forest sites would be reached. With this pattern of non-declining total C stock per ha, however, the MAI (i.e., the C sequestration) would decline after the maximum level was reached. Consequently, a complete protection strategy could maintain the existing C stock in forest landscapes but might not be the best strategy to enhance C sequestration.

16.3.2 C stock Dynamics Under Different Harvest Regimes in the Miyaluo Area

Our results indicated that with the increase of AAC, the total C stock would decrease (Fig. 16.5B). When AAC reached 200, 000 m^3, the total C stock could not be maintained as to current conditions. The total C stock could be vanished by the end of the 200-year simulation period if AAC reaches 280, 000 m^3. Figure 16.5A also shows that the mean forest age could decrease once the AAC reaches 200, 000 m^3. The old forest area would decrease once harvest was conducted; however, lower AAC levels could still result in increased old forest area but the speed to succeed current old forest area would be slowed with the increase of AAC. When AAC was larger than 150, 000 m^3, the old forest area would have difficulty reaching the current level. The old forest area could decrease in size or even become extinct when AAC is higher than 200, 000 m^3. MAI curves (Fig. 16.5C) indicated that all harvest levels could result in a short-term increase and then decline. However, a surprising result was that moderate levels of AAC could result in a C sequestration higher than that under a complete protection strategy. This was probably due to the regeneration after a harvest that increases the area of forests in higher C sequestration stages.

16.3.3 C stock Dynamics Under Different Combinations of Fire and Harvest Regimes

Not all combinations of fire and harvest regimes generated meaningful results since some simulations were infeasible before completing the 20 planning periods. The infeasible simulations were caused by a forest extinction resulting from either high AAC levels or high annual areas burned by fire. In all of the simulations, high AAC's and short fire cycles tended to contribute to the unsuccessful simulations. Therefore, all of the simulation results from this experiment will contain missing values in the unsuccessful simulations of the combinations of fire and harvest regimes.

Simulation results in Fig. 16.6A showed that the mean forest age would be the highest with the longest fire cycle and no harvest. The forest age dropped with a shortening fire cycle and increasing harvest rates. The mean C stock per ha (Fig. 16.6B) would have a similar surface with the mean forest age, suggesting a positive correlation between the two variables.

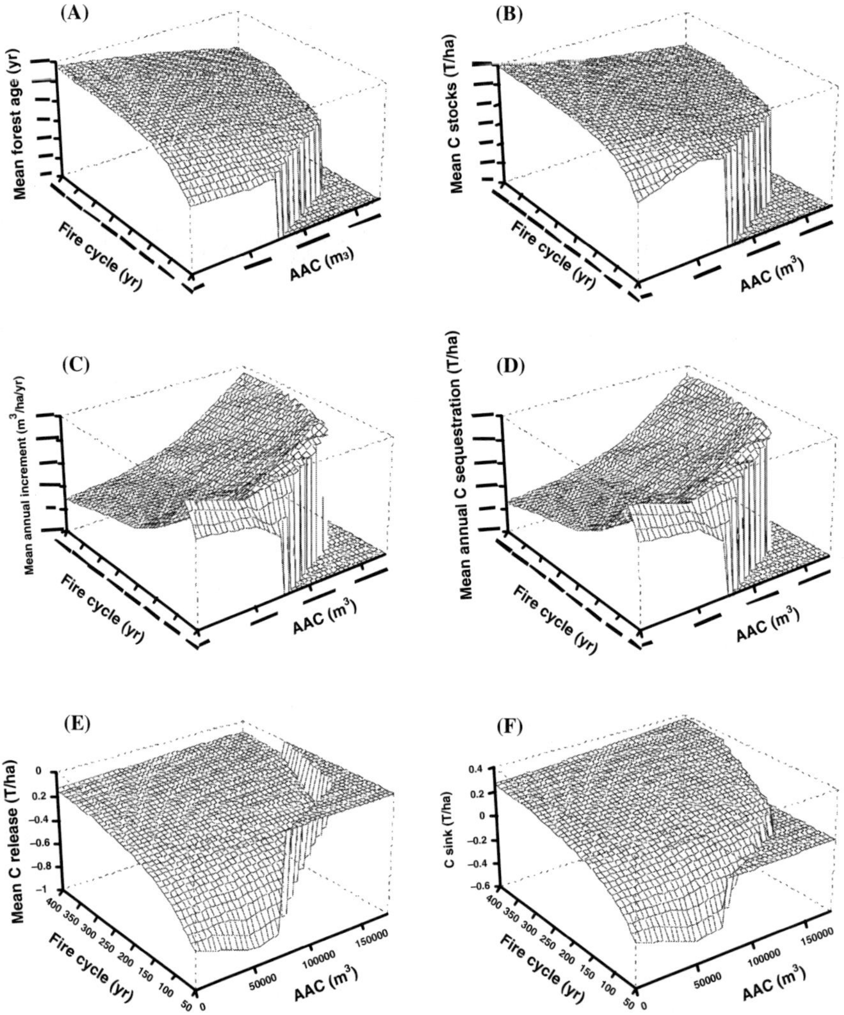

Fig. 16.6 Simulated forest and C dynamics under different combinations of fire cycles and annual allowable cuts. (**A**) Mean forest age. (**B**) Mean C stock. (**C**) Mean annual increment (MAI). (**D**) Mean annual C sequestration. (**E**) Mean C release. (**F**) Mean C sink size

MAI increased with the disturbance rate and this was true for both harvest and fire regimes (Fig. 16.6C). This is because stand-replacement disturbance can reduce the mean forest age thus pushing the MAI to a higher value. The MAI surface can be converted into a mean annual C sequestration surface (Fig. 16.6D).

A surface of mean C released from the fire disturbance was generated based on the method of Von Mirbach (2000) (Fig. 16.6E). By subtracting the C release (Fig. 16.6E) from the sequestration (Fig. 16.6D), a surface of the mean C sink size (i.e., the mean net change in C dynamics) under various combinations of fire cycles and AAC's (Fig. 16.6F) was also obtained. This surface represents a mix of positive and negative C sink sizes. Short fire cycles tended to make the C sink size negative thus representing a C source.

16.4 Discussion

The role of terrestrial ecosystems in the global C budget has been extensively discussed in the literature (e.g., Watson et al. 2000; Ciais et al. 1995; Houghton 1996). A number of estimates at national or regional scales have concluded that temperate and boreal forests are C sinks (e.g., Kauppi et al. 1992; Sedjo 1992; Dixon et al. 1994), while other studies reported that the forests function as a C source (e.g., Harmon et al. 1990). Kurz and Apps (1999) suggested that the role of forests in the C budget could change over time, depending upon changes in the prevailing disturbance regimes. For example, the forest ecosystems in Canada have been a sink of atmospheric C from 1920 to 1980 and a source during the 1980s because of a sharp increase in forest fire and insect disturbances starting around 1970 (Kurz and Apps 1999). The disparate results are also attributed to different geographical regions, climate and weather patterns, vegetation and soil conditions, disturbance regimes and different methods of estimating C stock.

The high-latitude forests were estimated to be a C sink increasing by $0.48 \pm 0.2\,\mathrm{Pg\,yr^{-1}}$ (Brown 1997), larger than the C sink in mid-latitude forests ($0.26 \pm 0.1\,\mathrm{Pg\,yr^{-1}}$) and the low-latitude forests (in which a relatively large net C source of $1.6 \pm 0.4\,\mathrm{Pg\,yr^{-1}}$ was reported). Goodale et al. (2002) summarized forest sector C budgets for Canada, the United States, Europe, Russia and China and concluded a general agreement that terrestrial systems in the Northern Hemisphere provided a significant sink for atmospheric CO_2; however, estimates of the magnitude and distribution of this sink vary greatly. Together, these suggest that northern forests and woodlands provided a total sink of 0.6–0.7 $\mathrm{Pg\,yr^{-1}}$ of C per year during the early 1990s, consisting of 0.21 $\mathrm{Pg\,yr^{-1}}$ in living biomass, 0.08 $\mathrm{Pg\,yr^{-1}}$ in forest products, 0.15 $\mathrm{Pg\,yr^{-1}}$ in dead wood and 0.13 $\mathrm{Pg\,yr^{-1}}$ in the forest floor and soil organic matter.

Using the vertical profiles of atmospheric CO_2 measured during midday at 12 global locations, an opposite result was proposed by Stephens et al. (2007) who concluded that "northern terrestrial uptake of industrial CO_2 emission plays a smaller role than previously thought and that, after subtracting land-use emissions, tropical ecosystems may currently be strong sinks for CO_2".

Diverse research results can be attributed to the data (e.g., sources, types, locations, variables, time and method in data collection, accuracy and representatives of the data) and methods (e.g., types of models in analytical and/or statistical estimates) used. Most of the studies and analyses mentioned above were based on large-scale categorized or statistical inventory data (Tier I and II data and C estimate methods for international reporting purpose approved by the IPCC, see Penman et al. 2003) and our comparative case study was based on the operational forest inventory that was spatially explicit, representing the most detailed and updated forest conditions in the two forest landscapes (Tier III data and C estimate methods for international reporting purpose approved by the IPCC, see Penman et al. 2003). Since the resolution of the operational forest inventory is much finer than the 100 km^2 minimum spatial resolution of the national forest inventory data used in the analysis of Goodale et al. (2002) and the biomass C dynamics in each forest stand was tracked in a spatially explicit way in the simulations, we expect that this study could provide a method for operational forest level C stock estimation that captures a more realistic estimate of C stock dynamics for the study areas. Furthermore, since the same model (Woodstock) was used in the C stock dynamics estimation in two forest types in two countries, the results are comparable and meaningful under the standard of the international reporting framework.

Because of the inclusion of the fire disturbance effect (Fig. 16.6E), our results (Fig. 16.6F) reflected the observed average changes in total C sink size per ha. Consequently, our results on the living biomass C stock dynamics in the FALC suggested that the area could generally function as a C sink as long as the fire cycle can be maintained longer than 100 years (i.e., the annual area burned can be managed under 1,325 ha). In boreal forests of Canada, fire regimes are the major factor determining the dynamics of forest age distribution (Van Wagner 1978; Li and Barclay 2001; Ryu et al. 2006) and our results in Fig. 16.6C further indicated that the mean forest age determined the strength of the C sink. Therefore, an understanding of regional fire patterns and management effects is necessary in order to estimate changes in C sink size (Figs. 16.5A and 16.5B). Our results are consistent with Pan et al. (2004) who estimated the different biomass densities in five tree development stages of various forest types in China and demonstrated that the forest age structure could have a significant effect on C sequestration estimation. Similar results were also reported in Euskirchen et al. (2002) from a net ecosystem productivity (NEP) perspective using the LandNEP model, Pregitzer and Euskirchen (2004) for forests across the world and Noormets et al. (2007) for five managed forest stands in northern Wisconsin, USA. By incorporating the effect of forest age structure, one could obtain a more accurate estimate on biomass because the volume-biomass relationship is not always linear.

Different patterns in simulated C sink dynamics in the two study areas are probably related to the different forest growth patterns displayed in Fig. 16.3. In the FALC area, forest growth patterns are consistent with the observations of age-related forest production decline phenomena (Gower et al. 1996; Ryan et al. 1997; Gower and McMurtrie 1999; Binkley et al. 2002; Berger et al. 2004), supporting the theoretical expectation from a diagram of C flux relative to gross primary production

(Barnes et al. 1998). Our results also suggested that the simplified assumption that younger forests had lower C sequestration and older forests had higher C sequestration might not always true. However, the complexity also allowed forest managers to have the opportunity to enhance forest C sink sizes through management options. For example, our simulation results suggest that by keeping mean forest age within a certain range (Fig. 16.6C), the C sequestered could have a higher probability of offsetting the C release from fires. In the Miyaluo sub-alpine forest area in China, forest dynamics were closer to the conceptual climax forest where no major disturbances would disrupt the continuous forest growth and much smaller-scale gap dynamics could sustain the steady state of the forest for a very long period. Therefore, no stand mortality could be identified. Even with this non-declining forest growth pattern, moderate disturbances could still contribute to increased C sequestration in the long run (Fig. 16.4C), which is consistent with our results from the FALC area. Therefore, keeping forests within a certain range of age classes is a better strategy than purely enhancing the C sequestration perspective. The management application of this result is that proper forest resource management operations can serve both increasing wood production and enhancing C sequestration.

Over-exploitation of forest resource in the Miyaluo area from the 1950s to the 1970s has resulted in the significant reduction of forest resource availability. Owing to the slow growth of trees, wood supply recovery takes a long time, even when harvest activities are completely stopped. Furthermore, there is a possibility that forests might not recover, due to soil erosion and reduced water availability.

Our results can be useful in estimating the C dynamics of managed forests in a nation's GHG accounting using a bottom-up approach from an operational forest management perspective, as our research is closely related to the forest management practices in west-central Canada using operational forest inventory data and harvest planning software. This study suggests that under a wide range of combinations of forest fire and harvest regimes, this FALC area could produce a C sink size that is large enough to offset the C released by fire disturbances. Our results thus provide an operational level case study of how the living biomass C stock at the landscape scale could be influenced by the inclusion of various forest management options. Similarly, in the Miyaluo area in sub-alpine China, a moderate level of harvest (ranging from 150,000 to 200, 000 km^3/yr) could maintain volume and keep MAI at a higher level (ranging from 2 to 3.3 m^3 per ha/yr). A lower level of harvest might have an increasing trend of total volume but MAI might be unable to maintain at a higher level and a higher level of harvest could lead to reduced total volume later on.

All of the fire-released C estimates presented in this case study are the means across all of the planning periods and replications. This treatment might not be able to capture some of the exceptional conditions such as the period with the highest forest growth rate, during which C sequestration was high and the extreme fire disturbance conditions in which the amount of C released is very high. The understanding of C dynamics under these exceptional conditions, nevertheless, would provide opportunities for forest managers to design the best strategy for taking advantage of the highest forest growth rate periods and to avoid extreme fire disturbance conditions. This is particularly meaningful if it is linked to current and future forest management strategies.

16.5 Conclusions

From this study, we found that forest C sink size cannot always be sustained without disturbances, which is consistent with the traditional understanding of boreal forests destroyed by and created by fire. Our results are supported by the fact that overmature forests tend to have a very small MAI and net C sequestration capacity but fuel loads (that are positively correlated with the tree age or time-since-last-fire in general) increase to a level that is more prone to fire. Also, native fire-dependent tree species such as pines cannot be sustained without fire to open their cones. Therefore, complete protection from any disturbance may not be the best strategy for enhanced C stock in forest lands and the existence of disturbances might not necessarily be bad for C dynamics.

We also confirmed that forest age distribution and its dynamics are important in determining a forest growth rate such as MAI and thus a certain range of mean forest age (Fig. 16.6C) will result in values higher than the default and average values recommended by IPCC. Clearly, keeping forests within age classes with a high MAI can serve as targets for forest managers. The FALC forests can function as C sinks except when a fire cycle becomes very short (<100 years); therefore, the area has the potential to contribute positively to the C sink. For the Miyaluo forests, a moderate level of harvest could maintain volume and keep MAI in a higher level.

Acknowledgments The authors thank Harinder Hans of the Canadian Forest Service for his GIS assistant and Brenda Laishley of the Canadian Forest Service and Lisa Delp of the University of Toledo for their critical reading and helpful comments on an earlier version of this manuscript. The funding support from the following sources at different stages of this research is much appreciated: Program of Energy Research and Development – Climate Change Impact on Energy Sector, Innovative Research Initiative and Mountain Pine Beetle Initiative of Natural Resources Canada, BIOCAP Canada, Outstanding Young Scientist Program of National Natural Science Foundation of China (No. 30125036), China National Major Fundamental Science Program (No. G2002cB111504), China National Scientific and Technical Key Project (No. 2006BAD03A04) and State Forestry Administration 948 Program of China (No. 2001-14, 2004-4-66 and 2005-4-26).

References

Alberta Sustainable Resource Development (2004) Annual report: forest health in Alberta, Department of Sustainable Resource Development, Public Lands and Forests Division, Forest Management Branch, Forest Health Section, Edmonton, Alberta, p 54

Barnes BV, Zak DR, Denton SR, Spurr SH (1998) Forest ecology, 4th edn. John Wiley & Sons Inc., New York, NY

BC Ministry of Forests (2003) Timber supply and the mountain pine beetle infestation in British Columbia. <http://www.for.gov.bc.ca/hts/pubs/beetledoc_oct29LO.pdf> Accessed on Feb. 15, 2006

Berger U, Hildenbrandt H, Grimm V (2004) Age-related decline in forest production: modelling the effects of growth limitation, neighbourhood competition and self-thinning. J Ecol 92:846–853

Binkley D, Stape JL, Ryan MG, Barnard HR, Fownes J (2002) Age-related decline in forest ecosystem growth: an individual-tree, stand-structure hypothesis. Ecosystems 5:58–67

Brown S (1996) The world's forest resources. Unasylva 43:3–10

Brown S (1997) Forests and climate change: role of forest lands as carbon sinks. Proceedings of the 11th World Forestry Congress, held in Antalya, Turkey, October 1997, pp 117–129

China Council for International Cooperation on Environment and Development (2002) Implementing the Natural Forest Protection Program and the Sloping Land Conversion Program: lessons and policy recommendations. Phase III (2002–2007). <http://www.harbour.sfu.ca/dlam/Taskforce/grassfindingindex.html> Accessed on Sept. 20, 2007

Ciais P, Tans PP, Trolier M, White JWC, Francey RJ (1995) A large northern hemispheric terrestrial CO2 sink indicated by the 13C/12C ratio of atmospheric CO2. Science 269:1098–1102

Clements FE (1916) Plant succession: an analysis of the development of vegetation. Carnegie Inst. Pub. 242. Washington, DC, p 512

Dixon RK, Brown S, Houghton RA, Solomon AM, Trexler MC, Wisniewski J (1994) Carbon pool and flux of global forest ecosystems. Science 263:185–190

Eggleston S, Buendia L, Miwa K, Ngara T, Tanabe K (eds) (2006) IPCC guidelines for National Greenhouse Gas Inventories, Institute for Global Environmental Strategies (IGES), Hayama, Japan on behalf of the IPCC

Euskirchen ES, Chen J, Li H, Gustafson EJ, Crow TR (2002) Modeling landscape net ecosystem productivity (LandNEP) under alternative management regimes. Ecol Model 154:75–91

Goodale CL, Apps MJ, Birdsey RA, Field CB, Heath LS, Houghton RA, Jenkins JC, Kohlmaier GH, Kurz W, Liu S, Nabuurs GJ, Nilsson S, Shvidenko AZ (2002) Forest carbon sinks in the northern hemisphere. Ecol Appl 2:891–899

Gower ST, McMurtrie RE (1999) An analysis of the age-related decline in aboveground net primary production; potential causes and stand-to-global scale implications. Santa Barbara (CA): NCEAS, University of Santa Barbara

Gower ST, McMurtrie RE, Murty D (1996) Aboveground net primary production decline with stand age: potential causes. Trends Ecol Evol 11:378–382

Harmon ME, Ferrell WK, Franklin JF (1990) Effects on carbon storage of conversion of old-growth forests to young forests. Science 247:699–702

Houghton RA (1996) Terrestrial sources and sinks of carbon inferred from terrestrial data. Tellus Ser. B Chem Phys Meteorol 48:430–433

IPCC (Intergovernmental Panel on Climate Change) (2001) Climate change 2001: the scientific basis. Contribution of Working Group I to the Third Assessment Report of the Intergovernmental Panel on climate change. In Houghton JT, Ding Y, Griggs KD, Noguer M, van der Linden PJ, Xiao D (eds) Cambridge Univ. Press, Cambridge, UK

Kauppi PE, Mielikainen K, Kuusela K (1992) Biomass and carbon budget of European forests, 1971 to 1990. Science 256:70–74

Kurz WA, Apps MJ (1999) A 70-year retrospective analysis of carbon fluxes in the Canadian forest sector. Ecol Appl 9:526–547

Li C (2004) Simulating forest fire regimes in the foothills of the Canadian Rocky Mountains. In Perera AH, Buse LJ, Weber MG (eds) Emulating natural forest landscape disturbances: concepts and applications. Columbia Univ. Press, New York, NY, pp 98–111

Li C, Barclay HJ (2001) Fire disturbance patterns and forest age structure. Nat Resour Model 14:495–521

Li C, Barclay H (2004) Simulation of Interactions among fire, mountain pine beetle and lodgepole pine forest. In Shore TL, Brooks JE, Stone JE (eds) Mountain pine beetle symposium: challenges and solutions, Can. For. Serv., Inf. Rep. BC-X-399, Victoria, BC, pp 257–266

Li C, Barclay HJ, Hawkes BC, Taylor SW (2005a) Lodgepole pine forest age class dynamics and susceptibility to mountain pine beetle attack. Ecol Complex 2:232–239

Li C, Barclay H, Liu J, Campbell D (2005b) Simulation of historical and current fire regimes in central Saskatchewan. For Ecol Manag 208:319–329

Li CP (1990) Ecological Research of Sichuan Forests. Sichuan Sci. & Tec. Press, Chengdu, PR, China

Li CW, Ge JP, Liu SR, Sun PS (2006) Landscape pattern and eco-hydrological characteristics at the upstream of Minjiang River, China. Front Biol China 1(4):455–462

Liu SR, Shi ZM, Chen LW, Liu XL (2002) Improving eco-environemnt in upper reaches of Yangze River by enhancing forest ecosystem services. Study report Vol 2 of forestry strategy of China sustainable development. In Jiang ZH (ed) China Forestry Publishing House, pp 163–170

Liu SR, Sun PS, Wang JX, Chen LW (2001) Hydrological functions of forest vegetation in upper reaches of Yangze river. J Nat Resour 16(6):451–456

Liu SR, Wang JX, Chen LW (2003) Ecology and restoration of sub-alpine ecosystem in western Sichuan, China. Inf Bot Ital 35(1):29–34

Metz B, Davidson O, de Coninck H, Loos M, Meyer L (eds) (2005) IPCC special report on carbon dioxide capture and storage. Cambridge Univ. Press, New York, NY, p 443

Noormets A, Chen J, Crow TR (2007) Age-dependent changes in ecosystem carbon fluxes in managed forests in northern Wisconsin, USA. Ecosystems 10:187–203

Odum EP (1969) The strategy of ecosystem development. Science 164:262–270

Pan Y, Luo T, Birdsey R, Hom J, Melillo J (2004) New estimates of carbon storage and sequestration in China's forests: effects of age-class and method on inventory-based carbon estimate. Climate Change 67:211–236

Penman J, Gytarsky M, Hiraishi T, Krug T, Kruger D, Pipatti R, Buendia L, Miwa K, Ngara T, Tanabe K, Wagner F (eds) (2003) IPCC good practice guidance for land use, land-use change and forestry. IPCC National Greenhouse Gas Inventories Programme

Pregitzer KS, Euskirchen ES (2004) Carbon cycling and storage in world forests: biome patterns related to forest age. Global Change Biol 10:2052–2077

Ryan MG, Binkley D, Fownes JH (1997) Age-related decline in forest productivity: pattern and process. Adv Ecol Res 27:213–262

Ryu SR, Chen J, Zheng D, Bresee MK, Crow TR (2006) Simulating the effects of prescribed burning on fuel loading and timber production (EcolFL) in managed northern Wisconsin Forests. Ecol Model 196:395–406

Saskatchewan Environment (1999) Fort A La Corne integrated forest land use plan – background information. http://www.se.gov.sk.ca/forests/landuse/fort/toc.htm. Accessed on Feb 15, 2006

Sedjo RA (1992) Temperate forest ecosystems in the global carbon cycle. Ambio 21: 274–277

Stephens BR, Gurney KR, Tans PP, Sweeney C, Peters W, Bruhwiler L, Ciais P et al (2007) Weak northern and strong tropical land carbon uptake from vertical profiles of atmospheric CO2. Science 316:1732–1735

Van Wagner CE (1978) Age-class distribution and the forest fire cycle. Can J For Res 8:220–227

Von Mirbach M (2000) Carbon budget accounting at the forest management unit level: an overview of issues and methods. Canada's Model Forest Program, Can. For. Serv., Nat. Resour. Can., Ottawa, ON, http://dsp-psd.pwgsc.gc.ca/Collection/Fo42-312-2000E.pdf. Accessed on Feb 15, 2006

Walters KR, Cogswell A (2002) Spatial forest planning: where did all the wood go? Forest Technology Group, Remsoft Inc., http://www.remsoft.com/docs/library/spatial.forest.planning.pdf. Accessed on Feb 15, 2006

Watson R, Noble IR, Bolin B, Ravindranath NH, Verado DJ, Dokken DJ (2000) Land use, land-use change and forestry. Special Report of the IPCC, Cambridge Univ. Press, p 377

Yang YP (1985) Guidelines for mountain forest management. Sichuan Science and Technology Press, Chengdu, p 395

Zhang YD, Liu SR, Ma JM (2006) Water-holding capacity of ground covers and soils in alpine and sub-alpine shrubs in western Sichuan, China. Acta Ecol Sin 26:2775–2782

Chapter 17

Emulating Natural Disturbance Regimes: an Emerging Approach for Sustainable Forest Management

Malcolm P. North and William S. Keeton

Abstract Sustainable forest management integrates ecological, social, and economic objectives. To achieve the former, researchers and practitioners are modifying silvicultural practices based on concepts from successional and landscape ecology to provide a broader array of ecosystem functions than is associated with conventional approaches. One such innovation is disturbance-based management. Under this approach, forest practices that emulate natural ecological processes, such as local disturbance regimes, are viewed as more likely to perpetuate the evolutionary environment and ecosystem functions of the forest matrix. We examine how this concept has been applied in three U.S. forest types: Pacific Northwest temperate coniferous, Western mixed-conifer, and Northeastern northern hardwood forests. In general, stand-level treatments have been widely used and often closely mimic historic disturbance because forest structure and composition guidelines have been well defined from reconstructive research. Disturbance-based landscape management, however, has not yet been closely approximated in the three forest types we examined. Landscape implementation has been constrained by economic, ownership, safety, and practical limitations. Given these constraints we suggest that disturbance-based management concepts are best applied as an assessment tool with variable implementation potential. Silviculture practices can be compared against the frequency, scale, and level of biological legacies characteristic of natural disturbance regimes to evaluate their potential impact on ecosystem sustainability.

17.1 Introduction

Recent landscape ecology texts (Turner et al. 2001; Lindenmayer and Franklin 2002) and some U.S. regional management plans (FEMAT 1993; SNFPA 2004) have proposed using natural disturbance as a model for sustainable forest management. This chapter examines how forest managers can use natural disturbance patterns and

M.P. North
Sierra Nevada Research Center, 1731 Research Park Dr. Davis, CA 95618
e-mail: mpnorth@ucdavis.edu

R. Lafortezza et al. (eds.), *Patterns and Processes in Forest Landscapes*,

processes as a coarse-filter model by manipulating forest structure and development, and the spatial distribution of treatments. Although management plans implementing these ideas vary regionally, most have the common goals of increasing forest structural complexity, maintenance of landscape connectivity and heterogeneity, and protection or restoration of riparian and watershed integrity. Most plans also focus on the matrix, the area between reserves that constitute most of the managed forest landscape. In this chapter we first summarize the concepts of disturbance-based management that are generally applicable to maintaining a sustainable forest landscape. Next we provide examples of disturbance-based management as applied in three distinct forest types: Pacific Northwest temperate coniferous forests, Western mixed-conifer forests, and Northeastern northern hardwood-conifer forests. For each of these examples, we describe existing forest conditions, summarize historic disturbance regimes, and then examine current management practices designed to emulate forest conditions produced by natural disturbance regimes. Finally we evaluate the strengths and weaknesses of using disturbance-based management in each of these forest types and identify lessons that may be useful for forest managers in other regions of the world.

17.2 Disturbance-Based Forest Management Concepts

17.2.1 Managing the Matrix

Concepts of forest sustainability have changed as the social perception of forests has shifted (Harrison 2002). Forest landscape sustainability, once measured as a constant supply of timber, has become a more complex concept where social, ecological, and biodiversity needs must be met in addition to economic revenues (Hunter 1999). This range of values cannot be fully sustained if forest landscapes are strictly segregated into reserves and production lands. Parks, wilderness areas, and reserves alone will never be able to sustain biodiversity and all the ecological services that society now demands of its forests. A significant majority of global forest lands, by one estimate about 88% (Dudley and Phillips 2006), have no formal protection. As the dominant element of the landscape, managed forestlands have a controlling influence on ecological processes, such as biological connectivity, watershed functioning, and carbon sequestration. Consequently, sustainable management of "matrix forests" is increasingly viewed as an essential complement to other conservation approaches (Lindenmayer and Franklin 2002; Keeton 2007). Matrix management incorporates concepts from the field of conservation biology. Lindenmayer and Franklin (2002) developed a framework for conserving forest biodiversity that we believe also provides appropriate metrics for assessing landscape sustainability, particularly if used in conjunction with protected areas based strategies. They list five core principles: (1) maintenance of stand structural complexity; (2) maintenance of connectivity; (3) maintenance of landscape heterogeneity; (4) maintenance of aquatic ecosystem integrity, and (5) risk spreading, or the application of multiple conservation strategies.

The first principle recognizes that intensive forestry practices usually simplify stand structure, resulting in less vertical complexity in the forest canopy, less horizontal variation in stand density, and fewer key habitat elements like large dead trees and downed logs (Swanson and Franklin 1992; Franklin et al. 1997). Thus, an alternative is to promote greater structural complexity (e.g. vertically differentiated canopies, higher volumes of coarse woody debris) in actively managed stands (Hunter 1999; Keeton 2006), which may benefit those organisms not well represented in simplified stands, as long as sufficient habitat is provided across multiple stands to support viable populations. The second principle, maintenance of connectivity, allows organisms to disperse, access resources, and interact demographically. Connectivity strategies include protection of terrestrial and riparian corridors, and restoration of linkage habitats. There are also non-corridor approaches, such as retention of well distributed habitat blocks and structures that provide "stepping stones" across harvested areas. Maintaining a diverse landscape, principle three, supports an array of ecological functions while also increasing ecosystem resilience to disturbance and stress (Perry and Amaranthus 1997). Principle four relates to minimizing deleterious forest management effects on riparian and aquatic ecosystem interactions (Naiman et al. 2005; Keeton et al. 2007b). Delineation of riparian buffers, riparian forest restoration, and ecologically informed forest road management are essential elements of matrix management (Gregory et al. 1997). Finally, "risk-spreading," principle five, deals directly with the scientific uncertainty associated with over-reliance on any one forest management approach. Uncertainty and risk are reduced if multiple management and conservation strategies are applied at different spatial scales and on different portions of the landscape (Lindenmayer and Franklin 2002).

17.2.2 Emulating Natural Disturbance

The central concept in disturbance-based management is that forest practices which are consistent with natural ecological processes, such as local disturbance regimes, are more likely to perpetuate the evolutionary environment and ecosystem functions of the forest matrix. Some of the negative ecological effects of forest management actions can be reduced if operations attempt to stay within the bounds of these natural disturbance regimes (Attiwell 1994; Bunnell 1995). Several useful indicators have been suggested as measures of differences between natural disturbance regimes and the effects of forest harvest. These include: (1) disturbance frequency, (2) disturbance magnitude (intensity and spatial attributes), and (3) the density and type of biological legacies persisting post-disturbance (Hunter 1999; Lindenmayer and Franklin 2002; Seymour et al. 2002). To evaluate the congruence between human and natural disturbances, managers need information on the frequency of historic disturbance events (e.g. local fire history; wind storm return interval), their patch size and distribution (e.g. fire extent; average sizes and formation rates of canopy gaps), and the number and arrangement of legacies structures (e.g. live and dead trees, and coarse woody debris left after natural disturbances). The effort to mimic natural disturbance regimes means disturbance-based forest management practices will vary, adapting to local and regional differences in disturbance patters (Fig. 17.1).

An important informational need in disturbance-based management is an understanding of ecosystem recovery following disturbances and long-term processes of stand development (Franklin et al. 2002). Research and evolving forest practices in Scandinavia (Vanha-Majamaa and Jalonen 2001), Canada (Beese and Bryant 1999), and several regions of the U.S., including the Pacific Northwest (Franklin et al. 2002; Keeton and Franklin 2005), upper Midwest (Palik and Robl 1999), Southeast (Palik and Pederson 1996; Mitchell et al. 2002), and New England (Foster et al. 1998; Seymour et al. 2002), have fostered a growing appreciation for the role of biological legacies in ecosystem recovery following disturbances. Biological legacies are "the organisms, organic materials, and organically-generated patterns that persist through a disturbance and are incorporated into the recovering ecosystem"

Fig. 17.1 Examples of disturbance-based silvicultural practices. The *upper right* is an example of both dispersed and aggregated retention in the U.S. Pacific Northwest (photo credits: Jerry F. Franklin). Shown on the *left* is a group selection cut with retention (both live and dead trees) within small (0.05 ha) harvested patches on the Mount Mansfield State Forest in Vermont (northeastern U.S.) (photo credit: Jeremy Stovall). Shown on the *bottom right* is mixed conifer in which understory trees were first thinned to reduce fuels and then the stand was prescribed burned to mimic historic low-intensity fire (photo credit: Malcolm North)

(Franklin et al. 2000). Biological legacies "lifeboat" organisms through the post-disturbance recovery period, ameliorate site conditions in stressed, post-disturbance environments, and promote accelerated and complex recolonization and successional pathways. To emulate these functions in managed forest stands, structures can be retained in varying densities/volumes and in different spatial patterns (e.g. aggregated vs. dispersed, Aubry et al. 1999). Retention schemes can mimic the landscape level patterns created by natural disturbances, such as, in some cases, greater tree survivorship within riparian zones in areas burned by wildfire (Keeton and Franklin 2004). Permanent retention of legacies, such as living trees, can influence (Zenner 2000) and even accelerate (Keeton and Franklin 2005) long-term stand development processes and recovery from disturbance.

An extension of this research has investigated effects of natural disturbances in mediating late-successional stand development (Abrams and Scott 1989; Lorimer and Frelich 1994). The objective is to develop silvicultural systems that provide a broader range of stand development stages, including old-growth forest habitats and associated functions (Franklin et al. 2002; Keeton 2006). These systems accelerate rates of stand development in young, mature, and riparian forests through underplanting, variable density thinning, crown release, and other methods (Singer and Lorimer 1997; Harrington et al. 2005).

One method of assessing disturbance-based practices has been to compare managed forests to their "historic range of variability" (HRV). Although ecosystem structure and function vary over time and space, HRV suggests there is a bounded range to these conditions that can be compared against the range of conditions produced in managed forests (Aplet and Keeton 1999). There are examples of forest management plans based on reconstructions of HRV (e.g. Cissel et al. 1999; Moore et al. 1999). In practice, however, HRV-based management is difficult to implement. To begin with, the feasibility of quantifying HRV for a given landscape varies greatly depending on data availability and modeling requirements (Parsons et al. 1999). There is the added difficulty of finding appropriate historical reference periods (Millar and Woolfenden 1999). Third, forest managers must determine whether HRV offers a realistic target for management, considering the extent to which conditions within the HRV are compatible with contemporary management objectives as well as altered ecosystem conditions and dynamics attributable to land use history. HRV, however, can provide an informative benchmark or reference for understanding landscape change.

The concepts of disturbance-based forestry have intuitive appeal because they take a cautious, less intrusive approach to management, one that attempts to stay within the bounds of historic conditions and "natural" variability. A central concern, however, is whether these concepts can be implemented in practice. Managers' best efforts to mimic natural disturbance regimes will inevitably involve tradeoffs between economic, social, and ecological objectives. The case studies that follow explore the basis, evolution, and limitations of disturbance-based forest management in the U.S, beginning with the Pacific Northwest where many disturbance-based forestry concepts were first developed.

17.3 Case Studies

17.3.1 Pacific Northwest Forests

Distribution and Current Condition – Temperate coniferous forests in the U.S. Pacific Northwest (PNW) and Canada extend over 2000 km from southeastern Alaska to northern California in a narrow band ranging from 60 to 200 km in width (Franklin and Halpern 2000). Low to mid elevation forests in this region are dominated by large conifers, including most commonly Douglas-fir (*Pseudotsuga menziesii*), western hemlock (*Tsuga heterophylla*), western red cedar (*Thuja plicata*), Sitka spruce (*Picea sitchensis*), Pacific silver fir (*Abies amabilis*), noble fir (*Abies procera*), and in northern California, coast redwood (*Sequoia sempervirens*). The climatc is strongly maritime influenced, having very wet (80–300 cm annual precipitation) mild winters, and warm, dry summers. The forests are noted for having some of the greatest biomass accumulations and highest productivity of any forests in the world. Historically, landscapes in Pacific Northwest were dominated by large areas of continuous forest cover. By some estimates roughly 60–70% of forests were in an old-growth condition (greater than 150 years of age) at any one time (Vogt et al. 1997). Stand structure in PNW forests changes dramatically in response to disturbance and with processes of stand development (Franklin et al. 2002), yielding an array of different biodiversity values and ecosystem functions (Hunter 1999). Therefore, the initial focus of disturbance-based forestry was on managing stand structure and age class distributions in this region (e.g. FEMAT 1993). Young to mature forests, especially in managed stands, tend to have single-layered canopies and low structural complexity, although young stands may have a high carryover of coarse woody debris if they originated from natural disturbances (Spies et al. 1988). Old-growth stand structure is typified by a range of tree sizes, including very large trees, exceptionally high volumes of coarse woody debris (both standing and downed), and vertically continuous canopies which have very high leaf area index values (Gholz 1982) (Fig. 17.2, upper left). The largest trees can reach diameters over 300 cm and heights over 90 m. Understory light availability can be limited beneath closed canopy forests, often producing a sparse or patchy herb and shrub community, extensive moss mats, and saplings and mid-canopies dominated by shade-tolerant tree species (Van Pelt and Franklin 2000). Tree mortality processes shift from density-dependent competition during early stand development to density-independent or disturbance-related mortality late in stand development. Thus, horizontal complexity associated with gap dynamics is a defining characteristic of old-growth forests in the PNW (Franklin et al. 2002; Franklin and Van Pelt 2004).

In the 1980s and '90s, controversy over the PNW's declining late-successional/old-growth (LS/OG) forests and associated biological diversity eventually led to changes in forest management both there and across much of the United States. After several decades of widespread clearcut logging (Fig. 17.2, lower left) and replanting, the majority of LS/OG forest was converted into short rotation (e.g. <60 year) plantations. Today less than 10% (or about 1.8 million ha) remains of the late-successional forest cover extant at the time of European settlement (FEMAT 1993). Loss and fragmentation of habitat at landscape scales has contributed to significant population declines in northern spotted owls (*Strix occidentalis caurina*), marbled murrelets (*Brachyramphus marmoratus*), and other LS/OG associated species. Loss of LS/OG and related high quality spawning and rearing habitats along headwater streams has been one of several factors causing declines in anadromous salmonid populations. By one estimate (FEMAT 1993), over a thousand species of plants, animals, and fungi are associated with LS/OG forests in the PNW.

Historic Disturbance Regimes - Wind and wildfire are the main disturbance agents in PNW forests, although floods, insects and pathogens are important at smaller scales. Though infrequent, large intense fires exert a strong influence on the age-class structure and development patterns of these forests. Historically fire return intervals generally increased along precipitation gradients varying, for example, from about 200 years in central Oregon to over 1000 years in coastal Washington (Agee 1993). Under the right weather conditions, tens of thousands of hectares can burn within a short period. Typically not all trees are killed even during extreme, large-scale wildfires (Morrison and Swanson 1990; Gray and Franklin 1997). Fires usually leave small groups of survivors on landforms providing refugia or dampening effects on fire intensity and spread (Camp et al. 1997). Standing dead trees and scattered live trees, varying by species-specific fire resistance traits, are often widely distributed throughout burn areas, depending on fire intensity, and stand age and structure at the time of disturbance (Keeton and Franklin 2004).

Wind is also an important disturbance in PNW forests at two scales and intensities. Large, catastrophic windstorms strongly influence coastal forests in particular. These storms can blow down large swaths of forests, particularly when soils are saturated after weeks of winter rain. For example, the 1962 Columbus Day windstorm caused a timber blow down in excess of 25 million cubic meters in western Oregon and Washington (Lynott and Cramer 1966). Another windstorm in 1921 blew down approximately 19 million cubic meters of timber along a 110 km long, 50 km wide swath on the west side of Washington's Olympic Peninsula (Guie 1921). Wind is also a chronic disturbance creating small- to moderate-sized gaps within closed canopy forests (Spies et al. 1990; Lertzman et al. 1996). Fine-scaled wind disturbance interacts with trees weakened by fungal pathogens, such as stressed trees, opening up the canopy and increasing understory light availability. Wind disturbances in the PNW typically leave fewer standing trees, compared to wildfires, and greater densities of snapped and up-rooted trees (Franklin et al. 2000).

Disturbance-based management - forests in the PNW have been extensively altered by over 100 years of logging and clearing for development. Following World War II clearcut logging became the dominant type of regeneration harvesting in the region. Clearcutting removes nearly all aboveground structure, whereas wind and fire typically leave abundant biological legacies, including live trees and very large accumulations of coarse woody debris (Kohm and Franklin 1997; Franklin et al. 2000). Studies have documented many differences in plant succession (Halpern and Spies 1995; Turner et al. 1998), soil erosion and nutrient loss

(Sollins and McCorison 1981; Martin and Harr 1989) and biodiversity responses (Hansen et al. 1991) in clearcuts compared to wind and fire created openings. The frequency of large disturbances also differs considerably from harvesting, which is generally practiced on 40–60 year rotations in the Douglas-fir region (Curtis 1997). At the landscape level, dispersed patch clearcutting practiced by the U.S. Forest Service on national forest lands left much of the PNW's forests highly fragmented, with a significant increase in forest edge (Franklin and Forman 1987) and a reduction in interior forest microclimate and habitat conditions (Chen et al. 1990) (Fig. 17.2 bottom left). In response to these changes, some researchers proposed a "new forestry," one which significantly lengthens rotations (Curtis 1997) and retains large green trees, snags, and logs in harvest areas to more closely mimic historic disturbance (Swanson and Franklin 1992; Franklin et al. 1997). With the implementation of the Northwest Forest Plan (NFP) in 1994, redevelopment of LS/OG within reserves established by the plan became a central objective, requiring innovative silvicultural approaches that would accelerate rates of stand development and promote eventual recovery of LS/OG structure and functional conditions (DeBell et al. 1997). Researchers are testing silvicultural systems designed to meet this need, such as variable density thinning (Harrington et al. 2005) and creation of variably sized gaps (Wilson and Puettmann 2005) in young and mature stands. These approximate and accelerate stand development processes, such as spatially variable density-dependent and disturbance related tree mortality, that reduce stand densities, increase light availability, and allow for understory reestablishment of shade-tolerant conifers (Keeton and Franklin 2005). Collectively these processes influence both overstory tree growth rates and redevelopment of the vertically and horizontally complex structure characteristic of late-successional temperate forests (Franklin et al. 2002). Another experimental study, called the "Demonstration of Ecosystem Management Options" (DEMO), is testing the "Variable Retention Harvest System" proposed by Franklin et al. (1997). DEMO is evaluating variable levels of post-harvest retention (ranging from 15 to 70% of basal area) in two spatial patterns, aggregated vs. dispersed (Aubry et al. 1999) (Fig. 17.1). Trees are retained permanently to provide legacy functions and multi-aged structure; biodiversity and regeneration responses will be monitored over the long-term (Aubry et al. 2004). The NFP requires management practices that increase the level of biological legacies which historically were associated with natural disturbance regimes. For instance, where regeneration harvests are employed (i.e. in 1.6 million ha of "matrix" areas), the NFP requires retention forestry practices that leave individual large trees and forest patches within harvest units. In addition, 15% of each 5th field watershed must be left in intact patches of mature and old-growth forest to provide residual structure across large matrix areas. The intent is to provide biological legacies and some degree of habitat connectivity (also achieved using riparian buffers) across managed landscapes. In late-successional reserves created by the NFP, development of late-successional forest structure is the management objective and thus regeneration harvests are prohibited. Only thinnings in stands less than 80 years of age are allowed to accelerate rates of stand development. This strategy addresses the need for large, well distributed, and connected blocks of habitat across the landscape,

Fig. 17.2 *Top left* is a typical Pacific Northwest old growth forest. *Bottom left* is a Pacific Northwest landscape fragmented by clearcut logging. *Middle top* is mixed conifer forest in Yosemite Valley, California in 1890. *Bottom middle* is the same forest in 1970 after many years of fire suppression with an inset photograph of the forest in 1990 (pictures from Gruell 2001). At *top right* is a structurally complex, old-growth northern hardwood stand in New York's Adirondack State Park. *Bottom right* is a young, structural simple secondary northern hardwood forest in Vermont's Green Mountains

in which natural disturbance dynamics will play a formative role. The NFP also encourages development of innovative approaches, particularly in Adaptive Management Areas. In this spirit Cissel et al. (1999) proposed an alternative management plan for one watershed covered by the NFP. Rotation periods and harvesting patterns were based on reconstructions of spatially-explicit fire return intervals, including stand replacement events in riparian forests. The projected result was a less fragmented landscape pattern over time compared to the harvesting pattern required by the NFP, in which placement of harvest units is constrained by the extensive network of riparian reserves.

17.3.2 Western Mixed-Conifer Forests

Distribution and current condition – the classification "mixed conifer" has been loosely applied to many coniferous forest types in North America that have a combination of species in which no one species clearly dominates. In the western United States, mixed conifer usually has a combination of shade-tolerant (e.g. cedars and true firs) and -intolerant (e.g. pines) conifers and is often a mid-elevation forest type, bounded at lower elevation by ponderosa pine (*Pinus ponderosa*) and at higher elevation by fir (e.g. *Abies magnifica*, and *A. lasiocarpa*), spruce (e.g. *Picea engelmanii*) or lodgepole pine (*Pinus contorta*) forests. Mixed conifer is widely distributed in the western U.S. but is most prevalent in the northern Rockies (northeastern Oregon, central Idaho and western Montana), the western slopes of California's Sierra Nevada, central Colorado, and the southern Rockies (northern Arizona and

New Mexico). Stands that were not heavily harvested can contain 300 500 year old trees and some species, such as sugar pine and Douglas-fir, can reach diameters of over 250 cm and 75 m in height.

Across a landscape, mixed-conifer conditions are highly heterogeneous not only due to historic fire regimes (Fig. 17.2 top center) but also because they occupy an elevational band where significant changes in precipitation form (rain vs. snow) and availability (immediate soil wetting vs. snow pack banking) occur over small scales. Spatially variable physiographic and microclimatic conditions can have strong influences on the size of vegetation patches, patch complexity and pattern, and horizontal fuel continuity, which collectively influence fire spread (Taylor and Skinner 2004). A century of fire suppression has homogenized forest patterns at landscape scales making delineation of patches and restoration of patch complexity a central challenge for disturbance-based management.

Historic disturbance regime – historically fire was the key disturbance agent with an average return interval of 15–35 years (Arno 1980; Agee 1991; McKelvey et al. 1996). Across much of the western U.S. this fire regime changed in the late 19th century concurrent with a cooling trend in global climate, an increase in grazing (which reduces herbaceous fuels), and a reduction in Native American ignitions. Beginning in the late 1930s with increased forest road construction and development of effective fire fighting methods, fire suppression also contributed to the reduction in burned acreage. Many mixed-conifer forests have not burned in the 20th century and one study, using the amount of acreage annually burned by wildfire in different forest types, estimated California's mixed conifer now has a fire return interval of 644 years (McKelvey and Busse 1996). Historically, mixed-conifer fires were usually low-intensity surface fires that consumed surface litter and fine fuels, and killed small, thin-barked trees. Researchers have found some evidence of higher intensity burns in the past but it appears these crown fires were infrequent events (>400 years) possibly driven by extreme weather (Stephenson et al. 1991).

Historically fire produced a highly heterogeneous landscape. Within a watershed, riparian areas and valley bottoms had longer fire return intervals, developing higher stem densities and fuel loads than adjacent upland forest (Bisson et al. 2003; Dwire and Kauffman 2003; Stephens et al. 2004). Midslope forests generally experienced frequent fires (8–20 years) and forest conditions were strongly influenced by slope, aspect and soil conditions. Ridgetops characterized by shallow soils and open forest conditions often slowed or contained surface fires because of low fuel loads. Reconstruction of past landscape patterns (Hessburg et al. 2005, 2007) suggest this high degree of heterogeneity was a defining characteristic of low-intensity fire regimes. This heterogeneity is self-reinforcing. The behavior of each successive fire is influenced by the spatially variable fuel conditions left by previous fires, thereby perpetuating patchy stand structures and patterns.

In the absence of fire, current forest conditions have become more homogeneous at all scales (Fig. 17.2 bottom center). When wildfires do occur in these conditions they are often higher severity than they would have been historically, because increases in surface and ladder fuels can sustain crown fires across large areas. Over the last 8 years, Arizona, Colorado, and Oregon have had the largest fires in their recorded histories, with much of each burn area experiencing crown fire and high tree mortality (>75%). The frequency of large high intensity fires is predicted to increase further over the 21st century in mixed-conifer forests due to climate change (Keeton et al. 2007a).

Other disturbance agents (i.e., wind, avalanches and flooding) are present in mixed conifer but historically their impacts have been localized or infrequent. In the absence of fire pests have become the principal mortality agent in mixed conifer attacking high-density, moisture-stressed stands (Ferrell 1996). As an ecological process, however, pests do not replace fire because their mortality is more clustered and does not select for smaller, thin barked trees (Smith et al. 2005). Pest mortality has reduced the number of large, old-growth trees, and increased fuel loading in many forests, exacerbating the potential for high-intensity wildfire.

Disturbance based management – Management of mixed-conifer forests has evolved as desired conditions have changed and research has demonstrated the importance of maintaining critical ecological processes, such as fire. This evolution, however, has produced hybrid management approaches, including practices that reflect past priorities while incorporating new concepts. For example, another subspecies of spotted owl is found in Californian and Southwestern mixed-conifer forests, where logging has reduced the extent of old growth. Consequently management became focused on retaining old-growth structures and providing suitable owl habit. Unlike the Pacific Northwest, however, western mixed-conifer forests are characterized by frequent, low to moderate intensity disturbance rather than long periods of old-forest conditions. Managers often find it difficult to reconcile the emphasize on providing undisturbed habitat for spotted owls and developing large, old trees, because increasing fuel loads threaten to eliminate both if high-severity wildfires burn across the landscape. Fire history studies have long established the frequency of historic burns (Biswell 1973; Agee 1991; McKelvey et al. 1996), and research has identified low-intensity fire as a "keystone" process for restoring and maintaining the ecological functions associated with forest "health" (Falk 2006; North 2006). Low-intensity fire shapes mixed-conifer ecosystems by reducing the understory canopy, slash, litter, and shrub cover, all of which open growing space, provide pulses of soil nutrients, and increase the diversity of plants and microhabitat conditions (Wayman and North 2007; Innes et al. 2006; North et al. 2007).

In mixed conifer, disturbance-based management has begun to focus on process restoration and the importance of influencing fire behavior (Fig. 17.1). In fire-dependent forests management practices are evaluated based on what kind of fuel conditions they create. Modeling software is used to estimate how different post-treatment fuel loads and weather could affect local fire intensity (Stephens 1998; Stephens and Moghaddas 2005). Fuels are reduced until the crowning and torching index (the wind speed needed to produce an active and passive crown fire) for the treated stand are higher than conditions that are likely to occur even under extreme weather events. With air quality regulations, increasing wildland home construction, and limited budgets, many forests cannot be prescribed burned, at least as an initial treatment. Yet restoration of these forests is still dependent on modifying fuels because they control wildfire intensity when the inevitably fire does occur, and in the mean time can produce stand conditions that simulate some of fire's ecological effects.

Disturbance-based management with a focus on process has two potential benefits that traditional silvicultural practices often lack: variability and adaptation to current conditions. Managers have often focused on structural targets, such as thinning all trees up to a maximum diameter limit, consistently applied throughout a treated area. This uniform application, however, is unlikely to produce the variable stand structures and composition that fire would have in the past (Hessburg et al. 2005). Management keyed to manipulating disturbance processes, however, produces different stand structures across a landscape because thinning prescriptions, designed to affect fire behavior, vary depending on a locale's slope position (i.e., riparian, midslope or ridgetop), aspect, and moisture conditions. A second benefit of process-based management is that forest structure and composition are allowed to re-establish to modern dynamic equilibrium by using fire under current climate and ignition conditions (Stephenson 1999; Falk 2006). Annual fluctuations in temperature and precipitation are expected to increase with global warming (Field et al. 1999). Process-focused management lets forests reach their own equilibrium in response to the interaction of fire with current climate conditions.

Landscape level management in mixed conifer is focused more on fire control than strictly mimicking historic disturbance patterns. Mechanical treatments of fuels vary depending on slope position. Riparian areas are usually left alone. Midslope forests are often thinned following process-focused management. Stands are thinned from below (removing the smallest trees first), and ladder and surface fuels are reduced until a wildfire burning through the stand is likely to stay on the ground rather than climbing into the overstory canopy. The location of treatment units, called "Strategically Placed Area Treatments" or SPLATs, follows model predictions about how a fire might move through a burnshed (Finney 2001). Treated units are placed in a stepped herringbone pattern, like speed bumps designed to reduce the rate of fire spread. Ridgetops and forests near wildland urban interfaces (WUIs) are considered control points and are heavily thinned to defensible fuel profile zone (DFPZ) standards to dramatically reduce fuels.

These landscape treatments were largely developed from fire simulation models (Finney 2002, 2003) and do not necessarily match historic landscape patterns. For example, current management practices that avoid riparian areas do not replicate natural fire patterns, because historically fire often reduced fuels and thinned stand structure, albeit not as frequently as adjacent upslope areas (Olson 2000; Dwire and Kauffman 2003; Everett et al. 2003). Another departure from historic landscape patterns is thinning prescriptions along ridgetops. Thinning in these areas reduces canopy cover to 40% by evenly spacing leave trees and separating their crowns. Research, however, has suggested there is limited reduction in crown fire potential through overstory thinning and tree crown separation (Agee et al. 2000; Butler et al. 2004, Agee and Skinner 2005). Furthermore, studies in active fire regime forests (Stephens and Fry 2004; Stephens and Gill 2005), and stand reconstructions (Bonnicksen and Stone 1982; North et al. 2007) indicate forest structures (live trees, snags, logs and regeneration) were highly clustered in forests with frequent low-intensity fire. Even spacing of leave trees produces a regular distribution which significantly departs from historic spatial patterns (North et al. 2004, 2007). Managers, however, have not attempted to reproduce historic conditions because even a small potential gain in fire intensity reduction is considered a priority in these key control areas. Disturbance-based management in mixed conifer is generally mimicking historic stand conditions but failing to replicate landscape-level patterns because of concern over fire containment.

17.3.3 Northern Hardwood Region

Distribution and current condition – the northern hardwood region of eastern North America[1] is characterized by evenly distributed annual precipitation and relatively fertile soils on post-glacial landscapes. The region's forests are thus both generally productive and diverse, comprised primarily of two dominant forest groups, the northern hardwood forest (beech-birch-maple) and the northern coniferous forests (spruce-fir-hemlock, but also white-red-jack pine). Central hardwood forests (oak-hickory) finger northwards through major valleys and along a transition zone in southeastern portions of the region. In New York and the New England states these major formations have been classified into 40 different cover types (Eyre 1980) and four type groups that collectively cover approximately 89% of the northeastern U.S. (Seymour 1995). The later include the northern hardwood or American beech-yellow birch-sugar maple (*Fagus grandifolia-Betula alleganiensis-Acer saccharum*) type; the red spruce-balsam fir (*Picea rubens-Abies balsamea*) type; the eastern white pine-eastern hemlock (*Pinus strobus-Tsuga canadensis*) with mixed hardwoods type; and the oak type (mostly red oak [*Quercus rubra*], but also white oak [*Quercus alba*], black oak [*Quercus velutina*], and others).

A post-European settlement history of land-use exceeding 300 years creates a unique and complex context for application of disturbance-based forestry concepts. Forest cover, composition, age class distribution, and structure in the northern hardwood region have changed dramatically since the 17th and 18th centuries (Cogbill et al. 2002; Lorimer and White 2003). Geophysical heterogeneity, climate variability, and disturbances, which included aboriginal clearing and burning, maintained a dynamic and diverse landscape in which forest structure and composition were spatially and temporally variable (Foster and Aber 2004). The landscape was nevertheless dominated by late-successional and old-growth forests (uneven-aged, >150 years in age), with young forests (up to 15 years old) representing <1–13% of the landscape on average (Lorimer and Frelich 1994; Lorimer and White 2003). Nineteenth century clearing, followed by land abandonment,

---

[1] Includes all or portions of Minnesota, Wisconsin, Michigan, New York, Vermont, New Hampshire, and Maine in the United States, and Ontario, Quebec, New Brunswick, and Nova Scotia in Canada. Delineations sometimes also include portions of Pennsylvania and the southern New England states.

secondary forest redevelopment on old-fields, and 20th century forest management, resulted in the current predominance of young to mature forests.

Research in remnant eastern old-growth over the last two decades has substantially broadened our understanding of structure and composition in pre-settlement forests. These studies have been conducted across a wide range of sites representing a significant portion of the region's biophysical diversity (see review in Keeton et al. 2007b). They tell us that forest structure, both in terms of landscape level patch complexity (Mladenoff and Pastor 1993) and stand structure (Tyrell and Crow 1994; Dahir and Lorimer 1996; McGee et al. 1999) (Fig. 17.2 top right) differs considerably between old-growth forests and the young to mature forests which currently dominate the landscape. Forest management has tended to convert landscapes with complex patch mosaics shaped by wind and other disturbances to simpler configurations (Mladenoff and Pastor 1993). Forest patches are now less diverse in size and less complex in shape. At the stand level younger, secondary forests tend to have less differentiated canopies, lower densities of large trees (both live and dead), lower volumes and densities of downed logs, smaller canopy gaps, and less horizontal variation in stand density (Fig. 17.2 bottom right). These relate both to the limited time over which secondary forest development has occurred, through predominately old-field succession, and forest management practices which tend to set back or hold in check late-successional stand development processes (Keeton 2006). The relative abundance of dominant tree species and their landscape position have also shifted as a result of land use history (Cogbill et al. 2002).

With changes related to land-use history have come shifts in the types of ecosystem goods and services provided by forested landscapes. For instance, young to mature northern hardwood forests provide lower quality habitats for late-successional species (see reviews in Tyrell and Crow 1994; Keddy and Drummond 1996; McGee et al. 1999), lower levels of biomass and associated carbon storage (Krankina and Harmon 1994; Strong 1997; Houghton et al. 1999), and reduced riparian functionality in terms of effects on headwater streams (Keeton et al. 2007b). Interest in disturbance-based forestry has developed as managers look for new approaches offering a broader array of ecosystems goods and services. Rehabilitation of forestlands degraded (e.g. poor stocking and genetic vigor) through intensive high-grade logging, a practice particularly widespread on former industrial timberlands, is another major concern (Kenefic et al. 2005). Disturbance-based approaches have great potential for restoring structural complexity at both landscape and stand scales. This would be achieved using harvesting approaches that emulate both natural disturbance effects and their interaction with processes of stand development, leading to provision of a range of stand structures, developmental stages, and associated ecosystem functions.

Historic disturbance regimes – development of disturbance-based forestry practices begins with an understanding of natural disturbance dynamics and their influence on ecosystem structure and function. In the northern hardwood region, a variety of disturbance agents, including wind, ice, insects, fungal pathogens, beavers (*Castor canadensis*), floods, and fire, have shaped forested landscapes for centuries. Wind disturbances are generally dominant, occurring most frequently as low intensity wind storms that result in fine-scaled canopy gaps. The region also experiences a variety of other types of wind events, including hurricanes, straight line winds and microbursts, and tornadoes. In New England, hurricane frequency and intensity decrease along a gradient running inland from the southeast to the northwest (Boose et al. 2001). Susceptibility to wind disturbance varies with topographic position and orientation relative to wind direction (Foster and Boose 1992), adding to patch complexity at landscape scales. High intensity wind events leave significant accumulations of downed wood debris as well as standing biological legacies, primarily snapped and uprooted stems (Foster 1988). Retention of legacy structure is, therefore, an appropriate way to emulate this type of disturbance.

Seymour et al. (2002) reviewed the literature and found a discontinuity in both frequency and spatial extent of natural disturbances in the northeastern U.S. They concluded that natural disturbances have been either relatively high frequency (e.g. returns intervals of 100 years) with small extent (e.g. 0.05 ha) or very low frequency (e.g. return intervals approaching or exceeding 1000 years) with large extent (e.g. >10 ha). However, recent studies suggest that intermediate intensity disturbances, such as ice storms and microburst wind events, may be more prevalent than previously recognized (Ziegler 2002; Millward and Kraft 2004; Woods 2004; Hanson and Lorimer 2007). These events tend to produce partial to high canopy mortality across a moderate to large sized area, but they can leave abundant residual live and dead or damaged trees (Keeton unpublished data). Remnant trees together with regeneration and release effects, can result in multi-aged stand structures. Multi-cohort silvicultural systems are thus analogous, in some respects, to the age structure produced by intermediate intensity disturbances.

The important role of canopy gap forming disturbances in stand dynamics and related ecosystem functions is well established (Dahir and Lorimer 1996; Runkle 2000). Disturbance gaps usually involve death or damage to individual or small groups of trees. Depending on size and orientation, gaps can result in regeneration of intermediate to shade tolerant species, release of advanced regeneration, and/or competitive release and accelerated growth in proximate overstory trees. In mesic, late-successional forest types, disturbance gaps form at the rate of about 1% of stand area per year on average (Runkle 1982). Gap patterns in northern hardwood stands are often highly diffuse, with individual gaps having irregular form and encompassing scattered residual or legacy trees, both live and dead. Sequential disturbance events can cause gap expansion over time (Foster and Reiners 1986). Gap phase processes are important drivers of both vertical and horizontal structural diversification, particularly late in stand development. Consequently, many late-successional habitat attributes depend on disturbance originated canopy gaps (Keddy and Drummond 1996). Hence, disturbance based forestry practices are often designed to emulate gap processes, especially where management objectives include regeneration of intermediate to shade-tolerant species and maintenance of multi- or uneven-aged structure.

Fire was far less prevalent, historically, in the northern hardwood region in comparison to western coniferous forests, although there were important exceptions. There are a number of fire dependent/fire maintained plant associations, such as pine barren, pitch pine (*Pinus rigida*)/oak communities, and the jack pine (*Pinus banksiana*) seral type in the upper Midwest. Many of these have declined as a result of fire exclusion. Restoration of stand structure and species composition characteristic of historic fire regimes remains an important management objective on appropriate sites. There is debate regarding the geographic extent of Native American burning prior to European settlement, with some authors stressing the amount of grassland and early successional shrubland/forest maintained for berries, game, and agriculture (DeGaaf and Yamasaki 2001, 2003). However, historical evidence suggests that aboriginal fire in the northeastern U.S. was primarily restricted to the vicinity of settlements and travel routes (Russell 1983).

Native insects and pathogens, such as defoliators (e.g. eastern spruce budworm [*Choristoneura fumiferana*]) and root rots (e.g. *Armillaria* spp.), historically had important influences on stand dynamics and habitat complexity at gap and stand scales. Introduced organisms, including beech bark disease (*Nectria* spp.), ash yellows (caused by a mycoplasma-like organism), pear thrips (*Taeniothrips inconsequens*), and hemlock woolly adelgid (*Adelges tsugae*), are among the greatest current threats to forest ecosystem sustainability in the northern forest region. Two exceedingly important species, American chestnut (*Castanea dentate*) and American elm (*Ulmus americana*), were functionally extirpated by exotic pathogens in the 20th century, although efforts are underway to reintroduce hybrid varieties bred for disease resistance. Declines in native tree species impacted by exotic organisms, together with a changing global environment, limits our ability to manage within the HRV and necessitates an adaptive, forward looking approach.

Disturbance-based management – application of disturbance based forestry concepts in the northern hardwood region has a number of things working in its favor. First, the region has had long experience with partial cutting and selection harvesting that in many ways mirrors the relatively frequent and low intensity, fine-scaled disturbances endemic to northern hardwood systems. Secondly, many of the commercially valuable hardwood species, and some of the commercial conifers, have intermediate to high shade tolerances and thus respond favorably, both in growth and regeneration, to low intensity harvests that might emulate natural disturbance effects. However, closer examination of the region's disturbance regime indicates a far greater degree of structural and compositional complexity – with respect to the range of effects associated with different disturbance types, frequencies, intensities, spatial patterns, etc. – than is afforded through conventional silvicultural systems. Hence, developing systems that produce and maintain complexity becomes a central objective of disturbance based forestry.

There are several examples of disturbance based silvicultural systems developed in the northern hardwood and southeastern boreal forest regions (e.g. Harvey et al. 2002; Seymour 2005; Keeton 2006; Seymour et al. 2006). These share a number of concepts that may have broader relevance outside the region. First, some of these systems emulate gap processes, but strive for variety in gap size and shape in a manner similar to heterogeneous disturbance effects. Secondly, they stress retention of biologically legacies to maintain and enrich stand structural complexity over multiple management entries. Restoration and management for stand structural complexity in general is an explicit objective. Thirdly, management for multi- or uneven-aged structure best emulates the dominant structural condition associated with natural disturbance regimes in these regions. Fourth, harvest entry cycles, desired stand age distributions, and percent of stand area harvested at each entry can be modeled on natural disturbance frequencies and scales. And fifth, carefully designed intermediate treatments can emulate the accelerating effect of low intensity natural disturbances on rates of stand development. This is true so long as they maintain and promote development of structural complexity (vertical, horizontal, dead and dying trees, etc.) rather than homogenizing structure, as is typical of conventional thinnings.

To guide disturbance-based forestry in the northeastern U.S. Seymour et al. (2002) proposed a "comparability index" based on their analysis discussed in the preceding section. The index depicts the correspondence between conventional harvest systems and natural disturbance frequencies and scales. Conventional even-aged approaches, such as clearcut logging, are not in synch with natural disturbance frequencies for northern hardwoods if practiced on short rotations (e.g. < 100 years). Extended rotations (see Curtis 1997) would move closer to this benchmark. Entry periods associated with uneven-aged forestry did show a close correspondence with natural frequency; scales were similar but typical group selection openings are generally slightly larger than natural gaps. While Seymour et al. (2002) identified two general regimes using frequency and scale (see preceding section), the various studies reviewed showed considerable variation around the means. This supports the need to vary opening sizes, levels of canopy retention, and spatial patterns to emulate the complexity inherent to natural disturbance regimes.

The principles described above primarily address stand level management. Yet in the northern hardwood region there are questions regarding whether landscape scale age class distributions should be shifted closer to that associated with natural disturbance regimes (Lorimer and White 2003; Keeton 2006). Given the current over abundance of young to mature stands, an artifact of land-use history, this would require a greater emphasis on management for late-successional forest characteristics. Late-successional forests are dramatically under-represented relative to HRV (Lorimer and White 2003). Others have advocated managing for early-successional forest habitats due to declines in some disturbance dependent wildlife species. Proponents of this approach favor patch-cut or large-group selection harvesting methods (Hunter et al. 2001; King et al. 2001; DeGaaf and Yamasaki 2003). Although early successional habitats represented something less than 10% of the landscape historically, there are concerns that grassland/shrubland habitats may be approaching this level in some locales (DeGaaf and Yamasaki 2003). Thus, a disturbance based approach in this region will require consideration of these differing, though not mutually exclusive, proposals for managing age class distributions.

Two examples of experimental research help illustrate the application of disturbance-based forestry concepts to the northern forest region. The first is a project called the "Acadian Forest Ecosystem Research Program" (Seymour 2005; Saunders and Wagner 2005; Seymour et al. 2006). It provides an example of "area based" prescriptions. The study is testing two silvicultural systems, an irregular

Fig. 17.3 Simulated view of the Acadian Forest Ecosystem Research Project areas on the Penobscot Experimental Forest, Maine. Shown is year 11 following treatment for group shelterwood with retention (*left*) and group selection with retention (*right*). The first group expansion has just occurred for the group shelterwood. Gaps are positioned based on actual GPS locations. Visualization of regeneration and reserve trees is based on tally data. Overstory structure is averaged across the blocks. Figure courtesy of Robert Seymour, University of Maine

group shelterwood with reserves (or retention) and a "small gap" group selection with reserves (Fig. 17.3). Both systems emulate "natural disturbance rates, patterns, and structural features of natural forests" by adjusting cutting cycles, removal rates, and reserve tree retention levels (Seymour 2005: 45). They approximate the 1% annual disturbance rate and partial mortality (i.e. persistence of biological legacies) typical of gap dynamics in this region. The first (large gap) treatment is modeled after the German *Femelschlag* or "expanding gap," in which large group harvests (each about 0.2 ha in size) expand previously created openings at each entry. This emulates observed natural gap dynamics (Runkle 1982). Under this approach 20% of stand area is cut every 10 years over 5 entries, followed by 50 years with no harvesting. If advanced regeneration is lacking, 30% of overstory basal area is retained within gaps; at the next entry this is reduced to 10% for permanent retention. The second (small gap) system is a half speed version of the first. It harvests and regenerates 10% of stand area in roughly 0.1 ha patches every 10 years. Individual gaps are expanded every 20 years; the within group retention prescription matches the first treatment. Both systems shift initial single cohort structures to "diverse, irregular within-stand age structures." Long-term retention of reserve trees within groups ensures that legacy large tree structure is maintained throughout the management unit.

A second example is provided by the Vermont Forest Ecosystem Management Demonstration Project (FEMDP) (Keeton 2006). This study is evaluating the ability of modified uneven-aged silvicultural approaches to accelerate rates of stand development. Prescriptions are based primarily on tree diameter distributions. Biodiversity responses (McKenny et al. 2006; Smith et al. 2008) and economic tradeoffs (Keeton and Troy 2006) are of key interest. Modified single-tree selection and group-selection are compared against an alternative approach called "structural complexity enhancement" (SCE). Both of the selection systems include higher levels of post-harvest retention than is typical for the region. The group selection treatment employs small (mean 0.05 ha) but variably sized groups, with light retention of individual live and dead trees within groups, to emulate the scale and structural diversity associated with natural gap dynamics (Fig. 17.1). Compliance with worker safety regulations is maintained by topping large snags within groups and through the use of fully enclosed harvesting machinery. SCE is a restorative approach that promotes development of old-growth structural characteristics (Keeton 2006). It combines a number of disturbance-based silvicultural approaches, including variable density marking to create small gaps, crown release to promote development of large trees, enhancement of coarse woody debris (standing and downed) densities, including pushing or pulling trees over to create tip-up mounds, and an unconventional marking guide based on a rotated sigmoid diameter distribution. The latter reflects the growing appreciation for the disturbance history-related diversity of diameter distributions found in late-successional forests (Goodburn and Lorimer 1999; O'Hara 2001).

Application of disturbance-based forestry at the landscape scale is complicated in the northern forest region because the majority (93%) of forests are privately owned and held in small parcel sizes (now averaging < 4 ha). Mean parcel sizes have been trending downward for several decades due to increasing rates of subdivision and exurban housing and commercial development. This contrasts with many regions of the western U.S., where large proportions of the landscape are in public ownership and can be managed holistically, for instance to plan patch dynamics at large scales. Meeting large scale objectives in highly parcelized landscapes, such as management of age class distributions and scheduling the frequency and spatial pattern of harvests to achieve desired patch configurations, can only be achieved through the collective or combined actions of many individual landowners operating on a parcel by parcel basis. Public land holdings in the region, including national and state forests, offer larger contiguous forest tracts where disturbance-based forest management is directly applicable.

There are, however, policy instruments that could be used to promote broader adoption of disturbance-based management objectives. Increasingly forest conservation on private lands in the Northeast, including large blocks of former and current industrial timberland, is achieved through a combination of incentive based and market mechanisms as well as limited acquisition of high conservation value forests. Conservation easements and tax incentive programs, such as current use value appraisal, provide a means for conserving working forests and promoting sustainable management practices. As former industrial timberlands are transferred to new ownerships under easement, there is the potential to build disturbance-based forestry requirements into conservation agreements and revised management plans.

Forest certification offers another potential avenue for explicate incorporation of disturbance-based forestry concepts into management planning. Finally, community based forestry can help achieve disturbance based objectives through the aggregate contribution of multiple landowners. Community-based initiatives involving multiple landowners provide strength in numbers. Landowners, in effect, voluntarily pool their resources and, to some degree, coordinate management across a larger area. This gives participants access to market opportunities not readily available to individuals. If conducted under a set of agreed upon standards there is an opportunity for disturbance-based forestry through community forestry.

17.4 Lessons

Disturbed-based forest management is increasingly used in forest types across North America to enhance the range of ecosystem goods and services provided by managed forests. Although specific silvicultural systems and implementation vary depending on regional disturbance regimes (Table 17.1), several common advantages and limitations to disturbance-based forestry have emerged.

17.4.1 Limitations

Before regionally specific disturbance-based management systems can be implemented, researchers need to provide comprehensive information on historic and current disturbance regimes, including disturbance frequencies, intensities, patterns, and associated biological legacies. With this information, managers may find that efforts to closely emulate natural disturbance regimes face social and economic constraints. For example, in the Pacific Northwest, large tracts of contiguous forest would need to be treated to emulate the scale of historic wind and fire disturbances. Management has been able to extend the rotation period between harvests and leave more structural legacies, but the public is not receptive to treating large (>400 ha) blocks of forest at one time. This would also carry significant ecological risk due to the current scarcity of late-successional forests (Aplet and Keeton 1999). In mixed-conifer forests, fuels need to be reduced every 15–30 years with either repeated applications of prescribed fire or service contracts that hand thin and pile burn small unmerchantable trees that have accumulated with fire suppression. Both practices can be expensive (e.g. > $200 and >$1000/ha, respectively). Managers are also constrained from mimicking historic landscape patterns because past practices (logging in riparian areas), public health (prescribed fire smoke), and safety concerns (rural homes) limit options. In Northeastern forests, extensive private ownership and a general skepticism of land use regulation makes coordination of landscape-level management difficult.

Table 17.1 Historic disturbances, disturbance-based silviculture, example projects, and management challenges for three regional forest types in the United States

	Pacific Northwest coniferous forests	Western mixed conifer forests	North hardwood forests
Dominant historic disturbances:			
Stand scale	Fine-scaled canopy gaps	Low-moderate intensity wildfire	Low intensity wind; fine-scale canopy gaps
Landscape scale	Infrequent, high-intensity fire and windstorm	Moderate intensity wildfire	Intermediate intensity microbursts and ice storms
			Infrequent, high-intensity hurricanes
Disturbance-based silvicultural systems:	Variable density thinning and underplanting	Fuels reduction that varies by landscape topographic position:	Variable density thinning; crown release
	Group selection/gap creation	Ridgetop: remove understory fuels and leave overstory trees with widely separated crowns	Selection harvesting with structural retention within variably sized groups
	Regeneration harvesting with aggregated and dispersed green tree retention	Midslope: thin from below up to 50–75 cm dbh	Expanding gap systems
	Variable retention harvest system	Riparian: no entry	Multi-cohort systems
Examples of experimental projects:[1]	Demonstration of Ecosystem Management Options	Fire and Fire Surrogate Study	Acadian Forest Ecosystem Research Program
	Olympic Habitat Development Study	The Teakettle Ecosystem Experiment	Vermont Forest Ecosystem Management Demonstration Project
	Montane Alternative Silvicultural Systems	Southern Utah Fuel Management Demonstration Project	
	Variable Retention Adaptive Management Experiments		
Challenges:	Large scale of dominant disturbances	Human constraints on treatment types and intensities	Extensive private ownership of small parcels

[1] For literature describing the project examples see Peterson and Maguire (2005).

Disturbance-based forestry practices have been legitimately criticized for carrying significant uncertainty when it comes to producing the process effects induced by natural disturbances (Lindenmayer et al. 2007). For instance, foresters can approximate the structural legacies and patterns associated with wind throw, but they may not achieve (or may only achieve in part) the same effects on soil turnover, soil carbon dynamics, and nutrient cycling. Similarly, thinning can restore the stand and landscape structures that historically supported low to moderate intensity fire regimes, but may fall short when it comes to the full range of effects on ecosystem processes associated with frequent natural fire (North 2006). Lindemayer et al. (2007) point out that the specific sequence of disturbances over time, their timing, intensity, type, and pattern, can result in complex process effects that may be hard to approximate through management.

These limitations, however, do not mean that disturbance-based forest management is fundamentally impractical or scientifically flawed. But they do suggest that forest managers often cannot fully or directly emulate historic disturbance patterns at the stand level, and are particularly limited at landscape scales. Rather, knowledge and inferences based on natural disturbance regimes can be used to guide and modify silvicultural manipulations to achieve a more limited set of objectives.

17.4.2 Modifying Silviculture to Better Match Disturbance Regimes

Silviculture has traditionally focused on manipulating stands (Oliver and Larson 1996) to influence forest succession while extracting wood products (Smith 1986). Thinning guidelines are developed to achieve a desired age structure, diameter distribution, species composition, and spatial pattern. This approach can attempt to engineer forest structure to fit a concept of stand dynamics that may not match disturbance processes. For example, to produce "semi-natural" forest conditions silviculturists have sometimes relied on the principles of uneven-aged silviculture (Smith 1986), which suggest cutting to a negative exponential or reverse-J shaped diameter distribution to produce a multi-aged structure. This was the shape of the diameter distribution North et al. (2007) found in unmanaged, fire-suppressed mixed conifer (Fig. 17.4, pretreatment bar) and which was maintained with diameter-based thinning prescriptions (Fig. 17.4, understory and overstory thinning bars). However, a reconstruction of the same forest in 1865, when it had an active fire regime, found an almost flat diameter distribution (Fig. 17.4, 1865 reconstruction bar), probably resulting from pulses of mortality and recruitment associated with fires and wet El Niño years (North et al. 2005). O'Hara (2001; O'Hara and Gersonde 2004) has pointed out that seral development and local disturbance patterns can produce a wide variety of diameter distributions in natural stands. Similar variability in age class structure has been documented in the Pacific Northwest (Zenner 2005) and the northern hardwood region (Goodburn and Lorimer 1999). Thus,

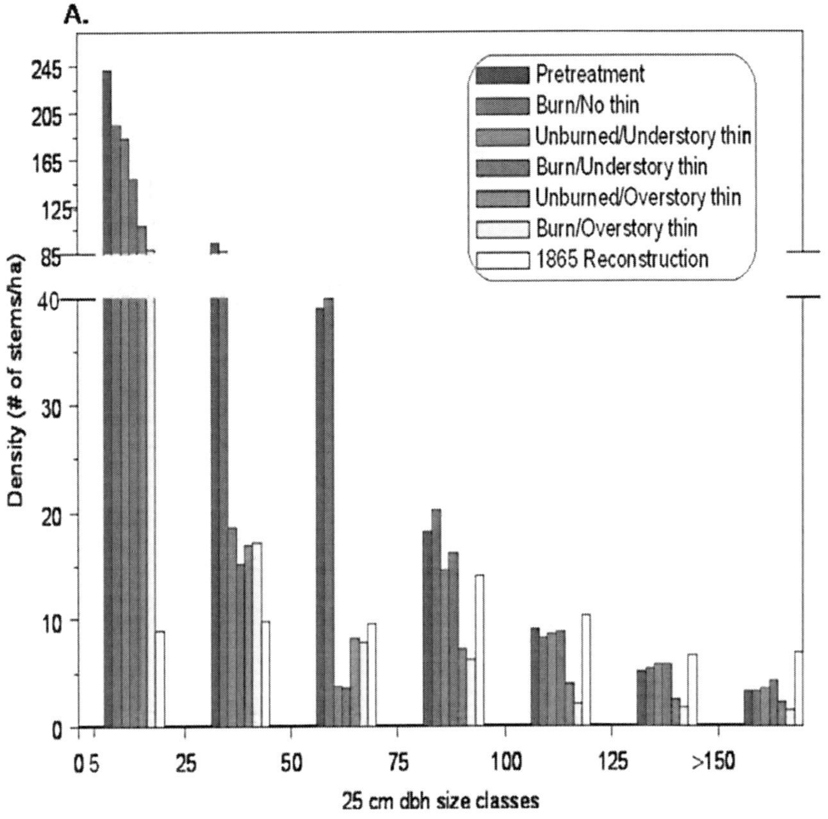

Fig. 17.4 Density of trees in 25 cm diameter classes in old-growth, mixed conifer at the Teakettle Experimental Forest. The pretreatment forest (fire suppressed modern conditions, *blue bar*) has a reverse-j shaped diameter distribution, as do the five silvicultural treatments used in an effort to reduce fuels and restore historic stand conditions. The reconstruction of stand conditions in 1865 (*white bar*), however, indicates a fairly flat diameter distribution and a greater number of large trees. Figure from North et al. 2007

modified silvicultural practices might manage for a broader range of diameter distributions and age-class structures more characteristic of local disturbance regimes (O'Hara 2001; Keeton 2006).

17.4.3 Comparing Management Practices to Natural Disturbances

One potential method for evaluating silvicultural practices is to examine their congruence with historic disturbance events. For example Seymour et al.'s (2002) comparability index evaluates the size and rotation length of management treatments against the scale and frequency of regional natural disturbance patterns. This interesting approach builds on two of the three characteristics of disturbance that

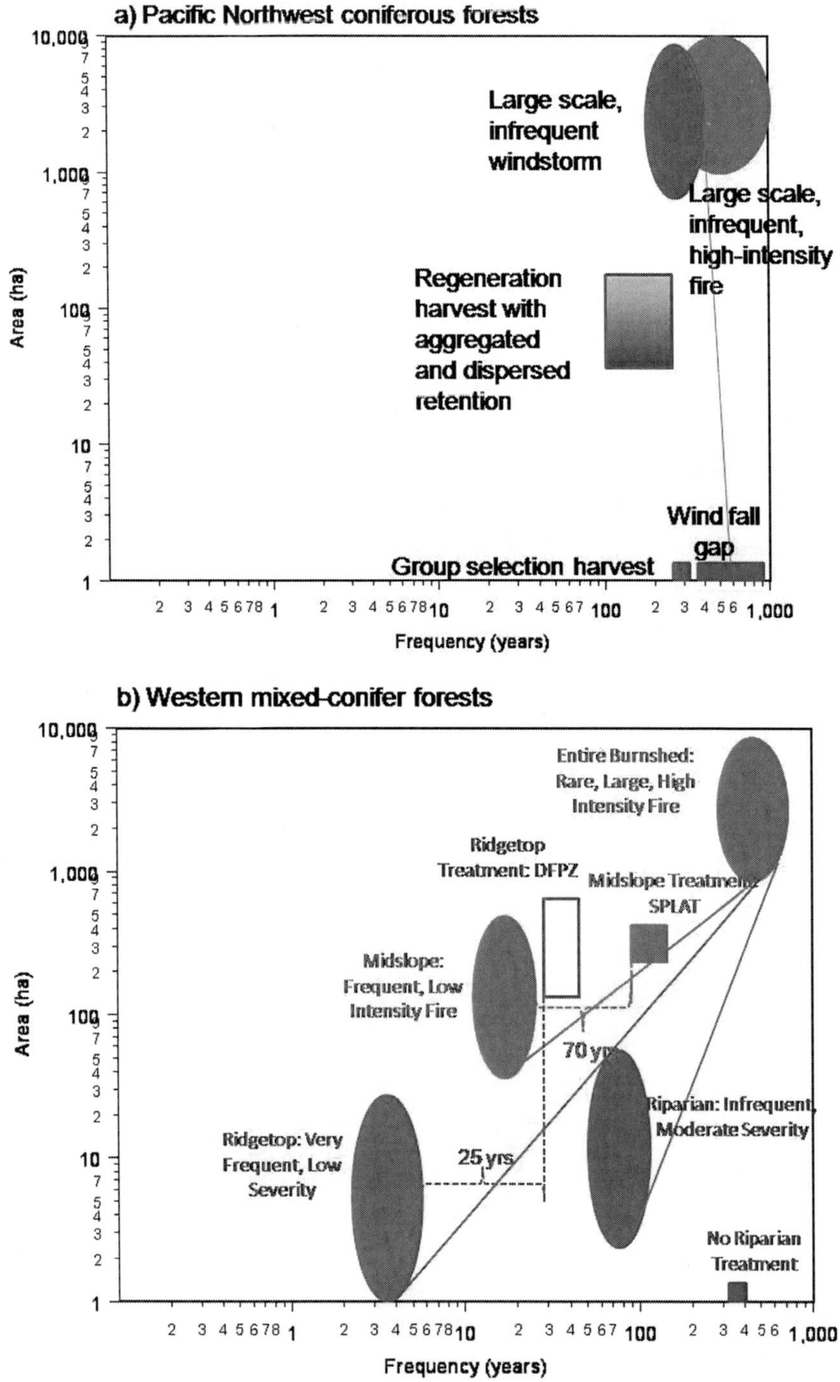

Fig. 17.5 (continued)

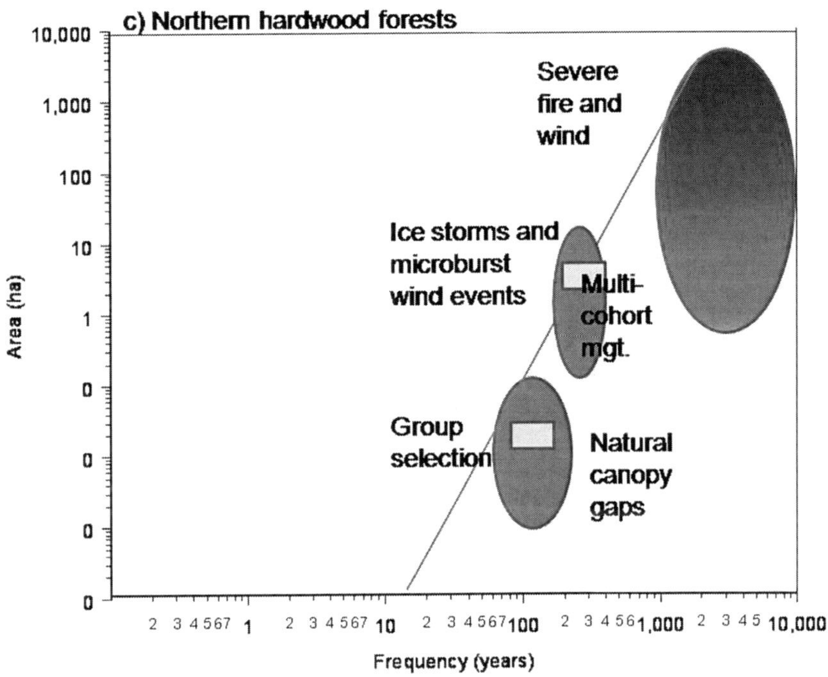

Fig. 17.5 A comparison of natural disturbance regimes and management treatments based on concepts in Seymour et al. (2002) in (**a**) Pacific Northwest coniferous forests; (**b**) Western mixed-conifer forests; and (**c**) Northeastern hardwood forests. The *x* axis is a logorithmic scale of the frequency of events in years and the *y* axis is a logorithmic scale of the size of events in hectares. Ovals represent historic disturbance regimes and rectangles represent management practices. For each oval and rectangle, shape width is the frequency range (*in years*) for a disturbance type, shape height is the range of scales (*in hectares*) and shape fill (*shaded* for aggregated, *non-shaded* for dispersed) is the pattern of biological legacies. The diagonal lines between the rare, large-scale and more frequent, small-scale ovals are a reference for the bounds (longest return interval and smallest scale) of each forest type's natural disturbance regime. The Northeastern hardwood diagram modifies one in Seymour et al. (2002), adding a hypothesized intermediate disturbance regime suggested by recent research (Millward and Kraft 2004; Woods 2004; Hanson and Lorimer 2007)

some researchers (Hunter 1999; Lindenmayer and Franklin 2002) have suggested using to evaluate management activities. In addition to Seymour et al.'s (2002) choice of scale and frequency, we suggest a third evaluation criterion, the level of biological legacies left by historic disturbances. We compared current silvicultural practices in the three regional case studies against the historic disturbance regimes for those forest types (Figs. 17.5a, 17.5b, 17.5c) using Seymour et al.'s (2002) concept. The Pacific Northwest case study illustrates the difficulty in using disturbance-based management at the landscape level. Fire and high-intensity wind disturbances generally affected large areas (>1000 ha), which managers cannot directly emulate due to competing management objectives (Fig. 17.5a). In mixed conifer, managers are attempting to vary their treatments across the landscape, depending on topographic position, but with varying success (Fig. 17.5b). The most significant management departure from historic disturbance patterns is for riparian zones which are currently not being treated and may act like wicks to spread crown fire throughout the landscape. Ridgetop treatments, the creation of defensible fuel profile zones, are conducted on a much larger scale than historic ridgetop fire sizes and are leaving trees regularly spaced rather than grouped together. Group selection cutting in Northeastern hardwood forests approximates fine-scaled gap disturbances but there is little opportunity to coordinate this approach at landscape scales because of extensive private, small-scale ownership (Fig. 17.5c).

Our case studies suggest that social values, competing ecological objectives, and encroaching human settlement sometimes constrain our ability to emulate natural disturbance dynamics at landscape scales. Although managers may not be able to meet all landscape objectives, by comparing silvicultural treatments against the scale, frequency, and biological legacies characteristic of historic disturbances they can understand where compromises are made and risks accrue.

Disturbance-based forest management is a conceptual approach where the central premise might be summarized as "manipulation of forest ecosystems should work within the limits established by natural disturbance patterns prior to extensive human alteration of the landscape" (Seymour and Hunter 1999). Although such an objective seems like a simple extension of traditional silviculture, it fundamentally differs from past fine filter approaches that have manipulated forests for specific objectives such as timber production, water yield, or endangered species habitat. Some critics have argued that this approach leaves managers without clear guidelines because the scale and processes of ecosystems are poorly defined, making it difficult to directly emulate the ecological effects of natural disturbances (Oliver and Larson 1996). Disturbance-based management, however, readily acknowledges these uncertainties. It emphasizes a cautious approach, targeted at those specific management objectives, such as provision of complex habitat structures, reduced harvesting impacts, and landscape connectivity, that can be achieved. Although this approach will require changes in how management success is evaluated, disturbance-based management is likely to minimize adverse impacts on complex ecological processes that knit together the forest landscape.

References

Abrams M, Scott M (1989) Disturbance-mediated accelerated succession in two Michigan forest types. Forest Sci 35:42–49

Agee J (1991) Fire history along an elevational gradient in the Siskiyou Mountains, Oregon. Northwest Sci 65:188–199

Agee J (1993) Fire ecology of Pacific Northwest forests. Island Press, Washington, DC

Agee J, Bahro B, Finney M et al (2000) The use of shaded fuelbreaks in landscape fire management. Forest Ecol Manag 127:55–66

Agee J, Skinner C (2005) Basic principles of forest fuel reduction treatments. Forest Ecol Manag 211:83–96

Aplet G, Keeton W (1999) Application of historical range of variability concepts to biodiversity conservation. In: Baydack R, Campa H, Haufler J (eds) Practical approaches to the conservation of biological diversity. Island Press, Washington, DC

Arno S (1980) Forest fire history of the northern Rockies. J Forest 78:460–465

Attiwell P (1994) Ecological disturbance and the conservative management of eucalypt forests in Australia. Forest Ecol Manag 63:301–346

Aubry K, Amaranthus M, Halpern C et al (1999) Evaluating the effects of varying levels and patterns of green-tree retention: Experimental design of the DEMO study. Northwest Sci 73:12–26

Aubry, K, Halpern C, Maguire D (2004) Ecological effects of variable-retention harvests in the northwestern United States: The DEMO study. For Snow Landsc Res 78:119–134

Beese W, Bryant A (1999) Effect of alternative silvicultural systems on vegetation and bird communities in coastal montane forests of British Columbia, Canada. Forest Ecol Manag 115:231–242

Bisson P, Rieman B, Luce C et al (2003) Fire and aquatic ecosystems of the western USA: Current knowledge and key questions. Forest Ecol Manag 178:213–229

Biswell H (1973) Fire ecology in ponderosa pine-grassland. Tall Tim Fire Ecol Conf 12:69–96

Bonnicksen T, Stone E (1982) Reconstruction of a presettlement giant sequoia-mixed conifer forest community using the aggregation approach. Ecology 63:1134–1148

Boose E, Chamberlin K, Foster D (2001) Landscape and regional impacts of hurricanes in New England. Ecol Monogr 71:27–48

Bunnell F (1995) Forest-dwelling fauna and natural fire regimes in British Columbia: Patterns and implications for conservation. Conserv Biol 9:636–644

Butler B, Finney M, Andrews P et al (2004) A radiation-driven model for crown fire spread. Can J For Res 34:1588–1599

Camp A, Oliver C, Hessburg P et al (1997) Predicting late-successional fire refugia pre-dating European settlement in the Wenatchee Mountains. Forest Ecol Manag 95:63–77

Chen J, Franklin J, Spies T (1990) Microclimatic pattern and basic biological responses at the clearcut edges of old-growth Douglas-fir stands. Northwest Environ J 6:424–425

Cissel J, Swanson F, Weisberg P (1999) Landscape management using historical fire regimes: Blue River, Oregon. Ecol Appl 9:1217–1231

Cogbill C, Burk J, Motzkin G (2002) The forests of presettlement New England, USA: Spatial and compositional patterns based on town proprietor surveys. J Biogeogr 29:1279–1304

Curtis R (1997) The role of extended rotations. In: Kohm K, Franklin J (eds) Creating a forestry for the twenty-first century: The science of ecosystem management. Island Press, Washington, DC

Dahir S, Lorimer C (1996) Variation in canopy gap formation among developmental stages of northern hardwood stands. Can J For Res 26:1875–1892

DeBell D, Curtis R, Harrington C et al (1997) Shaping stand development through silvicultural practices. In: Kohm K, Franklin J (eds) Creating a forestry for the twenty-first century: The science of ecosystem management. Island Press, Washington, DC

DeGaaf R, Yamasaki M (2001) New England wildlife: Habitat, natural history, and distribution. University of New England Press, Hanover, NH

DeGaaf R, Yamasaki M (2003) Options for managing early-successional forest and shrubland bird habitats in the northeastern United States. Forest Ecol Manag 185:179–191

Dudley N, Phillips A (2006) Forests and protected areas: Guidance on the use of IUCN protected area management categories. IUCN – The World Conservation Union, Gland, Switzerland

Dwire K, Kauffman J (2003) Fire and riparian ecosystems in landscapes of the western USA. Forest Ecol Manag 178:61–74

Everett R, Schellhaas R, Ohlson P et al (2003) Continuity in fire disturbance between riparian and adjacent sideslope Douglas-fir forests. Forest Ecol Manag 175:31–47

Eyre F (ed) (1980) Forest cover types of the United States and Canada. Society of American Foresters, Bethesda, MD

Falk D (2006) Process-centred restoration in a fire-adapted ponderosa pine forest. J Nat Conserv 14:140–151
FEMAT (1993) Forest ecosystem management: An ecological, economics, and social assessment. USDA Forest Service, Portland, OR
Ferrell G (1996) The influence of insect pests and pathogen on Sierra forests. In: Sierra Nevada Ecosystem Project: Final report to congress, vol II. University of California, Centers for Water and Wildlands Resources, Davis, CA
Field C, Daily G, Davis F et al (1999) Confronting climate change in California: Ecological impacts on the Golden State. 1999. Union of Concerned Scientists, Cambridge, MA and Ecological Society of America, Washington, DC
Finney M (2001) Design of regular landscape fuel treatment patterns for modifying fire growth and behavior. Forest Sci 47:219–228
Finney M (2002) Fire growth using minimum travel time methods. Can J For Res 32:1420–1424
Finney M (2003) Calculation of fire spread rates across random landscapes. Int J Wildland Fire 12:167–174
Foster J, Reiners W (1986) Size distribution and expansion of canopy gaps in a northern Appalachian spruce-fir forest. Plant Ecol 68:109–114
Foster D (1988) Species and stand response to catastrophic wind in central New England, U.S.A. J Ecol 76:135–151
Foster D, Boose E (1992) Patterns of forest damage resulting from catastrophic wind in central New England, USA. J Ecol 80:79–98
Foster D, Knight D, Franklin J (1998) Landscape patterns and legacies resulting from large, infrequent forest disturbances. Ecosystem 1:497–510
Foster D, Aber J (eds) (2004) Forests in time: The environmental consequences of 1,000 years of change in New England. Yale University Press, New Haven, CT
Franklin J, Forman R (1987) Creating landscape patterns by forest cutting: Ecological consequences and principles. Landscape Ecol 1:5–18
Franklin J, Berg D, Thornburgh D et al (1997) Alternative silvicultural approaches to timber harvesting: Variable retention harvest systems. In: Kohm K, Franklin J (eds) Creating a forestry for the twenty-first century: The science of ecosystem management. Island Press, Washington, DC
Franklin J, Halpern C (2000) Pacific northwest forests. In: Barbour M, Billings W (eds) North American terrestrial vegetation. 2nd edn. Cambridge University Press, New York City, NY
Franklin J, Lindenmayer D, MacMahon J et al (2000) Threads of continuity: Ecosystem disturbance, recovery, and the theory of biological legacies. Conserv Biol Pract 1:8–16
Franklin J, Spies T, Van Pelt R et al (2002) Disturbances and the structural development of natural forest ecosystems with some implications for silviculture. Forest Ecol Manag 155:399–423
Franklin J, Van Pelt R (2004) Spatial aspects of structural complexity in old-growth forests. J Forest 102:22–28
Gholz H (1982) Environmental limits on above-ground net primary production, leaf area, and biomass in vegetation zones of the Pacific Northwest. Ecology 63:469–481
Goodburn J, Lorimer C (1999) Population structure in old-growth and managed northern hardwoods: An examination of the balanced diameter distribution concept. Forest Ecol Manag 118:11–29
Gray A, Franklin J (1997) Effects of multiple fires on the structure of southwestern Washington forests. Northwest Sci 71:174–185
Gregory S, Swanson F, McKee W (1997) An ecosystem perspective of riparian zones. BioScience 40:540–551
Gruell G (2001) Fire in Sierra Nevada Forests: A photographic interpretation of ecological change since 1849. Mountain Press Publishing, Berkeley, CA
Guie H (1921) Washington's forest catastrophe. American Forest 27:379–382
Halpern C, Spies T (1995) Plant species diversity in natural and managed forests of the Pacific Norhtwest. Ecol Appl 5:913–934

Hansen A, Spies T, Swanson F et al (1991) Conserving biodiversity in managed forests: Lessons from natural forests. BioScience 41:382–392
Hanson J, Lorimer C (2007) Forest structure and light regimes following moderate wind storms: Implications for multi-cohort management. Ecol Appl 17:1325–1340
Harrington C, Roberts S, Brodie L (2005) Tree and understory responses to variable-density thinning in western Washington. In: Peterson C, Maguire D (eds) Balancing ecosystem values: Innovative experiments for sustainable forestry. USDA Forest Service General Technical Report PNW-GTR-635
Harrison R (2002) Forests: The shadow of civilization. The University of Chicago Press, Chicago, IL
Harvey B, Leduc A, Gauthier S et al (2002) Stand-landscape integration in natural disturbance-based management of the southern boreal forest. Forest Ecol Manag 155:369–385
Hessburg P, Agee J, Franklin J (2005) Dry forests and wildland fires of the inland Northwest USA: Contrasting the landscape ecology of the pre-settlement and modern eras. Forest Ecol Manag 211:117–139
Hessburg P, Slater B, James K (2007) Re-examining fire severity relations in pre-management era mixed conifer forests: Inferences from landscape patterns of forest structure. Landscape Ecol 22:5–24
Houghton R, Hackler J, Lawrence K (1999) The U.S. carbon budget: Contributions from land-use change. Science 285:574–578
Hunter M Jr (ed) (1999) Maintaining biodiversity in forest ecosystems. Cambridge University Press, New York City, NY
Hunter W, Buechler D, Canterbury R et al (2001) Conservation of disturbance dependent birds in eastern North America. Wildlife Soc B 29:425–439
Innes J, North M, Williamson N (2006) Effect of thinning and prescribed fire restoration treatments on woody debris and snag dynamics in a Sierran old-growth mixed-conifer forest. Can J For Res 36:3183–3193
Keddy P, Drummond C (1996) Ecological properties for the evaluation, management, and restoration of temperate deciduous forest ecosystems. Ecol Appl 6:748–762
Keeton W (2006) Managing for late-successional/old-growth characteristics in northern hardwood-conifer forests. Forest Ecol Manag 235:129–142
Keeton W (2007) Role of managed forestlands and models for sustainable forest management: Perspectives from North America. George Wright Forum 24:38–53
Keeton W, Franklin J (2004) Fire-related landform associations of remnant old-growth trees in the southern Washington Cascade Range. Can J For Res 34:2371–2381
Keeton W, Franklin J (2005) Do remnant old-growth trees accelerate rates of succession in mature Douglas-fir forests? Ecol Monogr 75:103–118
Keeton W, Troy A (2006) Balancing ecological and economic objectives while managing for late-successional forest structure. In: Zahoyska L (ed) Ecologisation of economy as a key prerequisite for sustainable development. Proceedings of the international conference, Sept. 22–23, 2005, Ukrainian National Forestry University, L'viv, Ukraine
Keeton W, Franklin J, Mote P (2007a) Climate variability, climate change, and western wildfire with implications for the suburban-wildland interface. In: Troy A, Kennedy R (eds) Living on the edge: Economic, institutional and management perspectives on wildfire hazard in the urban interface. Advances in the Economics of Environmental Resources, vol 6. Elsevier Sciences, New York, NY
Keeton W, Kraft C, Warren D (2007b) Mature and old-growth riparian forests: Structure, dynamics, and effects on Adirondack stream habitats. Ecol Appl 17:852–868
Kenefic L, Sendak P, Brissette J (2005) Comparison of fixed diameter-limit and selection cutting in northern conifers. Northern J Appl For 22:77–84
King D, DeGraaf R, Griffin C (2001) Productivity of early-successional shrubland birds in clearcuts and groupcuts in an eastern deciduous forest. J Wildlife Manag 65:345–350
Kohm K, Franklin J (eds) (1997) Creating a forestry for the 21st century: The science of ecosystem management. Island Press, Washington, DC

Krankina O, Harmon M (1994) The impact of intensive forest management on carbon stores in forest ecosystems. World Resour Rev 6:161–177

Lertzman K, Sutherland G, Inselberg A et al (1996) Canopy gaps and the landscape mosaic in a coastal temperate rain forest. Ecology 77:1254-1270

Lindenmayer D, Franklin J (2002) Conserving forest biodiversity: A comprehensive multiscaled approach. Island Press, Washington, DC

Lindenmayer D, Hobbs R, Montague-Drake R et al (2007) A checklist for ecological management of landscapes for conservation. Ecol Lett 10:1–14

Lorimer C, Frelich L (1994) Natural disturbance regimes in old-growth northern hardwoods: Implications for restoration efforts. J For 92:33–38

Lorimer C, White A (2003) Scale and frequency of natural disturbances in the northeastern U.S.: Implications for early-successional forest habitats and regional age distributions. Forest Ecol Manag 185:41–64

Lynott R, Cramer O (1966) Detailed analysis of the 1962 Columbus Day windstorm in Oregon and Washington. Mon Weather Rev 94:105–117

Martin C, Harr R (1989) Logging of mature Douglas fir in western Oregon has little effect on nutrient output budgets. Can J For Res 19:35–43

McGee G, Leopold D, Nyland R (1999) Structural characteristics of old-growth, maturing, and partially cut northern hardwood forests. Ecol Appl 9:1316–1329

McKelvey K, Busse K (1996) Twentieth-century fire patterns on forest service lands. In: Sierra Nevada Ecosystem Project: Final report to congress, vol II. Wildland Resources Center Report No. 37, University of California, Davis

McKelvey K, Skinner C, Chang, C et al (1996) An overview of fire in the Sierra Nevada. In: Sierra Nevada Ecosystem Project: Final report to congress, vol II, Assessments and scientific basis for management options. Centers for Water and Wildland Resources, University of California, Davis, CA

McKenny H, Keeton W, Donovan T (2006) Effects of structural complexity enhancement on eastern red-backed salamander (Plethodon cinereus) populations in northern hardwood forests. Forest Ecol Manag 230:186–196

Millar C, Woolfenden W (1999) The role of climate change in interpreting historical variability. Ecol Appl 9:1207–1216

Millward AA, Kraft, CE (2004) Physical influences of landscape on a large-extent ecological disturbance: The northeastern North American ice storm of 1998. Landscape Ecol 19:99–111

Mitchell R, Palik B, Hunter Jr M (2002) Natural disturbances as a guide to silviculture. Forest Ecol Manag 155:315–317

Mladenoff D, Pastor J (1993) Sustainable forest ecosystems in the northern hardwood and conifer forest region: Concepts and management. In: Aplet G, Johnson N, Olson J et al (eds) Defining sustainable forestry. Island Press, Washington, DC

Moore M, Covington W, Fule P (1999) Reference conditions and ecological restoration: A southwestern ponderosa pine perspective. Ecol Appl 9:1266–1277

Morrison P, Swanson F (1990) Fire history and pattern in a Cascade Mountain landscape. USDA Forest Service General Technical Report PNW-GTR-254

Naiman R, Bilby R, Bisson P (2005) Riparian ecology and management in the Pacific Coastal Rain Forest. BioScience 50:996–1011

North M, Chen J, Oakley B, et al (2004) Forest stand structure and pattern of old-growth western hemlock/Douglas-fir and mixed-conifer forest. Forest Sci 50:299–311

North M, Hurteau M, Fiegener R et al (2005) Influence of fire and El Nino on tree recruitment varies by species in Sierran mixed conifer. Forest Sci 51:187–197

North M (2006) Restoring forest health: Fire and thinning effects on mixed-conifer forests. USDA PSW Science Perspective 7

North M, Innes J, Zald H (2007) Comparison of thinning and prescribed fire restoration treatments to Sierran mixed-conifer historic conditions. Can J For Res 37:331–342

O'Hara K (2001) The silviculture of transformation—a commentary. Forest Ecol Manag 151:81–86
O'Hara K, Gersonde R (2004) Stocking control concepts in unevenaged-silviculture. Forestry 77:131–143
Oliver C, Larson B (1996) Forest stand dynamics. Updated edn. John Wiley and Sons, New York, NY
Olson D (2000) Fire in riparian zones: A comparison of historical fire occurrence in riparian and upslope forests in the Blue Mountains and Southern Cascades of Oregon. Thesis, University of Washington
Palik B, Pederson N (1996) Natural disturbance and overstory mortality in longleaf pine ecosystems. Can J For Res 26:2035–2047
Palik B, Robl J (1999) Structural legacies of catastrophic windstorm in a mature Great Lakes aspen forest. USDA Forest Service, Research Paper, NC-337
Parsons D, Swetnam T, Christensen N (1999) Uses and limitations of historical variability concepts in managing ecosystems. Ecol Appl 9:1177–1178
Perry D, Amaranthus M (1997) Disturbance, recovery, and stability. In: Kohm K, Franklin J (eds) Creating a forestry for the twenty-first century: The science of ecosystem management. Island Press, Washington, DC
Peterson C, Maguire D (eds) (2005) Balancing ecosystem values: Innovative experiments for sustainable forestry. USDA Forest Service General Technical Report PNW-GTR-635
Runkle J (1982) Patterns of disturbance in some old-growth mesic forests of eastern North America. Ecology 63:1533–1546
Runkle J (2000) Canopy tree turnover in old-growth mesic forests of eastern North America. Ecology 81:554–567
Russell E (1983) Indian-set fires in the forests of the northeastern United States. Ecology 64:78–88
Saunders M, Wagner R (2005) Ten-year results of the Forest Ecosystem Research Program: Successes and challenges. In: Peterson C, Maguire D (eds) Balancing ecosystem values: Innovative experiments for sustainable forestry. USDA Forest Service General Technical Report PNW-GTR-635
Schulte L, Mitchell R, Hunter Jr M et al (2006) Evaluating the conceptual tools for forest biodiversity conservation and their implementation in the U.S. Forest Ecol Manag 232:1–11
Seymour R (1995) The northeastern region. In: Barrett J (ed) Regional Silviculture of the United States. 3rd edn. John Wiley and Sons Inc., New York, NY
Seymour R, Hunter M (1999) Principles of ecological forestry. In: Hunter M (ed) Maintaining biodiversity in forest ecosystems. Cambridge University Press, Cambridge, MA
Seymour R, White A, deMaynadier P (2002) Natural disturbance regimes in northeastern North America—evaluating silvicultural systems using natural scales and frequencies. Forest Ecol Manag 155:357–367
Seymour R (2005) Integrating natural disturbance parameters into conventional silvicultural systems: Experience from the Acadian forest of northeastern North America. In: Peterson C, Maguire D (eds) Balancing ecosystem values: Innovative experiments for sustainable forestry. USDA Forest Service General Technical Report PNW-GTR-635
Seymour R, Guldin J, Marshall D et al (2006) Large-scale, long-term silvicultural experiments in the United States. Allg Forst Jagdztg 177:104–112
Singer M, Lorimer C (1997) Crown release as a potential old-growth restoration approach in northern hardwoods. Can J For Res 27:1222–1232
Smith D (1986) The practice of Silviculture. 8th edn. John Wiley and Sons Inc., New York, NY
Smith T, Rizzo D, North M (2005) Patterns of mortality in an old-growth mixed-conifer forest of the Southern Sierra Nevada, California. Forest Sci 51:266–275
Smith K, Keeton W, Twery M et al (2008) Understory plant responses to uneven-aged forestry alternatives in northern hardwood-conifer forests. Can J For Res 38:1303–1318
SNFPA (2004) Sierra Nevada Forest Plan amendment: Final environmental impact statement, vols 1–6. USDA Forest Service, Pacific Southwest Region, Vallejo, CA

Sollins P, McCorison F (1981) Nitrogen and carbon solution chemistry of an old-growth coniferous forest watershed before and after clearcutting. Water Resour Res 17:1409–1418
Spies T, Franklin J, Thomas T (1988) Coarse woody debris in Douglas-fir forests of western Oregon and Washington. Ecology 69:1689–1702
Spies T, Franklin J, Klopsch M (1990) Canopy gaps in Douglas fir forests of the cascade mountains. Can J For Res 20:649–658
Stephens S (1998) Evaluation of the effects of silvicultural and fuels treatments on potential fire behaviour in Sierra Nevada mixed-conifer forests. Forest Ecol Manag 105:21–35
Stephens S, Fry D (2004) Spatial distribution of regeneration patches in an old-growth Pinus jeffreyi-mixed conifer forest in northwestern Mexico. J Veg Sci 16:693–702
Stephens S, Meixner M, Poth M et al (2004) Prescribed fire, soils, and stream water chemistry in a watershed in the Lake Tahoe Basin, California. Int J Wildland Fire 13:27–35
Stephens S, Gill S (2005) Forest structure and mortality in an old-growth Jeffrey pine-mixed conifer forest in north-western Mexico. Forest Ecol Manag 205:15–28
Stephens S, Moghaddas J (2005) Silvicultural and reserve impacts on potential fire behavior and forest conservation: Twenty-five years of experience from Sierra Nevada mixed conifer forests. Biol Conserv 125:369–379
Stephenson N, Parsons D, Swetnam T (1991) Restoring natural fire to the sequoia-mixed conifer forest: Should intense fire play a role? In: Hermann S (ed) Proceedings of the 17th tall timbers fire ecology conference: High-intensity fire in wildlands: Management challenges and options. Tall Timbers Research Station, Tallahassee, FL
Stephenson N (1999) Reference conditions for giant sequoia forest restoration: Structure, process, and precision. Ecol Appl 9:1253–1265
Strong T (1997) Harvesting intensity influences the carbon distribution in a northern hardwood ecosystem. USDA Forest Service Research Paper NC-329
Swanson F, Franklin J (1992) New forestry principles from ecosystem analysis of Pacific Northwest forests. Ecol Appl 2:262–274
Taylor A, Skinner C (2004) Spatial patterns and controls on historical fire regimes and forest structure in the Klamath Mountains. Ecol Appl 13:704–719
Turner M, Baker W, Peterson C, Peet R (1998) Factors influencing succession: Lessons from large, infrequent natural disturbances. Ecosystems 1:511–523
Turner M, Gardner R, O'Neill R (2001) Landscape ecology in theory and practice, pattern and process. Springer-Verlag, New York City, NY
Tyrell L, Crow T (1994) Structural characteristics of old-growth hemlock-hardwood forests in relation to age. Ecology 75:370–386
Van Pelt R, Franklin J (2000) Influence of canopy structure on the understory environment in tall, old-growth, conifer forests. Can J For Res 30:1231–1245
Vanha-Majamaa I, Jalonen J (2001) Green tree retention in Fennoscandian forestry. Scand J For Res 3:79–90
Vogt K, Gordon J, Wargo J et al (1997) Ecosystems: Balancing science with management. Springer-Verlag, New York, NY
Wayman R, North M (2007) Initial response of a mixed-conifer understory plant community to burning and thinning restoration treatments. Forest Ecol Manag 239:32–44
Wilson D, Puettmann K (2005) Density management and biodiversity in young Douglas-fir forests: Challenges of managing across scales. Forest Ecol Manag 246:123–134
Woods K (2004) Intermediate disturbance in a late-successional hemlock-northern hardwood forest. J Ecol 92:464–476
Zenner E (2000) Do residual trees increase structural complexity in Pacific northwest coniferous forests? Ecol Appl 10:800–810
Zenner E (2005) Development of tree size distributions in Douglas-fir forests under differing disturbance regimes. Ecol Appl 15:701–714
Ziegler S (2002) Disturbance regimes of hemlock-dominated old-growth forests in northern New York, USA. Can J For Res 32:2106–2115

Chapter 18

Conserving Forest Biodiversity: Recent Approaches in UK Forest Planning and Management*

Kevin Watts, Christopher P. Quine, Amy E. Eycott, Darren Moseley, Jonathan W. Humphrey and Duncan Ray

Abstract The need to combat woodland loss and fragmentation are key objectives for forestry and biodiversity conservation strategies in the UK. Conservation action has often been centred on the protection and management of individual sites with limited, often *ad hoc*, action within the surrounding landscape. However, woodland biodiversity conservation efforts, including restoration and re-creation measures, are beginning to be scaled-up to the landscape level in an attempt to address habitat loss and fragmentation. There is also a need to integrate biodiversity goals with other objectives which are planned at the landscape scale, marking a significant shift from segregated to integrated planning. An assessment of landscape structure and function is needed to target conservation action and to evaluate landscape change. This will ensure that the appropriate action is applied in the most effective location. It will also contribute to the development of multi-use landscape plans ensuring biodiversity needs are adequately represented. The aim of this chapter is to present examples of recent approaches to landscape-scale forest planning in the UK. These will illustrate the application of both functional approaches, utilising focal species and estimates of functional connectivity, and also structural approaches, based on the use of landscape metrics, to target and evaluate potential biodiversity conservation action. These examples have been used to target strategic conservation action at a country scale, target specific locations for woodland planting schemes and assess the performance of a woodland planting policy to combat habitat fragmentation. They also demonstrate that the appropriate choice of a functional or structural approach is dependent upon the issue being addressed.

K. Watts
Forest Research, Alice Holt, Wrecclesham, Farnham, Surrey, England, UK
e-mail: kevin.watts@forestry.gsi.gov.uk

R. Lafortezza et al. (eds.), *Patterns and Processes in Forest Landscapes*,

18.1 Introduction

18.1.1 Woodland Loss and Fragmentation

The forests and woodlands of the UK, in common with many habitats throughout the world, have undergone considerable loss and fragmentation through a long history of human activity (Bailey et al. 2002; Kirby and Thomas 1994; Wade et al. 2003). Woodland once covered the majority of the UK landscape and represented the climax vegetation community (Peterken 1993) following the last glaciation, reaching an estimated high of around 75% cover around 6,000 years ago. Significant woodland clearance, with the advancement of agriculture, reduced woodland cover to only 12% by the 16th Century with barely 5% of woodland remaining by the start of the 20th Century (Forestry Commission 2003; Rackham 1990). Much of this woodland is modified remnants of the original forest cover, classed as ancient in origin (*sensu* Peterken 1993) and semi-natural, and is of high biodiversity value (Peterken 1996). The majority of ancient and more recent semi-natural woodland is currently dominated in the lowlands by mixed broad-leaved woodland and in the uplands by oak (*Quercus*) woodland, with wet woodland in particular being relatively uncommon (Table 18.1). Many such woods have been managed (e.g. by coppicing and wood pasture), in some cases for many centuries, without compromising their biodiversity.

During the latter part of the 20th Century, there was widespread conversion of ancient (primarily broad-leaved) woodland to plantations of introduced conifer species such as Sitka spruce *Picea sitchensis*, Norway spruce *P. abies* and Corsican pine *Pinus nigra* var. *maritima* (Table 18.2 and Harmer et al. 2005). This resulted in serious loss and modification of biodiversity, and in response to this, a large-scale programme of restoration of ancient woodland was initiated in the 1990s (Thompson et al. 2003).

In tandem with conversion of semi-natural woodland to non-native conifer plantations, there has also been extensive afforestation of previously non-wooded ground with exotic coniferous species in the last 90 years (Table 18.2), increasing woodland cover to around 12% (Forestry Commission 2003). This was motivated primarily by the need to augment timber production. Although these planted forests can be poor in biodiversity, recent research has emphasised the positive effect that good management can have (Humphrey 2003). In recent decades, management practices have evolved away from large scale felling and replanting towards lower impact silviculture (e.g. progressive thinning and small-scale felling) and retention of stands beyond normal felling age to benefit a range of species groups such as lichens and bryophytes (Humphrey 2005). Despite the increase in woodland cover the majority of semi-natural woodlands remain small and isolated within a primarily agricultural landscape (Fig. 18.1). For example, 75% of all woodlands are under 2 ha in size, with non-native conifer plantations accounting for the relatively few larger forests (Fig. 18.2). Agricultural activities within the surrounding landscapes would have initially produced complex and diverse habitats and landscapes, but following the post-war intensification of agriculture and subsequent

Table 18.1 Area of semi-natural woodland in Great Britain by type. HAP type = Habitat Action Plan woodland type recognised within the UK Biodiversity Action Plan (UK Biodiversity Steering Group 1995a). The HAP types are related to the CORINE Land cover classification (Moss and Davies 2002) and the EU Habitats and Species Directive Annex 1 types (European Community 1992)

Woodland HAP type	CORINE	Habitats Directive Annex 1 Type	GB area (1000's Ha)
Lowland Beech and Yew	42.A71 41.13 41.12 41.16	*Taxus baccata* woods *Asperulo-Fagetum* beech forests Beech forest with *Ilex* and *Taxus*, rich in epiphytes	30
Lowland Mixed Broadleaved	41.23, 41.32 41.24 41.51 41.52	*Tilio-Acerion* ravine forests *Stellario-Carpinetum* oak-hornbeam forests Old acidophilous oak woods with *Quercus robur* on sandy plains	250
Upland Mixed Ash	41.31, 41.32, 41.41 42.A71 62.3	*Tilio-Acerion* ravine forests *Taxus baccata* woods limestone pavement	68
Upland Oak	41.53, 41.52	Old oak woods with *Ilex* and *Blechnum* in the British Isles	70–100
Upland Birch	41.53, 41.52	Old oak woods with *Ilex* and *Blechnum* in the British Isles	30–40
Native Pine	42.51 44.A2	Caledonian forest Bog woodland	16
Wet	44.A1 44.31 44.13 44.92	Bog woodland Residual alluvial forests	50–70

Table 18.2 Amount of each coniferous species present as a percentage of total conifer area in each country (Forestry Commission 2003)

Species	England	Wales	Scotland	GB
Scots pine (*Pinus sylvestris*)*	24.0	3.1	15.0	15.9
Corsican pine (*P. nigra* ssp. *laricio*)	12.1	2.3	0.2	3.3
Lodgepole pine (*P. contorta*)	2.1	4.2	13.4	9.7
Sitka spruce (*Picea sitchensis*)	24.1	55.9	57.8	49.6
Norway spruce (*P. abies*)	9.2	7.6	3.8	5.5
European larch (*Larix decidua*)	4.0	0.4	1.0	1.6
Japanese larch (*L. kaempferi*)†	9.4	14.8	6.1	7.8
Douglas fir (*Pseudotsuga menziesii*)	7.3	7.4	1.1	3.3
Other conifers	5.4	4.0	0.6	2.1
Mixed conifers	2.4	0.3	0.9	1.2
Total Area (ha)	329832	145523	904155	1379510

* Native to Scotland.

† Includes hybrid larch (*Larix* x *eurolepis*).

Figures are for woodlands >2 ha.

Fig. 18.1 Aerial photograph showing a typical fragmented UK lowland landscape with semi-natural woodland (very dark green) surrounded by agriculture

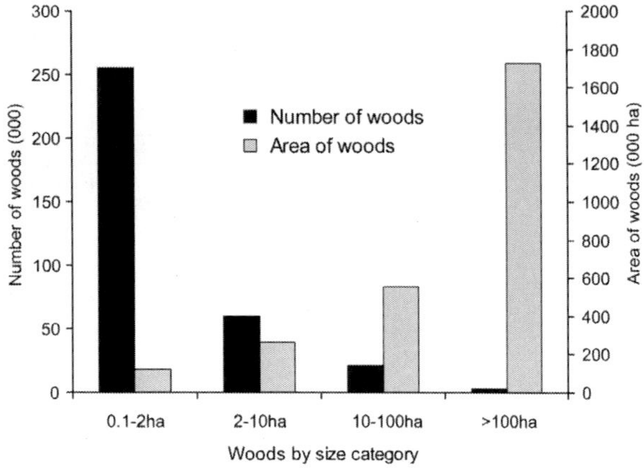

Fig. 18.2 Total number and area of woods by size category (Forestry Commission 2001a, 2002b, c, d, 2003)

loss of landscape diversity the impact on biodiversity has been profoundly negative (Robinson and Sutherland 2002; Sheail 1995). The loss and fragmentation of woodland coupled with this intensification of agriculture, with species being more restricted to fragmented patches, has resulted in further reductions in UK biodiversity (Humphrey 2003; Robinson and Sutherland 2002).

18.1.2 Conservation of Woodland Biodiversity

In spite of this loss and fragmentation, UK woodlands still contain considerable biodiversity interest. Over 40% of species within the UK Biodiversity Action Plan are associated with woodlands, and nearly 15% of priority habitats are specific woodland types, as detailed in Table 18.1 (Simonson and Thomas 1999; UK Biodiversity Steering Group 1995a). In the past decade, the need to conserve and enhance woodland biodiversity and combat habitat fragmentation has become a key element of the forestry and biodiversity conservation strategies for the UK (Forestry

Authority 1998; Forestry Commission 1999, 2000, 2001b; Quine et al. 2004; UK Biodiversity Steering Group 1995b).

The main emphasis of conservation activity has been on the protection and management of individual protected sites or biodiversity 'hotspots' (Kirby et al. 2002), the restoration of semi-natural woodland and habitats (Thompson et al. 2003), and measures to enhance the ecological value of the surrounding agricultural environment (Donald and Evans 2006). Over recent decades in the UK, a wide range of forestry and agri-environment incentives have been introduced in an attempt to address the impacts from habitat loss, fragmentation and agricultural intensification (Kleijn et al. 2006; Kleijn and Sutherland 2003). Such incentives are regarded as a key conservation mechanisms within a non-statutory planning system within the UK landscape (Gilg 1996; Watts and Selman 2004).

Many of these conservation measures have been applied within a relatively *ad hoc* manner with little consideration of the interactions with the surrounding landscape and what the overall consequences for biodiversity might be (Poiani et al. 2000). A site-based conservation focus also fosters a binary view of the landscape as either habitat or non-habitat, and fails to capitalise on the wider conservation, restoration and connectivity opportunities within the existing and future landscape (Haila 2002; Kupfer et al. 2006).

18.2 Towards Landscape-Level Planning and Management

18.2.1 Landscape-Level Conservation

There is a growing realisation that conservation and enhancement of woodland biodiversity cannot be achieved purely by a stand level or protected site approach (Margules and Pressey 2000), nor can it be guaranteed by concentrating on landscape aesthetics and assuming that ecological benefits are linked. Organisms and ecological processes are not constrained by management or ownership boundaries, and hence there has been the need to develop broader approaches. There is now an increasing focus on combating the effects of fragmentation through combining site protection, management and restoration measures with landscape scale approaches which improve connectivity and wider landscape quality (Frelich and Puettmann 1999). As a result an increasing proportion of forestry and agri-environment measures with a conservation focus are likely to be spatially targeted to capitalise on potential landscape impacts and cumulative benefits.

In order to support this shift to landscape scale conservation there is a need to incorporate biodiversity conservation objectives into the planning and design of multi-use sustainable landscapes, including land allocated to both protection and production (Donald and Evans 2006; Margules and Pressey 2000). Other landscape activities and resources (e.g. forestry, agriculture and water) are similarly being planned at this broader landscape scale in recognition that they cannot be managed exclusively at the level of habitat units or local sites (Liu and Taylor 2002).

As one approach to integrated conservation at the landscape scale, conservation plans and strategies within the UK and beyond are starting to focus on the development of ecological/habitat networks (Bennett 2002, 2004; Catchpole 2006; Jongman and Pungetti 2004; Latham et al. 2004; Opdam 2002; Opdam et al. 2006; Ray et al. 2004; Watts et al. 2005). These networks are considered especially important for fragmented, and formerly widespread, habitat systems such as woodland within the UK, as many species may need them to operate across multiple sites e.g. as metapopulations (Hanski and Ovaskainen 2000; Opdam 1991; Thomas et al. 1992; Verboom et al. 1993). Ecological networks are also being proposed as an adaptation measure to mitigate the impacts of climate change by allowing species to potentially track their changing climate space (Opdam and Wascher 2004; Pearson and Dawson 2003).

18.2.2 Implementing Landscape Approaches

The need to take a landscape approach has been part of many recent statements of UK forest policy reflecting a trend towards broader scale spatial and temporal planning (DEFRA 2007; Forestry Commission 2000, 2001b). There is also evidence within these statements of a significant, necessary but challenging shift from segregated to integrative landscape and biodiversity planning evident in the wider conservation policy arena (Bennett 2004; Bissonette and Storch 2003; Hobbs and Lambeck 2002; Turner et al. 2002).

In order to implement an integrated landscape approach there is a need to target conservation and restoration activities in the most effective areas, and to influence the development of multi-use landscape plans. There is also a complementary need to evaluate planned landscape change, which may entail a balance or compromise between the various environmental, economic and social objectives, to ensure biodiversity needs are adequately represented (Fig. 18.3).

As current and future landscape changes will impact on both the structure and function of the landscape there may be a need to assess the effect of both on

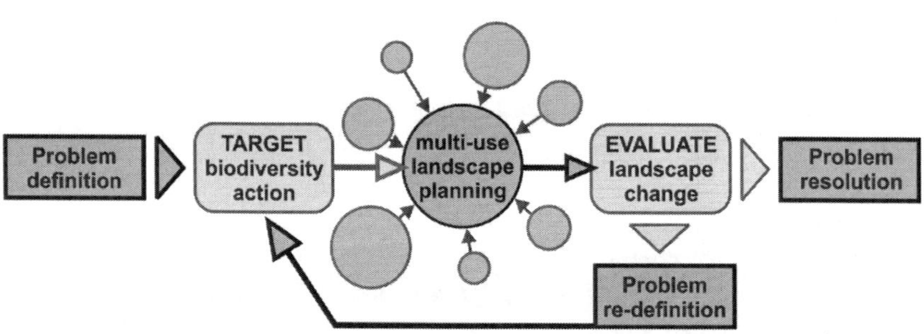

Fig. 18.3 Roles of targeting and evaluation of land use change for biodiversity conservation. Actions must always be integrated with other objectives, represented by the light blue circles

biodiversity, depending on the particular issue being addressed. Landscape structure refers to the spatial arrangement and organisation of distinct landscape elements. An assessment of landscape structure, through the use of landscape metrics, may be appropriate where the aim of the planned action is to change landscape structure. For instance, the Ancient Woodland Policy for England (Forestry Commission 2005b, p. 3) has a specific aim to 'promote woodland creation which extends, buffers and links ancient woodland'. The success of this policy aim could be evaluated by assessing change to the total area of woodland, the number of individual woodlands, their size distribution and the amount of core habitat.

Landscape function refers to the extent to which the landscape supports ecological processes such as reproduction, dispersal, and the transfer of resources through the food chain. In terms of biodiversity conservation landscape function is often related to the movement and viability of particular species. The Ancient Woodland Policy (Forestry Commission 2005a, pp. 10–11) also has an aim related to landscape function: 'The landscape context of woodland should be improved. . .create new native woodland to extend, link or complement existing woodland and other habitats. . .work towards creating landscapes that are "ecologically functional"'. This policy aim may require a more complex functional approach to target action and evaluate change.

18.3 Recent Applications

In order to illustrate the value of landscape ecology to UK forest landscape planning and management, this chapter now presents three recent applications which aid the conservation of woodland biodiversity in fragmented landscapes (Fig. 18.4).

Example 1 in Wales uses a relatively simple functional assessment of landscape fragmentation, utilising focal species and estimates of functional connectivity. This work has been used to strategically target and prioritise conservation action at a country scale.

Example 2 in the Scottish Highland uses a similar functional approach to the Welsh study but focuses on targeting and evaluating specific locations for new woodland planting.

Finally, Example 3 uses a simple metrics-based approach to evaluate the performance of a two different woodland planting schemes in combating structural fragmentation on the Isle of Wight.

These three examples attempt to illustrate the role of landscape ecology as an applied, problem-oriented science (Bissonette and Storch 2003; Gutzwiller 2002; Hobbs 1997; Turner et al. 2002). They demonstrate that the choice of a functional or structural approach to assess the impact of landscape change is dependent upon the specific issue being addressed and the availability of species and spatial data. These also illustrate how landscape ecology can be used to guide and support both strategic policy and operational management at a variety of scales.

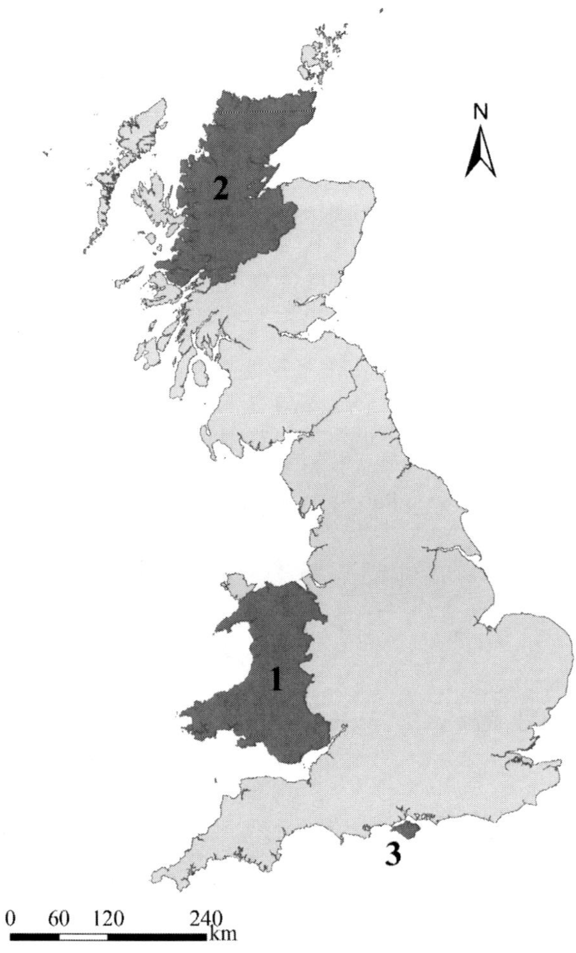

Fig. 18.4 Location of UK example applications

18.4 Example 1 – Targeting Strategic Conservation Action in Wales

18.4.1 Introduction

Semi-natural woodland habitats in Wales (Area 1 in Fig. 18.4) have undergone serious fragmentation over a sustained period. Welsh woodlands still contain many rare, threatened and characteristic species and there is considerable political momentum in Wales to protect them through the reduction of fragmentation and improvement of connectivity (Countryside Council for Wales 2004; Forestry Commission 2001b). This is in addition to site-based measures which already affect around 25% of land in some way. Some landscape scale action in Wales is now being targeted using habitat network maps based on functional connectivity. The research report *Towards a Woodland Habitat Network for Wales* (Watts et al. 2005) provided the foundation for this continuing research.

18.4.2 The Approach: Functional Connectivity

Key ecological theories and approaches formed the basis for the development of this strategy, particularly species-area relationships and island biogeography (MacArthur and Wilson 1967; Vellend 2003), and metapopulation theory (Hanski 1999). Pattern-based networks (e.g. Good et al. 2000) and greenways had initially been considered as possible options, particularly those based on neutral landscape models. However, many landscape ecologists now consider functional connectivity models that take account of the landscape matrix as being more robust and realistic for many species, particularly in agriculturally dominated landscapes (Crooks and Sanjayan 2006). It was therefore considered more useful to develop a focal species based habitat network which would explicitly express functional connectivity (Tischendorf and Fahrig 2000) in relation to woodland biodiversity.

Functional connectivity can only be measured in terms of species processes, e.g. of dispersal distance and response to different matrix elements. The matrix is the mosaic of non-habitat land between habitat patches, which may contain elements which promote or inhibit movement. Many functional approaches have been based therefore on the needs of particular species. However, a single species rarely represents the needs of the majority. Lambeck (1997) developed the 'focal species approach' where landscape needs are set by those species with the most demanding requirements in terms of patch area, patch isolation, habitat management and resource management. The drawback to both approaches is the absence of adequate species and spatial data. In addition, Lambeck's (1997) approach would possibly result in too demanding a plan for modern, multi-objective landscape planning.

In response to this, a set of generic focal species (GFS) profiles were developed. A GFS is a conceptual or virtual species, whose profile consists of a set of ecological requirements reflecting likely needs of real species where species data are unavailable. GFS are selected to represent particular species, groups of species, habitats, important landscape features or specific policy objectives. These are best developed with key stakeholders involved in strategic planning and management and relevant habitat and species experts. These GFS are similar to the ecoprofiles used within the work of Alterra and the LARCH model (Opdam et al. 2006; van Rooij et al. 2003; Vos et al. 2001).

The GFS profiles developed to study fragmentation in Wales reflected species habitat area requirements and dispersal preferences (maximum distance and permeability of different matrix land uses; Table 18.3). Two profiles were developed to represent one more demanding and one more generalist species. The former profile, termed 'core species', had a minimum patch size of 10 ha and a maximum dispersal distance of 1 km. This covers a range of woodland species' requirements, including woodland specialist butterflies and birds (Bailey 1998). The more generalist profile,

Table 18.3 Ecological permeability scores of matrix land uses for a broad-leaved woodland generic focal species (GFS). Lower scores equate to higher permeability

Score	Type	Examples
1	Semi-natural habitats with a high broad-leaved tree component	Planted broad-leaved woodland, dense scrub
3	Semi-natural habitats with some vertical structure	Heath
5	Semi-natural habitats with little vertical structure	Unimproved grassland
10	Some habitat modification, or extremely wet semi-natural habitats with little vertical structure	Conifer plantation, semi-improved grassland, bog
20	Highly modified with little or no vertical structure	Improved grassland, arable crops
50	Highly modified or impermeable	Urban, open water

named the 'focal' profile, had a minimum patch size of 2 ha and a maximum dispersal distance of 5 km, representing for example some birds (Bellamy et al. 1996) and vagile plants. The focal profile was so named as it outlines the area which conservation action could be focussed in, where some connectivity, species and processes may already be in place.

Both profiles had the same permeability scores for intervening landscape matrix (Table 18.3), with lower costs equating to higher permeability. Scores were assigned based on a combination of semi-naturalness and degree of vertical structure as a way of measuring similarity to woodland. Matrix permeability is difficult to quantify but in general terms, semi-natural and extensive habitats are considered to be more conducive or permeable to species movement, whereas intensive land uses are predicted to reduce connectivity and increase ecological isolation (Donald and Evans 2006; Ricketts 2001).

For example, arable land, which is known to be hostile to woodland species as it has little habitat diversity and vertical structure, was assigned a relatively high cost of 20. Whereas, semi-natural heathland was assigned a cost of 3 as it offers a semi-natural habitat with a degree of vertical structure.

The GFS profiles were combined with a simple land cover map based on the Welsh Phase 1 habitat survey (Howe et al. 2005), Land Cover Map 2000 (Centre for Ecology and Hydrology 2000) and elevation. Accumulated cost-distance modelling (Adriaensen et al. 2003) was used to generate habitat networks where the distance travelled from the source patch was moderated by the surrounding landscape. For example, the 'core' GFS may pass through 20 m of improved grassland with a permeability score of 20, using up 400 m of dispersal distance, then 60 m through conifer plantation, with a permeability score of 10, using up 600 m, and this point would delineate the edge of the network.

The result of the analysis was the production of a series of maps that showed the potential extent of networks for broad-leaved woodland throughout Wales. There were 1254 core networks and 1655 focal networks identified, with a mean area of 69 ha and 271 ha respectively. Figure 18.5 illustrates networks for both core and focal GFS profiles for a small area of Wales.

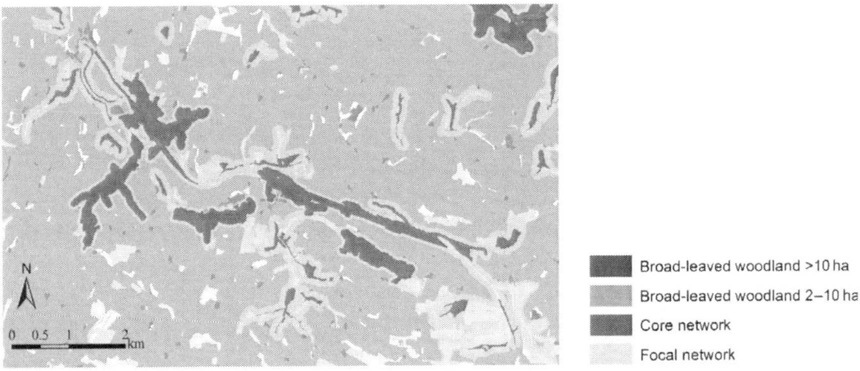

Fig. 18.5 Identification of broad-leaved woodland core and focal networks

18.4.3 Priorities for Developing Woodland Habitat Networks

Priorities must be established for the future development of woodland habitat networks, both in terms of actions to be taken and where to apply them.

18.4.3.1 Priorities for Actions

Halting the continuing process of fragmentation must be the first priority. This can be achieved through the protection and formal management of existing high quality broad-leaved woodland habitat. The national statutory designation for woodland to protect biodiversity in the UK is Site of Special Scientific Interest (SSSI). However, SSSIs are only examples, not the total area, of important habitats, so not all woodlands are included (for example, only 25% of ancient, semi-natural woodland is designated SSSI in Wales).

Once the process of fragmentation has been halted, the pattern of fragmentation can be changed, through measures of restoration, improvement, changes to the matrix and habitat creation, prioritised in that order (McIntyre and Hobbs 1999). The challenge is to find mechanisms for implementing such changes in a sustainable manner. In this case study, strategic and national-scale targeting of action can be achieved using woodland planting grant schemes and agri-environment plans, as well as forestry planning (Table 18.4). In areas selected as being of strategic importance, specific actions can be targeted on a site basis and then their impact evaluated before decisions are made.

18.4.3.2 Priorities for Spatial Targeting

The continuing decline of woodland biodiversity means that the focus should be upon consolidation and expansion of existing robust areas. Action can be combined or extended in the future to wider networks. In this example the core networks are

Table 18.4 Ranked priorities for the protection of woodland biodiversity at a landscape scale

Priority	Action	Mechanism – example
1	Protect and manage existing woodland resource	SSSI and SAC designations Management grants for private woodlands (Better Woodlands for Wales scheme) Management policy for state-owned forest
2	Restore degraded habitat, e.g. Plantations on Ancient Woodland Sites (PAWS); sites invaded by *Rhododendron ponticum*	Management policy for state-owned forest Management grants for private woodlands (Better Woodlands for Wales scheme)
3	Improve degraded secondary habitat	Management policy for state-owned forest Management grants for private woodlands (Better Woodlands for Wales scheme)
4	Improve the matrix by reducing land-use intensity	Agri-environment schemes Conversion of even-aged conifer plantation to Low Impact Silvicultural Systems (LISS)
5	Create new habitat	Planting grants for private woodlands (Better Woodlands for Wales scheme)

considered to be existing robust areas, and the focal networks to be the framework for future expansion. Individual sites should therefore be ranked in the following (descending) order (Fig. 18.6): Large core woods in large core networks (1), isolated core woodland in core networks (2), small woods in core networks (3), small woodlands in large to small focal networks (4), and isolated small woodlands (5). Figure 18.7 illustrates how the conservation actions in Table 18.4 could be spatially targeted.

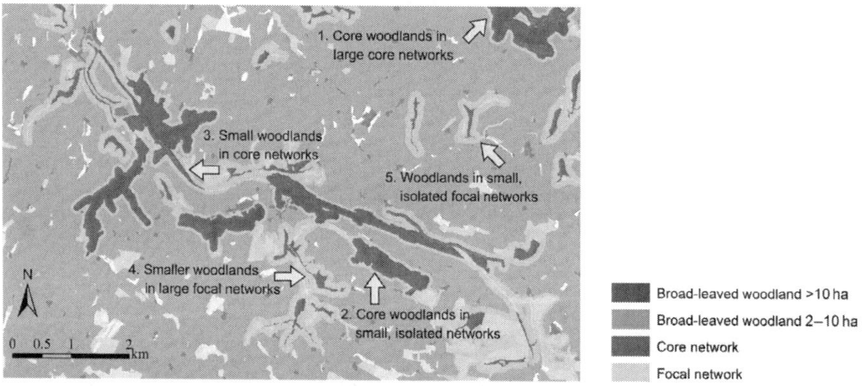

Fig. 18.6 Spatial prioritising of woodlands within the landscape

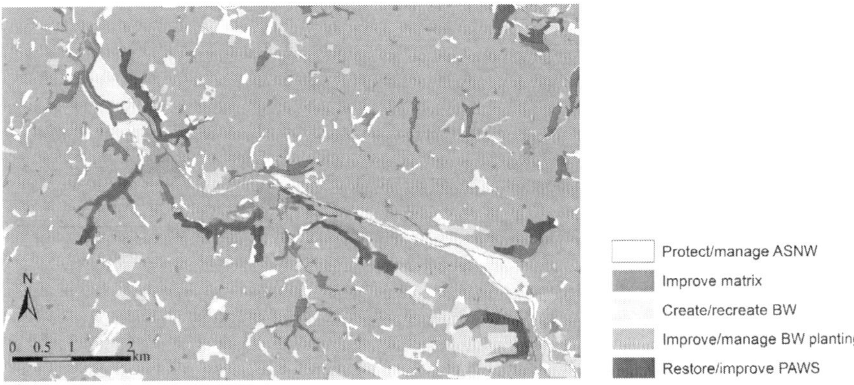

Fig. 18.7 Targeting and prioritising biodiversity conservation action to tackle habitat fragmentation within existing core and focal networks (ASNW – Ancient Semi-Natural Woodland; PAWS – Plantations on Ancient Woodland Sites; BW – native Broad-leaved Woodland)

18.4.4 Current Application

Including a connectivity element into grant schemes can reduce fragmentation more than reactive planting (See Example 3), and various state agencies are now beginning to use these habitat networks maps in their work. For example, grant schemes for planting new woodland in Wales typically had no incentive for spatial targeting but the new *Better Woodlands for Wales* scheme incorporated an incentive for planting in core or large focal networks. Another scheme (Forestry Commission 2005c) used the focal networks as part of the prioritising process for the restoration of Plantations on Ancient Woodland Site (PAWS) – former semi-natural woodland sites that had been felled and planted with non-native conifers. The Countryside Council for Wales have used the habitat networks as part of the basis of a vision for a national-scale network of semi-natural habitats (Latham 2007).

18.4.5 Further Development

Future development of this work will be based on validating the assumptions underlying the model, particularly the permeability of different matrix land uses for particular fragmentation sensitive species. Reaching the people who apply policy, and those who make it, must be a key part of a habitat network strategy, and communication to those groups should be further developed. Further work is required to produce complementary plans for open habitats, and to consider the impacts of climate change.

18.5 Example 2 – Targeting New Woodland Planting in the Scottish Highlands

18.5.1 Introduction

Scotland's woodland has also been fragmented with as little as 4% remaining by the 17th Century. Afforestation throughout the 19th Century and particularly the 20th Century increased woodland cover to 16.8% of the total land area of Scotland (Forestry Commission 2002b). However, the ancient semi-natural remnant woodlands remain largely fragmented, because new planting has been spatially unconstrained and consisted largely of exotic conifer plantations established on poor grazing land.

Policy in Scotland had advocated the use of networks for woodland planning since the mid 1990s (Peterken et al. 1995), when the rationale was a network of physical 'nodes' and 'links' at a range of scales e.g. Forestry Commission (2000, 2003) at the national scale, Towers et al. (1999), Peterken (1999) and Worrell et al. (2003) at the local scale. Since then, the functional habitat network approach using accumulated cost-distance modelling methods (see Section 18.4.2) has been adopted. Initial national scale networks (Moseley et al. 2007a) using generic focal species (see Section 18.4.2) have been followed up by smaller scale studies tailored to local conditions (Grieve et al. 2006; Moseley and Ray 2006, 2007; Moseley et al. 2005).

18.5.2 Approach and Analyses

18.5.2.1 National Scale, Strategic Networks

National scale, strategic networks (Moseley et al. 2007a) were based on the use of Generic Species Profiles (GFS) to represent habitats that are commonly found in Scotland to allow consistency across the country:

- Woodland Generalists – representing species which may disperse easily, and are not specifically associated only with woodland, but they may need woodland for a part of their life cycle, or partly within their range.
- Broad-leaved specialist – representing species specifically associated with broad-leaved woodland, may be found in mixed woodland to a lesser degree and occasionally in conifer. The term specialist signifies a rather reduced dispersal and a more exacting habitat requirement. This GFS profile was approximately equivalent to the core network GFS in Wales (Example 1).

18.5.2.2 Local Scale Networks

The regional analyses reflected important woodland types associated with each of the Scottish regions, e.g. pinewood in Scottish Highlands (Moseley et al. 2005), riparian and wet woodland in Grampians (Moseley and Ray 2007), ancient broad-leaved woodland in the Scottish Borders (Moseley and Ray 2006).

- Pinewood specialists – representing species specifically associated with pine woodland.
- Riparian woodland specialist – representing those species only associated with riparian woodlands. Sites are located adjacent to rivers and streams. Species in this category are generally limited to riparian areas.
- High quality broad-leaved woodland specialist – representing those species only associated with ancient and long established woodlands. The species may additionally be present in conifer plantations on ancient woodland sites (PAWS), but PAWS have not been classified as habitat in the analysis. The important issue is antiquity, which provides a long period of woodland cover. Species in this category are less mobile than broad-leaved specialists.

18.5.2.3 Use of the Networks

The analyses produced national and regional maps indicating the extent of networks for these habitats, which can then be used in conjunction with the series of strategic priorities for reducing fragmentation in a similar way to the Wales study (see Section 18.4). The protection and improvement of the important woodland habitats have

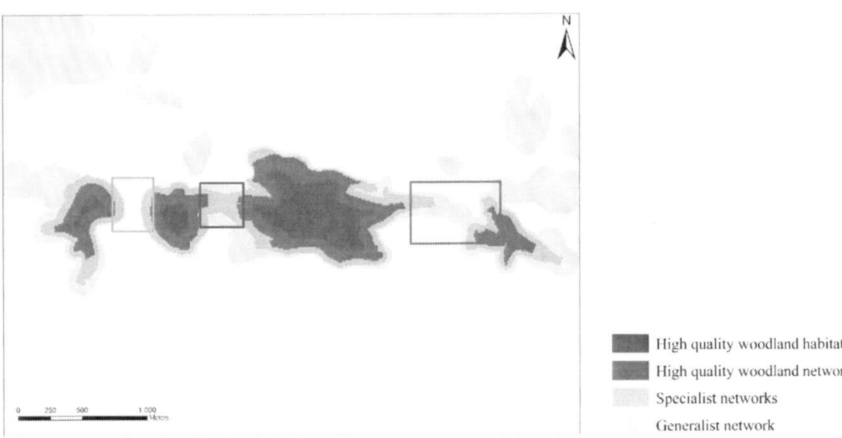

Fig. 18.8 Example of where ancient or high quality woodland patches (*green*) within their networks (*bright red*), are nested within lower quality specialist networks (*pale red*), which in turn are situated within the woodland generalist networks (*pale blue*). The *blue box* indicates where the woodland may be improved to increase the high quality network. The *red box* indicates where the woodland may be restored to become part of the specialist network, allowing dispersal to occur between the central and right hand side network. The *orange box* indicates where reducing the intensity of open ground management or a woodland 'stepping stone' may be introduced to functionally connect the existing networks

the highest priorities, followed by approaches that enhance existing network connectivity (Fig. 18.8). The connectivity of woodlands may be addressed by improving the permeability of the matrix, e.g. by reducing intensive agricultural practices or creating new woodland. The latter approach has been undertaken on a regional basis by using a locational premium scheme to spatially target planting to functionally connect existing woodland networks.

18.5.3 Highland Locational Premium Scheme (HLPS)

Previous woodland planting grant schemes in Scotland have been spatially unconstrained and have not addressed habitat fragmentation. The regional network

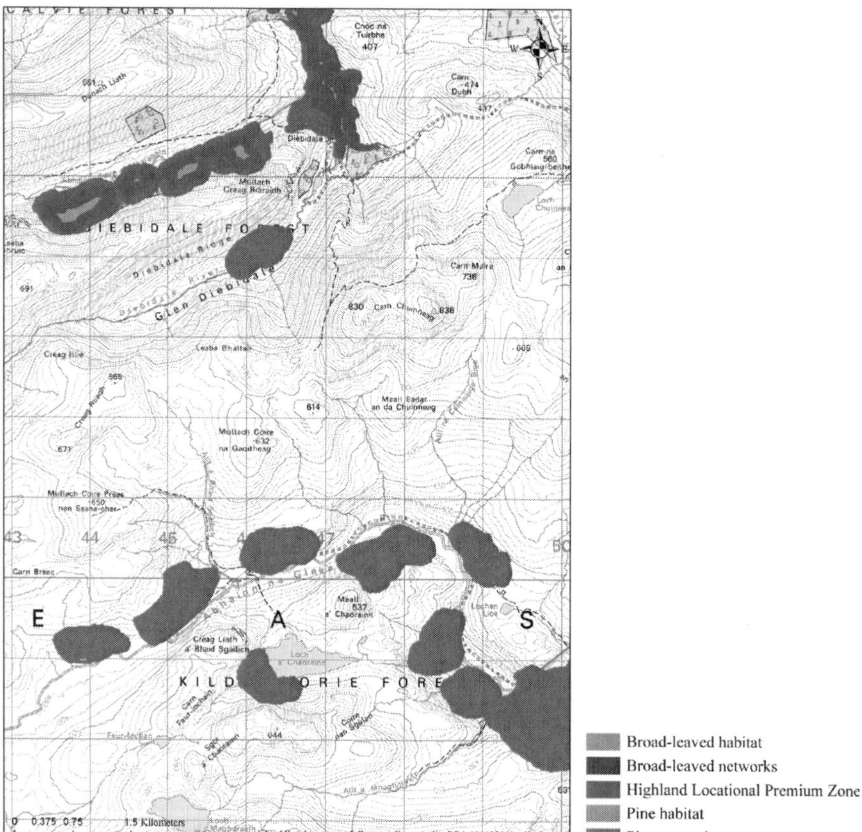

Fig. 18.9 Zone around existing networks, beyond which new planting schemes are not considered able to contribute towards addressing habitat fragmentation

analysis of the Scottish Highlands (Moseley et al. 2005) were used by Forestry Commission Scotland Highland Conservancy in 2006 to spatially direct a new Scottish Forestry Grant Schemes which offered a premium for targeted planting which would functionally connect pine or broad-leaved woodland habitat fragments (Moseley et al. 2007b).

The methodology involved the identification of areas where woodland expansion could link existing networks by creating an additional cost-distance buffer (see Section 18.4.2 and Adriaensen et al. (2003)) around the existing networks (Fig. 18.9). Schemes were required to intersect the buffer around two or more networks; outside this area it was assumed unlikely that that they would be close enough for dispersal events to occur between the existing and potential habitat.

The next step was to construct a user-friendly geographic information systems (GIS) tool to analyse how well the proposed new planting scheme would improve network connectivity for pinewood or broad-leaved woodland specialists. The automated GIS-based procedure was designed for speed and simplicity, but is transparent and open to public scrutiny. The analysis tool first checked that there was sufficient internal forest habitat to support species that are sensitive to woodland edge by removing a 50 m internal buffer. A habitat network analysis was then performed using two different generic focal species; pinewood specialists and broadleaf specialists, to determine whether the proposed new scheme would succeed in functionally connecting the two woodlands. Map outputs and statistics showing network connectivity formed by the proposal were produced.

This methodology quickly determined whether a proposal met the scheme objective and if so, applicants applied for grant aid through a series of one-to-one meetings with agents appointed by Forestry Commission Scotland. A scoring system (Table 18.5) was used to determine the contribution the proposal would make to the existing networks, allowing calculation of the amount of locational premium the proposed scheme would be eligible for. This was done by examining the area of habitat, rather than the area of network, linked together by the proposed scheme, to avoid bias towards schemes adjacent to semi-natural types of habitat (relatively more permeable to woodland species dispersal) over schemes adjacent to more modified and managed habitat (relatively less permeable to woodland species). The scoring was also weighted by new woodland size, to encourage small schemes that link together larger networks rather than unnecessarily large schemes that would link small networks. This approach allowed the amount of grant aid to be used carefully to achieve more benefit for woodland biodiversity.

The HLPS proved to be very popular. The final allocation for funding was made in September 2006, comprising 25 new planting schemes, covering just over 1000 ha. This will potentially link 50 networks, comprising approximately 13500 ha of existing network area.

Table 18.5 Scoring system developed with Highland conservancy to award locational premium

New habitat score[1]	Proportionality[2]	Size of new woodland eligible for premium (ha)[3]	Comment
5	1.5	None	If these minimum scores are not reached, then the scheme is not eligible
5–50	≥ 1.5	Up to 20	Premium payable on up to 20 ha of new woodland planted, but proportionality means that 'new habitat score' needs to be one and a half times as big as new area planted*
50–200	≥ 3	Up to 50	Premium payable on up to 50 ha of new woodland planted, but proportionality means that 'new habitat score' needs to be three times as big as new area planted*
200+	≥ 4	Up to 100	Premium payable on up to 100 ha of new woodland planted, but proportionality means that 'new habitat score' needs to be four times as big as new area planted*

[1] the total area of habitat in new network (ha) minus largest area of habitat in existing networks (ha).

[2] the 'new habitat score' divided by area of new woodland planted (ha).

[3] The 'size of the new woodland eligible for premium' was based on an amount that contributes towards the Forest Habitat Networks. This was agreed with the Woodland Officer during the consultation.

18.6 Example 3 – Evaluating Woodland Planting Schemes on the Isle of Wight

18.6.1 Introduction

The Isle of Wight (Area 3 in Fig. 18.4) is an island of 380 km^2 with a population of approximately 125000 located off the south coast of England. The land use history of the island, like that of much of lowland England, has resulted in an intensive agricultural landscape with small remnant woodland areas of considerable conservation interest. Approximately 66% of the landscape is agricultural, 12% is woodland (Forestry Commission 2002a), and only 2% is of ancient semi-natural woodland (ASNW); the latter is judged to be of particularly high conservation value with many species entirely dependent upon ASNW. Our case study on the Isle of Wight demonstrates the use of landscape metrics to assess the relative success, in terms of structural connectivity, of two contrasting woodland creation schemes.

18.6.2 New Woodland Planting

The re-creation of woodland though new planting schemes is one of the key mechanisms (see Table 18.4) in combating woodland loss and fragmentation. The first England Forestry Strategy (Forestry Commission 1999) declares that 'a priority will be to work towards reversing this fragmentation' (p. 23); promoting the need to 'target grants... to reverse the fragmentation of existing native woodlands' (p. 26). Wood re-creation has been encouraged through a number of financial incentives from small scale measures encouraging any contribution to increase woodland cover, to grant schemes that have set out to restore connectivity to existing woodlands.

This study on the Isle of Wight assessed the relative success of two contrasting grant aid schemes, WGS (Woodland Grant Scheme) and JIGSAW (Joining and Increasing Grant Scheme for Ancient Woodland), in improving the structural connectivity of woodland habitats (Quine and Watts under revision). WGS is a broad-based scheme to encourage general woodland expansion with little spatial targeting, whereas JIGSAW is a proactive, spatially targeted scheme that offers a premium for woodlands that expand, buffer or join existing woodland habitats (Fig. 18.10). The JIGSAW scheme was deployed on the island in the period 2001–2005, during which time approximately 200 ha of tree-planting was undertaken. The Woodland Grant Scheme (WGS) commenced in 1988 and resulted in the planting of 200 ha on the Island by 2005.

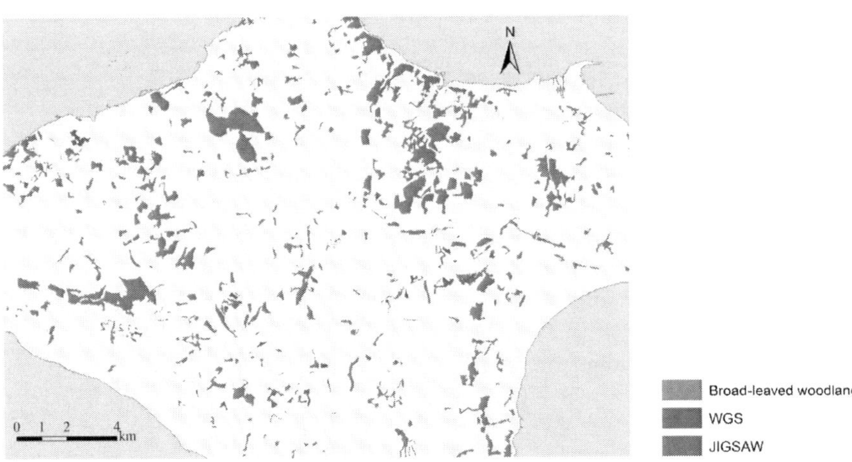

Fig. 18.10 Spatial distribution of non-targeted WGS woodland and spatially targeted JIGSAW woodland in relation to existing broad-leaved woodland in the east of the Isle of Wight

Table 18.6 Summary of selected indicators/metrics, and their interpretation, used to test the success of de-fragmentation of landscape structure

Indicator/metric	Underlying assumption	Relative increase in measure	Relative decrease in measure
Area	Habitat availability	Favourable – more habitat	Unfavourable – less habitat
No. of patches	Habitat composition	Unfavourable – more fragmented	Favourable – less fragmented
Total edge	Edge impacts	Unfavourable – bad for core species	Favourable – good for core species
Patch size	Habitat availability	Favourable – good for core species	Unfavourable – bad for core species
Core area (50 m edge)	Core habitat availability	Favourable – good for core species	Unfavourable – bad for core species
Nearest neighbour	Habitat configuration	Unfavourable – bad for connectivity	Favourable – good for connectivity

18.6.3 Evaluating Landscape Structure

There is a need to assess the effectiveness of spatially targeted approaches in comparison to untargeted woodland expansion, and this study was an opportunity to compare their relative success, based on spatial measures that reflect the aim of de-fragmentation.

As both JIGSAW and WGS grant schemes were focused on landscape structure, their relative impact in combination with existing woodland was assessed separately, using a selected number of indicators based on landscape metrics. These metrics were developed by reviewing the landscape ecology literature and selecting those with clear and appropriate assumptions (Li and Wu 2004) and whose context was consistent with the concerns expressed by local planners and managers (Failing and Gregory 2003). In all, six metrics were selected with clear assumptions and interpretation (Table 18.6) and were computed within FRAGSTATS (McGarigal et al. 2002).

18.6.4 Comparison Between Woodland Grant Schemes

The untargeted WGS created a substantial number of new woodlands, while the JIGSAW woodlands actually reduced the total number by extending and joining existing broad-leaved woodlands (Fig. 18.11a). The total edge of woodland produced by WGS was almost twice as much; on average WGS created 527 m of edge for each new hectare of woodland, whilst JIGSAW created only 255 m (Fig. 18.11b). The tendency for new small WGS woodlands also reduces median patch size and mean core area while JIGSAW provides a slight improvement over the semi-natural baseline (Fig. 18.11c & d).

Only with the indicator of mean nearest neighbour distance does WGS appear to have a more favourable impact than JIGSAW. However this suggests that WGS

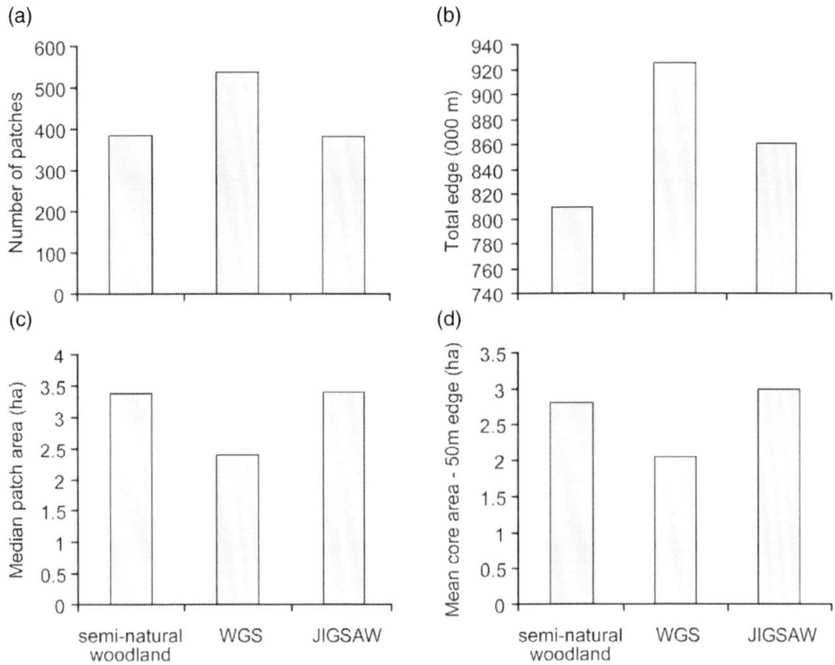

Fig. 18.11 Isle of Wight landscape metrics

woodlands were planted in between existing woodlands. The lower standard deviation also suggests these woodlands were planted in areas particularly far from existing woodland, increasing the regularity of the landscape pattern. This is unlikely to make a significant contribution to local connectivity (see Fig. 18.10).

18.7 Summary and Conclusions

These examples illustrate how these approaches to landscape ecology can provide useful support for forest planners and managers in taking decisions across a range of spatial scales.

- The application in Wales demonstrated a functional approach to define woodland habitat networks, acknowledging the importance of the surrounding landscape matrix. These networks are now providing a basis for strategic conservation action at national and regional scales.
- The habitat network approach developed for the Scottish Highlands provided a specific decision-support tool to enable the planning and assessment of new woodland planting to improve functional connectivity. This work targets woodland creation but also recognises the importance of other landscape-scale conservation activities to protect, restore and improve habitat and the matrix (see

Table 18.4). This is the first time in Britain that landscape ecology tools have been incorporated within a spatially-targeted forestry grant scheme.
- The JIGSAW scheme on the Isle of Wight aimed to encourage woodland expansion that would contribute to woodland de-fragmentation. Our assessment, using a variety of metrics, suggest that this targeted approach has been more successful in achieving the desired structural changes than more general incentives to expand woodland area; future schemes could be guided by tools such as that implemented in the Scottish Highlands.

Each of these examples demonstrated an application of landscape ecology to guide the potential defragmentation of woodland habitats. Other advantages of these approaches include:

- Flexibility: While the conservation of woodland biodiversity was the primary objective, the information is compatible with and relevant to decisions that seek to balance other environmental, economic and social objectives to develop a multi-use landscape. In addition, the choice of which action to focus upon (protection, improvement, restoration or habitat creation) can be varied depending on policy objectives, stakeholder input and regional context.
- Applicability: The examples presented in this paper are focused primarily around woodland habitats but the methods are applicable wherever species are affected by fragmentation and sensitive to the composition of the matrix.
- Practicality: The examples are all based on tools that have practical application. Such tools, especially the HLPS, can be used with little training and put relatively complex landscape analysis within the reach of the forest conservation planner and managers.

Future application of these tools will benefit from enhanced ecological knowledge and more sophisticated forms of decision analysis. The lack of information on species responses to landscape change and spatial data necessitates caution in the application of the functional approach to habitat networks; a number of the important assumptions require further validation. However, due to the pressing need for urgent landscape-scale action and a continuing threat to woodland biodiversity, we consider that any improvement on untargeted conservation action is preferable.

References

Adriaensen F, Chardon JP, De Blust G, Swinnen E, Villalba S, Gulinck H, Matthysen E (2003) The application of 'least-cost' modelling as a functional landscape model. Landscape Urban Plan 64:233–247

Bailey S (1998) Habitat fragmentation in England's ancient woods. University of Nottingham, Nottingham

Bailey S, Haines-Young RH, Watkins C (2002) Species presence in fragmented landscapes: modelling of species requirements at the national level. Biol Conserv 108:307–316

Bellamy PE, Hinsley SA, Newton I (1996) Factors influencing bird species numbers in small woods in south-east England. J Appl Ecol 33:249

Bennett A (2002) Linkages in the landscape: The role of corridors and connectivity in wildlife conservation. IUCN, Gland, Switzerland and Cambridge, UK
Bennett G (2004) Integrating biodiversity conservation and sustainable use: Lessons learned from ecological networks. IUCN, Gland, Switzerland and Cambridge, UK
Bissonette JA, Storch I (eds) (2003) Landscape ecology and resource management: Linking theory with practice. Island Press, Washington, DC, USA
Catchpole RDJ (2006) Planning for biodiversity – Opportunity mapping and habitat networks in practice: A technical guide. In: English nature research reports number 687. Natural England, Peterborough
Centre for Ecology and Hydrology (2000) Land Cover Map 2000. Centre for Ecology and Hydrology, Monks Wood, UK
Countryside Council for Wales (2004) Corporate Plan: 2005–2008 "Working together to create a better Wales". Countryside Council for Wales, Bangor, Wales
Crooks KR, Sanjayan M (2006) Connectivity conservation. Cambridge University Press, Cambridge, UK
DEFRA (2007) Strategy for England's trees, woods and forests. Department for the Environment, Food and Rural Affairs, London
Donald PF, Evans AD (2006) Habitat connectivity and matrix restoration: The wider implications of agri-environment schemes. J Appl Ecol 43:209–218
European Community (1992) Council Directive 92/43/EEC on the conservation of natural habitats and of wild flora and fauna. European Commission, Brussels
Failing L, Gregory R (2003) Ten common mistakes in designing biodiversity indicators for forest policy. J Environ Manag 68:121–132
Forestry Authority (1998) The UK forestry standard: The government's approach to sustainable forestry. Forestry Commission, Edinburgh
Forestry Commission (1999) England forestry strategy: A new focus for England's woodlands. Forestry Commission, Cambridge
Forestry Commission (2000) Forests for Scotland: The Scottish forestry strategy. Scottish Executive, Edinburgh
Forestry Commission (2001a) National Inventory of Woodland and Trees – England. Forestry Commission, Edinburgh
Forestry Commission (2001b) Woodlands for Wales: The National Assembly for Wales strategy for trees and woodlands. Forestry Commission, Aberystwyth
Forestry Commission (2002a) National Inventory of Woodland and Trees – County report for Isle of Wight. Forestry Commission, Edinburgh
Forestry Commission (2002b) National Inventory of Woodland and Trees – Scotland. Forestry Commission, Edinburgh
Forestry Commission (2002c) National Inventory of Woodland and Trees – Wales. Forestry Commission, Edinburgh
Forestry Commission (2002d) UK indicators of sustainable forestry. In: Economics and Statistics Unit, Forestry Commission, Edinburgh
Forestry Commission (2003) National Inventory of Woodland and Trees – Great Britain. Forestry Commission, Edinburgh
Forestry Commission (2005a) Keepers of time: A statement of policy for England's ancient and native woodland. Forestry Commission, Cambridge
Forestry Commission (2005b) Keepers of time: A statement of policy for England's ancient and native woodland – Action plan 2005–2007. Forestry Commission, Cambridge
Forestry Commission (2005c) Recliaming our forgotten inheritance. Forestry Commission, Aberystwyth, Wales
Forestry Commission Scotland (2003) Forest habitat networks and SFGS expansion grants. Forestry Commission, Edinburgh
Frelich LE, Puettmann K (1999) Restoration ecology. In: Hunter ML (ed) Maintaining biodiversity in forest ecosystems. Cambridge University Press, Cambridge, UK, pp 498–524

Gilg A (1996) Countryside planning: The first half century. 2nd edn. Routledge, London

Good JEG, Norris DA, Stevens PA (2000) Woodland habitat network for Wales: Ystwyth Valley. Countryside Council for Wales, Bangor

Grieve Y, Sing L, Ray D, Moseley DG (2006) Forest habitat networks Scotland – Broadleaved woodland specialist network for SW Scotland: contract report to Forestry Commission Scotland and Scottish Natural Heritage. Forest Research, Edinburgh

Gutzwiller KJ (ed) (2002) Applying landscape ecology in biological conservation. Springer-Verlag, New York

Haila Y (2002) A conceptual genealogy of fragmentation reserach: From island biogeography to landscape ecology. Ecol Appl 12:321–334

Hanski I (1999) Metapopulation ecology. Oxford University Press, Oxford

Hanski I, Ovaskainen O (2000) The metapopulation capacity of a fragmented landscape. Nature 404:755–758

Harmer RW, Thompson RN, Humphrey JW (2005) Great Britain: Conifers to broadleaves. In: Gardiner ES, Breland LJ (eds) Restoration of boreal and temperate forests: Documenting forest restoration knowledge and practices in boreal and temperate ecosystems. Proceedings of the IUFRO conference. CRC Press, Horshelm, pp 319–338

Hobbs RJ (1997) Future landscapes and the future of landscape ecology. Landscape Urban Plan 37:1–9

Hobbs RJ, Lambeck R (2002) An integrated approach to landscape science and management. In: Liu J, Taylor WW (eds) Integrating landscape ecology into natural resource management. Cambridge University Press, Cambridge, UK, pp 412–430

Howe L, Blackstock T, Burrows C, Stevens J (2005) The Habitat Survey of Wales. Brit Wildlife 16:153–162

Humphrey JW (2003) The restoration of wooded landscapes: Future priorities. In: Humphrey JW et al (eds) The restoration of wooded landscapes, Forestry Commission, Edinburgh, pp 153–157

Humphrey JW (2005) Benefits to biodiversity from developing old-growth conditions in British upland conifer plantations: A review. Forestry 78:33–53

Jongman RHG, Pungetti G (2004) Ecological networks and greenways: Concepts, design, implementation. Cambridge University Press, Cambridge

Kirby KJ, Latham J, Holl K, Bryce J, Corbett P, Watson R (2002) Objective setting and condition monitoring within woodland sites of special scientific interest. In: English Nature, Peterborough

Kirby KJ, Thomas RC (1994) Fragmentation patterns of ancient woodland in England. In: Dover JW (ed) Fragmentation in agricultural landscapes – Proceedings of the 1994 IALE (UK) Conference, held at Myerscough College, Preston, Lancashire, IALE (UK), 13–14 Sept 1994

Kleijn D, Baquero RA, Clough Y, Diaz M, Esteban J, Fernandez F, Gabriel D, Herzog F, Holzschuh A, Joehl R (2006) Mixed biodiversity benefits of agri-environment schemes in five European countries. Ecol Lett 9:243–254

Kleijn D, Sutherland WJ (2003) How effective are European agri-environment schemes in conserving and promoting biodiversity? J Appl Ecol 40:947–969

Kupfer JA, Malanson GP, Franklin SB (2006) Not seeing the ocean for the islands: The mediating influence of matrix-based processes on forest fragmentation effects. Global Ecol Biogeogr 15:8–20

Lambeck RJ (1997) Focal species: A multi-species umbrella for nature conservation. Conserv. Biol. 11:849–856

Latham J (2007) Ecological connectivity – Developing natural connections. Natur Cymru 22:33–35

Latham J, Watts K, Thomas C (2004) Development of a forest habitat network for Wales: Linking research with policy. In: Smithers RJ (ed) Landscape ecology of trees and forests – Proceedings of the 2004 IALE (UK) Conference, held at Cirencester Agricultural College, Gloucestershire, IALE (UK), 21–24 June 2004

Li HB, Wu JG (2004) Use and misuse of landscape indices. Landscape Ecol 19:389–399

Liu J, Taylor WW (eds) (2002) Integrating landscape ecology into natural resource management. Cambridge University Press, Cambridge, UK

MacArthur RH, Wilson EO (1967) The theory of island biogeography. Princeton University Press, Princeton

Margules CR, Pressey RL (2000) Systematic conservation planning. Nature 405:243–253

McGarigal K, Cushman SA, Neel MC, Ene E (2002) FRAGSTATS: Spatial pattern analysis program for categorical maps. University of Massachusetts, Amherst, www.umass.edu/landeco/research/fragstats/fragstats.html

McIntyre S, Hobbs RJ (1999) A framework for conceptualising human impacts on landscapes and its relevance to management and research models. Conserv Biol 13:1282–1292

Moseley DG, Ray D (2006) Forest habitat networks Scotland – Borders and Lothians: contract report to Forestry Commission Scotland and Scottish Natural Heritage. Forest Research, Edinburgh

Moseley DG, Ray D (2007) Forest habitat networks Scotland – Grampian: contract report to Forestry Commission Scotland and Scottish Natural Heritage. Forest Research, Edinburgh

Moseley DG, Ray D, Humphrey JW, Watts K (2007a) Forest habitat networks Scotland: contract report to Forestry Commission Scotland, Forestry Commission GB and Scottish Natural Heritage. Forest Research, Edinburgh

Moseley DG, Ray D, Watts K (2007b) Improving forest habitat networks with new woodland planting schemes. Forest Brit Tim 1:14–18

Moseley DG, Sing L, Ray D (2005) Forest habitat networks Scotland – Highland Conservancy: contract report to Forestry Commission Scotland and Scottish Natural Heritage. Forest Research, Edinburgh

Moss D, Davies CE (2002) Cross-references between the EUNIS habitat classification and nomenclature of CORINE Land Cover – CEH Project Report C00389. Centre for Ecology and Hydrology – European Environment Agency, Huntingdon, Cambridge

Opdam P (1991) Metapopulation theory and habitat fragmentation: A review of holarctic breeding bird studies. Landscape Ecol 5:93–106

Opdam P (2002) Assessing the conservation potential of habitat networks. In: Gutzwiller KJ (ed) Applying landscape ecology in biological conservation. Springer-Verlag, New York, pp 381–404

Opdam P, Steingrover E, van Rooij S (2006) Ecological networks: A spatial concept for multi-actor planning of sustainable landscapes. Landscape Urban Plan 75:322–332

Opdam P, Wascher D (2004) Climate change meets habitat fragmentation: Linking landscape and biogeographical scale levels in research and conservation. Biol Conserv 117:285–297

Pearson RG, Dawson TP (2003) Predicting the impacts of climate change on the distribution of species: Are bioclimate envelope models useful? Global Ecol Biogeogr 12:361–371

Peterken G (1996) Natural woodland: Ecology and conservation in northern temperate regions. Cambridge University Press, Cambridge

Peterken GF (1993) Woodland conservation and management. 2nd edn. Chapman and Hall, London

Peterken GF (1999) Clyde valley forest habitat network. Scotish Natural Heritage, Edinburgh

Peterken GF, Baldock D, Hampson A (1995) A forest habitat network for Scotland. Scotish Natural Heritage, Edinburgh

Poiani KA, Richter BD, Anderson MG, Richter HE (2000) Biodiversity conservation at multiple scales: Functional sites, landscapes, and networks. Bioscience 50:133–146

Quine C, Humphrey JW, Watts K (2004) Biodiversity in the UK's forests – Recent policy developments and future research challenges. In: Andersson F, Birot Y, Päivinen R (eds) Towards the sustainable use of Europe's forests – Forest ecosystem and landscape research: Scientific challenges and opportunities (EFI Proceedings No. 49). European Forest Institute, Joensuu, pp 237–248

Quine C, Watts K (in press) Successful de-fragmentation of woodland by planting in an agricultural landscape? An assessment based on landscape indicators. J Environ Mana

Rackham O (1990) Trees and woodlands in the British Landscape. J. M. Dent, London

Ray D, Watts K, Hope JCE, Humphrey J (2004) Developing forest habitat networks in lowland Scotland. In: Smithers RJ (ed) Landscape ecology of trees and forests – Proceedings of the 2004 IALE (UK) Conference, held at Cirencester Agricultural College, Gloucestershire, IALE (UK), 21–24 June 2004

Ricketts TH (2001) The matrix matters: Effective isolation in fragmented landscapes. AmNat 158:87–99

Robinson RA, Sutherland WJ (2002) Post-war changes in arable farming and biodiversity in Great Britain. J Appl Ecol 39:157–176

Sheail J (1995) Nature protection, ecologists and the farming context: A UK historical perspective. J Rural Stud 11:79–88

Simonson W, Thomas R (1999) Biodiversity: Making the links. English Nature, Peterborough

Thomas CD, Thomas JA, Warren MS (1992) Distributions of occupied and vacant butterfly habitats in fragmented landscapes. Oecologia 92:563–567

Thompson RN, Humphrey JW, Harmer R, Ferris R (2003) Restoration of native woodland on ancient woodland sites – Forestry commission practice guide, Forestry Commission, Edinburgh

Tischendorf L, Fahrig L (2000) On the usage and measurement of landscape connectivity. Oikos 90:7–19

Towers W, Malcolm A, Hester A, Ross I, Baird E (1999) Cairngorms forest and woodland framework. Cairngorms Partnership, Grantown-on Spey. 32pp

Turner MG, Crow TR, Liu J, Rabe D, Rabeni CF, Soranno PA, Taylor WW, Vogt KA, Wiens JA (2002) Bridging the gap between landscape ecology and natural resource management. In: Liu J, Taylor WW (eds) Integrating landscape ecology into natural resource management. Cambridge University Press, Cambridge, UK, pp 433–460

UK Biodiversity Steering Group (1995a) Biodiversity: The UK steering group report – Volume II: Action plans (Annex F – Lists of key species, key habitats and broad habitats). HMSO, London

UK Biodiversity Steering Group (1995b) Biodiversity: The UK steering group report – Volume II: Action plans (Annex F & G). HMSO, London

van Rooij SAM, van der Sluis T, Steingröver EG (2003) Networks for LIFE: Development of an ecological network for Persiceto (Emilia-Romagna, Italy). Report 729. Alterra, Wageningen

Vellend M (2003) Island biogeography of genes and species. Am Nat 162:358–365

Verboom J, Metz JAJ, Meelis E (1993) Metapopulation models for impact assessment of fragmentation. In: Vos C, Opdam P (eds) Landscape ecology of a stressed environment. Chapman and Hall, London, pp 172–191

Vos CC, Verboom J, Opdam PFM, Ter Braak CJF (2001) Toward ecologically scaled landscape indices. Am Nat 157:24–41

Wade TG, Riitters KH, Wickham JD, Jones KB (2003) Distribution and causes of global forest fragmentation. Conserv Ecol 7:7

Watts K, Griffiths M, Quine C, Ray D, Humphrey JW (2005) Towards a woodland habitat network for Wales. Countryside Council for Wales, Bangor, Wales

Watts K, Selman P (2004) Forcing the pace of biodiversity action: A force-field analysis of conservation effort at the 'landscape scale'. Local Environ 9:5–20

Worrell R, Taylor CMA, Spittal JJ (2003) Highland perthshire forest habitat network. Scot Forest 57:151–157

Synthesis

Ecology-Based Landscape Planning and Management

Thomas R. Crow

1 The Three Faces of Sustainability

The words "sustain" or "sustainable" are commonly found in the mission statements of resource management agencies. The mission of the USDA Forest Service, for example, is to "sustain the health, diversity, and productivity of the Nation's forests and grasslands to meet the needs of present and future generations." Sustaining the health, diversity, and productivity of a resource has certain fundamental requirements (Thayer 1989; Fedkiw et al. 2004). They start with a commitment to manage land and the water resources for the long term. It requires connecting the people living in the landscape with the natural resources that support their lives. It has to be inclusive of all sectors and functions of society by embracing meaningful civic involvement. And finally, sustaining the health, diversity and productivity of natural resources must create opportunities and preserve choices for people. These requirements necessitate a comprehensive approach in which economic, environmental, and social sustainability are given equal weight and are considered concurrently. Landscape ecology provides a conceptual as well as an operational framework for considering the three faces of sustainability.

2 Moving from Concept to Practice

The five chapters in this section support moving from concept to practice. Practicing sustainability within the context of landscapes is the unifying theme among the chapters. Each chapter illustrates the utility of a landscape perspective for considering sustainability in a variety of social and environmental settings. In Chapter 14, Azevedo et al. evaluated the changes in landscape structure and function that occur due to the application of a forest certification program, the Sustainable Forestry Initiative (SFI), in Texas, USA. They address the question: how is the widely applied SFI changing the

T.R. Crow
Research and Development, Environmental Sciences Staff, USDA Forest Service, Stop Code 1113
1400 Independence Avenue, SW Washington DC 20250-1113 USA
e-mail: tcrow@fs.fed.us

R. Lafortezza et al. (eds.), *Patterns and Processes in Forest Landscapes*,

landscape patterns in the intensively managed forests of East Texas? As always, there are trade-offs. There are winners and losers, but in general, the application of SFI at the landscape level improves hydrologic function and habitat suitability.

Authors for several of the chapters in this section of the book present landscape-level strategies for conserving biological diversity. The first of these, Pfund et al. (Chapter 15), explored the shift in paradigm from emphasizing reserves for conserving biodiversity to considering multifunctional landscapes with a spectrum of land uses ranging from protection to intensive utilization. The value of this new paradigm is the ability to integrate the livelihood of the people living in the landscape with biodiversity conservation. Pfund et al. provide valuable guidelines for connecting people with their natural resource – starting with creating the institutional partnerships and ending with regular monitoring. In contrast to the tropical setting presented in Pfund et al. where preventing forest fragmentation is a likely goal, in Chapter 18 Watts et al. had as their goal restoring connectivity among existing woodlands in the UK that are highly fragmented. As others have done (e.g., Gustafson 1996; Finney 2000), Watts and his co-authors stress the need to evaluate the effectiveness of spatially targeted treatments on the landscape. A challenge in doing so is the lack of information about species responses to change in landscape structure.

Turner (1989) defines landscape ecology as the study of the effect of pattern on process where "pattern" refers specifically to landscape structure. Two chapters in this section, Li et al. (Chapter 16) and North and Keeton (Chapter 17), dealt with the relation between landscape pattern and ecological processes. Li et al. compared and contrasted the patterns created by forest fire and timber harvesting on carbon stocks in two locations, boreal forests in central Saskatchewan, Canada, and sub-alpine forests in Miyaluo, Sichuan Province, P.R. China. Their study focused on carbon dynamics in living biomass and the ability to increase carbon pools through management. The strategy, while depending somewhat on the biotic and physical characteristics of the ecosystem, is basically the same – managing the age-class structure of the forest on the landscape.

Emulating natural disturbance regimes in management has been presented as a basis for practicing sustainable resource management (e.g., Palik et al. 2002; Crow and Perera 2004). North and Keeton used case studies from three U.S. forest types to explore the utility of this concept. A common approach is to compare managed forests to their "historic range of variability" (HRV), assuming that management is sustainable if the boundary conditions defined by HRV are not exceeded. The concept, however, as North and Keeton suggest, is difficult to implement. One problem is that the trade-offs among the three faces of sustainability – economic, environmental, and social – are rarely evaluated in a comprehensive way.

3 Finis

Landscape ecology has much to offer for developing sustainable social institutions and environmental practices. But as the authors of these chapters suggest, much remains to be learned about applying the concepts, theories, and methods from landscape ecology to the practice of sustainability. We do know, however, that a piecemeal approach in which one species, a forest stand, or a single ownership is considered in isolation is neither desirable nor tenable. Instead, a comprehensive, integrated approach needs to be applied at large scales – and this is exactly the strength of landscape ecology.

References

Crow TR, Perera AH (2004) Emulating natural landscape disturbance in forest management – an introduction. Landscape Ecol 19:231–233

Fedkiw J, MacCleery DW, Sample VA (2004) Pathways to sustainability, defining the bounds on forest management. Forest History Soc, Durham, NC

Finney MA (2000) Design of regular landscape fuel treatment patterns for modifying fire growth and behavior. Forest Sci 47:219–228

Gustafson EJ (1996) Expanding the scale of forest management: allocating timber harvests in time and space. Forest Ecol Manag 87:27–39

Palik BJ, Mitchell RJ, Hiers JK (2002) Modeling silviculture after natural disturbance to sustain biodiversity in the longleaf pine (*Pinus palustris*) ecosystem: balancing complexity and implementation. Forest Ecol Manag 155:347–356

Thayer RL Jr (1989) The experience of sustainable landscapes. Landscape J 8:101–110

Turner MG (1989) Landscape ecology: the effect of pattern on process. Annu Rev Ecol Syst 20:171–197

Appendix A

Evaluating Forest Landscape Connectivity through Conefor Sensinode 2.2

Santiago Saura

Abstract Maintaining and restoring landscape connectivity is currently a central concern in ecology and biodiversity conservation, but there is yet a lack of solid, operational and user-driven tools available for integrating connectivity in forest landscape planning. Here we describe the new Conefor Sensinode 2.2 (CS22) software, which quantifies the importance of forest habitat patches for maintaining or improving landscape connectivity and is conceived as a tool for decision-making support in landscape planning and habitat conservation. CS22 is based on (1) graph structures, which have been suggested to posses the greatest benefit to effort ratio for some conservation problems regarding landscape connectivity, (2) the habitat availability concept, which considers a patch itself as a space where connectivity occurs, integrating intrapatch and interpatch connectivity in a single measure and (3) the new probability of connectivity index, which has been recently shown to present improved properties compared to other existing indices and can be partitioned in four fractions considering the different ways in which a certain forest patch can affect the habitat availability and connectivity of the landscape. We provide an example of application to a case study for the Tengmalm's Owl (*Aegolius funereus*) in the region of Catalonia (NE Spain) to illustrate the results provided by the software and their potential for integrating connectivity in forest landscape planning. The CS22 software and all the geospatial data used in this case study can be downloaded from the World Wide Web, which allows performing the entire analysis as an exercise with real-world data.

A.1 Introduction

Landscape connectivity can be defined as the degree to which the landscape facilitates or impedes the movement across the habitat existing in that landscape (modified from Taylor et al. 1993). Maintaining or restoring landscape connectivity is currently a central concern in ecology and conservation planning (Crooks

---

S. Saura
Department of Agroforestry Engineering, Higher Technical School of Agrarian Engineering, University of Lleida; Forest Technology Centre of Catalonia, Solsona, Lleida, Spain
e-mail: ssaura@eagrof.udl.es

and Sanjayan 2006) and a key strategy for the conservation of the biodiversity and the ecological functions of forests (Forman 1995; Rochelle et al. 1999; Crist et al. 2005). For example, in the resolution 4 ("Conserving and enhancing forest biological diversity in Europe") of the Fourth Ministerial Conference on the Protection of Forests in Europe held at Vienna in 2003, the Signatory States and the European Community commit themselves to "prevent and mitigate losses of forest biological diversity due to fragmentation and conversion to other land uses and maintain and establish ecological connectivity". Indeed, the fragmentation and isolation of forest patches lead to a spatially structured habitat pattern in which the movements of dispersing individuals may be constrained, hampering the conservation of forest-dwelling species. Connectivity is in addition particularly crucial in the current challenge of alleviating the effects of climate change on species and ecosystems, since it may allow species to adapt to changing environmental conditions and to accommodate natural range shifts due to climate change (Hannah et al. 2002; Opdam and Wascher 2004). In this context, the spatial scales traditionally considered in forest resource management and analysis should be broaden to adequately characterize these ecological processes and characteristics, which occur at the landscape level.

There is a wide consensus in the literature that connectivity is species-specific and should be measured from a functional perspective. That is, not only the spatial arrangement of the habitat (structural connectivity) but also the dispersal distances and/or the behavioral response of the focal species to the physical structure of the landscape (functional connectivity) should be taken into account (e.g. Tischendorf & Fahrig 2000; Theobald 2006). Calabrese and Fagan (2004) further differentiated between potential and actual functional connectivity metrics. Potential connectivity metrics are those that incorporate some basic (perhaps indirect) knowledge about an organism's dispersal abilities together with spatial relationships among landscape elements or habitat patches. Actual connectivity metrics go a step further, by quantifying the real movement of individuals through a habitat or landscape and thus providing a direct estimate of the linkages that exist among landscape elements or habitat patches (Fagan and Calabrese 2006). However, these measurements of actual connectivity may be difficult to obtain in practice for applications at the landscape scale that usually cover large areas.

Although many different indices have been proposed and used in this context, there is still a lack of comprehensive understanding of their sensitivity to pattern structure and their behavior to different spatial changes, which seriously limits their proper interpretation and usefulness. As noted by Fagan and Calabrese (2006) "while definitions and measurement might seem boringly technical, conservation scientists must work to identify clear, replicable and well-understood metrics of connectivity if conservation is to invest funds and efforts wisely and responsibly". Indeed, research on connectivity and on the metrics that quantify it is still in an early stage, and much remains to be learned (Fagan and Calabrese 2006). Recent comparative analyses (Pascual-Hortal and Saura 2006; Saura and Pascual-Hortal 2007) have shown the weaknesses of different commonly used connectivity indices for prioritizing the most important areas for the maintenance of landscape connectivity. Additionally, an assessment of the sensitivity to spatial scale (both minimum mapping unit and extent) of ten graph-based connectivity indices (Pascual-Hortal and Saura 2007) has provided complementary criteria for evaluating the appropriateness of the analyzed metrics for conservation planning applications. Most of the examined metrics did not match up to all the desirable properties for decision-making (which are adequately reacting to all relevant landscape changes, being effective in identifying the most critical forest patches for conservation, and robust to changing spatial scale within a reasonable range), with the exception of two new landscape connectivity indices, the integral index of connectivity and the probability of connectivity (Pascual-Hortal and Saura 2006, 2007; Saura and Pascual-Hortal 2007). The integral index of connectivity (IIC) is based on a binary dispersal model, in which two forest patches are just either connected or not, with no intermediate modulation of the strength or feasibility of the connection between two patches, while the probability of connectivity index (PC) is based on a probabilistic connection model, in which there is a certain probability of direct dispersal between each two patches. Therefore, PC provides a more realistic and detailed picture of connectivity than IIC, and is as well less sensitive to uncertainties or errors in the estimation of the dispersal distances of the analyzed species, being in general preferable to IIC for forest landscape planning applications (Saura and Pascual-Hortal 2007), as noted as well by Bodin and Norberg (2007) for binary indices in general.

In many cases, the indices and methodologies developed for landscape connectivity analysis may fail to become widespread in practice because they may be too complex, too data intensive, not transparent enough or difficult to understand by land managers, or simply because they are not available or easy to implement in operational tools for real-world forest landscape planning. They may just remain as theoretical developments in the academic arena, having no real impact in the actual landscape planning or in improved forest biodiversity conservation. Land managers and forest planners may be expected to be aware of the general characteristics, limitations and scope of application of the different available approaches, but they rarely can develop or implement the operational tools that may derive from the conceptual and theoretical developments in this topic. More effort is required from the research community to provide end-user applications and practical recommendations for integrating connectivity considerations in forest landscape planning with a sound basis.

Here we describe the new Conefor Sensinode 2.2 (CS22) software, which includes new improved indices like the probability of connectivity, is intended to be easy to use for landscape and forest planners, and can be used free of charge for non-commercial purposes. We describe the major concepts, structures and indices in which the software is based (graphs, habitat availability indices) and provide an example of application to a case study for the Tengmalm's Owl (*Aegolius funereus*) in the region of Catalonia (NE Spain) to illustrate the results provided by the software and its potential for integrating connectivity in forest landscape planning.

A.2 Landscape Graphs and Habitat Availability Indices

A.2.1 Why Use Graphs to Analyze Forest Landscape Connectivity?

Graphs are mathematical structures made up by a set of nodes and links that may be used for quantitatively describing a forest landscape as a set of spatially or functionally interconnected patches. They are a powerful and effective way of overcoming computational limitations that appear when dealing with large data sets and performing complex analysis regarding forest connectivity. Their convenience for broad-scale studies in ecology, and specifically for the assessment of landscape connectivity, it is notable and it is being innovatively addressed for particular applications with threatened species. From the seminal papers by Bunn et al. (2000) and Urban and Keitt (2001), the applications and development of graph theory indices to the analysis of landscape connectivity have increased rapidly in recent years (Jordan et al. 2003; Brooks 2006; Pascual-Hortal and Saura 2006, 2008; Theobald 2006; Bodin and Norberg 2007; Ferrari et al. 2007; Saura and Pascual-Hortal 2007).

Nodes represent the spatial units that are considered for the connectivity analysis. They will typically be forest habitat patches or forest cells, but they may as well correspond to any other forest unit discriminated in the landscape based on ecological, administrative or management criteria (e.g. ownerships, public forests, forest blocks, natural parks, etc.), depending on the scale and objectives of the analysis. Links represent the functional connection between a pair of nodes; the existence of a link implies the potential ability of an organism to directly disperse between these two nodes. Links may be characterized by a probability of dispersal (in the probabilistic connection model), which is typically obtained as a function of distance. Distances between nodes (patches) can be obtained as Euclidean (straight-line) distances or, preferably, as minimum cost (effective) distances that take into account the variable movement preferences and abilities of the animal species through different land cover types (Adriaensen et al. 2003; Chardon et al. 2003; Nikolakaki 2004; Theobald 2006).

Graph-based connectivity metrics (like IIC or PC) have been suggested to "possess the greatest benefit to effort ratio for conservation problems that require characterization of connectivity at relatively large scales. These measures provide a reasonably detailed picture of potential connectivity, but have relatively modest data requirements" (Calabrese and Fagan 2004). Some simpler indices that only measure structural connectivity (e.g. nearest neighbour measures) are too crude to be considered as ecologically realist, while more complex metrics and models, such as those of metapopulation theory (e.g. Hanski 1994, 1998), may be too data-intensive and too difficult to parameterize for landscape-level planning applications and are generally limited to small study areas and scientific experiments (Calabrese and Fagan 2004). Indeed, it is important to balance metric performance with data requirements for operational landscape forest management, and connectivity metrics must be pragmatic and based upon data that might actually be attained on a regular basis (Fagan and Calabrese 2006).

A.2.2 Habitat Availability: Connectivity Between and Within Forest Patches

On the other hand, it has been suggested that connectivity should be considered within the broader concept of habitat availability in order to be successfully integrated in landscape conservation planning (Pascual-Hortal and Saura 2006). The habitat availability concept consists in considering a patch itself as a space where connectivity occurs, integrating intrapatch connectivity (habitat patch area) and interpatch connectivity (connections between different habitat patches) in a single measure (Pascual-Hortal and Saura 2006). For a forest habitat being easily available for an animal or population, it should be both abundant and well connected. Therefore, habitat availability for a species may be low if forest habitat patches are poorly connected, but also if the forest habitat is very connected but highly scarce. For example, in Fig. A.1 landscape A may be considered to be more connected than landscape B because it presents a high number of connections between forest habitat patches (with all the patches interconnected in a single component or connected region), while in landscape B the only two forest habitat patches are completely isolated from each other. However, this conclusion is misleading because only one of the isolated patches in landscape B provides much more connected area (within that patch) than all the patches in landscape A together, no matter how strongly connected they are (Fig. A.1). In fact, landscape A may be the result of a fragmentation and habitat loss process in just one of the big patches in landscape B. Many connectivity indices that are not based on the habitat availability concept fail for landscape conservation planning applications by considering that the loss of one of the small patches in landscape A is more detrimental for landscape connectivity than the loss of one of the patches in landscape B, since the former reduces the number of connections between forest habitat patches while the later does not (Fig. A.1), as has been shown in detail by Pascual-Hortal and Saura (2006) and Saura and Pascual-Hortal (2007). The new IIC and PC have been developed as habitat availability indices and are free of these limitations. Other graph-based indices have been

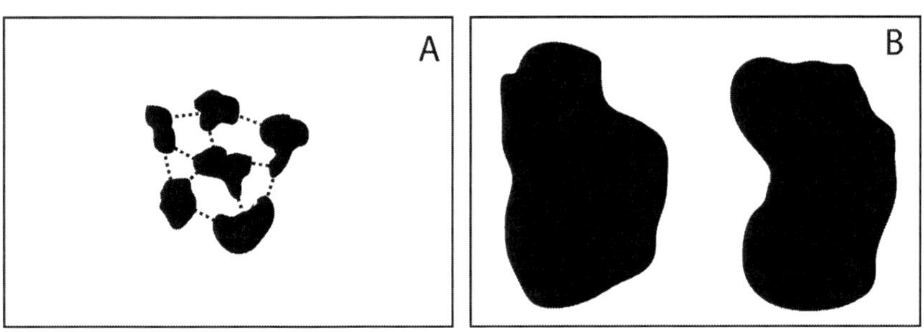

Fig. A.1 Two simple landscapes (**A** and **B**) to illustrate the concept of habitat availability and the potential pitfalls of some connectivity indices for forest landscape planning purposes, as described in Section A.2.2. Forest habitat patches in each landscape are shown in *black*, while the connections existing between the patches are indicated by *dashed lines*

as well proposed recently to quantify the vulnerability of the landscape to the loss of individual patches and to identify patches that significantly contribute to landscape connectivity, such as the normalized betweenness centrality index (Bodin and Norberg 2007). However, this index does not measure connectivity from the habitat availability perspective and only considers connectivity between (and not within) habitat patches, suffering from the same type of problems just illustrated through the two landscapes in Fig. A.1.

A.3 The Probability of Connectivity Index Implemented in the Conefor Sensinode 2.2 Software

A.3.1 Definition and Computation of the Probability of Connectivity Index

The Conefor Sensinode 2.2 (CS22) software includes nine different connectivity indices. Some of them are based in the habitat availability concept (IIC an PC), while some others are basic and classical indices that provide complementary information on the landscape and its degree of connectivity (e.g. number of links, number of components, etc.). However, the new probability of connectivity (PC) is the index that has been shown to present the best performance for forest landscape planning applications (Saura and Pascual-Hortal 2007) and is the one most recommended for use among those available in CS22.

The probability of connectivity index (PC) ranges from 0 to 1, increases with higher connectivity, and is defined as the probability that two points randomly placed within the landscape fall into habitat areas that are reachable from each other (interconnected) given a set of n habitat nodes (e.g. forest patches) and the connections among them. It is given by the following expression (Saura and Pascual-Hortal 2007):

$$PC = \frac{\sum_{i=1}^{n}\sum_{j=1}^{n} a_i \cdot a_j \cdot p_{ij}^*}{A_L^2} = \frac{PC_{num}}{A_L^2}, \tag{A.1}$$

where $a_i$and a_j are the attributes of the nodes i and j, typically habitat patch area or some other attribute that may be considered relevant for the analysis (quality-weighted habitat area, habitat suitability, core area, etc.). A_L is the maximum landscape attribute; in the case that the node attribute is patch area A_L corresponds to total landscape area (area of the study region, comprising both forest and non-forest patches). PC is based on the probabilistic dispersal model, with p_{ij} being the probability of a step between nodes i and j (where a step is a direct dispersal between i and j without passing by any other intermediate habitat nodes). The product probability of a path (where a path is made up of a set of steps in which no node is visited more than once) is the product of all the p_{ij} belonging to each step in that path. p_{ij}^* is defined as the maximum product probability of all possible paths between patches i and j (including single-step paths). If nodes i and j are close enough, the maximum probability path will simply be the step (direct movement) between nodes i and j ($p^*_{ij} = p_{ij}$). If nodes i and j are more distant, the "best" (maximum probability) path will probably comprise several steps through intermediate stepping stone nodes yielding $p^*_{ij} > p_{ij}$. When two nodes are completely isolated from each other, either by being too distant or by the existence of a land cover impeding the movement between both nodes (e.g. a road), then p^*_{ij}=0. When $i = j$ then p^*_{ij}=1 (it is sure that a patch can be reached from itself); this relates to the habitat availability concept that applies for PC, in which a patch itself is considered as a space where connectivity exists. The numerator in the equation for the PC index (PC_{num}, Equation A.1) varies depending both on the spatial arrangement and characteristics of the forest patches and on the dispersal abilities of the analyzed species (functional connectivity), while the denominator (A_L^2) is just a constant that only depends on the extent of the study area and is included to normalize the range of variation of PC from 0 to 1 as for the degree of coherence by Jaeger (2000).

PC is a habitat availability index that measures functional connectivity and is based on graph structures and algorithms for its computation (e.g. determination of the maximum probability paths). The PC index is general enough and may be measuring either potential or actual connectivity depending on how the probabilities of dispersal have been quantified (p_{ij}). However, for landscape-level applications the functional aspect of connectivity is usually quantified through an estimation of the average dispersal distance of the analyzed species (as in the case study presented in Section A.5), therefore being more typically applied as a potential connectivity metric.

As implemented by default in CS22 it is possible to easily compute the importance of each individual node (*dPC*) for forest landscapes comprising up to about 2000 nodes in a standard personal computer. In addition, CS22 includes the possibility of specifying a minimum probability to discard from the analysis the direct connections (p_{ij}) and the maximum probability paths (p_{ij}*) with a probability equal or lower than that minimum (that will be treated as completely unconnected). Moderate values of this minimum probability may have a very minor effect on the results and accuracy of the analysis, while they may considerably reduce the processing time by decreasing the number of connections and paths to be analyzed by CS22. This minimum probability option will typically provide faster results in very big or sparsely connected landscapes, therefore increasing the total number of nodes that can be processed at a time with CS22.

A.3.2 From Overall Landscape Connectivity to the Importance of Individual Forest Patches for Connectivity

When analyzing forest landscape connectivity through an index such as PC, two different types of outcomes are possible. On one hand, a single index value may characterize the degree of connectivity of the whole landscape; this provides an idea of the current status of the landscape, but is simply descriptive and not particularly relevant for decision making. On the other hand, an operational connectivity analysis oriented to forest planning would pursue identifying the most critical landscape elements for the maintenance of overall connectivity (Keitt et al. 1997; Jordán et al. 2003; Pascual-Hortal and Saura 2006). Most critical landscape elements (typically forest patches) are those whose absence would cause a larger decrease in overall landscape connectivity. The ranking of landscape elements by their contribution to overall landscape connectivity according to the PC index can be obtained by calculating the percentage of importance (*dPC*) of each individual element (Keitt et al. 1997; Urban and Keitt 2001; Pascual-Hortal and Saura 2006; Rae et al. 2007), as implemented in CS22:

$$dPC(\%) = 100 \cdot \frac{PC - PC_{remove}}{PC}, \tag{A.2}$$

where *PC* is the index value when the landscape element is present in the landscape and PC_{remove} is the index value after removal of that landscape element (e.g. after a certain forest patch loss). The *dPC* values for all the patches in the landscape are very useful for decision-making in forest landscape planning, since they allow identifying the most critical nodes (forest patches) for the maintenance or improvement of landscape connectivity, in which a forest management oriented to the conservation of the forest habitat should be implemented. Note that the *dPC* values are only affected by PC_{num} and not by the denominator in Equation (A.1) for the PC index, since A_L only depends on the extent of the study area and remains constant after the removal of any forest patch. On the other hand, although the formula for the PC index (Equation A.1) only depends on the 'best' (maximum product probability) path between two patches, the existence of alternative paths different from that one is considered through the importance analysis (*dPC*); when the loss of a patch breaks the only path existing between other patches this will result in a large *dPC*, while when there are many other good paths between those patches (nearly as good as the one that has been broken, as quantified by p^*_{ij}), this will result in a comparatively much lower *dPC*.

A.3.3 On the Different Ways a Forest Patch Can Contribute to the Probability of Connectivity

To better understand what the PC index is really measuring and integrating in a single measure, and to adequately interpret the results provided by CS22, it is interesting to note that the equation for the PC index (Equation A.1) can be partitioned in four distinct fractions considering the different ways in which a certain patch *k* can affect the value of PC (PC_{num}) for the entire landscape, as follows:

$$PC_{num} = PC_{intra} + PC_{flux} + PC_{connector} + PC_{indep} \tag{A.3}$$

PC_{intra} is the intrapatch connectivity or the available habitat area provided by the area of patch k itself (or other patch attribute, $PC_{intra} = a_k^2$), as related to the habitat availability concept. Note that the value of this fraction is fully independent on how patch k may be connected to other patches and would be the same even if patch k was completely isolated ($p_{ij} = 0$ to any other patch in the landscape). This corresponds to the degree of coherence proposed by Jaeger (2000) as a fragmentation index (not considering connections or the possibility of dispersal between patches).

PC_{flux} is the area-weighted flux of the connections of patch k with all the other patches in the landscape when k is either the starting or ending node of that connection (dispersal flux starting from or ending in that patch k). This is similar to an area-weighted version of the index of dispersal flux by Urban and Keitt (2001) but considering the maximum product probability (p_{ij}^*) instead of the probability of direct dispersal between patches (p_{ij}). This fraction depends both on the area (attribute) of node k (bigger patches producing more flux, being the rest of the factors equal) and in its location in the landscape network (topological position, interpatch connectivity). It results from summing $a_i \cdot a_j \cdot p_{ij}^*$ for all the pairs of patches in the landscape in which either $i = k$ or $j = k$. This fraction measures how well this patch is connected to other patches in the landscape (in terms of the amount of flux), but not necessarily how important that patch is for maintaining the connectivity between the rest of the patches, as quantified by the next fraction.

$PC_{connector}$ is the contribution of patch k to the connectivity between other habitat patches, as a connector or stepping stone between them. This fraction is independent of the area or any other attribute of patch k (a_k), and does only depend on the topological position of the patch in the landscape network. A certain patch k will only contribute to PC through $PC_{connector}$ when it is part of the best (maximum product probability) path for dispersal between other two patches i and j. It results from summing $a_i \cdot a_j \cdot p_{ij}^*$ for all the pairs of patches i and j in which $i \neq k$, $j \neq k$ and k is part of the maximum probability path ($p_{ij}*$).

PC_{indep} is the part of the PC value that is fully independent of patch k, not affected neither by the topological position nor by the value of the attribute of patch k. This fraction corresponds to the habitat area made available by the rest of the patches by themselves (sum of a_i^2 when $i \neq k$) and by the connections between the rest of the patches i and j (sum of $a_i \cdot a_j \cdot p_{ij}^*$) when $i \neq k$, $j \neq k$ and k is not part of the maximum probability path between i and j.

Depending on the intrinsic characteristics (node attribute) and on the topological position within the landscape network (interpatch connectivity) of a certain forest patch, it will present a higher or lower importance (*dPC*) coming from one or more of the different fractions described above. When a patch is completely isolated it will only contribute to PC through PC_{intra}. When a patch is connected to some degree to some other patches (and $a_k > 0$), it will surely contribute to PC through PC_{intra} and PC_{flux} and, depending on the cases (topological position in the landscape network), it may also contribute through $PC_{connector}$ (only if it is a stepping stone in the paths between other two nodes). When a patch is lost from the landscape, PC_{intra} and PC_{flux} will be entirely lost from the new resultant PC value ($PC_{remove} \cdot A_L^2 < PC_{num}\text{-}PC_{intra}\text{-}PC_{flux}$), while $PC_{connector}$ may be lost only partially, depending on the alternative paths between the remnant patches that are available after loosing patch k (as quantified by the decrease in $p_{ij}*$ produced by the loss of patch k), as described above.

A.4 The Conefor Sensinode 2.2 Software: Inputs, Outputs, and User Settings

A.4.1 What is Conefor Sensinode 2.2?

Conefor Sensinode 2.2 (CS22) is a simple program that allows quantifying the importance of forest habitat patches for maintaining or improving landscape connectivity through graph structures and habitat availability indices, as described in previous sections. CS22 is conceived as a tool for decision-making support in landscape planning and habitat conservation. CS22 includes several connectivity indices, but the best performing and recommended index is the probability of connectivity (PC) (Saura and Pascual-Hortal 2007), as described above.

CS22 has been developed by Josep Torné and Santiago Saura at the University of Lleida (Spain) by modifying, reprogramming and including new indices and features in the Sensinode 1.0 version (LandGraphs package) developed by Dean L. Urban (Duke University, USA). It is distributed free of charge for non-commercial use, with the only condition of citing the software and the two most-related references (Pascual-Hortal and Saura 2006; Saura and Pascual-Hortal 2007). The last available version of the software and the user's manual can be directly downloaded from http://www.udl.es/usuaris/saura/cs22.htm or from http://www.conefor.udl.es/cs22.htm. CS22 only requires a standard computer running a Windows operative system and about 8 MB of free space in the hard disk.

A.4.2 Which Information is Needed to Run Conefor Sensinode 2.2?

CS22 quantifies functional connectivity; that is, it requires as an input the information necessary for quantifying both the structural (spatial arrangement of forest patches) and the functional (dispersal abilities of the analysed species) aspects of connectivity (Fig. A.2). The information required by CS22 consists of two input files (the node file and the connection file) and some other specific user settings, all of them shown in a simple interface in the main screen of the software (Fig. A.3). The user should specify the connectivity indices to be calculated, up to nine different ones including PC (Fig. A.3). Both input and output files are ASCII text or DBF formats which may be easily obtained from or incorporated into any GIS, word processor or spreadsheet program.

The node file contains a list of the forest habitat nodes existing in the landscape and their attributes (one value per node corresponding to the a_i variable in Equation (A.1) for the PC index). Nodes may represent forest patches, forest cells

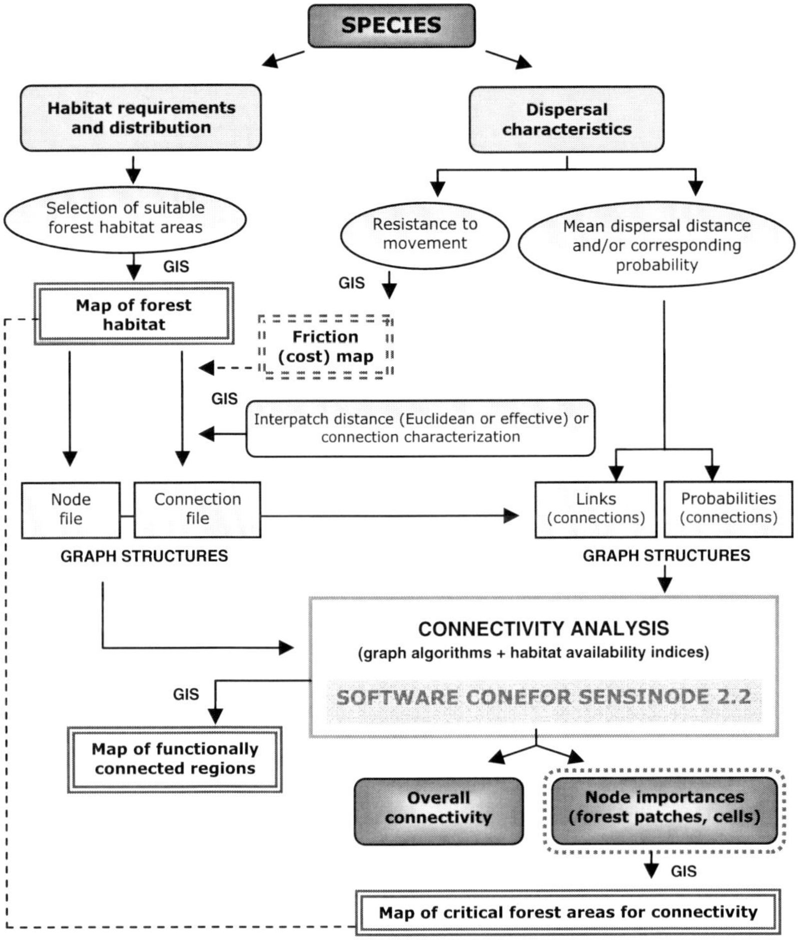

Fig. A.2 Schematic outline of the methodology for the analysis of forest landscape connectivity through the Conefor Sensinode 2.2 software

or other bigger spatial units (e.g. ownerships, public forests, forest blocks, natural parks, etc.), while the node attribute is the characteristic of the node that is considered relevant for the analysis, such as forest habitat area, habitat quality, quality-weighted area or some other attributes where appropriate (e.g. population density, core area, carrying capacity, habitat suitability, etc.). The definition and determination of the nodes and their attributes depends on the objectives and characteristics of particular applications, the spatial scale of the analysis, and the available information on the species and the forest landscape.

The connection file contains the information necessary for characterizing the connections (probability of direct dispersal, p_{ij}) between every two nodes in the landscape as required by the PC index, and may be entered in two ways, either as a

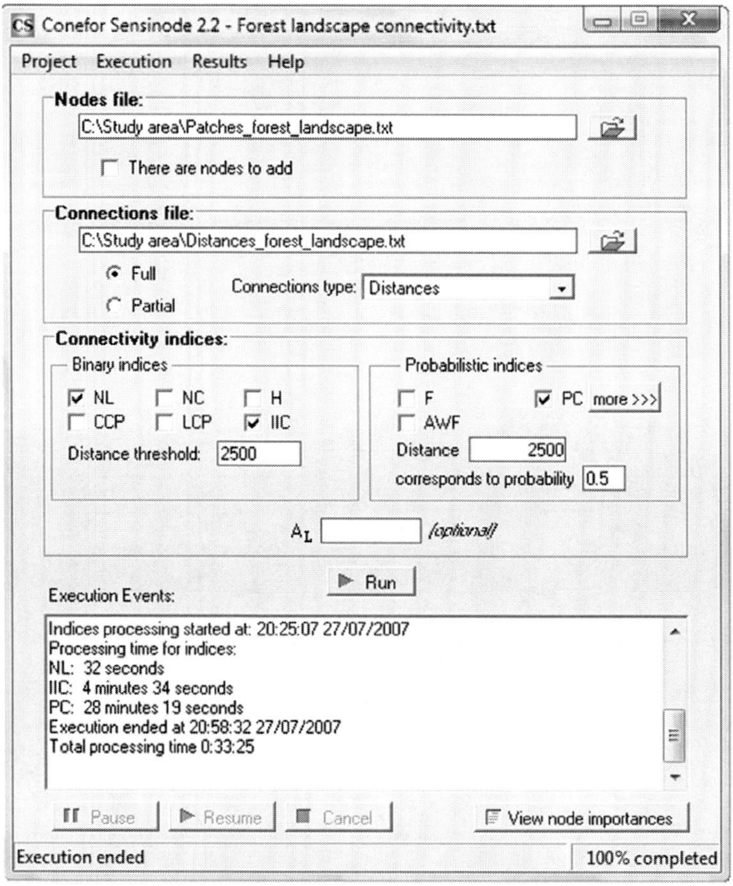

Fig. A.3 Main screen of the Conefor Sensinode 2.2 software

distance file or as a probability file. The former is the most common way, with the connection file reporting the distance, either Euclidean or effective (minimum cost), between every two nodes. In addition the user needs to specify the distance that the analyzed species disperses with a certain probability (see the box for the probabilistic indices in Fig. A.3). Typically, this corresponds to the median dispersal distance of the species that is assigned to a probability of 0.5 (Fig. A.3), although it may also be for example the maximum dispersal distance corresponding to a probability of 0.05 or 0.01 depending on the cases. From this information, CS22 models and computes automatically the probabilities of direct dispersal (p_{ij}) as a decreasing exponential function of the distance between each pair of nodes (e.g. Keitt et al. 1997; Bunn et al. 2000; Urban and Keitt 2001; Saura and Pascual-Hortal 2007) matching the distance-probability values entered by the user. The user may also compute these direct dispersal probabilities in a different way externally to CS22 or measure them directly through actual movement patterns monitoring or mark-release-recapture methods (if those data-intensive measurements can be carried out), and provided as an input probability file of already calculated p_{ij} to CS22. In this latter case, no other information apart from that probability file is required by CS22 regarding the connections in the landscape.

Obviously, the user is responsible of providing adequate data with a sufficient accuracy and degree of detail for the specific landscape, focal species and goals of the analysis. The "garbage in, garbage out" axiom applies here as everywhere else.

The node and connection files (already formatted for processing with CS22) for the bird species and study area described in Section A.5 can be downloaded at http://www.conefor.udl.es/form_springer.php by providing the password "springer1aegolius2tetrao". Further details on the format of the files can be found in the CS22 user's manual.

A.4.3 Which Results Are Provided by Conefor Sensinode 2.2?

CS22 provides different outputs (Fig. A.2), being the most outstanding one the importance of each individual node (forest habitat patch) for maintaining overall landscape connectivity (*dPC*). This allows ranking forest habitat patches by their contribution to landscape connectivity (patches prioritization), which provides objective criteria for the selection of the most critical forest areas for landscape conservation planning purposes. CS22 also allows including in the analysis potential new forest areas (nodes) that currently do not exist but that may be added in the landscape through forestation or habitat restoration. In this case, CS22 will also compute the contribution of these potential new nodes to the improvement of landscape connectivity.

Other results provide complementary information on the landscape and its degree of connectivity, such as the overall values of the different connectivity indices (e.g. PC, Equation A.1), the component to which each forest patch belongs (where a component is a set of nodes for which a path exists between every pair of nodes), the probabilities of direct dispersal (p_{ij}) and the maximum product probabilities (p^*_{ij}) between every two nodes, etc. Further details can be found in the user's manual of CS22.

A.4.4 Which Other Software is Needed to Analyze the Connectivity of the Forest Landscape Together with Conefor Sensinode 2.2?

Although the core of the analysis (the connectivity analysis itself) is performed by CS22 (Fig. A.2), the user will at least need a GIS to prepare the information required by CS22 and to visualize the results provided by CS22 and integrate them with other geospatial data for further analysis.

The node file can be very easily prepared from the attribute table of any GIS layer, by simply exporting the two columns of that table containing the feature ID and attribute table.

The connection file (usually a distance file) may require some more processing than the node file, but it can also be easily obtained from the basic capabilities of any GIS such as ArcView/ArcGis. However, the processing will be different depending on which type of distance is used to characterize the connections between nodes. Euclidean (straight-line) distances may be used for those species that are not much affected by the land cover types (matrix) between the forest habitat patches (e.g. some bird species), for landscapes with a more or less homogeneous matrix, or simply as a first level of analysis that can be refined later with some more detailed considerations. The last version of CS22 includes the free extensions for ArcView 3.x ("ID Within Distance: Conefor") and ArcGis 9.x ("Conefor Inputs") that allow measuring the edge-to-edge Euclidean distances between all polygon features of a theme, producing a text distance file that is directly usable (with no other change) in CS22. For other forest species estimating effective (minimum-cost) distances may improve the results provided by a connectivity analysis performed through straight-line distances (e.g. Adriaensen et al. 2003; Theobald 2006) by taking into account the variable movement abilities and mortality risk of a species through different land cover types. Effective distances are typically obtained through least-cost path algorithms, requiring a considerably larger processing time than the Euclidean distances. One option to calculate the effective distances is to use the PathMatrix extension for ArcView 3.x that needs to be used in conjunction with the Spatial Analyst module; this extension is freely available for download from http://cmpg.unibe.ch/software/pathmatrix/ (accessed July 2007). Another option is to use the cost distance tools in ArcGis 9. Note that these are only some alternatives to calculate distances among the variety of GIS and image processing software and extensions to them, and you may use any other existing application (different from ArcView/ArcGis) to get the same result and provide it to CS22.

Several of the results are provided by CS22 in DBF format (node importance, components), which allows easily joining these results within a GIS layer for visualization and further analyses. Other applications such as spreadsheet software may also be used for further processing of the numerical results provided by CS22.

A.5 Example of Application to a Case Study

To illustrate the use and effectiveness of the methodology, the CS22 software and the probability of connectivity index for forest landscape planning applications, we analyzed the connectivity of the forest habitat of the Tengmalm's Owl (*Aegolius funereus*) in the region of Catalonia, located in the Northeast of Spain (Fig. A.4). Catalonia is a heterogeneous region comprising the provinces of Barcelona, Girona, Lleida and Tarragona and with a total extension of 32,107 km^2, including mountainous areas like the Pyrenees (with an altitude up to 3,143 m) and a long coastline

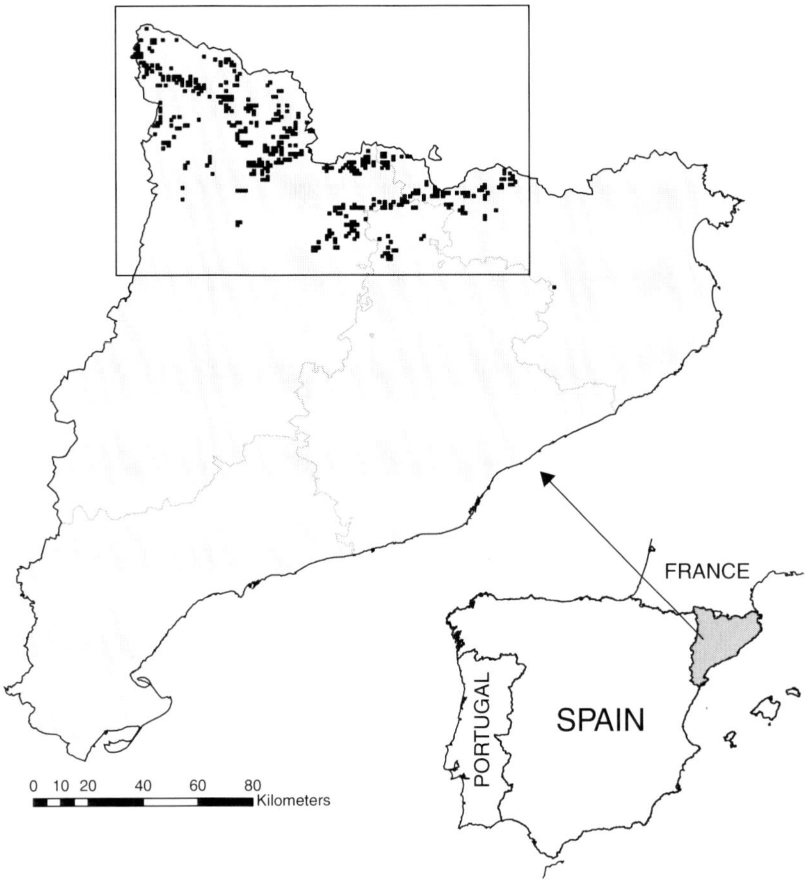

Fig. A.4 Location of the study area (Catalonia) in the map of Spain and distribution of Tengmalm's Owl habitat in the map of Catalonia. Black squares correspond to forest habitat patches (UTM 1 × 1 km squares) with a probability of occurrence equal or above 0.2 in the Catalan Breeding Bird Atlas 1999-2002 (Estrada et al. 2004)

along the Mediterranean Sea. The climate is, according to Papadakis classification, mostly Mediterranean temperate, with presence also of maritime temperate climate in the coast and temperate cold climate in the Pyrenees. According to the Third Spanish National Forest Inventory, about half of the total area of Catalonia is covered by forests with a canopy cover above 20%, with *Pinus halepensis*, *Pinus sylvestris*, *Quercus ilex* and *Pinus nigra* as the most abundant forest tree species. In Catalonia the Tengmalm's Owl (*Aegolius funereus*) is at the southern border of its distribution range, and only occurs in subalpine Scots Pine (*Pinus sylvestris*), Mountain Pine (*Pinus uncinata*) and Silver Fir (*Abies alba*) forests in the Pyrenees, usually between 1700 and 2100 m, and in areas where the maximum summer temperatures remain below 18° (Estrada et al. 2004). It is a scarce and vulnerable species in Catalonia that breeds in mature open forests with presence of young trees, clearings, available cavities, fallen trees, stumps and little understory (Mariné and Dalmau 2000).

Forest habitat distribution data for this species were obtained from the Catalan Breeding Bird Atlas 1999-2002 (Estrada et al. 2004), which provides the estimated probability of occurrence for the Tengmalm's Owl in 1 × 1 km UTM cells covering all Catalonia, as a result of field sampling and niche-based modeling (Estrada et al. 2004). Further details on this atlas can be as well obtained at http://www.ornitologia.org/monitoratge/atlesa.htm (accessed July 2007). All cells with a probability of occurrence equal or greater than 0.2 were selected in this study as the forest habitat patches (nodes) to be analyzed, resulting in a total of 436 1 × 1 km cells (Figs. A.4 and A.5). The probability of occurrence in each cell was considered as a measure of habitat quality and as the relevant patch attribute for the analysis (a_i variable in the PC index, see Equation (A.1), indicating that patches with higher probability of occurrence are more suitable for the Tengmalm's Owl. To quantify the species dispersal ability we considered a median dispersal distance of 34 km, as reported in previous studies (Korpimäki and Lagerström 1988, Sutherland et al. 2000). We set to 0.5 the direct dispersal probability (p_{ij}) associated to that distance of 34 km, and calculated all the remaining interpatch p_{ij} by applying a negative exponential function (as implemented in CS22) of the edge-to-edge Euclidean distance between forest patches (1 × 1 km forest cells).

The application of the probability of connectivity index (PC) and the CS22 software to the analysis of the importance of forest patches (1 × 1 km cells) allowed identifying and prioritizing the forest patches and public forests that most contribute to overall landscape connectivity for the Tengmalm's Owl (Fig. A.5), as evaluated by *dPC*. This outcome is particularly useful for forest landscape planning, as it allows concentrating conservation efforts and adapting forest management to the species requirements in those areas that are most important for the maintenance of connectivity, in which an eventual habitat loss or degradation would have more critical impact on the remnant habitat network and on the habitat availability for this species.

Some of the forest habitat patches attained *dPC* values as high as 1.27% (Fig. A.5), more than five times the importance that would result if all the 436 patches would have the same importance from a total of 100% (0.23%), reflecting the existence of key areas for connectivity (such as stepping stones) in the analyzed landscape. The analysis showed that about 41% of the critical areas for connectivity were located outside the Natura 2000 Network in Catalonia (as measured by the sum of *dPC* for all the 1 × 1 km cells or portions of them located outside this network, relative to the sum of *dPC* for all the habitat cells), indicating the specific locations where the current network could be expanded to considerably improve its efficiency for protecting the most important habitat areas for the Tengmalm's Owl. The most important public forests for this species resulted to be "Bandolèrs, Dossau, Beret, Ruda e Aiguamòg" (municipality of Naut Aran), "Muntanya de Lles" (municipality of Lles) and "Aubas-Portilhon" (municipality of Bossòst), concentrating 6.4%, 5.1% and 4.5% of accumulate connectivity importance respectively (quantified as the sum of the dPC of each 1 × 1 km habitat cell or portions of it falling within each

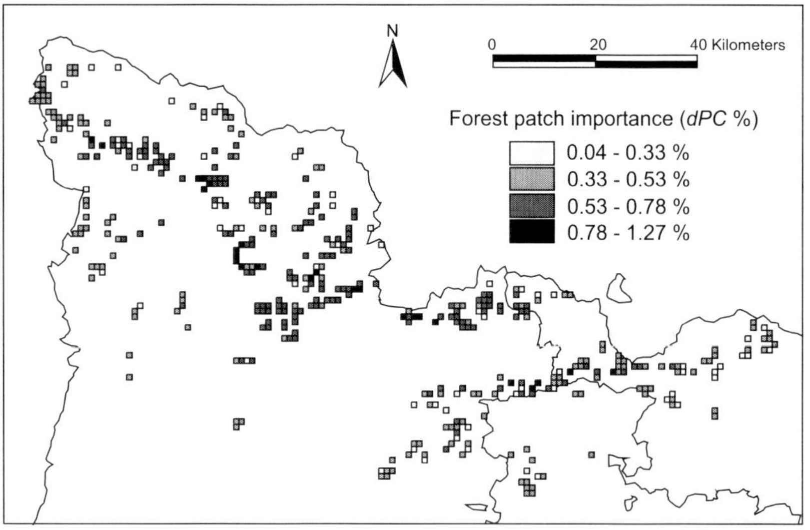

Fig. A.5 Importance of forest habitat patches (UTM 1 × 1 km cells) for the maintenance of overall landscape connectivity for the Tengmalm's Owl in Catalonia according to the PC index (dPC). The geographical extent of this figure corresponds to the rectangle over Catalonia indicated in Fig. A.4.

public forests). In these most important forests, a management specifically oriented to the conservation of the Tengmalm's Owl habitat should be applied. This concerns attaining structurally heterogeneous forests, preferably uneven aged stands, and avoiding excessive stem densities and too closed canopies through intense thinning and clear-cutting in small areas if necessary. Management should retain a large amount of dead wood, old-growth and standing dead trees, stumps, and other structural elements that favor both the presence and the hunting of small mammals by the Tengmalm's owl (Mariné and Dalmau 2000; Estrada et al. 2004). Cavities are also extremely important for the nesting and reproductive success of this species, which are associated to the presence of old-growth trees in the forest.

All the geospatial data used in this case study for the *Aegolius funereus* can be downloaded from http://www.conefor.udl.es/form_springer.php (by providing the password "springer1aegolius2tetrao"), and can be opened directly with ArcView/ArcGIS or other GIS software. These include the distribution of the forest habitat in 1 × 1 km cells provided by the Catalan Institute of Ornithology (which can be downloaded from http://www.ornitologia.org/scoc/, accessed July 2007), the administrative boundaries of Catalonia, the public forests managed by the Catalan Department of the Environment (official version produced on 28/02/2007), and the Natura 2000 Network in this region (official version produced by the government of Catalonia on 29/09/2006), the latter three available from http://mediambient.gencat.net/cat/el_departament/cartografia (accessed July 2007). This allows performing the entire analysis as an exercise with real-world data to get familiar with CS22 and the input and output information, obtaining the same results that have been presented above for the Tengmalm's Owl. In addition, the equivalent information can also be found in the indicated webpage for the Capercaillie (*Tetrao urogallus*), for which a median dispersal distance of 2.3 km has been reported by Hjeljord et al. (2000). Although the results for this species are not presented in this chapter, they have been reported previously performing the analysis through the IIC index (Pascual-Hortal and Saura 2008), which allows confirming the results obtained from these spatial data for the Capercaillie (although if analyzed through the PC index the results will vary from those reported for IIC). Finally, the geospatial data for the habitat distribution for these and many other bird species in Catalonia can be obtained freely from http://www.ornitologia.org/scoc/, to which the same methodology and type of analysis may be as well applied.

A.6 Conclusions and Further Development of the Software

The need for maintaining ecological fluxes in the landscape and the natural dispersal routes for the movement and survival of wildlife species call for a more integrated management of the land in which connectivity considerations should be necessarily incorporated. The Conefor Sensinode 2.2 software and the methodology in which it is based (graph structures, habitat availability concept, and the probability of connectivity index) may be a helpful decision support tool for integrating connectivity in forest landscape planning. It presents several improved characteristics compared to other approaches available for analyzing connectivity, and at the same time it is conceived as a user-driven application that is easy to understand by land managers and forest planners. The software allows identifying which forest patches are more relevant (critical) for the maintenance of overall landscape connectivity, which may provide valuable guidelines for orienting forest management and focusing conservation efforts and further analyses directly on those forest areas that are critical due to their attributes and specific network location within the landscape mosaic. Given scarce funding and limited capacity, it is crucial to know which of the potentially many forest areas represent the best conservation investment.

Further development of the software may include, among others, (1) a specific evaluation of the importance for connectivity of individual corridors and linkages in the landscape (and not only of the forest patches as currently implemented), (2) a better integration with the most common GIS and geospatial data formats, (3) a separated quantification of the different fractions in which the loss of a forest patch may affect the habitat availability and connectivity of the landscape and (4) an improved processing capabilities to allow the analysis of landscapes with larger sets of nodes.

Acknowledgments Funding was provided by the Ministerio de Educación y Ciencia (Spain) and the European Union (FEDER funds) through projects CONEFOR (REN2003-01628) and IBEP-FOR (CGL2006-00312/BOS). Special thanks to Begoña de la Fuente Martín for her valuable assistance in the preparation of this manuscript. We thank Dr. Dean Urban (Duke University) for providing the source codes of the Sensinode software (Landgraphs package version 1.0), as the starting point for the Conefor Sensinode 2.2 software. We thank all volunteers that made possible to collect the information for the Catalan Breeding Bird Atlas, which was provided by the Instituto Catalán de Ornitología (ICO). ICO provided the digital spatial data on the distribution of the Tengmalm's Owl and the Capercaillie in Catalonia, and the rest of the GIS layers were provided by the Catalan Department of Environment (Generalitat de Catalunya). All these data and layers can be downloaded from http://www.conefor.udl.es/form_springer.php (by providing the password "springer1aegolius2tetrao").

References

Adriaensen F, Chardon JP, De Blust G, Swinnen E, Villalba S, Gulinck H, Matthysen E (2003) The application of "least-cost" modelling as a functional landscape model. Landsc Urban Plan 64(4):233–247

Bodin O, Norberg J (2007) A network approach for analyzing spatially structured populations in fragmented landscape. Landsc Ecol 22:31–44

Brooks CP (2006) Quantifying population substructure: extending the graph-theoretic approach. Ecology 87(4):864–872

Bunn AG, Urban DL, Keitt TH (2000) Landsc connectivity: a conservation application of graph theory. J Environ Manage 59(4):265–278

Calabrese JM, Fagan WF (2004) A comparison-shopper's guide to connectivity metrics. Front Ecol Environ 2(10):529–536

Chardon JP, Adriaensen F, Matthysen E (2003) Incorporating landscape elements into a connectivity measure: a case study for the Speckled wood butterfly (*Pararge aegeria* L). Landsc Ecol 18(6):561–573

Crist MR, Wilmer B, Aplet GH (2005) Assessing the value of roadless areas in a conservation reserve strategy: biodiversity and landscape connectivity in the northern Rockies. J Appl Ecol 42(A.1):181–191

Crooks KR, Sanjayan M (eds) (2006) Connectivity conservation. Cambridge University Press, New York

Estrada J, Pedrocchi V, Brotons L, Herrando S (eds) (2004) Atles dels ocells nidificants de Catalunya 1999–2002. Lynx Edicions, Barcelona

Fagan WF, Calabrese JM (2006) Quantifying connectivity: balancing metric performance with data requirements. In: Crooks KR, Sanjayan M (eds) Connectivity conservation. Cambridge University Press, New York, pp 297–317

Ferrari JR, Lookingbill TR, Neel MC (2007) Two measures of landscape-graph connectivity: assessment across gradients in area and configuration. Landsc Ecol DOI 10.1007/s10980-007-9121-7

Forman RTT (1995) Land mosaics: the ecology of landscapes and regions. Cambridge University Press, Cambridge

Hannah L, Midgley GF, Lovejoy T, Bond WJ, Bush M, Lovett JC, Scott D, Woodward FI (2002). Conservation of biodiversity in a changing climate. Conserv Biol 16(A.1):264–268

Hanski I (1994) A practical model of metapopulation dynamics. J Anim Ecol 63:151–162

Hanski I (1998) Metapopulation dynamics. Nature 396:41–49

Hjeljord O, Wegge P, Rolstad J, Ivanova M, Beshkarev AB (2000) Spring-summer movements of male capercaillie Tetrao urogallus: a test of the "landscape mosaic" hypothesis. Wildl Biol 6:251–256

Jaeger JAG (2000) Landsc division, splitting index, and effective mesh size: new measures of landscape fragmentation. Landsc Ecol 15(A.2):115–130

Jordan F, Baldi A, Orci KM, Racz I, Varga Z (2003) Characterizing the importance of habitat patches and corridors in maintaining the landscape connectivity of a *Pholidoptera transsylvanica* (Orthoptera) metapopulation. Landsc Ecol 18:83–9

Keitt TH, Urban DL, Milne BT (1997) Detecting critical scales in fragmented landscapes. Conserv Ecol [online] 1(A.1):4. URL: http://www.consecol.org/vol1/iss1/art4

Korpimäki E, Lagerström M (1988) Survival and natal dispersal of fledglings of Tengmalm's owl in relation to fluctuating food conditions and hatching date. J Anim Ecol 57:433–441

Mariné R, Dalmau J (2000) Uso del hábitat por el mochuelo boreal *Aegolius funereus* en Andorra (Pirineo Oriental) durante el periodo reproductor. Ardeola 47:29–36

Nikolakaki P (2004) A GIS site-selection process for habitat creation: estimating connectivity of habitat patches. Landsc Urban Plan 68:77–94

Opdam P, Wascher D (2004) Climate change meets habitat fragmentation: linking landscape and biogeographical scale levels in research and conservation. Biol Conserv 117(A.3):285–297

Pascual-Hortal L, Saura S (2006) Comparison and development of new graph-based landscape connectivity indices: towards the priorization of habitat patches and corridors for conservation. Landsc Ecol 21(7):959–967

Pascual-Hortal L, Saura S (2007) Impact of spatial scale on the identification of critical habitat patches for the maintenance of landscape connectivity. Landsc Urban Plan 83 (2-3):176–186

Pascual-Hortal L, Saura S (2008) Integrating landscape connectivity in broad-scale forest planning through a new graph-based habitat availability methodology: application to capercaillie (*Tetrao urogallus*) in Catalonia (NE Spain). Eur J Forest Res 127:23–31

Rae C, Rothley K, Dragicevic S (2007) Implications of error and uncertainty for an environmental planning scenario: a sensitivity analysis of GIS-based variables in a reserve design exercise. Landsc Urban Plan 79(3–4):210–217

Rochelle JA, Lehman LA, Wisniewski J (1999) Forest fragmentation: wildlife and management implications. Die Deutsche Bubliothek, CIP, Holland

Saura S, Pascual-Hortal L (2007) A new habitat availability index to integrate connectivity in landscape conservation planning: comparison with existing indices and application to a case study. Landsc Urban Plan 83 (2-3):91–103.

Sutherland GD, Harestad AS, Price K, Lertzman KP (2000) Scaling of natal dispersal distances in terrestrial birds and mammals. Conserv Ecol [online] 4(A.1):16. URL: http://www.consecol.org/vol4/iss1/art16

Taylor P, Fahrig L, Henein K, Merriam G (1993) Connectivity is a vital element of landscape structure. Oikos 68(A.3):571–573

Theobald DM (2006) Exploring the functional connectivity of landscapes using landscape networks. In: Crooks KR, Sanjayan M (eds) Connectivity Conservation. Cambridge University Press, New York, pp 416–443

Tischendorf L, Fahrig L (2000) On the usage and measurement of landscape connectivity. Oikos 90(A.1):7–19

Urban D, Keitt T (2001) Landsc connectivity: a graph-theoretic perspective. Ecology 82(5): 1205–1218

Index